THE WORLD SCIENTIFIC
HANDBOOK OF ENERGY

World Scientific Series in Materials and Energy* ISSN: 2335-6596

Series Editor: Leonard C. Feldman *(Rutgers University)*

Published

Vol. 1   Handbook of Instrumentation and Techniques for Semiconductor Nanostructure
Characterization
edited by Richard Haight (IBM TJ Watson Research Center, USA),
Frances M. Ross (IBM TJ Watson Research Center, USA),
James B. Hannon (IBM TJ Watson Research Center, USA) and
Leonard C. Feldman (Rutgers University, USA)

Vol. 2   Handbook of Instrumentation and Techniques for Semiconductor Nanostructure
Characterization
edited by Richard Haight (IBM TJ Watson Research Center, USA),
Frances M. Ross (IBM TJ Watson Research Center, USA),
James B. Hannon (IBM TJ Watson Research Center, USA) and
Leonard C. Feldman (Rutgers University, USA)

Vol. 3   The World Scientific Handbook of Energy
edited by Gerard M. Crawley (University of South Carolina, USA)

Forthcoming

Batteries
edited by Jack Vaughey (Argonne National Lab., USA) and David Schroeder
(Northern Illinois, USA)

Encyclopedia of Practical Semiconductors (In 4 Volumes)
edited by Eugene A. Fitzgerald (Massachusetts Institute of Technology, USA)

Graphene
edited by Michael G. Spencer (Cornell University, USA)

Handbook of Green Materials: Properties, Technologies and Applications
(In 4–5 Volumes)
edited by Kristiina Oksman (Lulea University of Technology, Sweden)

Handbook of Silicon Surfaces and Formation of Interfaces: Basic Science in the
Industrial World (2nd Edition)
Jarek Dabrowski and Hans-Joachim Mussig (Institute for Semiconductor
Research, Germany)

Handbook of Solid State Batteries and Capacitors (2nd Edition)
edited by Nancy J. Dudney (Oak Ridge National Lab., USA), William C. West
Caltech, USA) and Jagjit Nanda (Oak Ridge National Lab., USA)

*The complete list of titles in the series can be found at
http://www.worldscientific.com/series/mae.

MATERIALS AND ENERGY – Vol. 3

THE WORLD SCIENTIFIC
HANDBOOK OF ENERGY

Editor

Gerard M. Crawley
University of South Carolina, USA

Senior Scientific Advisor

Richard Haight
IBM T. J. Watson Research Center

World Scientific

NEW JERSEY · LONDON · SINGAPORE · BEIJING · SHANGHAI · HONG KONG · TAIPEI · CHENNAI

Published by

World Scientific Publishing Co. Pte. Ltd.

5 Toh Tuck Link, Singapore 596224

USA office: 27 Warren Street, Suite 401-402, Hackensack, NJ 07601

UK office: 57 Shelton Street, Covent Garden, London WC2H 9HE

Library of Congress Cataloging-in-Publication Data
The World Scientific handbook of energy / edited by Gerard M. Crawley (University of South Carolina, USA).
    pages cm -- (World Scientific series in materials and energy : v. 3)
    Includes bibliographical references and index.
    ISBN 978-9814343510
    1. Power resources--Handbooks, manuals, etc.   I. Crawley, Gerard M., editor of compilation
II. Title: Handbook of energy.
    TJ163.235.W65 2013
    621.042--dc23
                               2012034832

British Library Cataloguing-in-Publication Data
A catalogue record for this book is available from the British Library.

In-house
Snr Consulting Editor: Juliet Lee Ley Chin
Production Editor: Divya Srikanth

Typeset by Stallion Press
Email: enquiries@stallionpress.com

Printed in Singapore by Mainland Press Pte Ltd.

Contents

vi   *Contents*

6   Other Coal Uses   29
7   Challenges in Coal Production   30
8   Challenges in Coal Usage   31
     8.1   Worldwide Coal Usage   31
     8.2   Coal Usage Projections for OECD Nations 32
     8.3   Coal Usage Projections for Non-OECD Nations 35
9   Carbon Dioxide   37
     9.1   Carbon Dioxide Produced per Kg of Coal   37
     9.2   Geologic Storage of Carbon Dioxide   38
           9.2.1   Saline-bearing formations   38
           9.2.2   Natural gas and oil-bearing formations   38
           9.2.3   Unmineable coal seams   38
           9.2.4   Organic-rich shale basins   39
           9.2.5   Basalt   39
     9.3   Carbon Dioxide Utilization   39
     9.4   Cost of Carbon Storage   39
References   40

4.   Petroleum Liquids   41
     William L. Fisher

1   Introduction   41
2   Production and Consumption   42
3   Reserves and Resources   44
     3.1   Reserves   46
     3.2   Resources   47
4   Petroleum Refining   52
     4.1   Combustion of Gasoline and Diesel Fuel   53
5   Future Production   54
6   Oil Production Costs   54
References   57

5.   Natural Gas   59
     John B. Curtis

1   Introduction   59
2   Why is Natural Gas Important?   60
3   How Natural Gas Forms   61
4   Exploration   62
5   Development   63
6   Production   64
     6.1   Gas Fields   64
     6.2   Stranded Gas   64

6. Nuclear Power   83

Bertrand Barré

7.   Magnetic Fusion Energy   111

R.J. Goldston and M.C. Zarnstorff

8.   Progress Toward Inertial Fusion Energy   131

Erik Storm

Foreword

Human beings are now facing many kinds of crises, concerning issues such as energy resources, food scarcities and global warming. The scarcity of energy resources is an especially serious problem. Sooner or later, humanity will exhaust the supply of fossil fuels, and in any case carbon dioxide produced from fossil fuels may cause global warming. To deal with this problem, we must know the trend of world energy consumption, analyze and understand the present state of affairs, and forecast many things about the future. All we really know is that there are finite resources and the issues are many. What are the fossil fuel reserves, the consumption trends, and the development status of alternative energy resources including renewable energy resources? What is the current status and particularly the future of nuclear fission power generation including safe conditions of operation and the future plans for storage of nuclear waste? What is the likelihood of fusion nuclear reactors, and the present status and the future of energy conservation? It is beyond the capability of a single person to understand all these questions in detail.

Fortunately, *The World Scientific Handbook of Energy* is just being published, edited by Professor G.M. Crawley, who is an excellent nuclear physicist and educator. In each chapter of this handbook, representative experts in each area analyze the current status and discuss future prospects, using the most reliable data. I have read a number of the chapters very carefully including those on Coal, Nuclear Energy and Wind Power, and strongly believe that this is the very Handbook we need for an understanding of energy issues. We are provided with an overall view of the complex and serious energy problem by reading this book. By doing so, each individual person becomes capable of responding to the issues rationally and thoughtfully.

On 11 March 2011, the Pacific coast of the North East region of Japan suffered tremendous damage from a 9.0-magnitude earthquake and the consequent gigantic tsunami. The tsunami proved to be 13 meters high and destroyed the power generators for cooling at Fukushima Daiichi. As a result, four nuclear reactors had cooling problems and three suffered hydrogen explosions. In addition, radioactive isotopes were released into the environment and about 150 thousand people were forced to move. This accident produced a sense of mistrust in nuclear power to the extent that all nuclear reactors in Japan, except two, were shut down on 31 October 2012. At present, Japan has not decided on its future plans for nuclear energy. In addition, the Fukushima accident triggered the German government's decision to have a nuclear power phase-out by 2022.

The Japanese people have lost a feeling of trust not only in nuclear researchers and engineers but also more generally in scientists and engineers including earthquake researchers. Besides, they have lost faith in science and technology. Japanese scientists and engineers have to take this matter seriously, and reflect on what they should do in order to regain the public's trust. In order to do so, scientists and engineers should have a full understanding of energy issues around the world in addition to their own small specialized fields.

Renewable energy is expected to play a much greater role in Japan in the future. It is notable that Germany succeeded in expanding renewable energy use, especially wind power use, by adopting feed-in tariffs for renewable energy in 2000. As a result, Germany's total production of electricity by renewable resources, excluding hydropower, reached 82.95 billion kWh in 2010, which was 13% of the annual total electrical production that year. This is a remarkable achievement but still the amount was only about 9% of Japan's total electrical production of 956.5 billion kWh in 2009, and was also only 30% of Japan's total nuclear generation in that year. Of course, renewable energy resources must be developed. However, following the German approach would only bring Japan to about 10% of its total electricity production in 10 years. At this same rate, we would need 30 years to replace nuclear power generation, which is 30% of Japan's total. In addition, Japan does not have any more sites for large hydropower plants, so Japan must rely on thermal power plants. An increase in the Japanese trade deficit would be inevitable as a result of increasing the import of fossil fuels. Aside from that, we need an immediate plan for how to compensate for the current power shortage until we can get sufficient electricity from renewable energy resources.

I believe that the Japanese nuclear power plants should be restarted, since the operators are now sufficiently aware of the importance of disaster prevention, especially from earthquakes and tsunamis. More and more Japanese should gain a deeper understanding of the energy issues when making this kind of decision. From this point of view, the present Handbook should be appreciated because of its timeliness.

World population reached seven billion on 31 October 2011. This is a huge increase from the eighteenth century when it was at most about one billion people, and it is still increasing. Additionally, the average energy consumption per person has risen from 2,000 kcal/day for primitive man to about 45,000 kcal/day currently. This is an increase of 22.5 times. In the developed countries, the average equivalent oil use per person per year is 7.5 tons in Canada, 7.1 tons in the United States, about 3.5 tons on average in the European Union, 4.7 tons in Korea and 3.7 tons in Japan.

The total population of China and India will likely reach three billion in 2030. Alternatively, the population of developed countries will be about 1.2 billion. Suppose now we assume that the energy use in the developed countries remains at about 4 tons of oil equivalent per year, and China and India increase to a similar value of 4 tons/year. Leaving aside all the other countries in the world, the energy consumption in the US, EU, Japan, Korea, China and India would increase to about

3.5 times larger than the present total energy use of these countries. This is a rough calculation so we may need a more exact forecast. However, there is no doubt that the world's energy consumption will rapidly increase over the next 30 years. According to one calculation by the Institute of Energy Economics of Japan, the energy use of 11.2 billion tons of oil equivalent in 2009 will increase to 17.3 billion tons in 2035. This is an increase of 1.5 times. What should we do to deal with this rapid increase? The newly developed fossil fuels such as shale gas and methane hydrate may slow down the depletion of fossil fuels, but the rapid increase of energy consumption will compensate for these new fuels. Almost all fossil fuels will be used up within at most one hundred years. We are responsible for developing the promising new technologies which will need to be deployed as the fossil fuels are used up. From this point of view, we need to have a full understanding of the present situation and the future possibilities of energy production and use, and to promote technologies with the assurance of energy safety. Therefore, we must make progress using all our energy resources including renewable energy resources, safety-assured nuclear fission power generation, nuclear fusion power generation, and so forth. I have very little doubt that this excellent *World Scientific Handbook of Energy* will help up us to examine this important issue of energy which is so critical for the future of the world.

Human beings have conquered periods of trial like the Little Ice Age, especially the shortage of food, and epidemics like the plague. I am convinced that we can deal with the current crises thoughtfully and solve them with wisdom.

Akito Arima
November 2012

Professor Akito Arima is a prominent theoretical physicist, who has received many awards and prizes for his research into nuclear structure, and he continues to be actively engaged in nuclear physics research. In addition, with a strong commitment to policy in education and research he has served the scientific community in a number of policy and administrative roles. Professor Arima received his Doctor of Science in 1958 from the University of Tokyo. His positions include Associate Professor in the Department of Physics at the University of Tokyo; Visiting Professor, Rutgers University, New Jersey; Professor, State University of New York at Stony Brook; and Professor, Department of Physics, University of Tokyo. He was President of Tokyo University from 1989 to 1993 and President of the Institute of Physical and Chemical Research (RIKEN) from 1993 to 1998. From 1998 to 1999 he was Minister

of Education, Science, Sports and Culture, and Minister of State for Science and Technology for Japan. Professor Arima has received many notable awards including the Nishina Prize, Japan; Honorary Professorship, the University of Glasgow, UK; the Humboldt Award, Germany; the Prize of the Haiku Society, Japan; the Wetherill Medal of the Franklin Institute, USA; Order Das Grosse Verdienstkreuz, Bonn, Germany; Kanselarij der Netherlandse Orden's Gravenhage, Amersterdam; the Bonner Prize of the American Physical Society; the Japan Academy Prize; and Honorary Doctorates from Drexel University, USA, the University of Science and Technology, China, Chung Yuan Christian University, Taiwan, the State University of New York at Stony Brook, USA, the University of Groningen, the Netherlands and the University of Birmingham, UK.

Among his other activities Professor Arima played an important role in supporting the public understanding of science as Chairman of the Japan Science Foundation. As Co-Chairman of the Board of Governors of the international graduate university, the Okinawa Institute of Science and Technology, he continues to advance science worldwide.

Chapter 1

Introduction

Gerard M. Crawley

Marcus Enterprises, Columbia SC 29206, USA
marcusenterprise@gmail.com

The Handbook of Energy is intended to gather, in one place, a brief description of the current state of knowledge about a broad range of energy sources and energy issues, including comprehensive, reliable, understandable, and timely sets of data on resource availability and use worldwide.

The importance of energy cannot be overstated. The development of new sources of energy has enabled humans to expand their abilities enormously and has led quite directly to our technological civilization. People are now able to produce food, manufacture goods, travel, and communicate in ways that would have been unthinkable a thousand years ago. Based on per capita energy use, citizens in many countries have their human efforts multiplied by more than 100 times because of the use of external energy sources such as electrical energy.

Historically, the growth of civilization has also been the story of expansion in the use of various sources of energy. Animals can plow fields and grind corn more easily than human beings acting on their own. Later on, the energy from falling water and wind was also used to augment human efforts. Even more recently, as our understanding of the nature of matter and energy has increased, we have greatly expanded the possible sources of energy that have come into use. It might be said that we now live in the age of fossil fuels since, for the past 100 years or so, coal, petroleum liquids, and natural gas have been the primary sources of energy. However, fossil fuels were all laid down by nature over millennia and are not replaceable on the scale they are being used. Therefore, we have begun to expand the use of ever more sophisticated energy sources such as nuclear fission and perhaps fusion, and we are exploring many possible renewable resources, often making use of the abundant but diffuse energy from the sun.

The growing appetite for energy has enormous consequences both economically and politically. Defense alliances and confrontations have taken place and continue

1

to take place over access to energy resources. The price of food worldwide has been affected by the growing use of crops as energy sources rather than for food. Even within countries, there are often wide disagreements about energy policy, pitting expanding energy resources against environmental hazards and safety risks. One thing seems clear: almost no energy resource is without some risk, and the challenge is to balance risk versus benefit in some reasonable way. Efforts to respond to the growing need for energy have triggered a veritable avalanche of new programs to develop both further existing energy sources and new renewable approaches such as solar, wind, bio-, and tidal energies. Therefore, it has become imperative to compile this cornucopia of energy sources and related data in a single volume.

In comparing various energy options, it is important to have a quantitative approach. The issues are complicated, and a solution for one region may not be appropriate for another. The numbers do indeed tell a story. This Handbook is intended therefore to gather, in one place, a brief description of the current state of knowledge about a broad range of energy sources and energy issues, including comprehensive, reliable, understandable, and timely sets of data on resource availability and use worldwide. One advantage of such a compilation is that the information is readily accessible in one place. In addition, the individual chapters are all written by experts in that specific area. While it is possible to gather data from the Internet, the reliability of such information is often suspect. The expert authors, who are leading current practitioners in the field, provide reliable and credible information that is not available from anonymous sources. The authors present not only an understandable and comprehensive introduction to the various sources but also the current status of different technologies and their probability of success. The environmental or other risks associated with particular energy sources or storage technologies are also described. The emphasis is on presenting the data rather than advocating a particular approach. Different solutions may well be appropriate in different situations. In cases where there are critical material resources or economic limitations for particular technologies, this volume provides data relevant to these issues. The Handbook is intended to provide information that helps readers to first understand the issues and then to use their judgment in formulating answers.

The fossil fuels, coal, petroleum liquids, and natural gas, are discussed in Chapters 3, 4, and 5, where current resources and use is described, along with the best current estimates of the future lifetimes of these finite resources. Nuclear fission is discussed in Chapter 6, including different types of reactors currently in use and plans for the next generation of reactors (generation 4). Two very different approaches to nuclear fusion are being tried: inertial fusion using very high energy laser beams striking a tiny pellet of deuterium–tritium fuel, and magnetic confinement fusion. The timeframe for these two approaches toward first scientific demonstration and later commercial exploitation are described in Chapters 7 and 8. Various approaches to tapping into the abundant but diffuse energy from the sun are described in Chapters 9 through 11. The use of photovoltaic cells to transform electromagnetic energy from the sun directly into electricity is becoming more and more economically viable, and technological and scientific advances will expedite this process. Using energy from the sun to heat a working fluid and then using more conventional turbine technology to produce electricity is also becoming more widespread. Cultivation of non-food crops to produce fuel is described, including specially engineered varieties to optimize solar energy capture.

Tapping into the geothermal energy of the earth is discussed in Chapter 12. This technology has been used in various countries, particularly Iceland, very successfully, but new methods will be needed for wider application. Another well-established technology, hydropower, the energy from falling water, is discussed in Chapter 13. While this technology is quite mature, new advances in tapping into smaller sources in an economic fashion may allow this resource to continue its expansion. Energy from the wind has also been used for many centuries, but advances in material science and the ability to model wind turbine behavior has led to a huge renaissance in the use of wind as an energy resource. These advances are discussed in Chapter 14. Various methods for extracting energy from the ocean, including waves, tidal currents, and ocean thermal gradients, are described in Chapters 15 and 16. While the economic exploitation of these latter resources is likely limited to particular regions where they are viable, the potential resources are large.

In addition to the basic sources of energy, there are also issues related to the storage of energy. Most renewable sources such as solar and wind energy are intermittent so that there is a need for storing energy produced by these resources for later use. Pumped storage of water, as described in Chapter 13, is one example of a storage solution employed to smooth out energy supply and demand requirements. Another energy storage option, the capacitor, which is used in specific applications, is discussed in Chapter 17. Batteries are increasingly being used both as standalone energy sources for transportation and as supplements to the internal combustion engine to increase fuel efficiency, as discussed in Chapter 18. Fuel cells, powered by either hydrogen or natural gas, provide additional options for mobile power sources. Their current status is described in Chapter 19.

The use of electrical energy has become almost universal and many different energy sources from coal-fired plants, gas turbines, nuclear fission plants, wind turbines, and, recently, solar farms are all being used to produce electricity which is then fed to the electric grid. The transport of electricity by the grid both within countries and beyond national boundaries, as well as the variety of energy sources producing electricity for the grid, all bring both advantages and challenges as discussed in Chapter 20.

Another extremely important factor in the use of energy is efficiency. The amount of energy used per capita is discussed in Chapter 21 along with its relation to GDP. The chapter also describes the important elements in improving energy efficiency in many processes. One of the most important aspects, as well as one of the most controversial, at least in the US, has been the impact of the growing use of energy, particularly fossil fuels, on the temperature of the earth and its potential effect on climate. There is a growing scientific consensus on this subject, which is discussed in Chapter 22.

The situation *vis*-à-*vis* energy resources is changing rapidly, both on the supply side and the demand side. Large countries like China, India, and Brazil are demanding an increasing share of energy resources to drive their rising economies. The price of petroleum is extremely volatile but is only likely to increase in the long term. New methods of extracting natural gas from shale and new ways of using coal that are cleaner and produce fewer carbon emissions are also being considered. Generation-4 nuclear fission reactors provide a new level of safety. There is also the promise of breakthroughs in fusion energy using both inertial fusion and magnetic confinement techniques. Solar energy is steadily becoming a more competitive resource, and there are possibly new approaches to the use of solar energy. Energy from the wind, particularly offshore, is becoming a rapidly increasing resource in many countries. The Handbook provides up to date information from authoritative sources on all of these issues.

What kind of energy resources should the world focus on? The answer may likely vary by country, or even by region within a country, but the impact will be felt worldwide. There are no simple answers. The best solution is likely to be a mix of energy sources, with the mix changing from country to country and especially with time. Continuing to explore scientific and technological options for energy resources, for energy storage and for energy conservation are all necessary strategies. Hopefully, the Handbook of Energy will help both researchers and policy makers to better understand the science and engineering behind the development of energy resources, storage and conservation as well as their economic implications.

Chapter 2

Energy, Power, Units, and Conversions

Gerard M. Crawley

Marcus Enterprises, Columbia SC 29206, USA
marcusenterprise@gmail.com

Many different units are used to measure the amount of energy in a system. These units can be related to one another so that different kinds of energy can be compared. There is also sometimes confusion between the usage of the terms "energy" and "power". Units of power will also be related to one another and contrasted with energy units.

1 Introduction

Energy comes in many forms and therefore many different units are used in various situations to describe the energy content of fuels or the amount of energy used in particular processes. Just as energy in one form (e.g. electrical energy) can be transformed into another form (mechanical energy), the units of energy can also be converted from one to another as convenient, using the relationships described in the tables presented in this chapter. There are many sources available for energy conversion, including online calculators.[1−3] Some of the most common conversions are given in Table 1.

Energy is one of the most basic concepts in science and has very broad and powerful uses in describing the physical world. The simplest definition of energy is the capacity to do **work**. Thus energy and work have the same units. Work is formally defined as the scalar product of force times the distance over which the force acts. Thus, whereas force and distance (or displacement) are vectors, work and therefore energy are scalar quantities.

The most basic unit of energy in the SI (International System) or MKS system[a] is the **joule (J)**, which can be thought of as the unit of force (Newton) times the unit of displacement (meter). All other units of energy can be related to the joule.

---

[a]In the SI and MKS systems, displacement is measured in meters, mass in kilograms, and time in seconds.

Table 1:   Conversion of Energy Units.

1 kilowatt-hour (kWh)	3.6×10^6 J
1 kilowatt-hour (kWh)	3412 BTU[b]
1 calorie[a]	4.184 J
1 kilocalorie (Calorie)	4184 J
1 BTU[b]	1054 J
1 BTU[b]	252 calories
1 eV	1.602×10^{-19} J
1 MeV	1.602×10^{-13} J

[a]There are other definitions of the calorie such as the International Table calorie defined as 4.1866 J.[2]
[b]Slightly different definitions of the BTU make these relationships approximate.

Power is the rate of doing work or the rate of using energy. Therefore the unit of power in the SI system is the joule per second or **watt (W)**. Again, many different units of power are used in different situations and these can all be related to one another.

2 Different Forms of Energy

The various forms of energy, which are useful in different situations, can be related to one another by transforming the units into a common unit. This is illustrated below for different forms of energy.

2.1 *Mechanical Energy*

An object can have energy because of its motion, called kinetic energy (KE).

$$\mathrm{KE} = \tfrac{1}{2}\, mv^2,$$

where m is the mass of the object and v is its velocity.

An object is also said to possess potential energy (PE) because of its position or situation. One example is the PE of an object lifted to a height h above some reference level in a constant gravitational field.

$$\mathrm{PE} = mgh,$$

where m is the mass of the object, raised to a height h, and g is the (constant) acceleration of gravity.

Another example of PE is the energy possessed by a mass attached to the end of a compressed or stretched spring. This PE is converted into KE as the spring is released.

$$\mathrm{PE} = \tfrac{1}{2}\, kx^2,$$

where k is the spring constant and x is the displacement from the equilibrium position of the spring.

2.2 *Thermal Energy*

A body also possesses energy because it contains heat. The basic unit of heat energy in the SI system is the **Calorie** (or sometimes written as a kilocalorie or Kcal), which is the amount of heat energy required to raise the temperature of 1 kilogram of water by 1 degree Celsius.

Note that it is necessary to distinguish the "big" calorie defined above from the "small" calorie defined in the CGS system,[b] as the amount of heat energy required to raise the temperature of 1 gram of water by 1 degree Celsius. 1 Calorie = 1,000 calories.

Another commonly used energy unit for heat energy is the **British Thermal Unit** (BTU). This is defined as the amount of heat required to raise the temperature of 1 pound of water by 1 degree Fahrenheit. Obviously, the BTU can be related to the Calorie (see Table 1).

The chemical energy contained in food is also measured in Calories. Humans consume about 2,000 Calories each day, although clearly this number varies widely from person to person. These Calories are used to maintain body temperature and to convert into kinetic energy as we move around. If we exercise vigorously, we use (burn) more Calories. While the number of Calories required for exercise depends upon body weight and on how vigorously we exercise, a few guidelines are given in Table 2.

2.3 *Electrical Energy*

Another very important form of energy is electrical energy. When a charge of 1 coulomb moves through a potential difference of 1 volt, the energy gained (or lost) by the charge is 1 joule.

$$1 \text{ joule} = 1 \text{ coulomb} \times 1 \text{ volt.}$$

A common unit of electrical energy is the **kilowatt-hour** (kWh). Since the watt is a unit of power, the corresponding unit of energy (power × time) is the watt sec,

Table 2:   Calories/Hour Used in Different Forms of Exercise.[4,5]

	Weight 120 lbs (55 kg)	Weight 160 pounds (73 kg)	Weight 200 pounds (91 kg)
Running (8 mph)*	740	986	1,229
Swimming laps**	384	511	637
Walking (3–4 mph)	208	277	346
Rowing, stationary**	384	511	637
Tennis Singles**	439	584	728

*Mph is miles per hour, and 8 mph equals 1 mile in 7.5 min or 1 km in 4 min and 40 sec.
**The precise numbers will depend on how vigorously one exercises.

[b]In the CGS system, distance is measured in centimeters, mass in grams, and time in seconds.

equivalent to the joule. A watt is also the power from a current of 1 ampere flowing through a potential difference of 1 volt. Kilowatt-hours can be related to other energy units as shown in Table 1.

Energy can also be stored in an electrical capacitor and in an electrical inductor.

Energy stored in a capacitor $(J) = \frac{1}{2}\,CV^2$,

where C is the capacitance in farads and V is the potential difference, in volts.

OR

Energy stored in a capacitor $(J) = \frac{1}{2}\,Q^2/C$,

where Q is the charge on the capacitor, in coulombs.

Energy stored in a coil $(J) = \frac{1}{2}\,LI^2$,

where L is the inductance of the coil in henrys (H) and I is the current in the coil in amperes (A).

2.4 *Atomic and Nuclear Energy*

When dealing with the energy at the atomic and molecular scale, a convenient unit of energy is the electron-volt (eV): 1 eV is the energy gained by an electron in moving through an electrical potential difference of 1 volt.

Since the charge on an electron is 1.602×10^{-19} coulomb, this means that the energy gained is 1.602×10^{-19} J. So the eV is a very small energy unit (Table 1).

In some cases, a unit of 1,000 electron volts *viz.* a kiloelectron volt (keV) is used. In nuclear processes, where greater amounts of energy are released than in atomic processes, the common measurement unit is a million electron volts (MeV).

2.5 *Chemical Energy*

As mentioned earlier, there is chemical energy in food, usually measured in Calories. Different fuels such as coal, oil, and natural gas also contain chemical energy, which is used to drive the energy needs of our civilization. There are many different units used to describe the energy content of fuels. There is an in-built imprecision in such units because the actual energy content of a particular ton of coal or barrel of oil depends upon its detailed composition, which varies from place to place. For example, the energy content of a barrel of crude oil originating in different countries can vary from about 5.6 million BTU to about 6.3 million BTU. For coal, the difference in energy content can be even greater, ranging from 10 million BTU per ton to 30 million BTU per ton, depending on the type of coal and its country of origin. However, an average or nominal value of the energy content of different fuels is often assumed in order to facilitate comparisons between different types of fuel, as shown in Table 3.

Table 3:   Energy Content (Nominal) of Different Fuels.[1,2]

1 ton of coal	26.6×10^9 J	27.8×10^6 BTU
1 tonne of coal[a]	29.3×10^9 J	25.2×10^6 BTU
1 barrel of oil (bbl)	6.11×10^9 J	5.80×10^6 BTU
1 tonne of oil (toe)	41.87×10^9 J	39.7×10^6 BTU
1 cubic foot (ft^3) of natural gas	1.05×10^6 J	1000 BTU
1 therm (natural gas)	1.05×10^8 J	1×10^5 BTU

[a]1 tonne (metric) = 1,000 kg.

Table 4:   Large Units of Energy.

1 quad	1×10^{15} BTU	1.055×10^{18} J	
1 TWyr	8.76×10^{12} Kwh	31.54×10^{18} J	29.89 quad
1 Mtoe	3.97×10^{13} BTU	4.187×10^{16} J	
1 Gigatoe	3.97×10^{16} BTU	4.187×10^{19} J	39.68 quad
1 CMO	153×10^{15} BTU	162×10^{18} J	153 quad

3 Large Energy Units

In many situations, where large amounts of energy are considered, it is useful to use units of energy that describe large quantities such as a quadrillion BTU (quad), a terawatt-year (TWyr), megatonne of oil equivalent (Mtoe), or a Cubic Mile of Oil (CMO).[6] Some examples are given in Table 4.

4 Power

As mentioned earlier, the unit of power (rate of doing work or rate of using energy) is the watt (W). Other power units can be derived from the equivalent energy units by dividing the energy unit by time. One rather peculiar unit of power is the **horsepower**, perhaps because of its historical significance.

$$1 \text{ horsepower (hp)} = 745.7 \text{ W or } 0.7457 \text{ KW.}$$

In generating electricity, the output power is often denoted as Megawatt electric (MWe) or Gigawatt electric (GWe), as opposed to the input power in the form of heat, which is designated as Megawatt thermal (MWt) or Gigawatt thermal (GWt). The difference is because of the efficiency of the power generating device. For example, a power plant operating at 30% efficiency would require an input thermal power of 2,000 MWt to produce an electrical power output of 600 MWe.

References

1. US Energy Information Administration. Available at: http://www.eia.gov/energy-explained/index.cfm?page=about_energy_units. Accessed November 2011.
2. American Physical Society. Available at: http://www.aps.org/policy/reports/popa-reports/energy/units.cfm. Accessed November 2011.
3. D. Silverman, UC Irvine. Available at: http://www.physics.uci.edu/~silverma/units.html. Accessed November 2011.

4. Mayo Clinic. Available at: www.mayoclinic.com/health/exercise/SM00109. Accessed November 2011.
5. Health Status. Available at: http://www.healthstatus.com/cgi-bin/calc/calculator. cgi. Accessed November 2011.
6. H. Crane, E. M. Kinderman and R. Malhotra, *A Cubic Mile of Oil* (Oxford University Press, New York, 2010).

Chapter 3

Coal

Thomas Sarkus
National Energy Technology Laboratory
US Department of Energy, USA
Thomas.Sarkus@netl.doe.gov

Adrian Radziwon and William Ellis
KeyLogic Systems Inc., USA

This chapter introduces and describes coal as a widely available and versatile energy source that is available in substantial quantities throughout the world. The origins of coal deposits are quickly visited, and the ranking of different types of coal by energy and carbon content is presented. The production and chemical analysis of coal from major US coalbeds (including representative values for some lignite and anthracite beds) are detailed. World coal resources, reserves, and production levels are discussed for coal types. Coal utilization is developed with a detailed focus on the major coal-based electrical power production technologies, which includes pulverized coal, fluidized bed, and gasification technologies. These technologies are presented in detail, including relevant emission control technologies. Carbon dioxide capture is also addressed. Liquefaction of coal and other uses of coal are briefly described. The safety and environmental challenges associated with coal production and its uses are also discussed. Current and projected usages of coal as well as other energy sources by world region and selected countries for electrical power production are presented. Selected internet-based coal information resources from the US Department of Energy, including the National Energy Technology Laboratory (NETL), are referenced for the reader's use.

1 Introduction

Coal is a rock of sedimentary origin, formed from decomposed and lithified biomass and is utilized widely as an energy resource. Coal is commonly termed a mineral resource, but according to a strict scientific definition, the term "mineral" denotes a fixed chemical composition, whereas the composition of coal varies widely. Indeed, coal composition varies so much that samples taken from adjacent parts of the same coal seam can and do vary significantly in their exact chemical composition. To be more precise, the complex composition of coal may vary both vertically and horizontally in a given coalbed or coal seam.

Biomass → Peat → Lignite → Sub-bituminous Coal → Bituminous Coal → Anthracite
(aka Brown   (aka Hard
Coal)   Coal)

Fig. 1.   Coal ranks (from least mature to most mature).

2 Coal Rank

Coal is most commonly classified by rank. Coal rank increases as heat and pressure are applied over time through geologic processes. Accordingly, the heating value (usually measured in Btu/lb or MJ/kg) tends to increase and volatile compounds tend to decrease with higher rank; however, these are only trends and do not necessarily apply in every instance. The ranks of coal are shown in Fig. 1 in order of ascending maturity.

Some experts do not consider peat, let alone biomass, to be a bona fide rank of coal but, at a minimum, they represent early precursors of the coalification process. Less mature ranks, such as lignite and sub-bituminous coal, often occur in very thick seams which can be surface mined. To be sure, underground sub-bituminous coal mines are not unknown, but surface mining of sub-bituminous coal is overwhelmingly predominant. More mature ranks such as bituminous coal and anthracite usually occur in comparatively thinner seams and, while surface mining is performed in some circumstances, underground mining is commonly utilized to obtain these higher heating value coals. Peat, lignite, sub-bituminous coal, and bituminous coal generally occur in layers, beds, or seams that are more or less horizontal. However, anthracite undergoes a higher degree of alteration or metamorphism, so anthracite seams can be inclined sharply, or folded, through geologic processes. The coal ranks can be divided into groups or sub-ranks, listed in Table 1, in descending order of maturity.

Table 1:   Coal Ranks and Sub-ranks, with Some Defining Parameters.

Coal group	Fixed carbon	Volatile matter	Calorific value (Btu/lb)
Meta-Anthracite	$98\% \leq$ fc	vm $\leq 2\%$	—
Anthracite	$92\% \leq$ fc $< 98\%$	$2\% <$ vm $\leq 8\%$	—
Semi-anthracite	$86\% \leq$ fc $< 92\%$	$8\% <$ vm $\leq 14\%$	—
Low-volatile bituminous	$78\% \leq$ fc $< 86\%$	$14\% <$ vm $\leq 22\%$	—
Medium-volatile bituminous	$69\% \leq$ fc $< 78\%$	$22\% <$ vm $\leq 31\%$	—
High-volatile A bituminous	fc $< 69\%$	$31\% <$ vm	$14{,}000 \leq$ cv
High-volatile B bituminous	—	—	$13{,}000 \leq$ cv $< 14{,}000$
High-volatile C bituminous	—	—	$11{,}500 \leq$ cv $< 13{,}000$
Sub-bituminous A	—	—	$10{,}500 \leq$ cv $< 11{,}500$
Sub-bituminous B	—	—	$9500 \leq$ cv $< 10{,}500$
Sub-bituminous C	—	—	$8300 \leq$ cv $< 9{,}500$
Lignite A	—	—	$6300 \leq$ cv $< 8{,}300$
Lignite B	—	—	cv $< 6{,}300$

Source: Modified from Geological Survey Circular 891, Coal Resource Classification System of the US Geological Survey (2003).

3 Coal Resources and Geographic Distribution

Coal is distributed broadly across much of the US and similarly around the world. However, many coal deposits are not economical for mining. For example, they may be too small, too thin, or too deep to mine economically. As technology improves, it may become possible to mine or otherwise utilize (e.g. through well-controlled underground combustion or gasification) some of these coal deposits. In other cases, coal seams may underlie cities, towns, or sensitive environmental areas.

3.1 *Coal Resources of the US*

Coal fields in the US are shown in Fig. 2, characterized by geologic age. The total coal resources in the US are currently estimated to be about 4 trillion tons. However, the estimate could be augmented substantially by additional coal deposits in Alaska, where Flores *et al.* posit a coal resource of more than 5.5 trillion tons. These resource estimates are somewhat hypothetical and carry a fair level of uncertainty. More germane is the demonstrated reserve base, which carries higher certainty and tallies with the coal that could be mined commercially. As of 1 January 2010, the estimated US coal reserve base is approximately 486 billion tons. It is estimated that approximately 54% or 261 billion tons of that coal can be recovered. This estimate of recoverable coal could increase as regional estimates are refined, mining technologies improve, and/or the costs of competing energy resources such as oil, gas, and

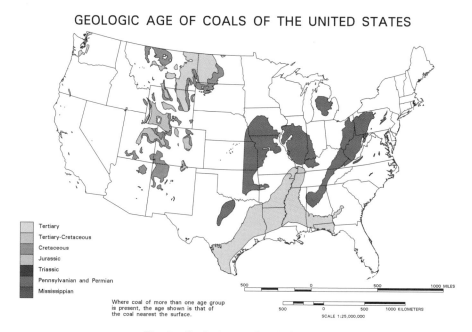

Fig. 2.   Geologic age of coal of the US.

Source: US Geological Survey Coal Fields of the Conterminous United States, Digital Compilation by John Tully, Open-File Report No. 96-92 (1996).

nuclear power increase. Alternatively, it could decrease as coal is mined and utilized over time, or if environmental restrictions impede the mining and utilization of coal.

3.2 *Coal Analyses*

Owing partly to its usage in a wide variety of applications over the past two centuries, there are numerous formats and techniques for analyzing coal. Two of the most prevalent types of coal analyses are proximate analysis and ultimate analysis. Proximate analysis includes fixed carbon, volatile matter, ash, and moisture contents — all on an as-received basis. Ultimate analysis includes calorific or heating value (on both as received and dry bases), as well as the amounts of carbon, hydrogen, nitrogen, oxygen, and sulfur. Other forms of coal analyses may include agglomerating or caking tendency, agglutinating or binding tendency, ash fusion temperature, ash softening temperature, free swelling index, hardgrove grindability, petrography (e.g. maceral contents; macerals are coal portions that can be identified with a petrographic microscope), trace element levels (often expressed as portions of either the parent coal or the ash fraction of the parent coal), and washability (or sink/float testing).

3.3 *US Coal Production and Representative Coal Analyses*

Table 2 lists the 10 top-producing bituminous and sub-bituminous coalbeds in the US, as well as a lignite coalbed in North Dakota, and averages for multiple lignite coalbeds in Texas and multiple anthracite beds in Pennsylvania.

Table 3 illustrates some typical proximate analysis and heating values for these same coalbeds (within the US states previously listed in Table 2). These data are merely for illustrative purposes; readers are reminded that coal quality can vary widely within any given coal seam or coal mine. Table 4 shows ultimate and trace element analysis by coalbed.

4 Worldwide Coal Resources, Reserves, and Production Levels

Coal resources are significant both domestically and internationally. Exploration and estimation of coal resources do not carry the same level of risk and uncertainty as oil and natural gas exploration. One must, however, exercise caution when discussing resources and reserves (e.g. comparing recoverable coal reserves to recoverable oil and/or natural gas reserves), as the terminology can have rather precise meaning(s). Worldwide recoverable coal reserves, as of 2008, are shown for world regions in Table 5, and for the top 20 countries in Table 6.

The US Energy Information Administration (EIA) defines recoverable coal as "Proved Recoverable Reserves, Coal: Defined by the World Energy Council as the tonnage within the Proved Amount in Place that can be recovered (extracted from the earth in raw form) under present and expected local economic conditions with existing available technology. It approximates the US term proved (measured) reserves, coal." These recoverable reserve statistics were drawn from the EIA

Table 2: US Coal Production by Selected Coalbeds and Mine Type, 2010.[1]

Coalbed Name	Coal rank[a]	Under-ground	Surface	Total
		(Thousand short tons)		
Wyodak	Sub-bituminous	—	382,805	382,805
Pittsburgh	Bituminous	78,969	5,148	84,116
No. 9	Bituminous	40,746	6,212	46,958
Herrin (Illinois No. 6)	Bituminous	29,018	2,612	31,630
Canyon	Sub-bituminous	—	29,405	29,405
Beulah-Zap	Lignite	—	27,493	27,493
Coalburg	Bituminous	4,841	16,979	21,819
Anderson-Dietz 1-Dietz 2	Sub-bituminous	—	21,254	21,254
Roland	Sub-bituminous	—	16,226	16,226
Rosebud	Sub-bituminous	—	15,403	15,403
Multiple Texas Coalbeds[b,c]	Lignite	—	40,982	40,982
Multiple Pennsylvania Coalbeds[d,e]	Anthracite	na	na	1,705[f]
Identified Coalbed Total		**153,574**	**564,519**	**718,093**
U.S. Total		**337,155**	**745,357**	**1,084,368**

Source: Unless otherwise identified, data from Report No: DOE/EIA-0584 (2010), Table 5, Revised: July 3, 2012
[a]Data from Report No: DOE/EIA-0584 (2010), Table 6, Revised: July 3, 2012.
[b]Production data not available by individual coalbed.
[c]Data from Report No: DOE/EIA-0584 (2010), Table 6, Revised: July 3, 2012.
[d]Production data not available by individual coalbed.
[e]Data from Report No: DOE/EIA-0584 (2010), Table 6, Revised: July 3, 2012.
[f]Production of Anthracite not available by mine type.

International Energy Statistics Database. In similar fashion, Tables 7 through 11 illustrate coal production by world regions and top 20 countries over the period 2006–2010. These production statistics were also drawn from the EIA International Energy Statistics Database. The US has the largest coal reserves in the world. Other nations that have large coal reserves include Russia, China, Australia, and India.

5 Coal Utilization

Historically, coal has served a variety of uses, some of which (e.g. coal-fueled loco-motives, ships, and residential furnaces) have become outmoded given the abundance, convenience, and affordability of oil, natural gas and, in the case of at least some ships, nuclear energy. Today, the largest use of coal, by far, is to generate electricity in large central power stations. Approximately one half of all the electricity generated in the US is derived from coal. The most common method of using coal to generate electricity is to burn pulverized coal (PC; in Europe it is termed

Table 3:   Coal Proximate Analysis, by Coalbed.

Coalbed name	Proximate analysis (weight percent)				Calorific value	
	Moisture	Volatile matter	Fixed carbon	Ash	Kcal/kg	Btu/lb
Wyodak	27.18	33.77	30.85	8.20	4,641	8,362
Pittsburgh	2.32	34.88	54.23	8.57	7,379	13,295
No. 9	8.06	36.55	44.79	10.60	6,496	11,705
Herrin (Illinois No. 6)[a]	5.98	35.11	48.42	10.63	6,441	11,605
Canyon	31.04	28.54	31.60	8.82	4,116	7,416
Beulah-Zap	37.61	26.27	28.35	7.78	3,563	6,420
Coalburg	3.08	34.01	52.93	9.97	7,262	13,084
Anderson-Dietz 1-Dietz 2[b]	28.06	28.80	35.88	7.26	4,509	8,124
Roland	23.59	32.59	38.21	5.62	4,518	8,140
Rosebud	25.28	27.95	35.83	10.95	4,611	8,308
Texas Lignite Coalbeds[c]	26.51	31.61	29.24	12.64	4,234	7,628
Pennsylvania Anthracite Coalbeds[d]	2.02	7.47	77.14	13.38	6,962	12,544

Note: Coal sample data from USGS US Coal Quality Database.

[a]Weighted average samples were used for analysis from the Illinois No 6, Springfield No 5, and Colchester coalbeds.

[b]Anderson-Dietz 1-Dietz 2 coalbed analysis data not available, all available data for Montana Sub-bituminous used instead.

[c]Analysis data for all available USGS Texas Lignite Coalbed samples used due to production data not being available by individual coalbed.

[d]Analysis data for all available USGS Pennsylvania Anthracite Coalbed samples used due to production data not being available by individual coalbed.

pulverized fuel or PF) in a boiler to generate steam that is used to drive a steam turbine/generator set. Coal is generally the least costly fossil fuel for electricity generation.

5.1 *Pulverized Coal (aka Pulverized Fuel) Combustion*

While relatively inexpensive, coal also contains several elements that present environmental problems, namely sulfur, which is emitted primarily as sulfur dioxide (SO_2); nitrogen, which is emitted as various nitrogen oxides (NO_x); and mercury. SO_2 and NO_x are both acid rain precursors. NO_x can be distinguished from SO_2 in at least two respects, however. While SO_2 is essentially fully derived from sulfur in the coal, NO_x is derived from both coal-bound nitrogen and combustion air (of which nitrogen is the main constituent). Also, while SO_2 is an acid rain precursor, NO_x is widely acknowledged as a precursor of both acid rain and ozone. Control equipment is required to control emissions of SO_2 and NO_x, as well as emissions of particulate matter, which is largely composed of finely divided ash particles that also contain a small portion of unburned carbon. The US Environmental Protection Agency is currently engaged in the process of regulating mercury emissions, and this

Table 4:   Coal Ultimate and Trace Element Analyses by Coalbed.

Coalbed Name	Ultimate analysis (percent)					Trace elements (ppm)
	Hydrogen	Carbon	Nitrogen	Oxygen	Sulfur	Hg
Wyodak	6.51	48.01	0.92	35.74	0.64	0.15
Pittsburgh	5.27	74.09	1.52	8.50	2.02	0.21
No. 9	5.37	64.46	1.33	14.29	3.94	0.07
Herrin (Illinois No. 6)[a]	5.12	65.14	1.17	14.60	3.48	0.09
Canyon	6.00	43.95	0.74	39.88	0.62	0.10
Beulah-Zap	6.78	38.76	0.56	45.28	0.85	0.14
Coalburg	5.12	73.34	1.37	9.41	0.79	0.06
Anderson-Dietz 1-Dietz 2[b]	6.31	47.65	0.86	37.01	0.72	0.08
Roland	5.00	47.20	1.40	40.50	0.78	0.08
Rosebud	6.13	48.00	0.73	32.35	1.85	0.10
Texas Lignite Coalbeds[c]	6.14	44.27	0.80	35.15	1.15	0.19
Pennsylvania Anthracite Coalbeds[d]	2.55	78.21	0.85	4.12	0.89	0.18

Note: Coal sample data from USGS US Coal Quality Database.

[a]Weighted average samples were used for analysis from the Illinois No 6, Springfield No 5, and Colchester coalbeds.

[b]Anderson-Dietz 1-Dietz 2 coalbed analysis data not available, all available data for Montana Sub-bituminous used instead.

[c]Analysis data for all available USGS Texas Lignite Coalbed samples used due to production data not being available by individual coalbed.

[d]Analysis data for all available USGS Pennsylvania Anthracite Coalbed samples used due to production data not being available by individual coalbed.

Table 5:   Recoverable Total Coal Reserves, by Region[a] (Million Short Tons), 2008.

Reserves (region)	Anthracite and bituminous[b]	Lignite	Total
	Coal reserves (Million short tons)		
Asia & Oceania	175,627	117,415	293,042
North America	123,912	145,431	269,343
Eurasia	95,598	155,767	251,364
Europe	6,906	77,296	84,202
Africa	34,743	192	34,934
Central & South America	7,595	6,193	13,788
Middle East	1,326	0	1,326
Total	445,706	502,293	948,000

Source: EIA International Energy Statistics Database.

[a]Totals may not be exact due to independent rounding.

[b]Includes sub-bituminous.

Table 6: Recoverable Total Coal Reserves by Country, 2008.

Reserves (Country)	Total	Anthracite and bituminous	Lignite
	Coal reserves (Million short tons)		
US	260,551	119,135	141,416
Russia	173,074	54,110	118,964
China	126,215	68,564	57,651
Australia	84,217	40,896	43,321
India	66,800	61,840	4,960
Germany	44,863	109	44,754
Ukraine	37,339	16,922	20,417
Kazakhstan	37,038	23,700	13,338
South Africa	33,241	33,241	—
Serbia	15,179	10	15,169
Colombia	7,436	7,017	419
Canada	7,255	3,829	3,426
Poland	6,293	4,782	1,511
Indonesia	6,095	1,676	4,419
Brazil	5,025	—	5,025
Greece	3,329	—	3,329
Bosnia	3,145	534	2,611
Mongolia	2,778	1,290	1,488
Bulgaria	2,608	2	2,606
Turkey	2,583	583	2,000

Source: EIA International Energy Statistics Database.

Table 7: Total Coal Production by World Region, 2006–2010.

| Producer (Region) | Coal production (thousand short tons) | | | | |
	2006	2007	2008	2009	2010
Asia & Oceania	3,859,826	4,126,453	4,483,134	4,757,761	5,136,770
North America	1,248,169	1,236,890	1,259,089	1,155,926	1,171,117
Europe	791,937	802,754	788,102	754,626	726,990
Eurasia	507,798	514,063	546,495	497,657	563,820
Africa	275,662	278,395	283,295	281,271	285,839
Central & South America	88,485	92,565	97,746	97,210	99,069
Middle East	1,685	1,801	1,754	1,294	1,294
Total	6,773,564	7,052,922	7,459,615	7,545,745	7,984,900

Source: EIA International Energy Statistics Database.

will likewise require control equipment. Another environmental issue often associated with coal is climate change, in that it produces the largest amount of carbon dioxide, per unit of heating value, of any fossil fuel. A number of CO_2 reduction methods are currently being researched at laboratories and power plants around the world.

A generic PC boiler system schematic diagram is presented in Fig. 3. The boiler is a large, usually rectangular vessel that contains a tube wall where the steam is produced and a series of heat exchange surfaces which further heat the steam,

Table 8:   Total Coal Production by Country, 2005–2009.

Producer (Country)	Coal production (thousand short tons)				
	2006	2007	2008	2009	2010
China	2,573,897	2,746,303	3,048,536	3,226,094	3,522,973
US	1,162,750	1,146,635	1,171,809	1,074,923	1,085,281
India	500,193	531,521	568,448	613,402	622,818
Australia	405,047	430,104	431,453	444,101	463,256
Indonesia	249,700	287,227	301,587	332,372	370,379
Russia	313,680	318,591	336,163	304,228	357,043
South Africa	269,817	273,005	278,017	276,219	280,788
Germany	220,554	225,526	214,268	202,410	200,955
Poland	171,135	159,773	157,993	148,356	146,237
Kazakhstan	106,555	107,837	122,437	111,173	122,135
Colombia	72,307	77,054	81,022	80,256	81,957
Turkey	70,829	83,075	87,526	87,633	79,090
Canada	72,757	76,461	74,680	69,375	74,840
Greece	71,415	73,092	72,444	71,532	62,303
Czech Republic	69,339	69,033	66,359	62,189	60,941
Ukraine	67,981	64,954	65,729	60,645	60,117
Vietnam	42,891	46,899	44,785	48,443	49,233
Serbia	40,543	40,949	42,669	42,438	41,169
Korea, North	38,698	33,442	35,641	34,785	34,785
Romania	38,496	39,441	39,530	37,436	33,985

Note: EIA International Energy Statistics Database.

Table 9:   Bituminous Coal Production by Country, 2006–2010.

Producer (Country)	Bituminous coal production (thousand short tons)				
	2006	2007	2008	2009	2010
China	1,985,981	2,147,264	2,383,624	2,519,422	2,751,709
US	1,076,964	1,066,482	1,094,413	1,000,524	1,010,165
India	465,707	494,065	532,710	575,845	586,348
Australia	329,095	356,895	358,446	368,710	388,356
Indonesia	246,458	282,511	293,425	324,108	360,845
South Africa	268,071	270,416	275,584	273,402	277,925
Russia	222,808	230,621	238,153	221,545	265,381
Kazakhstan	101,424	103,020	117,171	105,568	115,980
Poland	104,066	96,349	92,220	85,405	83,945
Colombia	72,307	77,054	81,022	80,256	81,957
Canada	61,249	64,841	63,746	57,746	63,526
Ukraine	49,806	51,836	52,309	48,824	48,288
Vietnam	39,608	44,187	42,798	46,455	47,201
Korea, North	27,205	22,179	24,771	23,810	23,810
United Kingdom	18,309	16,896	17,986	17,756	18,372
Mongolia	2,708	3,602	4,594	7,820	18,253
Czech Republic	14,754	14,213	13,959	12,127	12,605
Germany	24,008	24,470	18,920	13,090	12,050
Mexico	12,662	13,794	12,599	11,627	10,996
Venezuela	8,676	8,429	8,429	9,692	9,692

Note: EIA International Energy Statistics Database.

Table 10:   Anthracite Production by Country, 2006–2010.

Producer (Country)	Anthracite coal production (thousand short tons)				
	2006	2007	2008	2009	2010
China	487,255	491,657	546,352	577,526	630,754
Ukraine	17,919	12,912	13,240	11,778	11,664
Russia	9,138	9,548	7,036	6,612	7,920
Spain	4,168	3,842	3,474	4,476	3,538
Korea, North	2,734	4,123	2,895	3,309	3,309
South Africa	1,746	2,589	2,433	2,818	2,864
Korea, South	3,113	3,181	3,057	2,777	2,297
Germany	2,185	2,189	2,099	2,084	2,169
Vietnam	3,283	2,712	1,986	1,989	2,032
United States	1,538	1,568	1,712	1,921	1,940
United Kingdom	1,619	1,336	1,419	1,396	1,268
Australia	1,284	883	218	157	797
Swaziland	20	20	20	20	20

Note: EIA International Energy Statistics Database.

Table 11:   Lignite Production by Country, 2006–2010.

Producer (Country)	Lignite coal production (thousand short tons)				
	2006	2007	2008	2009	2010
Germany	194,361	198,867	193,250	187,235	186,735
China	100,661	107,382	118,560	129,146	140,511
Russia	81,734	78,422	90,974	76,072	83,741
Turkey	67,775	79,500	83,964	83,309	74,978
Australia	74,667	72,326	72,789	75,235	74,103
United States	84,248	78,585	75,684	72,479	73,177
Greece	71,415	73,092	72,444	71,532	62,303
Poland	67,069	63,425	65,773	62,951	62,292
Czech Republic	54,584	54,820	52,401	50,063	48,336
Serbia	40,507	40,940	42,662	42,429	41,160
India	34,486	37,457	35,738	37,557	36,470
Romania	38,496	39,441	39,530	37,436	33,985
Bulgaria	28,275	31,325	31,689	29,926	32,303
Thailand	21,022	20,105	20,157	19,606	20,346
Estonia	15,537	18,237	17,766	16,467	19,769
Canada	11,508	11,619	10,935	11,629	11,314
Hungary	10,970	10,822	10,366	9,905	10,006
Mongolia	6,192	6,581	6,507	8,101	9,578
Indonesia	3,242	4,716	8,162	8,264	9,534
Korea, North	8,759	7,141	7,975	7,665	7,665

Note: EIA International Energy Statistics Database.

first by radiant heat transfer then by convective heat transfer. As the combustion products approach the exit of the boiler they pass through the economizer which preheats the boiler feed-water and finally the air heater which preheats the combustion air.

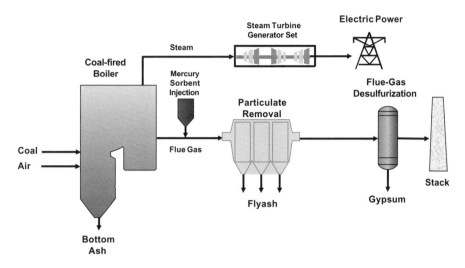

Fig. 3.   Pulverized coal fired boiler system.

As-received coal is first crushed to effect some level of size reduction and fed to mills which pulverize the coal to the proper size for use in the boiler — typically 70% passing through a 200-mesh screen (i.e. less than 0.074 mm). Air is fed to the mills to carry the coal into the boiler through an array of burners.

Additional air is fed to the burners to complete the combustion process. The burners are commonly mounted either on the walls of the boiler (wall-fired) or on the corners (tangentially fired). The latter are aimed slightly off the center of the boiler to impart a swirling motion to the flame, thus increasing the mixing of the air and coal. There is a variety of combustion practices designed to minimize NO_x formation in the boiler. High flame temperatures enhance NO_x formation (i.e. thermal NO_x). Most modern burners are low-NO_x burners that prolong the mixing of the air and fuel to minimize NO_x formation by lowering flame temperature. Some low-NO_x firing systems add insufficient (sub-stoichiometric) air at the burner to complete combustion while the required air (over-fire air) is added above the burners after the flame has cooled somewhat. Another technique to reduce NO_x is to add ammonia or urea to the boiler where the flue gas is at the appropriate temperature (1600°F to 2100°F) for the NO_x to react with the reagent to form molecular nitrogen and water vapor. This technique is known as Selective Non-Catalytic Reduction (SNCR). The flue gas can also be routed to a catalyst bed between the economizer and air heater, where the NO_x is catalytically reacted with ammonia to produce nitrogen and water vapor. This technique is referred to as Selective Catalytic Reduction (SCR). In order to achieve the desired level of NO_x reduction, two or three of these NO_x control methods are often used together.

Low-NO_x burners typically reduce NO_X by 40% to 60%, somewhat higher if used with overfire air. SNCR is generally able to effect a NO_X reduction of 30% to 75%, while SCR can achieve reductions ranging from about 70% to over 90%.

The precise level of reduction is dependent on individual boiler configuration and operating conditions. Most of the ash is removed from the bottom of the boiler as dry ash or as molten ash (slag) and much of it is sold for a variety of uses. Some of the fine ash particles are carried out of the boiler with the flue gas and must be removed. Two types of equipment are commonly used for this purpose — baghouses and electrostatic precipitators (ESPs). Baghouses consist of an array of filter bags hung in a steel casing. These filter bags, commonly about 6 inches in diameter by 20 feet or slightly longer are installed through holes in the roof of the baghouse. Rigid cages are inserted into the bags to maintain their shape.

The particle-laden flue gas flows into the casing and through the bags which filter out the particulate matter. Above the roof is the plenum into which the flue gas flows from the inside of the individual bags. The bags are periodically pulsed with high-pressure air to dislodge the collected particles, which fall into hoppers located at the bottom of the casing. The fly ash is removed from the hoppers for disposal or sale. Baghouses are capable of up to 99.99% particulate removal and are currently better at collecting the extremely small particles than are most ESPs.

ESPs are also installed inside a steel casing. An ESP consists of a series of grounded parallel steel plates with an electrified grid between each pair of plates. As the particle-laden flue gas enters the ESP, the electrified grid causes the particles to acquire a charge. The charged particles migrate to the grounded plates and adhere to them. The plates are periodically struck mechanically to dislodge the collected particles, and the particles fall into hoppers located at the bottom of the ESP for removal and sale or disposal. ESP efficiency is affected by an important ash property known as resistivity, which is the tendency of the ash particles to resist acquiring an electric charge. If the resistivity is too high, the particles are difficult to charge, which may require chemical conditioning or a larger ESP. If the resistivity is low, the particles acquire a charge more easily, but can also lose it just as easily and become re-entrained in the flue gas. Present-day ESPs are capable of up to 99.9% particulate removal.

SO_2 is commonly removed by contacting the flue gas with a material (sorbent) that reacts with the SO_2. Lime or limestone is commonly used as the sorbent. The contact normally occurs in one of two ways. Wet flue gas desulfurization systems use a vessel, usually a column, to achieve good contact between the flue gas and a water-slurry of the sorbent. The reaction product is initially a mixture of calcium sulfite and calcium sulfate, which is difficult to dewater and presents disposal problems. Air can be forced through the calcium sulfite/sulfate slurry to produce relatively pure calcium sulfate (gypsum), which is readily dewatered and is a valuable byproduct.

Alternatively, SO_2 can be removed in a spray dryer. In a spray dryer, a dense slurry of sorbent is sprayed into the vessel and the latent heat in the flue gas evaporates the water. The sorbent is carried by the flue gas to a baghouse, where it is removed. Since a relatively small amount of the sorbent reacts in one pass (typically lasting only a few seconds), it is recycled a number of times before its disposal. In retrofit applications, an existing ESP can be retained for the fly ash collection. If mercury removal is required, powdered activated carbon can be injected upstream of a baghouse or Flue Gas Desulphurization (FGD) system. If injected upstream of an ESP, the carbon is less readily captured due to its low resistivity. FGD systems capture some portion of the incoming mercury, and separate mercury removal equipment may not always be needed. In general, spray dryers are more effective for mercury capture than wet FGD systems. The vast majority of ESPs on coal-based systems are dry ESPs. While wet ESPs are more efficient at capturing very small particles such as fume, they are seldom used in utilities due to corrosion problems when used with flue gas.

CO_2 can be removed from the flue gas by contact with a material that absorbs CO_2, such as an amine. The CO_2 can then be separated from the amine, dried, compressed, and sent to an end use such as enhanced oil recovery or to a sequestration site. Due to the large volume of flue gas to be handled, primarily due to the nitrogen in the combustion air, such a process is expensive and requires substantial energy. A process that eliminates the need for handling these large volumes is oxy-combustion. An air separation unit (ASU) provides nearly pure oxygen for combustion, resulting in a flue gas that is predominantly CO_2. A portion of the flue gas is cooled and mixed with the oxygen to avoid excessive boiler temperatures. The rest can be used or sequestered geologically.

Power plants are classified as subcritical, supercritical, or ultra-supercritical based on the temperature and pressure of the steam. The critical point of water is 705.4°F and 3206.2 pounds per square inch absolute (psia). At and above these conditions, the vapor and liquid states of water are indistinguishable. From a practical standpoint, the pressure determines whether the boiler is sub-critical or supercritical since it is desirable to operate at the highest temperature/pressure practical for better efficiency. Construction materials generally limit operating temperatures to around 1100°F. While there are no set operating conditions, sub-critical boilers often operate at 2,500 psia, supercritical boilers around 3,500 psia, and ultra-supercritical boilers at significantly higher pressures. While efficiency improves with higher temperatures and pressures, cost/efficiency tradeoffs are required. The plant is also more prone to equipment/material and operational issues as operating conditions become more extreme.

5.2 *Fluidized Bed Combustion*

A second type of coal-fired boiler that is finding increased application is the fluidized bed combustion (FBC) boiler. It offers several potential advantages over the PC boiler. For example, much of the SO_2 is removed within the boiler itself by injecting limestone into the boiler, thus eliminating or greatly reducing the need for a separate FGD system. It operates at substantially lower temperatures (e.g. 1500–1700°F versus 2200°F–2300°F for PC combustion), thus nearly eliminating thermal NO_X formation. If the amount of NO_X remaining still presents a problem, only a polishing NO_X destruction system is needed. Another advantage is that the FBC boiler can burn a wide variety of properly sized solid fuels, allowing the operator to take advantage of lower priced/waste fuels such as waste coal or petroleum coke where available. In several regions that have long been major coal producers, there are numerous waste piles that still contain significant amounts of coal. These piles often pose environmental problems due to acid drainage. Waste coal from these piles can be burned in an FBC; however, it may or may not require some "fresh" coal as a supplemental fuel. Since the ash/limestone waste from the FBC contains some unreacted limestone, FBC waste can be used to mitigate acid mine drainage at the fuel supply source. Figure 4 shows a typical FBC arrangement.

FBC boilers can operate at atmospheric or elevated pressure, with the atmospheric systems heavily dominating the existing fleet. There are basically two types of atmospheric pressure fluidized bed boilers — bubbling bed and circulating bed. In both, air is injected into the bottom of a bed consisting of coal, lime, and ash. The bubbling version has a lower air flow so that the air essentially bubbles up through the bed, keeping it stirred while oxidizing the coal. In the circulating fluidized bed (CFB), the air flow is somewhat higher so that the bed is expanded and the particles separated, but still slow enough that the entire bed is not lifted out of the combustor. The circulating fluid bed is most commonly used. Again, fuel flexibility is a strength of FBC systems. They can be designed around a wide variety of solid fuels with heating values to 3,500 Btu/lb or lower (versus, say, 5,500 Btu/lb for PC boilers). This enables usage of a greater fraction of the coal resource, with less resultant waste.

Coal and limestone are fed into the boiler near the bottom of the bed. Fluidizing or primary air enters the bed from the bottom through a distribution system

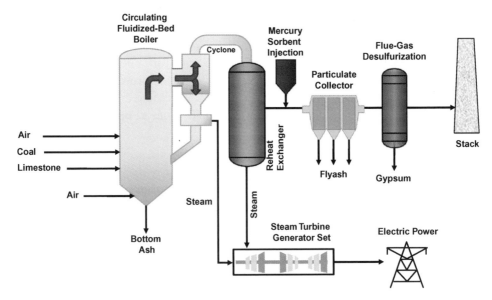

Fig. 4.   Typical fluid bed combustor.

that might consist of a perforated plate or an array of tubes configured to provide the fluidization. Secondary air is injected into the upper portion or above the bed to provide the balance of the combustion air. Bed temperatures are sufficiently high to convert the limestone ($CaCO_3$) to lime (CaO), which reacts with the SO_2 to produce calcium sulfate ($CaSO_4$) and/or calcium sulfite ($CaSO_3$). As the coal particles are combusted, they become smaller and rise through the bed and, along with the other bed material, are carried out of the bed with the flue gas. This material primarily consists of ash, reacted lime, and unreacted lime, along with some unburned carbon. It is removed from the flue gas in a cyclone and returned to the bed, possibly after it is cooled to recover energy as steam. This recycled material constitutes the bulk of the bed material, typically 80% to 90%. Maintaining this high level of basically inert material is necessary to maintain a stable bed temperature in the desired range, typically 1500°F to 1700°F. Steam is generated in tube walls within the boiler. Some of the ash/spent lime is removed from the bottom of the combustor for beneficial use or disposal.

Gas exiting the cyclone then passes through a series of heat transfer bundles located in a vessel downstream of the cyclone, where the steam generated in the boiler is superheated before it is used to generate electricity in a steam turbine/generator set. As with a PC boiler, additional energy is recovered from the flue gas in an economizer, then an air preheater. As a minimum, the flue gas will be treated in either a baghouse or ESP for removal of fly ash. If a polishing step for additional NO_X destruction is needed, SCR or SNCR can be used at the appropriate point. Ammonia or urea can be injected (SNCR) as the flue gas exits the cyclone or a small catalyst bed also requiring ammonia (SCR) can be installed between the economizer and air preheater. SO_2 removal within the FBC is typically over 90%. However, if that is insufficient, a polishing FGD system may also be required and would use the same technologies as would be used with a PC boiler. If mercury removal is required, activated carbon can be injected upstream of a baghouse.

There has been some commercial development of pressurized FBC systems. Typically, the combustor and FBC are installed in a larger pressurized vessel to avoid high pressure differential on the walls of the FBC. The major difference is that hot flue gas is passed through a combustion turbine after the bulk of the particulate matter has been removed, which increases the overall efficiency. The downstream equipment and processes are similar to an atmospheric FBC system. Their use is not currently common since some operational issues still need to be resolved.

5.3 *Gasification*

One use of coal that has received increased attention in recent years is gasification. The coal is fed to a gasifier to produce syngas (synthesis gas) — primarily carbon monoxide (CO) and hydrogen (H_2). After the impurities are removed, the syngas can be burned in a gas turbine/generator set to produce electric power or be further processed to produce a variety of chemicals or other fuels such as synthetic pipeline gas (methane; note that synthetic gas is to be distinguished clearly from synthesis gas) or, via Fischer–Tropsch synthesis, liquid transportation fuels. To recap, gasification can yield syngas to produce (1) electricity, (2) chemicals, and/or (3) fuels.

A worldwide gasification database is available online at: http://www.netl.doe. gov/technologies/coalpower/gasification/worlddatabase/index.html. According to this source, at the time of writing, there are 191 active gasification projects worldwide, consisting of over 500 gasifiers. Of the 191 projects, 137 are in operation and 54 are in some stage of development, ranging from preliminary engineering to construction. Of this total, 92 are (or will be) fed with coal. Fourteen are used to produce electricity and the balance produce syngas, which will be converted into a variety of products. Regardless of the end use, similar syngas cleanup steps are required. Figure 5 is a generic gasification system configured to produce electric power in an integrated gasification combined-cycle (IGCC) power plant and chemical byproducts.

Combined cycle refers to a system in which a gaseous fuel (e.g. methane, syngas) is used in a combustion turbine/generator set to produce electricity; heat is recovered from the turbine exhaust to generate steam, which is then used in a steam turbine/generator set to produce additional electricity. Although gasification technology dates back over 100 years, IGCC is a comparatively new application that couples a gasifier with a combustion turbine, in combined-cycle mode.

Although gasifiers can operate at atmospheric pressure, elevated pressure operation is currently most common. Gasifiers can be entrained flow, fluidized beds,

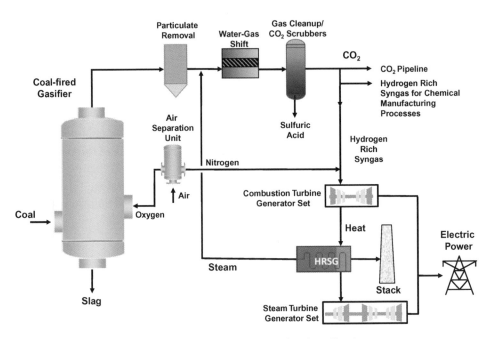

Fig. 5.   Simplified flow diagram of coal gasification.

or fixed beds, with most newer installations being entrained flow or fluidized bed designs. Gasification is the reaction of a hydrocarbon fuel with oxygen to form a mixture consisting primarily of carbon monoxide and hydrogen. In the simplest sense, it is a partial combustion reaction in which a portion of the coal is oxidized under extremely oxygen-deficient conditions to provide the energy for the gasification reactions. These reactions can be generalized as:

$$C_xH_y \text{ (coal) } + O_2 + H_2O => CO + CO_2 + H_2.$$

The primary feeds to the gasifier are coal and air or oxygen, with oxygen being more common in recent installations where the oxygen is supplied as a liquid by an Air Separation Unit (ASU). It is vaporized, compressed, and piped to the gasifier. The coal can be fed to a pressurized gasifier in one of two ways. It can be pressurized through a series of lockhoppers and discharged to the gasifier. Alternatively, it can be mixed with water and pumped into the gasifier as a dense-phase slurry. Some gasifiers inject the coal at two points. If a slurry feed system is used, the water takes part in the gasification process by reacting with the carbon in the coal. Steam may also be added to serve as a reactant, especially in dry feed systems.

The bulk of the mineral matter in the coal is removed from the gasifier base as either dry ash or as molten slag. If removed as slag, it flows from the bottom of the gasifier into a water bath and is removed through a series of pressure let-down lockhoppers for dewatering and subsequent sale. Dry ash is also removed through a series of lockhoppers.

Many gasifiers are equipped with an integrated cyclone to capture coarse, partially reacted solids and return them to the gasifier in order to completely consume the carbon. The raw syngas leaving the gasifier cyclone is cooled, generating steam in the process. In gasification systems that are used in an IGCC mode, this steam is used as part of the feed to the steam turbine and for other heat requirements in the plant. After some level of cooling, the syngas is treated to remove particulate matter. This can be accomplished by ceramic filters and/or cyclones. If extremely high efficiency removal is required, a wet particulate scrubber can be used.

The next step in the syngas treating process is water gas shift/carbonyl sulfide (COS) hydrolysis. In gasification, the bulk of the sulfur in the coal is transformed into hydrogen sulfide (H_2S), but some COS is also formed. While H_2S is readily removed in a number of processes, COS is not. It must therefore be hydrolyzed to H_2S for effective removal. The water gas shift reaction is a catalytic reaction in which CO and water vapor react to form hydrogen and CO_2. The water gas shift is carried out to provide the correct CO:H_2 ratio when the syngas will be used to create various chemical or fuel products, or to convert as much of the CO to CO_2 as possible, thereby allowing for efficient CO_2 separation prior to combustion of H_2 in an IGCC plant. (Although no coal-based IGCC power plants currently practice water gas shift and subsequent CO_2 capture at full scale, intensive studies are under way, and this approach constitutes an important potential pathway to control CO_2 emissions from future coal-fueled IGCC power plants.)

Returning to the gasification process description, the next step involves removal of pollutants (primarily H_2S) from the syngas. Sulfur and CO_2 are removed prior to combustion since the volume of the syngas is much smaller than the volume of the syngas combustion products; the smaller pre-combustion gas volume simply makes their removal easier. Since the syngas is relatively cool, activated carbon can be used to remove mercury, if needed. While it is possible to remove H_2S at higher temperature with a dry sorbent, most systems use a liquid that absorbs the H_2S, which is then stripped from the sorbent and recovered either as elemental sulfur or sulfuric acid. The liquids that are currently specified for this purpose include amines, glycols, and chilled methanol. These systems have the advantage that they can be designed as two-stage systems that offer the ability to separately remove and recover the CO_2 for beneficial use or sequestration, thus avoiding or reducing the emission of that greenhouse gas.

At this point, the syngas is ready for further processing into a chemical feedstock, or to fuel a combined cycle unit. In the latter case, it is diluted with byproduct nitrogen from the ASU or steam and fed to a combustion turbine/generator set where it is burned with air to generate electricity. The diluents add mass and maintain the correct temperature for the combustion turbine. The hot combustion products are passed through a heat recovery steam generator (HRSG) that uses the heat to generate steam. This steam is used to drive a steam turbine to generate additional electricity. Finally, the pollutant-free turbine exhaust is then discharged to the atmosphere through the plant stack.

Pulverized coal boilers first came on line in the 1920s and have long been the workhorse of the electric utility industry, with nearly a thousand boilers in the US alone, the largest rated at 1,300 megawatts electric (MWe). Significant work on coal-fired FBCs basically started in the 1960s resulting in over 100, mostly small, units in the US. The largest single unit is around 500 MWe. Coal gasification for syngas applications began in a small way over 50 years ago, but serious work in power generation did not start until the 1980s. A single gasifier train is limited to about 300 MWe. As stated earlier, there are 92 coal-fired gasification projects worldwide that are either operating or in different stages of development.

5.4 *Liquefaction*

Another form of coal conversion is coal liquefaction. Although the term "liquification" can be found in historical literature, it fell into disuse over a generation ago and has been supplanted totally by "liquefaction." Most recent projects for coal liquefaction use indirect liquefaction. The coal is gasified to produce syngas, the syngas is shifted to adjust the hydrogen/carbon ratio, and the syngas is catalytically reacted to yield a product slate that can range from light hydrocarbon gases to heavy waxes, which might be subjected to additional refinement. With proper catalyst selection and operating conditions, a crude diesel fuel can be produced by the catalytic reaction. The diesel fuel can then be subjected to further refining, to produce most liquid transportation fuels.

The hydrogen-to-carbon ratio of liquid fuels is higher than that of coal. In direct liquefaction, the coal is directly converted to a synthetic crude oil by the addition of hydrogen. It is then further processed to produce the desired end products. In a typical system, a recycle stream of raw synthetic oil is mixed with crushed coal and catalyst to form a slurry. The slurry is then fed to a fluidized bed reactor, which operates at elevated temperature and pressure. Hydrogen is separately fed to the reactor to react with the coal and produce a raw synthetic crude oil. The reactor effluent is then separated from the catalyst and the unreacted coal residue. After the solids are separated, a portion of the oil and all of the catalyst are recycled to the slurry preparation area. The balance of the raw oil is then subjected to typical refinery operations such as hydrotreating, cracking, and distillation. End products include both diesel fuel and gasoline. The unreacted coal that was separated from the reactor product can be gasified and processed, as described earlier, to supply the hydrogen for the liquefaction reactor and subsequent processing of the raw oil.

6 Other Coal Uses

Minor amounts of coal are used as a heat source in a variety of industrial applications such as the manufacture of steel, cement, glass, ceramics, and paper. The use of coal is declining in cement making, where it is being displaced by waste materials such as scrap tires and biomass. The conditions in a cement kiln are ideal for the combustion of scrap tires. If biomass is used in a cement kiln, co-firing some coal is still necessary to attain the required temperatures. Fly ash resulting from coal combustion can also be used in cement making.

Some of the coal used for steelmaking is used to produce metallurgical coke. Coke, limestone, and iron ore are charged into a blast furnace to produce crude molten iron (hot metal), which is further processed into various grades of steel. The coke serves as the reducing agent and heat source to reduce the iron oxides to elemental iron. Coke is produced by charging it into slot ovens, a connected series of which make up a battery. The coal is heated through the oven walls. The coal is heated to high temperatures in the absence of air to drive off the volatile materials which are recovered as valuable byproducts. Coal used for this purpose tends to be more expensive since certain properties are required. The coal must be strongly caking, since metallurgical coke must have sufficient mechanical strength to support the charge (burden) in the blast furnace. It must swell slightly during the coking process to maintain good contact with the walls to prevent them from overheating. If the tendency to swell is too great, the walls of the oven can be damaged and the coke will be difficult to push from the oven. The amount of coal used for this purpose in the US has declined to about 15 million tons in recent years. A major reason is that other fuels such as granulated or pulverized coal can now be substituted for a portion of the coke. However, a significant amount of coke is still needed to support the burden and to provide the necessary porosity for the hot air blast to pass through the burden.

7 Challenges in Coal Production

Any industrial activity that involves the use of heavy machinery has the potential to result in injury or death. The production of fossil fuels, including coal, is no exception. The production and use of fossil fuels also has the potential to adversely affect the environment. The preceding paragraphs discuss how the impacts of coal consumption are virtually eliminated through technology and operating techniques. The hazards and adverse environmental impacts peculiar to coal mining can likewise be largely mitigated through good operating practices and the application of the appropriate technology.

Coal is mined either by removing the rocks and dirt above the coal and digging the coal out (surface mining) or by removing the coal from the seam by constructing a shaft to the seam and removing the coal from underground (deep mining). The adverse environmental impact most associated with both types of mining is acid drainage, which results from water leaching due to sulfur compounds, primarily iron sulfides (pyrites), present in coal and thereby becoming acidified. This problem is solved by collecting the water and neutralizing it before discharge.

Since surface mining removes all dirt and rock above the coal seam(s) to be mined, the surface is essentially destroyed and this is unavoidable. However, the topsoil can be kept separate from the underlying rocks and dirt, and barriers made to avoid erosion during the mining process. At the conclusion of mining, the underlying rock and dirt are used to establish pre-mining surface contours then the topsoil is spread over the area. The area can then be planted with native plants and trees to restore the areas as close as possible to the pre-mining conditions. If desired, the area could actually be reclaimed to improve the area for wildlife habitat or other uses. While these techniques will eventually return the mine site to a good condition, it can take some number of years.

The risk of injury in surface mines is primarily due to human interaction with machinery. Training and the establishment/enforcement of safe work practices are the main methods used to prevent accidents.

Deep mining has several major challenges for environmental impact specific to deep mining in addition to the construction and operating activities at the mine mouth. One is the risk of subsidence, which can result in damage to structures, groundwater, and the surface environment. This can be mitigated to a degree by the mining technique. In some cases, even when subsidence occurs, the impact is minor. Beneficiation (washing) is often used to separate the mineral matter (ash) from coal. The result is a higher quality coal product and a waste portion that still contains some coal. Disposal of the solid waste from mining and beneficiation presents a second major environmental problem. It is often disposed of by simply piling it near the mine. The runoff from these waste piles is acidic and must be treated to avoid environmental consequences. Both old mines and waste piles occasionally catch fire. These fires can smolder for decades, emit noxious fumes, and are difficult or impossible to extinguish. Depending on the coal content of the waste piles, some of them can be burned in FBCs. The excess ash is then returned to the site of the waste pile and the unreacted lime in the ash serves to effectively neutralize any acidity in the runoff.

Safety challenges peculiar to deep mining include rock (roof) falls, methane and coal dust explosions/fires, breathing coal dust for prolonged periods, and pockets of CO_2 (sometimes referred to as "black damp"). The risk of rock falls is mitigated by properly supporting the roof in active mining areas by installing posts or bolting the roof to stable rock above the roof. A spray of limestone slurry is used to coat the mine walls to greatly reduce the problems associated with the coal dust. Further protection from the effects of breathing the dust is provided by requiring miners to wear respirators. Both methane and CO_2 tend to evolve from coal seams. Both gases can asphyxiate miners, and methane can cause gas explosions. Both hazards can be effectively mitigated by rigorous monitoring of the atmosphere and adequate ventilation.

Although safety and environmental challenges still exist in coal mining, most can be readily mitigated. Where these mitigation techniques are rigorously practiced, environmental and human costs have been greatly reduced since the early days of coal mining.

8 Challenges in Coal Usage

The abundance and low cost of coal have made it the fuel of choice for nations endowed with this natural resource and for those nations that choose to import it. Early use of coal in these countries has demonstrated the need, however, to investigate technologies that not only reduce the emissions associated with this fuel but also allow it to combust more efficiently, thereby conserving this energy source for future generations. Examples of innovative research, development, and deployment for clean and efficient coal utilization can be found at the US Department of Energy's National Energy Technology Laboratory Major Demonstration Projects website http://www.netl.doe.gov/technologies/coalpower/cctc/.

8.1 *Worldwide Coal Usage*

The largest use of coal worldwide, by far, is to provide steam to generate electric power. While coal is used in a variety of industries, to provide heat to various processes, the quantities involved are much less than the quantity used in power generation. There are nearly 6,400 operating units worldwide rated at over 25 MWe, including over 322 FBCs. Not surprisingly, countries or regions that rely on coal for a significant portion of their electrical generation are usually those that have significant coal reserves. For example, China has nearly 2,300 operating coal-fired generating units. The US is second in coal-based generation, with approximately 1,100 units. On a regional basis, the European Union has over 860 operating units, while the Commonwealth of Independent States (former Soviet republics) accounts for over 600 units. Of these over 400 are located in Russia.

T. Sarkus, A. Radziwon and W. Ellis

8.2 *Coal Usage Projections for OECD Nations*[a]

Sections 8.2 and 8.3 address international coal marketed energy projections based on the International Energy Outlook 2011 (IEO, 2011) prepared by the US Energy Information Administration (EIA) through 2035. The coal consumption projections are summarized by membership grouping in the Organization for Economic Cooperation and Development (OECD) and utilize projections from the Reference case. The IEO2011 report can be accessed at http://www.eia.gov/forecasts/ieo/. Data for the pie-chart graphics displayed in Secs. 8.2 and 8.3 are from the International Energy Agency (IEA) statistics database for 2009, which can be accessed at http://www.iea.org/stats/prodresult.asp?PRODUCT=Electricity/Heat. Hydroproduction includes pumped storage. Statistics are for gross generation.

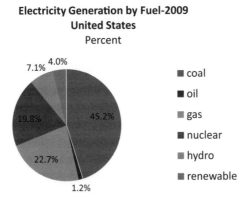

Electricity Generation by Fuel-2009
United States
Percent

Coal consumption in the US totaled 22.4 quadrillion Btu in 2008, which is anticipated to rise to 24.3 quadrillion Btu in 2035. Electricity generation from coal in the US is anticipated to account for 22% of power generation growth between 2008 and 2035. The increase in coal consumption is projected to result not only from increased coal-based power generation but also to the startup of several coal-to-liquids (CTL) plants.

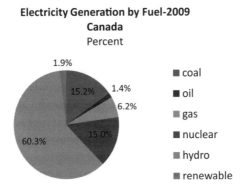

Electricity Generation by Fuel-2009
Canada
Percent

---

[a]Charts in Secs. 8.2 and 8.3 are read starting with coal, then oil, etc.

Electricity Generation by Fuel-2009
Mexico
Percent

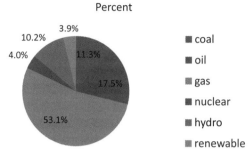

Canada is expected to experience a decrease of 0.2 quadrillion Btu in coal consumption from 2008 to 2035 due to the phasing out of Ontario's 6.1 GW of coal-fired generating capacity by 2014 (IEA Online Statistics ©OECD/IEA 2012). Mexico added 0.7 GW of coal-fired generating capacity in 2010.

Electricity Generation by Fuel-2009
United Kingdom
Percent

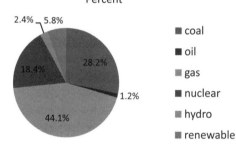

The coal consumption of European OECD member countries is projected to decline from 12.5 quadrillion Btu in 2008 to 10.4 quadrillion Btu in 2035. Consumption in this region is attributed almost in its entirety to the electricity and industrial sectors.

Electricity Generation by Fuel-2009
Republic of Korea
Percent

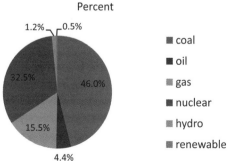

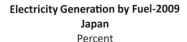

Electricity Generation by Fuel-2009
Japan
Percent

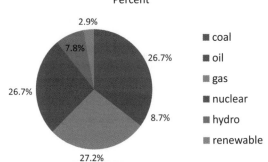

Electricity Generation by Fuel-2009
Australia
Percent

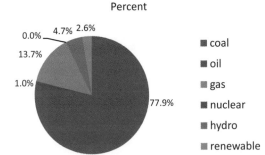

The OECD Asia region is anticipated to show no increase in coal consumption. Japan is anticipated to decrease coal use by 0.9 quadrillion Btu due to a shift to alternative sources of electricity generation, and South Korea is anticipated to increase coal use by 0.8 quadrillion Btu from 2008 to 2035. Australia and New Zealand are anticipated to maintain their current coal consumption through 2035. South Korea increases coal consumption from 2.6 quadrillion Btu in 2008 to 3.4 quadrillion Btu in 2035, primarily due to increased power production.

Electricity Generation by Fuel-2009
Germany
Percent

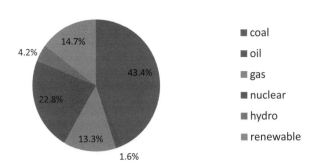

8.3 *Coal Usage Projections for Non-OECD Nations*

The non-OECD nations by contrast, are anticipated to experience significant growth in coal consumption, particularly among the Asian nations. These nations are expected to see coal consumption grow from the 2008 total of 92.2 quadrillion Btu to 162.5 quadrillion Btu in 2035, which is a 76% increase.

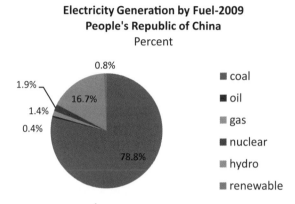

Electricity Generation by Fuel-2009
People's Republic of China
Percent

China's use of coal for electricity generation is projected to increase from 28.7 quadrillion Btu in 2008 to 63.4 quadrillion Btu in 2035. The growing demand for electricity is anticipated to require the addition of 485 GW by 2035 to their estimated existing capacity of 557 GW. Roughly 52% of coal use in China in 2008 was in their industrial sector, a significant portion of which was for iron and steel production.

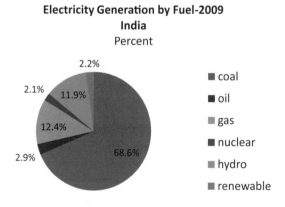

Electricity Generation by Fuel-2009
India
Percent

Over half (54%) of Indias anticipated growth in coal consumption is projected to be for electricity production and industrial use. In 2008, India consumed 6.7 quadrillion Btu of coal for electricity generation, which is anticipated to grow to 11.4 quadrillion Btu in 2035. This growth will require an increase in the national capacity of 99 GW in 2008 to an estimated 171 GW in 2035.

Electricity Generation by Fuel-2009
Russian Federation
Percent

0.3%

17.8%   16.5%   1.6%

16.5%

47.3%

■ coal
■ oil
■ gas
■ nuclear
■ hydro
■ renewable

Russia consumes the largest amount of coal (4.5 quadrillion Btu in 2008) of any nation of the non-OECD Europe and Eurasia regions. Coal provided 15% of the total energy requirements for Russia in 2008. Coal consumption projections for Russia are anticipated to be 4.9 quadrillion Btu in 2035 with a slight decrease in the percentage of electricity generated from coal.

Electricity Generation by Fuel-2009
South Africa
Percent

1.7%

0.0%   5.1%   0.1%
0.0%

93.0%

■ coal
■ oil
■ gas
■ nuclear
■ hydro
■ renewable

Coal consumption in Africa is projected to increase by 2.5 quadrillion Btu in the 2008 to 2035 period. South Africa consumes approximately 93% of the coal utilized by the African continent. Additionally, South Africa is anticipated to restart 3.8 GW of inactive capacity by 2012 and add additional generating capacity of 9.6 GW by 2017. Power shortages are thought to be driving the capacity additions.

Electricity Generation by Fuel-2009
Brazil
Percent

5.3% 2.1% 3.1%
             2.9%   2.8%

83.8%

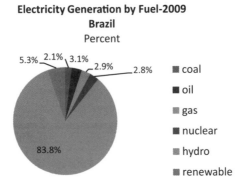

■ coal
■ oil
■ gas
■ nuclear
■ hydro
■ renewable

Central and South American countries consumed 0.8 quadrillion Btu of coal in 2008, with Brazil, consuming 61% of the region's demand for coal. Colombia, Puerto Rico, Argentina, and Peru accounted for most of the remaining coal demand. The Central and South American region is projected to increase coal consumption by 1.5 quadrillion Btu, from 2008 to 2035. Brazil is anticipated to account for most of the increased consumption for coke manufacturing.

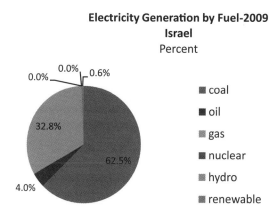

Electricity Generation by Fuel-2009
Israel
Percent

The Middle Eastern countries consumed an estimated 0.4 quadrillion Btu of coal in 2008, which is anticipated to remain near this level of consumption through 2035. Israel (an OECD member) consumes approximately 83% of the 0.4 quadrillion Btu total, with Iran accounting for most of the remaining consumption.

9 Carbon Dioxide

Coal is a dependable, cost-effective energy source. However, there are concerns over the impacts of concentrations of greenhouse gases (GHGs) in the atmosphere — particularly carbon dioxide (CO_2) resulting from burning coal as a fuel.

9.1 *Carbon Dioxide Produced per Kg of Coal*

The combustion of coal produces CO_2; however, the amount of CO_2 produced does vary with the rank of the coal. Table 12 illustrates the CO_2 calculated to be produced from several ranks of coal with an assumed carbon content.

Table 12: CO_2 Produced (Calculated) from Various Ranks of Coal.

Coal type	Percent carbon	Kg CO_2 per kg coal	HHV, Kcal/kg coal	kg CO_2/MJ
Lignite	44.27	1.62	4234	0.0915
Sub-bituminous	48.01	1.76	4461	0.0942
Bituminous	65.14	2.39	6441	0.0885
Anthracite	78.21	2.87	6292	0.1088

9.2 Geologic Storage of Carbon Dioxide

Geologic storage of CO_2 is a focus area of the NETL Carbon Storage R&D Program (http://www.netl.doe.gov/technologies/carbon_seq/corerd/storage.html) and involves the injection of CO_2 into deep geologic formations (injection zones) overlain by competent sealing formations and geologic traps that will contain CO_2. Current research and field studies are focused on developing better understanding of 11 major types of geologic storage reservoir classes, each having their own unique characteristics and challenges. The different storage formation classes include deltaic, coal/shale, fluvial, alluvial, strandplain, turbidite, eolian, lacustrine, clastic shelf, carbonate shallow shelf, and reef. Basaltic interflow zones are also being considered as potential reservoirs. These storage reservoirs contain fluids that may include natural gas, oil, and/or saline water; any of which may affect CO_2 storage differently. Following are brief descriptions taken from the NETL Carbon Sequestration Program R&D website of CO_2 storage capability for fluids that may be present in more conventional clastic and carbonate reservoirs (saline water, and oil and gas), as well as unconventional reservoirs (unmineable coal seams, organic-rich shales, and basalts):

9.2.1 Saline-bearing formations

Potential storage reservoirs may contain layers of porous rock saturated with brine (fluid containing >10,000 ppm total dissolved solids). This is the most widespread reservoir type and thus has the highest potential storage capacity of all geologic media. Saline water can contain minerals that could potentially react with the injected CO_2 to form solid carbonates. The reactions can increase permanence but could decrease the porosity in the immediate vicinity of an injection well. R&D efforts to investigate saline formations as potential storage media are ongoing to better understand their behavior under CO_2 storage conditions.

9.2.2 Natural gas and oil-bearing formations

Oil and natural gas can be present in permeable rock formations with an overlying low-permeability rock layer that forms a stratigraphic trap that holds the oil and gas in place. As a value-added benefit, when CO_2 is injected into an oil-bearing formation, it can produce additional oil (e.g. 10%–15%). This process is known as enhanced oil recovery (EOR). The ability of EOR to increase oil projection and provide storage for CO_2 is currently under investigation.

9.2.3 Unmineable coal seams

Unmineable coal seams include those that are too deep or too thin to be economically mined. All coals have varying amounts of methane (CH_4) adsorbed onto pore surfaces, and wells can be drilled into unmineable coal seams to recover coal bed methane (CBM). Initial CBM recovery methods leave a considerable amount of CH_4 in the formation, but this can be increased by injecting CO_2, which is adsorbed onto the surface of the coal, and results in the desorption of methane (enhanced coal bed methane [ECBM] recovery). Once the methane has been depleted from the formation, additional CO_2 may be injected for permanent storage.

9.2.4 *Organic-rich shale basins*

Shale, the most common sedimentary rock, is characterized by thin horizontal layers of rock with low permeability in the vertical direction. Many shales contain 1%–2% organic material in the form of hydrocarbons, which provide an adsorption substrate for CO_2 storage similar to CO_2 storage in coal seams. Modern drilling methods may allow the permeability of shales to be increased to provide economical CO_2 storage.

9.2.5 *Basalt*

Basalt formations are geologic formations of solidified lava. Basalt formations have a unique chemical makeup that could potentially convert much of the injected CO_2 to a solid mineral form, thereby isolating it from the atmosphere permanently.

9.3 *Carbon Dioxide Utilization*

CO_2 utilization is another focus area of the NETL Carbon Storage R&D Program (http://www.netl.doe.gov/technologies/carbon_seq/corerd/co2utilization.html), which investigates pathways and novel approaches for reducing CO_2 emissions by developing beneficial uses for the CO_2 that will mitigate CO_2 emissions in areas where geologic storage may not be an optimal solution. CO_2 can be used in applications that could generate significant benefits. It is possible to develop alternatives that can use captured CO_2 or convert it to useful products such as chemicals, cements, or plastics. Revenue generated from the utilized CO_2 could also offset a portion of the CO_2 capture cost.

The development of CO_2 utilization is ongoing. Specific applications of CO_2 utilization are further addressed at the readers' convenience in the following publication: "The US Department of Energy's R&D program to reduce greenhouse gas emissions through beneficial uses of carbon dioxide" (http://www.netl.doe.gov/technologies/carbon_seq/refshelf/project%20portfolio/2011/SelectedPubs/GHG35_final.pdf).

9.4 *Cost of Carbon Storage*

A cost and performance baseline study "Cost and Performance Baseline for Fossil Energy Plants Volume 1: Bituminous Coal and Natural Gas to Electricity, Revision 2, November 2010" (http://www.netl.doe.gov/energy-analyses/pubs/BitBase_FinRep_Rev2.pdf) commissioned by US DOE, NETL, provides a general cost and performance assessment for Pulverized Coal (PC) Combustion, Integrated Gasification Combined Cycles (IGCC), and Natural Gas Combined Cycles (NGCC) power plants. The cost estimates are presented for the plants both with and without CO_2 capture and storage. The study assumes that the plants use technology available today. The study shows the Levelized Cost of Electricity (mills/kWh, 2007 US$) for the Pulverized Coal (PC) Sub-critical Plant without Carbon Capture and Storage (CCS) to be 75.3 mills/kWh, PC Sub-critical Plant with CCS to be 139.0 mills/kWh, PC Supercritical Plant without CCS to be 74.7 mills/kWh, and PC Supercritical Plant with CCS to be 135.2 mills/kWh. Cost information for other power generation technologies such as IGCC and NGCC can also be found within the reference document if the reader desires additional information.

References

1. Modified from Geological Survey Circular 891, Coal Resource Classification System of the U.S. Geological Survey (2003).
2. J. Tully, U.S. geological survey coal fields of the conterminous United States, Digital Compilation, Open-File Report No. 96–92 (1996).
3. R. Flores, G. Stricker and S. Kinney, Alaska coal geology, resources, and coalbed methane potential, U.S. Geological Survey, Report DDS-77 (2004).
4. National Energy Technology Laboratory. Geologic storage focus area. U.S. Department of Energy, Available at: http://www.netl.doe.gov/technologies/carbon_seq/corerd/storage.html. Accessed 5 March 2012.
5. CO_2 Utilization Focus Area. U.S. Department of Energy, National Energy Technology Laboratory. Available at: http://www.netl.doe.gov/technologies/carbon_seq/corerd/co2utilization.html. Accessed 5 March 2012.
6. D. Damiani, J. Litynski, H. McIlvried, D.Vikara and R. Srivastava, The U.S. Department of Energy's R&D program to reduce greenhouse gas emissions through beneficial uses of carbon dioxide. Published online at Wiley Online Library (wileyonlinelibrary.com). DOI: 10.1002/ghg.35.
7. J. Black, Cost and performance baseline for fossil energy plants Volume 1: Bituminous coal and natural gas to electricity, Revision 2, November 2010. National Energy Technology Laboratory. Available at: http://www.netl.doe.gov/energy-analyses/pubs/Bit Base_FinRep_Rev2.pdf. Accessed December 2011.

Chapter 4

Petroleum Liquids

William L. Fisher

Jackson School of Geosciences
University of Texas at Austin, TX, USA
wfisher@jsg.utexas.edu

Petroleum liquids constitute the largest energy source in the global energy mix. However, these liquids have been declining as a percent of the energy mix for about 30 years and will continue to decline as alternative sources become commercial. The absolute volume of production will likely increase to levels 15% to 20% greater than that at present. The proven reserves of petroleum liquids, those quantities that can be produced at current prices and technology, range from 1.128 to 1.475 trillion barrels. The higher estimates include unconventional sources such as tar sands and oil shales. Estimates of ultimate petroleum liquid resources, yet to be discovered or developed, vary by a factor of three or more. Lower estimates, on the order of 2 trillion barrels, are based largely on projections of historical discovery trends. These estimates indicate early peaking of global production. Higher estimates where discovery is based on geologic factors, and which also include reserve growth and unconventional resources, range from 3 to 5 trillion barrels; these estimates indicate greater future production, commonly with long-term plateaus. The costs of petroleum liquids production vary by more than one order of magnitude. Certain large conventional fields, mostly situated in the Middle East and North Africa, have the lowest production costs. Future production will increasingly come from more costly sources, such as enhanced recovery, ultra deep water, Arctic areas, and unconventional sources.

1 Introduction

Petroleum liquids, as generally defined, include crude oil, condensate, and natural gas liquids. Petroleum liquids are used in a wide range of products and chiefly as transportation fuels (gasoline, diesel, jet fuel, propane, and other fuels). Globally, about 60% of liquids are used in transport, while in the US, the percentage is about 70%. Most projections of oil demand show a growing percentage will be devoted to transport fuels, with much of the growth in the emerging economies. In the US, the amount of petroleum liquids used in transport is steady or even declining slightly. In Europe, the decline is greater than that in the US. These trends will continue as the use of biofuels, compressed natural gas, and electricity for transportation increases as alternatives to petroleum.

The second largest use of petroleum liquids (about 17% globally) is for non-energy uses, mostly in the petrochemical industry, where petroleum is used to manufacture products such as plastics and fertilizers. The remaining use of petroleum liquids is in agricultural power equipment, power generation, and residential heating oil. The volume of non-transport use of petroleum liquids is projected to remain essentially at current levels, but with a decreasing percent as transport use increases.

2 Production and Consumption

Oil constitutes 40% of total energy consumed worldwide. In 1973, it contributed 48%, and in the early 1980s, reached its peak of about half of all energy used, but has declined since then. While absolute volumes of oil consumed will likely increase in the future, its decline in percent of total energy used will continue as natural gas and renewable energy use increase.

In the early part of the previous decade (2000–2010), global production and consumption of petroleum liquids was about 77 million barrels per day (mmb/d). In the middle years of the decade, production volumes rose to about 85 mmb/d and since then, have averaged about that same volume (Fig. 1). This leveling of production is due to the comparative stability or even a slight decline of oil consumption in the US and the continuing decline in Europe, offsetting increases in the world's emerging economies. Most recent projections of liquids production out to 2030 fall with a range of 95 to slightly more than 100 mmb/d, reflecting declining demand in most of the established economies and increasing demand in the emerging economies.

The top 15 producing countries provide 80% of total world production of petroleum liquids (Table 1), led by Saudi Arabia, Russia, and the US, which account for nearly 30% of the total.

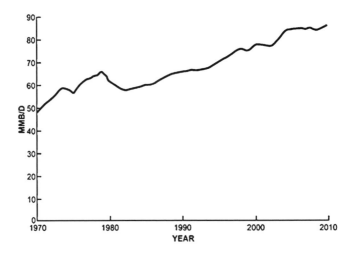

Fig. 1. Global production volumes of petroleum liquids during 1970–2010.

Table 1: Top World Producers, Consumers, Importers, and Exporters of Oil in 2009. Data in Thousands of Barrels per Day.[1]

Rank	Country	Production	Rank	Country	Consumption
1	Russia	9,934	1	US	18,771
2	Saudi Arabia	9,760	2	China	8,324
3	US	9,141	3	Japan	4,367
4	Iran	4,177	4	India	3,110
5	China	3,996	5	Russia	2,740
6	Canada	3,294	6	Brazil	2,522
7	Mexico	3,001	7	Germany	2,456
8	United Arab Emirates	2,795	8	Saudi Arabia	2,438
9	Brazil	2,577	9	Korea, South	2,185
10	Kuwait	2,496	10	Canada	2,151
11	Venezuela	2,471	11	Mexico	2,084
12	Iraq	2,400	12	France	1,828
13	Norway	2,350	13	Iran	1,691
14	Nigeria	2,211	14	United Kingdom	1,667
15	Algeria	2,086	15	Italy	1,528

Rank	Country	Imports	Rank	Country	Exports
1	US	9,631	1	Saudi Arabia	7,322
2	China	4,328	2	Russia	7,194
3	Japan	4,235	3	Iran	2,486
4	Germany	2,323	4	United Arab Emirates	2,303
5	India	2,233	5	Norway	2,132
6	Korea, South	2,139	6	Kuwait	2,124
7	France	1,749	7	Nigeria	1,939
8	United Kingdom	1,588	8	Angola	1,878
9	Spain	1,439	9	Algeria	1,767
10	Italy	1,381	10	Iraq	1,764
11	Netherlands	973	11	Venezuela	1,748
12	Taiwan	944	12	Libya	1,525
13	Singapore	916	13	Kazakhstan	1,299
14	Thailand	601	14	Canada	1,144
15	Belgium	597	15	Qatar	1,066

Countries rounding out the top 15 produce 2.2 to 4.2 mmb/d of petroleum liquids. Of these, Mexico and Venezuela are currently in decline, although these countries hold large potential resources. Brazil, with major new discoveries over the past decade, and Iraq, with large potential for reserve growth and new discovery, will have a substantial increase in production in the coming decade. Canada's role in the top 15 owes chiefly to production from extra-heavy oil and bitumen, which will likely increase in the future.

Among the oil-consuming countries, the US is by far the largest, consuming about 23% of the world's total production while producing about 10%, making it the largest importer of oil. China, Japan, India, and Russia immediately follow the US in oil consumption, together consuming about 22% of the world's total; China and India have shown substantial increase in rates of consumption over the past decade and are projected by most analysts to continue to increase oil consumption as their economies expand.

Following the US as the major importer of petroleum liquids are Japan and China; Japan imports essentially all of the petroleum liquids it consumes and China, like the US, imports half of its supply. Germany, South Korea, and India import most of the oil they consume. Future imports by China and India are projected to increase significantly as they compete for a growing share of the world's petroleum production.

Outside of Venezuela, Norway, and Russia, most of the major exporters of petroleum are in the Middle East, led by Saudi Arabia. Over the next couple of decades, Iraq, Brazil, and several countries of West Africa will provide ever-increasing volumes of petroleum exports.

3 Reserves and Resources

Two types of resources are normally considered: conventional, including both crude oil and natural gas liquids and both discovered and undiscovered volumes; and unconventional, including extra-heavy oils, tar sands, and oil from shales. Resources are additionally distinguished by discovery status, economic status, technical certainty, and development status.

A variety of terms is and has been used to define categories of oil reserves and resources. The most authoritative definitions are those of the Petroleum Resources Management System, developed in 2007 by the Society of Petroleum Engineers, the American Association of Petroleum Geologists, the World Petroleum Council, and the Society of Petroleum Evaluation Engineers (Fig. 2).[2] This system classifies total petroleum initially in place (PIIP) into discovered and undiscovered resources. Discovered resources are further divided into reserves and contingent resources, and undiscovered resources are considered as prospective resources. Each element is characterized by the degree of certainty associated with the assessment. Reserves thus are labeled as proved (1P, highest certainty), probable (2P), and possible (3P, lowest certainty). Contingent resources are likewise characterized as 1C, 2C, and 3C, reflecting decreasing certainty. The less certain reserve categories (2P and 3P), along with certain contingent resources, are commonly grouped under such terms as reserves growth, recovery growth, or inferred reserves. Prospective resources, commonly also referred to as yet-to-be discovered, are described with respect to probability, with a low estimate of resources (commonly also stated as having a 95% probability), a high estimate of resources (commonly stated as having only a 5% probability), and a best estimate of resources (commonly denoted as mean or 50% probability). Contingent and prospective resources, but not reserves, include unrecoverable portions (Fig. 2).

The total petroleum originally in place is the quantity of petroleum that is estimated to exist originally in naturally occurring accumulations, both known and estimated as yet to be discovered. Reserves are those quantities of petroleum anticipated

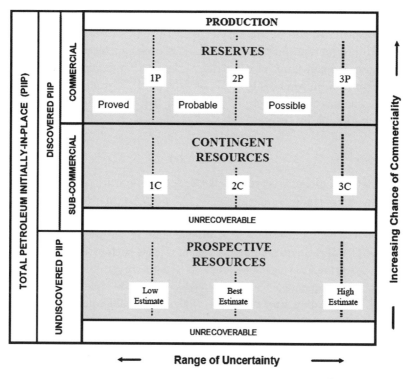

Fig. 2.   Petroleum resource classification framework.[2]

to be commercially recoverable from a given date forward under defined conditions; they must be discovered, recoverable, commercial, and remaining. Contingent resources are those quantities of petroleum estimated to be potentially recoverable but not yet considered ready for commercial recovery due to such contingencies as lack of viable market, technology under development, or high degree of uncertainty in evaluation. Prospective resources are those quantities of petroleum estimated to be potentially recoverable from accumulations yet to be discovered.

Some degree of uncertainty is inherent in any resource calculation or estimation, ranging from lowest for proved reserves to highest for low probability estimates of undiscovered resources. Furthermore, a variety of techniques are utilized in making calculations and estimates, commonly yielding significantly different results. And in the case of reserve estimations, not all countries follow the same standards. In many parts of the world, estimates of reserves and contingent resources are made by independently audited third parties. In other areas, notably among the Arab OPEC[a] nations, calculations are not externally audited and are considered propriety and

---

[a]OPEC stands for the Organization of Petroleum Exporting Countries. Currently it has 12 member countries: Algeria, Angola, Ecuador, Iran, Iraq, Kuwait, Libya, Nigeria, Qatar, Saudi Arabia, United Arab Emirates, and Venezuela.

an issue of national security. As such, some analysts consider many of the OPEC reserves as suspect. To add to this suspicion, in the early 1980s, OPEC tied volume of reserves to production quotas. Following this decision, between 1984 and 1988, the five largest Persian Gulf oil countries — led by Saudi Arabia — raised their estimates of proven oil reserves by 40%, or a total of 237 billion barrels. The fact that most of these estimates were relatively flat before and since 1984–1988 leads to further questioning by skeptics.

3.1 *Reserves*

Public estimates of global reserves of oil, condensate, natural gas, and natural gas liquids are made by trade organizations — *Oil and Gas Journal* (OGJ) and World Oil (WO), by one corporation (BP), and by governmental agencies — International Energy Agency (IEA) and the Central Intelligence Agency (CIA). Private consultancies make global estimates and on occasion release such estimates to the public. In addition, governmental agencies (such as the Energy Information Administration in the US, the Agencia Nacional do Petroleo, Gas Natural e Biocombustiveis in Brazil, and the Canadian Energy Board) make individual country estimates. And a number of independent authors make estimates, many of which discount OPEC revisions. Among the reporting entities, categories may differ. In some cases, oil and natural gas are reported together as barrels of oil equivalent; crude oil may be reported separately or considered as total liquids including condensate and natural gas liquids. Some estimates of oil include only conventional oil while others include unconventional oil (tar sands, heavy oil, and bitumen) as well. And finally, reserve estimates might be restricted to proven reserves (1P) or might include less certain categories. As a result the more conservative (or pessimistic) estimates currently range from 850 to 900 billion barrels while the more inclusive (or optimistic) estimates range from 1,150 to 1,350 billion barrels. Some recent published estimates are shown in Table 2.

When the above estimates are normalized to include only conventional crude oil and condensate, they range from 1,000 to 1,184 billion barrels; they also include some portion of the upgraded OPEC estimates. The several estimates by individual

Table 2:   Recent Published Reserve Estimates.[3–8]

Source	Year	Amount (billion barrels)	Includes
BP Statistical Review	2010	1,476.4	Crude, tar sands, condensate, and natural gas liquids
Oil & Gas Journal	2010	1,354	Tar sands and condensate
World Oil, year-end	2007	1,184	Crude and condensate
CIA	2010	1,349	Crude, tar sands, and condensate
International Energy Agency	2008	1,241	Crude, condensate, and natural gas liquids
IHS Energy	2009	1,128	Condensate and natural gas liquids

Table 3:   Proven Reserves (Includes Crude, Condensate,
Tar Sands, and Natural Gas Liquids (Top 20 countries).[6]

Country	Reserves (billion barrels)
Saudi Arabia	264.1
Canada	178.1
Iraq	143.1
Iran	137.6
Kuwait	101.5
Venezuela	98.6
United Arab Emirates	97.8
Russia	79.0
Libya	46.0
Nigeria	36.2
Kazakhstan	30.0
Qatar	27.2
USA	21.3
China	15.7
Algeria	15.2
Angola	13.5
Mexico	13.4
Brazil	12.6
Azerbaijan	7.0
Sudan	6.8

Source: CIA World Factbook, 2010.

authors, ranging from 850 to 900 billion barrels, differ from the earlier mentioned normalized estimates chiefly by the amount they discount OPEC upward revisions. Nehring[9] makes a compelling case for considering only 1P proved and developed reserves, putting the other categories of reserves in the category of reserve growth. Thus he carries 700 billion barrels as proved reserves, including crude oil, condensate, and natural gas liquids.

The proven reserves for the top 20 countries, including oil, tar sands, and natural gas liquids, are listed in Table 3. Apart from the significant tar sands in Canada, the other top five countries are all in the Middle East.

3.2 *Resources*

While the divergence in the estimates of proved reserves is not great, estimates of other parts of the total oil resource vary by a factor of three or more (Fig. 3). Some of this variance results from the resource components that are included or excluded. For example, some estimates ignore or do not include unconventional resources and reserve growth; some exclude natural gas liquids. Also, many components of the resource base simply are not as readily quantified as proved reserves and thus carry a wider range of uncertainty. These other components include undiscovered or yet-to-be discovered volumes, reserves growth, or recovery growth and recovery factors for so-called unconventional oil resources, including very heavy oil, tar sands, bitumen, and oil shale.

W.L. Fisher

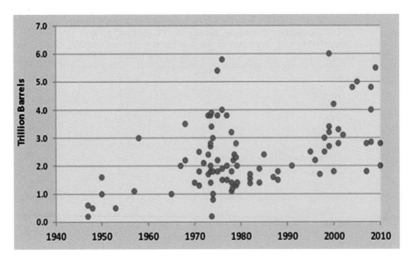

Fig. 3. Estimates of global petroleum resources over time.
Source: Updated from Ref. 11.

In estimating the several resource categories, undiscovered resources would seem to be the most difficult; however, they are the most extensive and thoroughly analyzed components. Assessment methodologies, while different, have been rigorously developed and have solid theoretical underpinnings.

One approach follows that originally established by Hubbert,[10] which postulates that discovery in any particular area increases exponentially until half the resource has been found and thereafter discovery declines in a fashion similar to its increase. A symmetrical life cycle is constructed with a mid-discovery peak. Production that follows discovery is postulated to follow a similar cycle whereby peak production is reached when half of the resource has been depleted. Hubbert's approach, or some modification of it, is used extensively today, especially by analysts predicting an eventual sharp peak in maximum production (Campbell and Siverstsson[12]; Deffeyes[13]; and Nashawi *et al.*[14]). It is a fairly simple model whereby historical data of discovery and/or production are plotted and projected. It assumes that the discovery and production profile is basically symmetrical when it is, in fact, commonly asymmetrical to the right, and it assumes that the total resource volume is known when in fact it can only be estimated. The approach is best suited for areas that are thoroughly explored and developed, have undergone major early development, are geographically limited, and are in an early phase of reserve growth. It is much more difficult to apply to areas not yet in decline, since it does not accommodate reserve growth (the major source of oil reserve additions over the past 25 years) and unconventional oil, and, being based on historical data, cannot anticipate entirely new areas of discoveries such as the ultra-deep waters of Brazil and the US Gulf of Mexico. It has no application to potential new frontiers, such as the Arctic, which lack discovery and production history. The approach thus gives inherently minimum

Table 4:   Estimates of Global Petroleum Resources, in Billions of Barrels.

End-1995 to 2025

Cumulative Production, end-1995, (BP, 2008)	786	
Proven Reserves, end-1995, (BP, 2008)	1,027	
Canadian Oil Sands, end-2007, (BP, 2008)	152	
Total Known Reserves		**1,965**

At 95% Confidence for end-2025

Undiscovered Oil (non-US)	334	
Undiscovered Oil + NG Liquids (US)	66	
Reserves Growth Oil (non-US)	192	
Reserves Growth Oil + NG liquids (US)	76	
Undiscovered NG liquids (non-US)	95	
Reserves Growth NG liquids (non-US)	42	
Total New Resources		**805**
Total Petroleum Liquids		**2,770**

At Mean Confidence for end-2025

Undiscovered Oil (non-US)	649	
Undiscovered Oil + NG Liquids (US)	83	
Reserves Growth Oil (non-US)	612	
Reserves Growth Oil + NG liquids (US)	76	
Undiscovered NG liquids (non-US)	207	
Reserves Growth NG liquids (non-US)	42	
Total New Resources		**1,669**
Total Petroleum Liquids		**3,634**

At 5% Confidence for end-2025

Undiscovered Oil (non-US)	1,017	
Undiscovered Oil + NG Liquids (US)	104	
Reserves Growth Oil (non-US)	1,031	
Reserves Growth Oil + NG liquids (US)	76	
Undiscovered NG liquids (non-US)	378	
Reserves Growth NG liquids (non-US)	42	
Total New Resources		**2,648**
Total Petroleum Liquids		**4,613**

Source: US Geological Survey (2000) and Ahlbrandt *et al.*[11]

values, with estimates of additional discovery based on these models on the order of only 100 to 300 billion barrels.

A second type of approach is that utilized by the US Geological Survey (Ahlbrandt *et al.*[11]) (Table 4), by Nehring[9] (Table 5) (Edwards[15] and Odell[16]) and some of the private consultancies (Stark and Chew[17] and Jackson[18]). For undiscovered conventional oil, the assessment of an area requires (1) choice of a minimum accumulation size; (2) assignment of geological and access risk; and (3) estimation of the number and sizes of undiscovered accumulations in the assessment area. The combination of these variables yields probability distributions for potential additions

Table 5: Ultimate World Oil Potential in Billions of Barrels (as of 12 December 2009).

	Low	Medium	High
Cumulative production	1,205	1,205	1,205
Proven developed reserves	700	700	700
Recovery growth	715	1,095	1,585
Future Discoveries	320	750	1,280
Unconventional	330	540	850
World total	3,270	4,290	5,620

to reserves. Results are controlled by geology-based input parameters supplied by knowledgeable geologists using total petroleum systems analyses, as opposed to estimates based chiefly on projection of historical discovery and production trends.

Ahlbrandt *et al.*[11] report undiscovered petroleum resources which they judge discoverable in a 30-year timeframe to range from 495 billion barrels (95% probability), 939 billion barrels (mean) to 1499 (5% probability). Nehring,[9,19] who basically followed the US Geological survey (USGS) estimates but with modifications and updates, estimates a low value (roughly 95% probability) at 320 billion barrels, a middle estimate of 750 billion barrels, and a high estimate of 1280 billion barrels (5% probability). IHS Energy,[8] a consultancy that provides a global database widely used by industry and energy analysts, estimates an undiscovered petroleum resource of 839 billion barrels, similar to the mean or mid-range estimates of the USGS and Nehring.

A second resource component is reserve growth, a widely recognized tendency whereby reserves once discovered tend to increase over time with greater understanding of the reservoir through additional characterization and from knowledge of the production history and patterns, generally leading to improvement in the percent recovery. While certain independent analysts ignore or dispute the significance of reserve growth and commonly include no or minimal values in their resources estimates, the major governmental entities such as the USGS, the Energy Information Administration (EIA), the International Energy Agency (IEA), as well as major consultancies such as IHS Energy (Table 6) consider reserve growth a major

Table 6: Estimates of Ultimate World Oil Potential in Billions of Barrels.[9,19]

Resource/Region	USA	Canada	Rest of world	Total
Cumulative production	234	31	813	1,076
Proved	19	7	1,042	1,066
Reserves Growth (EOR)	76	6	510	592
Undiscovered deepwater	8	3	50	61
Undiscovered Arctic	5	3	100	108
Undiscovered elsewhere	83	15	660	758
Total conventional				3,673
Extra-heavy oil	0	167	277	444
Oil Shale	500	4	200	704

part of the remaining resource base. In fact, reserve growth has been the most significant component of reserve additions over the past 25 years. The USGS[9] estimates reserve growth over the next 30 years to range from a low of 310 billion barrels to a high of 1,149 billion barrels, with a mean estimate of 730 billion barrels. Nehring[19] estimates reserve growth, or what he calls recovery growth, at a low of 715 billion barrels (assuming a 35% recovery factor), a mid-level at 1,095 billion barrels (40% recovery), and a high estimate of 1,585 billion barrels (45% recovery); Nehring's estimates exceed the USGS estimates. He reduces proved reserve values to 700 billion barrels and includes 270–350–450 billion barrels from fields yet to be developed. Unlike the USGS, which estimates for a 30-year timeframe, Nehring assumes his estimates of reserve growth may take 50 to 80 years to realize.

A third major component of petroleum resources, a significant portion now carried as proved reserves, is the so-called unconventional petroleum resources. While some analysts classify deep water and polar resources as unconventional, the generally accepted components of unconventional resources include extra-heavy oil/bitumen, oil from mature source rocks, and so-called oil (kerogen) shales. In-place volumes of these resources are very high. Nehring[19] reports in-place volumes of extra-heavy oil and bitumen at 4.1 trillion barrels, oil shale at 3.2 trillion barrels and oil from mature source rocks at 5 to 20 trillion barrels. Currently, the most extensively developed of the unconventional petroleum resources are the extra-heavy oil and bitumen. Oil from shales has historically been produced but only in modest amounts. Only in this decade has the potential of production from mature sources been realized, particularly in the US, with development of the Bakken Shale and the Eagleford Shale. Unconventional resources are generally expensive to recover (or at least more expensive than conventional resources), currently have low recovery values (10%–20%) and commonly utilize substantial levels of energy in their production, yielding a reduced net energy output. They are, however, subject to future technologies to enhance recovery. Nehring[19] estimates recoverable portions of the large resource base of unconventional resources as 330 billion barrels (low), 540 billion barrels (middle), and 850 billion barrels (high), based on assumptions about future recovery values.

Analysts basing future discoveries chiefly on production and discovery trends commonly also discount or ignore reserve growth and also exclude or include only small volumes of unconventional petroleum resources; furthermore, many analysts consider crude oil only and exclude natural gas liquids and many also discount reported proved reserves. Accordingly, their estimates of ultimate oil recovery are low, ranging from 1,825 to 2,130 billion barrels (Fig. 4). Analysts who take a more comprehensive, and in many cases, a probabilistic approach, estimate ultimate recoveries that range from 2,860 to 5,620 billion barrels. However, if reserve growth and unconventional resources are excluded, Nehring's estimate of 2,225 billion barrels and the USGS estimate of 2,308 billion barrels are comparable to the lower estimates derived from discovery and production trend analysis.

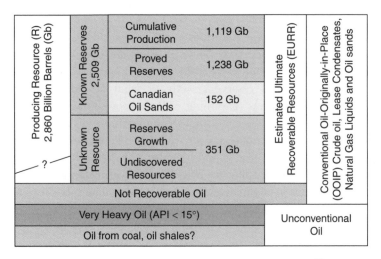

Fig. 4.   Estimate of world oil reserves and resources.[20]

4 Petroleum Refining

Petroleum liquids that are extracted from a well can be used only after processing. Crude oil contains many different long-chain hydrocarbon molecules, and the process of refining consists of separating these into groups that are useful for different purposes. Refining is primarily a distillation process, where the oil from the well is heated and then allowed to condense at different temperatures, which separates out different hydrocarbon products depending on their boiling points. Generally, the longer the carbon chain, the higher is the boiling point of the material. Table 7 shows the boiling points of some of the shorter chain molecules.

Gasoline has carbon chains typically of length 7–11 carbon molecules, and diesel fuel has carbon chains of around 15 carbon molecules. Since more transportation fuels, particularly gasoline, are needed than are present in the crude oil, a process called "cracking" is used to break up longer chain hydrocarbon molecules to produce smaller chains more appropriate for gasoline or diesel production. The cracking process typically involves catalysts such as zeolite, aluminum hydrosilicate, bauxite and silica-alumina. Other processes, also using catalysts, combine shorter

Table 7:   Boiling Points of Light Hydrocarbons.

Formula	Name	Boiling point (°C)
CH_4	methane	−161
CH_3CH_3	ethane	−89
$CH_3CH_2CH_3$	propane	−42
$CH_3(CH_2)_2CH_3$	butane	−0.5
$CH_3(CH_2)_3CH_3$	pentane	+36
$CH_3(CH_2)_6CH_3$	octane	+125

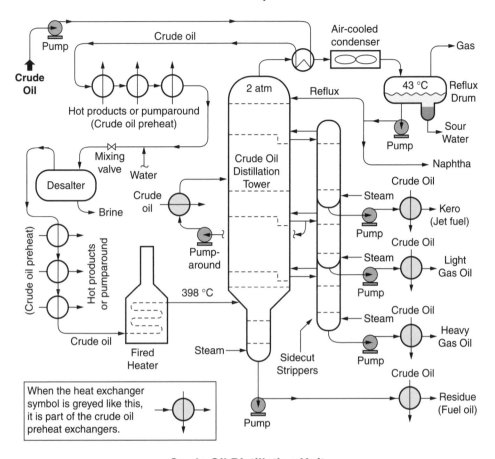

Crude Oil Distillation Unit

Fig. 5. Schematic of the refining of crude oil to produce usable products.

chain molecules to produce longer chain molecules ("reforming" or "unification") or change the chemical composition of shorter molecules to form longer chains (known as polymerization). All these processes are carried out online in a refinery and a schematic is shown in Fig. 5.[21]

4.1 *Combustion of Gasoline and Diesel Fuel*

Gasoline is one of the most commonly used products of crude oil distillation. Along with diesel fuel, it is usually about 50% or more of the output of a refinery. Table 8 shows the energy output of the combustion of gasoline and diesel fuel both as a function of mass and volume. The heating values per kilogram are very similar, but since diesel is denser, the heating value per liter of diesel is greater than that of gasoline. The CO_2 emitted from the combustion of both fuels is very similar.

Table 8:   Combustion of Gasoline and Diesel Fuel.

Fuel	Density (kg/l)	Heating Value (Mass) (MJ/kg)	Heating Value (Vol.) (MJ/l)	CO_2 emitted (g/MJ)
Gasoline	0.745	43.2	32.18	73.38
Diesel	0.832	43.1	35.86	73.25

5 Future Production

Historically, projections of future production of petroleum liquids were made primarily based on analysis of future demand, where ample production capacity existed or could be readily established. In recent years, many analysts, primarily those calculating a limited remaining resource base, have projected future production as supply constrained. Several project immediate or very near-term peaking of global oil production near current levels of about 84 mmb/d, followed by rapid decline. The analysts calculating a larger remaining resource base predict maximum production occurring anywhere from 2020 to 2050, with higher level of production at maximum and an extended plateau of 20 to 30 years where maximum production is maintained. Amongst the highest levels of projected future production is that of Odell,[16] with a peak of 180 mmb/d reached in 2050. Jackson[22] of Cambridge Energy Associates projects a 25-year undulating plateau of production at 115 mmb/d, starting about 2030.

Nehring,[9,19] while calculating a substantial resource base at the mid and high level, recognizes that much of the remaining resource other than proved reserves will require significant lead times in development. He judges, in his estimates of reserve growth, that it will add significantly to future production but will be added in modest increments and take 50 to 80 years to fully realize. New discovery volumes are chiefly in such frontier areas as ultra deep water and in Arctic basins, which are at the boundaries of current exploration and production technology, and many are 10 to 40 years away from contributing to world production in any significant amount. The unconventional part of the resource base generally has poor rock and/or fluid characteristics, which means more expensive and incremental development is needed, barring major technological breakthroughs. Nehring projects extended plateaus of future production, without distinct peaks. For his low estimate, he projects production will plateau at about 88.5 mmb/d starting in 2020 and extend to 2040 before declining. Mid-level estimates of the resource base result in a plateau of 92.1 mmb/d starting in 2025 and extending to 2050. Nehring's high resource base yields production of 95.3 mmb/d with a plateau beginning in 2036 continuing to 2065 (Fig. 6)

6 Oil Production Costs

The average cost of oil production is defined as all technical costs related to finding, development and lifting oil divided by the volume to be recovered. Jojarth[23]

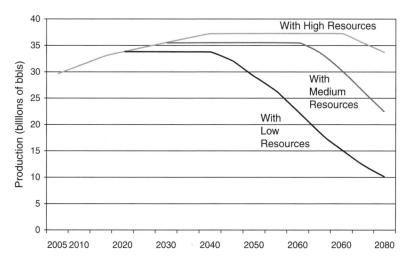

Fig. 6. Projected future oil production per year.[19]

indicates four elements or categories of cost. One is finding costs, which the Energy Information Administration[24] defines as "the costs of adding proven reserves of oil and natural gas via exploration and development activities and the purchase of properties that might contain reserves." Another is development costs, including drilling of production wells, installation of platforms and wellheads, and the construction of pipelines connecting the field to a trunk transport line or a processing plant. Lifting costs is a third category, commonly referred to simply as production or operating costs, including operation and maintenance as well as related equipment and facilities after hydrocarbons have been found, acquired, and developed (EIA, 2005).[23] A final cost element is capital expenditures on infrastructure, such as major transport pipelines, terminals, and processing plants. In addition, there are payments specific to operations, such as royalties and taxes.

Specific numbers on production costs are not commonly available in the public realm because most producers prefer to keep such information proprietary. Numbers that are published are commonly aggregated by country or large regions. Reuters[25] compiled representative costs in several countries worldwide based on conversations with traders and industry analysts. They report Saudi Arabian crude as the cheapest in the world to extract, given the size of fields and its shallow occurrence; simple lifting costs are $1 to $2 per barrel, and total cost (including capital expenditures) is $4 to $6 per barrel. Production costs in Iraq are likewise low, between $4 and $6 per barrel for the larger-sized fields. In the United Arab Emirates, operating and capital costs combined are around $7 per barrel.

Elsewhere among Arab OPEC countries, including Algeria, Iran, Libya, Oman, and Qatar, estimated production costs are somewhat higher, running between $10 and $15 per barrel. Inland production costs in Nigeria are estimated at about $15 per barrel. Costs are comparable in Kazakhstan, with costs for their largest operator

Table 9:   All-in Costs of Producing Oil from Various Parts of the
World.[19]

Oilfield/Source	Estimated costs per barrel (2008 US$)
Mideast/North Africa fields	6–28
Other conventional fields	6–39
Enhanced oil recovery	30–82
Deep-ultra deep oilfields	32–65
Arctic oil fields	32–100
Heavy oil/bitumen	32–68
Oil (kerogen) shales	52–115
Gas to liquids	38–113
Coal to liquids	60–113

about $10 to $12 per barrel and around $15 to $18 for the smaller state-owned companies. In South American OPEC countries, costs are higher, running about $20 per barrel for light oil in both Venezuela and Ecuador.

Offshore production is invariably more expensive than almost any inland production; the deeper the water the higher the cost. Operating costs in the British North Sea run about $30 to $40 per barrel. Comparable costs are encountered in the deeper waters of offshore Nigeria. In the ultra deep waters of Brazil, break-even costs are $35 to $45 per barrel. And in the Venezuela production of the heavy oil from the Orinoco oil belt runs around $30 per barrel.

The International Energy Agency in its 2008 World Energy Outlook[26] gave the following estimates for the all-in costs of producing oil from various types of hydrocarbons in different parts of the world (Table 9):

Size is a substantial factor in determining production costs. Large fields inland and at relatively shallow depths, common to most of the inland production in the Middle East and North Africa, lead to the lowest costs. Likewise, fields in hostile environments like the Arctic or in ultra deep waters are commercial only if they are large in size, as in the case of the Prudhoe Bay field (15 billion barrels) in the Alaska North Slope and the giant fields (5 to 15 billion barrels) in the ultra deep offshore of Brazil.

Recovery of in-place oil beyond primary and secondary stages is more costly, since tertiary or enhanced recovery calls for the injection of steam to reduce viscosity or chemicals or miscible gas to render oil mobile in the reservoir. Non-conventional oil is more expensive to produce if it is dense or has a low API gravity, such as the bitumen in western Canada and the heavy oil of Venezuela. The so-called shale oil, in reality kerogen, is expensive to produce because it occurs in very low permeability reservoirs and must be upgraded to crude oil by extensive and long-term heating. However, fully mature oil is now produced commercially from oil shale (Eagleford of South Texas) or from non-conventional source beds such as the Bakken (Williston Basin and North Dakota).

From a geopolitical standpoint, it must be noted that most of the low cost oil is confined to the OPEC countries of the Middle East and North Africa. Similar low-cost oil in the US, Russia, and elsewhere has mostly been produced, leaving inland potential largely in the more costly Arctic region, small fields or recovered through enhanced oil recovery. Also, much of the shallow shelf area of the world in such areas as the US, western Africa, and Brazil, is mature so that future potential resides mostly in the deep (greater than 1,000 feet) to the ultra deep (greater than 5,000 feet) waters.

References

1. US Energy Information Administration, Available at: http://www.eia.gov/petroleum/. Accessed December 2011.
2. Society of Petroleum Engineers, American Association of Petroleum Geologists, World Petroleum Congress and Society of Petroleum Evaluation Engineers (2007). *Petroleum Management System*, p. 47. Available at: http://www.spe.org/spe-app/spe/industry/reserves/index.htm. Accessed December 2011.
3. BP Statistical Review of World Energy, June 2010.
4. *Oil and Gas Journal* (2010).
5. World Oil Year-End (2007).
6. CIA World Factbook (2011). Available at: https://www.cia.gov/library/publications/the-world-factbook/index.html. Accessed December 2011.
7. World Energy Outlook (2008). Available at: http://www.iea.org/textbase/nppdf/free/2008/weo2008.pdf. Accessed December 2011.
8. IHS Energy (2009). Available at: http://www.ihs.com/products/oil-gas-information/index.aspx. Accessed December 2011.
9. R. Nehring, Traversing the mountaintop: World fossil field production to 2050, *Phil. Trans. Royal Soc.* **364** (2009), pp. 3067–3079.
10. M. K. Hubbert, Nuclear energy and the fossil fuels: In drilling and production practice, *Proc. Spring Meeting* (American Petroleum Institute, San Antonio, 1956).
11. T. S. Ahlbrandt *et al.*, Global resources estimates from total petroleum systems, *Amer. Assoc. Petrol. Geol.* **86** (2005), p. 324.
12. C. J. Campbell and A. Siverstsson, Updating the depletion model, *Second International Workshop on Oil Depletion* (Paris, 2003).
13. K. S. Deffeyes, *Hubbert's Peak: The Impending World Oil Shortage* (Princeton University Press, Princeton, 2001).
14. Nashawi *et al.*, Forecasting world crude oil production using multicyclic Hubbert Model, *Energy and Fuels* **24** (2010), pp. 1788–1800.
15. J. D. Edwards, Crude oil and alternate energy production forecasts for the twenty-first century: The end of the hydrocarbon era, *Am. Assoc. Petrol. Geol. Bull.* **81** (1997), pp. 1292–1305.
16. P. R. Odell, "Draining the World of Energy" in *World in Crisis*, eds. R. J. Johnston and P. J. Taylor (Blackwell, London, 1989), pp. 79–100.
17. P. H. Stark and C. Chew, Pillars of oil and gas supplies, *NAPE International Forum* (2009).
18. P. M. Jackson, Why the (peak oil) theory falls down — Myths, legends, and the future of oil resources, November 14 Press Release (Cambridge Associates-HIS, 2006).
19. R. Nehring, Peak oil: Why, when and how, Presented at Pikes Peak Economic Club, 2 February 2010.
20. M. Al-Husseini, World production of conventional petroleum liquids to 2030: A comparative overview, *GeoArabia* **14**(1) (2009), pp. 215–267.

21. Oil Refinery. Available at: en.wikipedia.org/wiki/oil_refinery. Accessed 2 April 2012.
22. P. M. Jackson, The future of global oil supply: Understandingxs the building blocks, IHS CERA Special Report, (2009), p. 13.
23. C. Jojarth, The end of easy oil: Estimating average production costs for oil fields around the world, CDDRL Working Paper No. 81, Stanford University (2008), p. 30.
24. Energy Information Administration, Performance profiles of major energy producers, DOE/EIA-0206 (04) Uc-959 (2005).
25. IEA, World Energy Outlook (2008). Available at: http://tonto.eia.doe.gov/ftproot/ financial/o20605.pdf. Accessed December 2011.
26. Reuters Press Release (July 2009).

Chapter 5

Natural Gas

John B. Curtis
Potential Gas Agency
Department of Geology and Geological Engineering
Colorado School of Mines, Golden, CO 80401, USA
jbcurtis@mines.edu

Worldwide production and consumption of natural gas has increased dramatically in the last two decades. As the lowest-carbon fossil fuel, there are societal implications for its increased usage compared to more carbon-rich (and hydrogen-poor) fuels. Natural gas requires significant infrastructure investment, including pipelines and ocean tankers, to produce, transmit, store, and distribute these growing volumes to the end user. While all energy sectors (residential and commercial heating, industrial and power generation) have historically used natural gas, greater emphasis is now being placed on using more gas for power generation, particularly to displace coal-fired plants and to be used as a transportation fuel. The application of new horizontal drilling and hydraulic fracturing techniques has allowed increased access to the vast amounts of gas contained in low porosity and permeability geological formations. Gas reservoirs in organic-rich shales, coal seams, and "tight gas" sands comprise these unconventional resources. While the US has been the leader in their production, with over 50% of gas now produced from such reservoirs, exploration and development activities are underway in Europe, China, India, Latin America, and Australia. Natural gas hydrates have a potential to provide even larger volumes of gas but will require significant future R&D investment.

1 Introduction

Natural gas is a combustible gaseous mixture of simple hydrocarbon compounds, usually found in deep underground reservoirs of porous and permeable rocks. Natural gas is a fossil fuel composed largely of methane. A molecule of methane, the simplest and lightest hydrocarbon, consists of one carbon atom surrounded by four hydrogen atoms. Natural gas also contains lesser amounts of "higher" hydrocarbon gases, i.e. those of greater molecular weight, namely ethane, propane, butane, isobutane, and pentane (together commonly referred to as C2+), as well as variable concentrations of non-hydrocarbon gases, namely carbon dioxide, nitrogen, hydrogen sulfide, and helium (Table 1). Natural gas is the cleanest burning fossil fuel, producing smaller amounts of combustion by-products than either coal or refined oil products. Petroleum is often defined to include both liquids and gases.

Table 1:   Composition of Natural Gas.

Hydrocarbon gases	Chemical formula	Typical composition	Non-hydrocarbon gases	Chemical formula	Typical composition
Methane	CH_4	80%–95%	Carbon dioxide	CO_2	1%–2%
Ethane	C_2H_6	2.5%–7%	Nitrogen	N_2	1%–4%
Propane	C_3H_8	1%–3%	Hydrogen sulfide	H_2S	variable
Butane	C_4H_{10}	1%–3%	Water vapor	H_2O	variable
Pentane	C_5H_{12}	trace	Helium	He	trace
Hexane	C_6H_{14}	trace	Argon	Ar	trace

Volumes are reported either as cubic feet (e.g. trillion cubic feet, Tcf) or cubic meters ($1\,\text{ft}^3 = 0.028316\,\text{m}^3$; $1\,\text{m}^3 = 35.3146667\,\text{ft}^3$).

2 Why is Natural Gas Important?

Natural gas has become the "fuel of choice" in many residential, commercial, and industrial applications. Furthermore, natural gas has become particularly important in electric power generation.

The first practical use of natural gas in the US dates to 1821 in Fredonia, New York, where a crudely drilled well and hollowed-out log pipes were used to deliver gas from a natural seep to nearby homes for lighting.[1] It was not until the 1880s, however, did natural gas for home heating and lighting and for industrial use become prevalent. By the late 1940s, natural gas had all but replaced the use of "illuminating" gas manufactured from coal and wood. The transition was facilitated in part by federal regulations that discouraged oil field operators from wasting natural gas by venting and flaring. An unknown but likely enormous volume of gas resource was lost through such practices. Natural gas became a marketable commodity, and production flourished. In the years following World War II, the interstate pipeline system, which began in 1925, was greatly expanded, thereby bringing natural gas service to consumers all over the lower 48 US states.

According to statistical data from the US Department of Energy's Energy Information Administration (EIA) for 2009, natural gas satisfied 25% of the nation's energy demand (consumption), moving ahead of coal, which accounted for 21%. Crude oil and natural gas liquids, while still accounting for the largest share, declined from a recent high of 40% (2004–2006) to 37% at present. Nuclear power rose slightly to 9%, and hydropower and renewables totaled 8%, including liquid biofuels.[3]

World consumption of natural gas also increased from 2004 to 2010 by just under 18%. However, there were major differences between countries. The top consumers of natural gas are shown in Table 2. China saw the largest percentage increase (175%) in that time period but Iran (58%), Mexico (29%), Saudi Arabia (28%), and Japan (22%) also saw significant increases.

Where does the US supply of natural gas come from? First, consider that well over one-half of all the crude oil Americans consume annually must be

Table 2: World Consumption of Natural Gas in Tcf per Year.[2]

Country	2004 Tcf	2006 Tcf	2008 Tcf	2010 Tcf	Share of total	Change 2010–2004
US	22.39	21.69	23.27	24.13	21.6%	7.79%
Canada	3.39	3.42	3.37	3.31	3.0%	−2.39%
Mexico	1.89	2.15	2.34	2.43	2.2%	29.03%
Germany	3.03	3.08	2.87	2.87	2.6%	−5.36%
Italy	2.61	2.73	2.75	2.69	2.4%	2.98%
Russia	13.92	14.43	14.69	14.62	13.1%	5.07%
U.K.	3.44	3.18	3.31	3.31	3.0%	−3.70%
Iran	3.05	3.84	4.21	4.83	4.3%	58.27%
Saudi Arabia	2.32	2.60	2.84	2.96	2.6%	27.70%
China	1.40	1.98	2.87	3.85	3.4%	174.56%
Japan	2.72	2.96	3.31	3.34	3.0%	22.73%
World Total	**95.2**	**100.4**	**108.1**	**111.9**	**100.0%**	17.61%

imported — from more than 35 countries — and 65% of those imports come from OPEC and Persian Gulf states. Canada, however, remains our largest single source of imported oil. In contrast, about 90% of all natural gas consumed in the US comes from domestic onshore and offshore sources — 21.6 trillion cubic feet (Tcf = 10^{12} cubic ft) out of 24.1 Tcf consumed in 2010. The remainder is imported from Canada via pipelines and to six to eight other foreign countries via ocean tankers carrying liquefied natural gas, or LNG (452 Bcf in 2009). There were 478,562 gas and gas condensate wells in production in the US in 2008.[4]

Using the US as an example, interdependent sectors of the natural gas industry are involved in delivering natural gas from the wellhead to consumers. *Exploration and production* (E&P) companies explore, drill, and extract natural gas from the ground and seabed. *Gathering and processing* ("midstream") companies connect the wellheads to *transmission* companies, which operate the pipelines linking the gas fields to major consuming areas. *Distribution* companies are the local utilities that deliver natural gas to customers. *Marketing* companies serve as intermediaries between production companies and ultimate customers.

3 How Natural Gas Forms

According to prevailing scientific theory, the natural gas that is produced commercially today formed millions of years ago when very small plant and animal remains were buried by mud and silt at the bottoms of oceans and lakes. Layers of sediment and plant and animal matter that slowly built up became deeply buried over time until the pressure and heat resulting from the weight of the overlying sediment eventually converted this organic matter into natural gas and crude oil. Bacteria also are intimately involved in this *generation* process, which continues to this day in modern swamps, peat bogs, wetlands, and lakes; large river deltas; and in some deep ocean basins. Through time, underground forces cause the buoyant hydrocarbons to move slowly, or *migrate*, out of their *source rocks* and into porous and permeable *reservoir* rocks, where they accumulate and become *trapped* if impermeable *seals* are present. Within a given basin or region, these four essential components — source rocks, reservoirs, trapping mechanisms, and seal rocks — comprise what geologists call a *petroleum system*, which may contain oil alone, gas alone, or oil and gas together.

Gas accumulates in two types of reservoirs. *Conventional* reservoirs include *clastic* sedimentary rocks such as porous sandstone, siltstone, and conglomerate, and *carbonates* (limestone and dolomite). *Unconventional* reservoirs include coalbeds, organic-rich marine and lacustrine (lake-formed) shales, and low-permeability sandstones. The grains in these "tight" sandstones have been so pervasively cemented together with silica or carbonate that high pressure, artificial *stimulation*, or *hydraulic fracturing* is required to create permeable pathways that allow the gas to flow out of the sand's intergranular pore spaces to the wellbore. Coalbeds and organic-rich shales, which are both reservoirs *and* self-generating source rocks, also usually require stimulation. *Although unconventional reservoirs present many technological challenges, they now account for greater than 50% of US natural gas production.* Recent drilling in Poland, Germany, Ukraine, China, and Australia is also targeting these unconventional reservoirs. In rare cases, natural gas has migrated into naturally fractured igneous and metamorphic rocks.

In industry terminology, natural gas occurring alone, without oil, is called *nonassociated* gas, whereas *associated* gas occurs with crude oil. When oil is pumped to the surface, the natural gas dissolved within it is released and is referred to as *casinghead* gas.

4 Exploration

Exploration prospects are the most risky. Tens of millions of dollars may be required to drill a single *wildcat* well in a previously untested area, where the chance of success may be no more than 5% to 20%. Target reservoirs may lie up to 30,000 ft below the ground surface or seabed. Explorers have adapted their drilling strategies to virtually every type of setting where natural gas may have accumulated — over 175 miles offshore in water as deep as 9,000 ft (2,743 m), below a busy Mid-Continent metropolitan area, on a remote stretch of Rocky Mountain rangeland, in the midst of coal-mining operations in Appalachia, or on desolate frozen tundra above the Arctic Circle in Alaska. The *drilling rigs* used to bore holes deep into the rock strata are technological marvels. Drilling tools and techniques are continually being improved in order to drill more quickly, more accurately, and more safely; to better access the reservoirs; to deal with the extreme temperatures and pressures encountered in extremely deep strata; to control costs; and to minimize the surface "footprint" of drilling operations, both onshore and offshore. Drilling multiple *horizontal* or *lateral* wells and *extended-reach* wells from a single drilling pad has become a desirable and even necessary option in many cases in order to reduce the surface impacts of drilling. Moreover, advanced drilling technologies such as these are proving inestimable in accessing remote and/or structurally complex reservoirs, including those offshore, and in improving production efficiencies. A number of lateral wells and even more elaborate *multilateral* wells are capable of tapping one or more reservoirs from several thousand feet to about 10,000 ft radially from one vertical production wellbore, thereby draining an area that would otherwise require drilling of several vertical wells. As their name implies, extended-reach wells can be driven even farther and involve remarkably tortuous but precisely guided underground paths. Early in 2011, Exxon Mobil reported a new world record for an extended-reach well drilled in eastern Russia — 40,502 ft (12,345 m), or nearly 7.7 miles (12.3 km) with the horizontal segment extending a record 37,648 ft (11,475 m) or 7.1 miles (11.5 km).

A successful well is termed a *discovery*. Nearly 2,800 gas exploration wells were drilled in 2007 in the US, and 2,900 were drilled in 2008. These wells achieved a record total drilled footage of 20 million ft in 2007, for an average depth of 5900 ft per well.[5] With better geology, geophysics, and geochemistry behind the selection of drill sites, the overall success rate for exploration wells has improved markedly over the last decade, from 35% to over 65%. The pace of new well permit applications and the number of new wells drilled in many areas of the country dropped off considerably during the second half of 2008 and into 2009 as a consequence of the economic recession.

The number of commercial rigs reported drilling for gas and oil across the country varies from week to week, depending on the season, availability of rigs, availability of funding, gas prices, and demand. The number of rigs drilling for natural gas each year since 1994 has exceeded those drilling for oil. At any given time, more than 80% of active rigs are drilling for natural gas. The average number of active gas-directed rigs totaled 1,466 in 2007 and 1,491 in 2008.

5 Development

Once the productive capacity of a well has been established through testing, which can last for months, it is typically *shut in* until it can be connected to a *gas-gathering system*, a network of small-diameter pipes that collect raw gas from a number of wells and deliver it to a treatment plant or, in some cases, directly to a pipeline. State regulations usually permit an operator to *flare*, or safely burn off, the gas from a new well during testing.

Much additional work and expense then are required to bring the discovery to market. *Delineation wells* are drilled to define the subsurface areal extent of the productive reservoir and to locate additional *pools* of hydrocarbons. The results of those wells are integrated with other technical information into a strategy for establishing commercial production. Less risky *development wells* then are drilled, *completed*, and *stimulated* with special surface and downhole equipment and materials to induce hydrocarbons to flow (*produce*) safely and efficiently from the reservoir into the wellbore, then up through the *wellhead* and into a gathering system.

Development wells now typically have a 92% chance of success. EIA estimates that more than 29,200 gas development wells were drilled in 2007 and 29,700 in 2008. These wells achieved record total drilled footages of nearly 193 million ft in 2007 and 195 million ft in 2008.[5]

6 Production

6.1 *Gas Fields*

Any number of development wells — from one to several hundred — producing gas from one or more rock formations in an area delineated by a controlling geologic structure or other type of trap comprise a *gas field*, whose operation and boundaries are governed by state regulations.

According to US industry's classification, the volume of gas recoverable from a developed field can vary from 187 Mcf (million cubic feet) to as much as 50 Tcf ("supergiant"), although discoveries of the latter are extremely rare. Most of the largest gas fields in the US are classified as "giants" — containing more than 3 Tcf — and were discovered between 1925 and 1950. Most of the giant fields discovered within the last decade have been found in the Gulf of Mexico. Statistically, only comparatively smaller accumulations likely will be discovered in the future in our mature gas-producing regions, as they are now, at least onshore. For this reason, industry must drill more wells each year — and often drill deeper — to find sufficient new supplies to meet consumer demand.

In standard oil field practice, the fluids produced from development wells are separated using treatment equipment into natural gas, liquid hydrocarbons (crude oil and condensate), non-hydrocarbon gases, and water. Oil and condensate are pumped to market through pipelines; oil can be stored in tanks for later transport by truck, rail, or ship. The gas flows into separate pipelines.

In some cases, the associated natural gas that is separated from the crude oil is more valuable for maintaining pressure levels within the oil reservoir than it is for sales. This gas is *reinjected* in order to enhance oil productivity. About 3.8 Tcf of the total 25 to 26 Tcf of US natural gas that is extracted and separated annually as *gross withdrawals* is reinjected for repressurization, principally on the North Slope of Alaska (approximately 3 Tcf), where no sales-gas transportation infrastructure exists yet, and to a lesser extent in the lower 48 States, mostly in Texas, California, Wyoming, Louisiana, and Colorado.[5,6]

6.2 *Stranded Gas*

Some natural gas, although technically producible, cannot be produced and delivered to market because of its low quality, remote location, or the fact that no large-volume market for it exists within a reasonable transport distance. If it cannot be produced and consumed for more than on-site lease use or for local domestic use, this *stranded* gas must be either shut in or, if associated with oil production, reinjected, or wasted by flaring.

Sub-quality or *low-Btu* natural gas contains excess concentrations of various non-hydrocarbon contaminants, principally carbon dioxide (CO_2), nitrogen (N_2), hydrogen sulfide (H_2S), and even helium (He). In large amounts, these undesirables act to lower the heat content of natural gas to less than 950 Btu/scf (standard cubic foot), the minimum specification generally required for pipeline transportation, thus rendering the gas unsuitable for nearly all purposes. Furthermore, these contaminants can cause serious operational problems in the field, such as health and fire hazards (H_2S) and corrosion-induced wellbore casing and pipeline leakage. The Gas Technology Institute defines sub-quality gas as having one or more of the following characteristics — $\geq 2\%$ CO_2 content, $\geq 4\%$ N_2 and ≥ 4 ppm (parts per million) H_2S. If such gas cannot be economically upgraded to pipeline quality or blended with higher quality natural gas, it remains "behind pipe."

What qualifies as low-Btu gas is substantial. Researchers speculate that high-nitrogen gas may constitute 60 Tcf, or 25% of proven US reserves.[7] In the larger picture, approximately one-fifth of the world's total gas reserves likely is highly contaminated with CO_2.[7,8]

Gas processors have developed several exotic technologies to treat and upgrade low-Btu gas in large-scale centralized facilities — cryogenic separation, membrane separation, adsorption, and absorption. Research is also under way to develop affordable, small-scale, mobile treatment units, for nitrogen rejection in particular, for field and well-site application.[9] Operators have even successfully demonstrated the direct use of sub-quality gas, without treatment, for small-capacity, on-site power generation via gas engines and turbine-driven generators modified to run on very lean gas mixtures.

Remotely located gas accumulations, onshore and offshore, remain stranded until such time that a gas pipeline can be built or the gas can otherwise be converted into a transportable and marketable form, either through *liquefaction* into LNG (see following section), *compression* into CNG (compressed natural gas) or CGL (compressed gas liquids, an emerging technology), or *reforming* via GTL (gas to liquids) into synthetic liquid hydrocarbon fuels (ultralow-sulfur diesel, naphtha) and petrochemical feedstocks (methanol, dimethyl ether). An intriguing new conceptual technology under development aims to pelletize gas into a solid natural gas hydrate (NGH) for marine tanker transport and then regasify (dissociate) it back into gaseous form at the delivery point.

6.3 *Producing Wells*

The world natural gas production for 2010 was 112.8 Tcf, while consumption for 2010 was 111.9 Tcf.[2] There is typically a small discrepancy each year due to withdrawal from storage, liquefaction, and measurement/definition differences. The top 10 producing countries are shown in Table 3. Together, these account for 62% of total world production of natural gas. Table 3 also shows that the world production from 2004 to 2010 increased by about 19%. The total number of US-producing

Table 3:   World Production of Natural Gas in Tcf per Year.[2]

Country	2004 Tcf	2006 Tcf	2008 Tcf	2010 Tcf	Share of total
US	18.59	18.50	20.16	21.58	19.1%
Russian Federation	20.25	21.02	21.25	20.80	18.4%
Canada	6.49	6.61	6.23	5.64	5.0%
Qatar	1.38	1.79	2.72	4.12	3.7%
Norway	2.77	3.09	3.51	3.76	3.3%
China	1.47	2.07	2.84	3.42	3.0%
Saudi Arabia	2.32	2.60	2.84	2.96	2.6%
Indonesia	2.48	2.48	2.46	2.90	2.6%
Algeria	2.90	2.98	3.03	2.84	2.5%
Malaysia	1.90	2.24	2.28	2.35	2.1%
World Total	**95.1**	**101.7**	**108.1**	**112.8**	**100.0%**

natural gas wells has risen steadily over the years but has jumped substantially since 1999; they now total nearly 480,000.

6.3.1 *Marketed production*

A total of 32 states produced natural gas in 2009. The Gulf of Mexico, together with Texas (onshore), Wyoming, Oklahoma, and New Mexico, accounted for 70% of total *marketed gas production* of 21.6 Tcf. Marketed production is gross withdrawals less the volumes that are reinjected, flared, and removed as impurities. *Dry gas* is the volume of gas — almost entirely methane — resulting from the removal of natural gas liquids from marketed production.

6.4 *Shale Gas*

Application of horizontal drilling and advances in hydraulic fracturing technology has allowed access to increasingly large amounts of natural gas from shale formations. Shales typically have low porosities in the range of 3%–8% and permeabilities in the micro-Darcy to nano-Darcy range. While these rocks have been known as source rocks for natural gas for over 100 years, direct production of gas in commercial quantities from shales is relatively new. In 2010, US shale gas production reached 4.87 Tcf, constituting 23% of US production. The recent release[10] by the US DOE-EIA indicates that 97.4 Tcf of the 317.6 Tcf of 2010 US met gas reserves (30.6%) are shale gas, a 9% increase from 2009. An initial assessment of 32 countries[10] suggests that these shale resources are also available in other regions of the world. Their study indicates that 5,760 Tcf of technically recoverable gas may exist, compared to 862 Tcf in the US. The top five countries with these postulated shale gas resources are shown in Table 4. This could alter the balance of world import/exports. For example, while China imports 5% of the 3.9 Tcf it consumes each year, South Africa now imports 63% of its natural gas needs. India, which imports 24% of the 1.9 Tcf consumed each year, has only a small assessed shale gas endowment of 63 Tcf.

Table 4:   Top Five Shale Gas Resource Countries.[11]

Country	Shale gas resources (Tcf)
China	1275
US	862
Argentina	774
Mexico	681
South Africa	485

Economic production of shale gas from horizontal wells requires hydraulic fracturing, whereby fluids (typically 99+% water) and proppant (typically sand or ceramics) are pumped under pressure to crack the rock and then prop open the created fractures after the pumping ceases. This process allows the wellbore to access more of the productive formation and results in increased flow rates. The US Department of Energy[12] has released a report of findings and recommendations, particularly concerning water and air quality, that will make the fracturing process more transparent to regulators and the general public. This is in direct response to concerns in the US and international community about the safety of the process and to a lack of readily available scientific and engineering data to ground the discussions.

6.5 *Natural Gas Hydrates*

Natural gas hydrates are solid, crystalline, ice-like substances composed of water, methane, and at times, small amounts of other gases. The gas molecules occupy the intermolecular cavities of a crystalline lattice of water molecules. These hydrates form under moderately high pressure and at temperatures near the freezing point of water. The estimated carbon content of these deposits, which occur onshore in permafrost regions and offshore in ocean-bottom sediments, exceeds that of all other fossil hydrocarbons combined. The "central consensus" estimate independently obtained using various estimation techniques is about 742,000 Tcf.[13]

Known and suspected deposits are unevenly distributed but occur near every continental landmass except Australia. Means of economically and safely producing this potential, but by no means assured, future energy source do not yet exist. Three possible production methods are depressurization (resulting in decomposition of the hydrate), thermal stimulation to decompose the hydrate, and chemical inhibition to destabilize the hydrate. Chemical inhibition is used in pipelines to inhibit the formation of hydrates, so-called water-ice.

7 Delivering Natural Gas from Producing Region to Market

7.1 *Processing*

Natural gas from completed wells within a producing field is delivered through the gas-gathering system to *processing plants,* where separators, dehydrators, membranes, fractionators, and other treatment techniques clean and condition the gas for safe and efficient transportation and end-use by consumers. Heavier hydrocarbon gases, primarily ethane, propane, and butane (*wet gas*), are typically removed, leaving principally methane or *dry gas*. At atmospheric pressure, propane and other wet gases can be liquefied into *condensates* or *natural gas liquids* (NGLs). Gas processors and oil refiners market these valuable byproducts as *liquefied petroleum gases* (LPG) for agricultural and rural residential fuel or as feedstocks for the petrochemical industry. Non-hydrocarbon impurities such as CO_2, H_2S, N_2, O_2, and water vapor must be reduced to levels that satisfy specifications for transport through interstate pipelines, which, together with transportation rates, are detailed in a pipeline company's *tariff*. More than 500 gas-processing plants, with a total daily capacity in excess of 68 Bcf, are in operation throughout the nation's oil and gas-producing regions.[14] Increasingly, these facilities are owned and operated by independent *midstream* companies rather than the *upstream* E&P companies.

7.2 *Transportation*

Conditioned gas that leaves the processing plant via the *tailgate* is delivered to a *receipt point*, where it is pressurized for transport to market via buried interstate and intrastate *transmission* pipelines or *trunk lines*, typically 12 to 42 inches in diameter. *Compressor stations* are required at intervals along a pipeline to maintain the high pressures needed to move the gas along. A significant proportion of gas consumed in Europe is supplied from Russia by Gazprom. Termed the Uniform Gas Supply System (UGSS) of Russia, it is the largest gas transmission system in the world and has a total length of 160,400 km.[15] A total of 152.8 Bm^3 (5.4 Tcf) was supplied to 23 countries in 2009. The greatest volumes were purchased by Germany (via the Jamal-Europe, WAG, and Transgas UGSS pipelines), Turkey, and Italy.

7.3 *Delivery*

The US interstate and intrastate transmission pipeline companies deliver about 40% of transported gas directly to large-volume *end users*, such as industrial gas consumers and independent power generators. The remainder is delivered by way of the *city gate*, where ownership of the gas changes (*custody transfer*) to *local distribution companies* (LDCs), who then deliver the gas to their residential and business customers through their own networks of lower pressure, smaller diameter *distribution lines* and still smaller *laterals*. More than 70% of the approximately 1,300 LDCs in the US are owned and operated by municipal governments, and about 20% are investor-owned and regulated by state public utilities commissions.[16] Investor-owned LDCs, who serve the greatest number of customers, are responsible for about 92% of the total volume of delivered LDC gas. The remaining distributors are privately owned (and state-regulated) or operated as non-profit cooperatives.

7.4 *Storage*

Some of the natural gas transported through the US interstate pipeline network is routinely diverted and temporarily placed in pressurized underground *storage fields*. These facilities are designed to deliver gas on short notice (daily and even hourly) to meet surges in consumer demand, such as for heating and power requirements during extremely cold weather, and to offset unexpected curtailments in production or delivery due to pipeline accidents, hurricanes, and other severe weather events. Thus they help avoid price spikes. Just as importantly, storage sites help *balance* or equalize the volumes of gas normally contracted to be withdrawn from a pipeline system with the volumes injected. Active storage sites currently number about 400 in 30 states and include depleted reservoirs in oil and gas fields (326), aquifers (43), and man-made salt caverns and salt mines (31).[17] Gas-storage sites are owned and operated by pipeline companies, LDCs, and independent storage developers. Storage capacity exists in other countries, either as manufactured storage sites or in underground geological formations. For example, the Wierzchowice facility, operated by the Polish Oil and Gas Company (PGNiG SA) can store $1.2\,\mathrm{Bm^3}$ in geological formations. The Rehden facilty in Lower Saxony, Germany, is the largest storage site in Europe, with a capacity of $4\,\mathrm{Bm^3}$. Rehden is operated by WINGAS, a joint German-Gazprom company. There are 25 gas storage facilities in the Russian Federation, 8 in water aquifers, and 17 in depleted gas fields.[18]

The total volume of gas in a given storage field consists of two components — a larger, relatively permanent volume of *base* gas, which is needed to maintain adequate pressure and deliverability rates, and a smaller cushion of *working gas* that can be withdrawn quickly during times of high demand and later replenished by injection during times of normal demand, a routine practice known as *cycling*. Storage inventories typically are rebuilt following the winter heating season (for the US, November 1 to March 31).

As a result of US expansions and new construction, the total demonstrated peak working-gas capacity rose in 2008 and early 2009 to 3,889 Bcf.[19] The total volume of working gas available at any given time, in relation to historical maxima and the running five-year average, strongly influences the price of natural gas that is set in futures contracts traded on the commodity market. With the higher output of natural gas in 2008, together with lower demand and prices due to the recession and to moderated temperatures in some areas, working-gas volumes reached an historical high of 3,565 Bcf.

Periodically, a transporter or storage operator may hold an *open season*, a period of time, usually several weeks, during which existing and prospective customers and marketers may bid, on an equal and often non-binding basis, for a specified amount of available or planned transport and/or storage capacity and other services.

7.5 *Commerce*

The place where a number of large pipelines interconnect to facilitate access to multiple regions of supply is termed a *market center* or *hub*. Located at strategic points

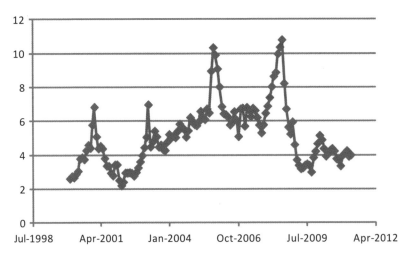

Fig. 1.   US natural gas wellhead price Jan 2000–Dec 2010 in dollars per thousand cubic feet. *Source*: EIA (2011).

on the pipeline grid, market centers offer a variety of essential transportation and administrative services to shippers, purchasers, and marketers — not only storage and compression but also short-term receipt/delivery balancing, volume aggregation, and title transfer.[20]

Hub configurations vary — some are associated with one major pipeline system while others operate through shorter bidirectional (flowing both directions) *header* systems connected to other pipelines or to storage sites and gas-processing plants.

The Henry Hub in southern Louisiana is the designated delivery point for natural gas futures contracts, which are traded on the New York Mercantile Exchange. Accordingly, Henry Hub is the market center to which gas prices at other hubs are compared daily. How much the short-term cash or *spot* price for gas (in US\$/Mcf or US\$/MMBtu) varies with respect to the futures price is referred to as the *basis differential*. In gas-supply regions that may have less outbound transport capacity or access to fewer markets, the differential can, in effect, be a relative price disadvantage under which producers, transporters and marketers must make decisions regarding sales, trades and other business.

The volatility of US gas prices (which are seen in other countries depending on international and regional factors) is shown in Fig. 1.

7.6 *The Integrated Delivery System*

For the US, 470,000 gas-producing wells, 500 gas-processing plants, 210 gas pipeline systems consisting of 300,000 miles of interstate and intrastate transmission pipelines, more than 1 million miles of gas-gathering lines, 1,400 compressor stations, 400 storage sites, 50 gas-import/export border pipeline interconnections,

8 LNG import and regasification terminals (including 2 offshore), 28 market centers, and more than 1,300 local distribution companies comprise the vast complex physical infrastructure that provides natural gas service to consumers in all 50 states.

Russian natural gas statistics are also available from Gazprom.[15] The Unified Gas Supply System of Russia, mentioned earlier, has 160,400 km of gas trunk lines and branches, 215 pipeline compressor stations, and 25 underground storage facilities. This accounted for 80% of total Russian production. The Gazprom infrastructure saw a throughput of 461.5Bm^3 (16.3 Tcf) in 2009, from a reserve base of 33.6Tm^3 (1186.6 Tcf).

7.7 *Liquefied Natural Gas*

LNG is an alternative method to transport natural gas in which gas is cooled to $-161.5°\text{C}$. The volume of the liquefied gas is approximately 610 times less than the equivalent volume of gas. The first LNG delivery by tanker occurred in 1959 between Lake Charles (Louisiana, US) and Great Britain. Commercial shipments between Algeria and the United Kingdom began in 1964. Commercial trade in Asia began in 1969 with LNG imports from Alaska, followed by exports from Libya, Brunei, and Algeria in 1970. Since 1979, LNG trade has grown at an average annual rate of 7%. LNG world trade in 2010 was approximately 297.6Bm^3 (10.5 Tcf).[2]

The top 10 exporting countries of LNG are shown in Table 5 and together these make up 88% of all LNG exports. Also shown in Table 5 are the 10 leading importers of LNG. This list, accounting for 87% of world total imports, is headed by Japan, which accounted for 31% of total world imports in 2010.[2]

LNG is scheduled to power trucks used to service the Haynesville Shale play in Louisiana and Texas. Encana Corp will develop mobile LNG refueling stations to be used by a heavy-duty truck fleet of Heckmann Water Resources. Heckmann would become the largest LNG fleet (> 200 trucks) operator in North America.

Table 5: Top 10 LNG Exporting and Importing Countries in 2010 (Bm^3).[2]

Exporting country	2010 (Bm^3)	Percentage of total	Importing country	2010 (Bm^3)	Percentage of total
Qatar	75.75	25%	Japan	93.48	31%
Indonesia	31.36	11%	South Korea	44.44	15%
Malaysia	30.54	10%	Spain	27.54	9%
Australia	25.36	9%	United Kingdom	18.67	6%
Nigeria	23.90	8%	Taiwan	14.90	5%
Trinidad & Tobago	20.38	7%	France	13.94	5%
Algeria	19.31	6%	China	12.80	4%
Russian Federation	13.40	5%	US	12.23	4%
Oman	11.49	4%	India	12.15	4%
Egypt	9.71	3%	Italy	9.08	3%
World Total Exports	**297.6**	**88%**	**World Total Imports**	**297.6**	**87%**

8 How Natural Gas is Used

Again using the US as an example, natural gas finds many important uses among the nation's four principal energy consumer sectors — residential, commercial, industrial, and transportation — and in electric power generation.

8.1 *Power Generation*

Throughout the 1990s, natural gas had become the preferred fuel for generating electricity at "peaker" plants, power plants that are smaller in capacity than the utilities' *base-load* plants (typically coal, nuclear, and hydro). Peakers can be brought on line at full capacity quickly to help meet peak electricity demands as well as to satisfy normal demands during scheduled and unscheduled shutdowns at base-load plants. Because of ever-increasing demands on base-load stations, many peakers now run longer than just during peak hours. Most peakers are operated by *independent power producers* (IPPs). In addition, because of its high reliability, natural gas is becoming important as a backup fuel for alternative and renewable energy projects.

Natural gas has accounted for a steadily increasing share of total net electricity generated (in kilowatt-hours, kWh) from fossil fuels in the power generation sector — from 13% in 1988 to 28% in 2007 and 2008. In the last two years, natural gas accounted for 20% of the net electricity generated from all energy sources (fossil fuels, nuclear, hydro, and other renewables).[3]

Of the total new generation capacity added between 1999 and 2002, 96% was natural gas fired. In its projection to 2030,[21] EIA forecasts that natural gas plants will account for more than one-half (53%) of all new generation capacity additions — 137 gigawatts (GW, or 1,000 MW) out of 259 GW net after retirement of older power plants.

Utilities and IPPs base their decisions regarding the choice of fuel for future capacity additions on such factors as electricity demand growth; the need to replace inefficient plants; capital, variable, and transmission costs; operating efficiencies; fuel prices; emissions constraints; and availability of federal tax credits for certain technologies. Compared to coal, nuclear, and wind, natural gas-fired plants have substantially lower capital costs but considerably higher variable costs. Fuel expenditures in particular, constitute the largest component ($\sim$75%) of total plant costs. Thus, while coal, nuclear, and wind plants are highly sensitive to construction costs, which are escalating over the near term, gas plant costs are sensitive to natural gas prices, which are tied closely to oil prices and to domestic gas production levels.

Natural gas also has the advantage that it releases less CO_2 than either coal or oil per unit of electrical energy produced. As shown in Table 6, producing one MWh of energy using a coal plant releases about twice the amount of CO_2 released by a plant using natural gas.[22]

Table 6: Release of CO_2 Per Unit of Electricity Produced by Different Fuels.

Fossil fuel	CO_2 released (kg)/MWh
Coal	974
Oil	726
Natural Gas	484

8.2 *Combined Heat and Power Generation*

Combined heat and power (CHP) generation includes two high-efficiency technologies for producing electricity — natural gas combined-cycle systems and cogeneration turbine systems. Both systems capture and reuse waste heat that normally is lost. A combined-cycle power plant uses waste heat from gas combustion (in a gas turbine or a battery of gas engines) to produce additional electricity by heating water to make steam, which, in turn, drives a steam turbine. A cogeneration system, on the other hand, uses the captured thermal energy to generate additional electricity, provide space heating, or fulfill other energy needs of a building, factory, industrial park, or campus. In most electric power plants, the waste heat from fuel combustion is lost, resulting in substantially lower operating efficiencies. Combined-cycle systems account for nearly 80% of installed CHP capacity.

8.2.1 *Combined-cycle generation*

Natural–gas-fueled combined-cycle systems offer attractive economic, environmental, and operating characteristics. For example, combined-cycle gas turbine plants generate electricity more efficiently than conventional fossil-fuel plants, with efficiencies approaching 60%, compared with 30% to 35% for typical boiler units.

In addition, gas-fueled combined-cycle units offer lower construction and maintenance costs, higher operating reliability, and shorter construction timeframes. Furthermore, compared to similarly sized coal-fired units equipped with pollution-control equipment, combined-cycle units produce no solid wastes, less than 1% of the sulfur oxides (SO_x) and particulate matter, and about 85% less nitrogen oxides (NO_x).

8.2.2 *Cogeneration*

Gas-fired cogeneration or "cogen" systems offer the same advantages as combined-cycle generation, in terms of capital and operating costs, efficiencies, and emissions. Large coal-fired cogeneration systems average from one and one-third to more than three times the capital costs of natural-gas–based systems because they require pollution control equipment, more land for the plant and fuel stockpiles, expensive fuel-handling equipment, greater boiler maintenance, and more personnel.

Many commercial and industrial facilities can take advantage of natural gas cogen systems because of their highly variable requirements for heat and electricity. Cogen systems are available in sizes from as small as 2.2 kW to as large as several hundred megawatts. Natural-gas–fueled cogeneration is being successfully applied in the pulp and paper, pharmaceutical, food processing, textile, oil refining, fertilizer, and other petrochemical industries, as well as in hospitals, universities, hotels, computer centers, and other commercial facilities.

Electricity generation and thermal output (Btu basis) from CHP systems have in recent years declined in the power sector as well as in the commercial and industrial sectors, the latter two reflecting overall declines both in consumption of combustible fuels for power generation and in total electricity output.

8.3 *Transportation*

Natural gas is quickly gaining recognition as the most readily available, environmentally acceptable, and economic alternative fuel for America's cars, trucks, and buses for the near term. About 120,000 natural gas vehicles (NGVs) in the US operate safely, cleanly, efficiently, and economically. California, Texas, and Arizona are home to one-half of all NGVs in use in the US and together accounted for nearly two-thirds of the natural gas consumed by NGVs in 2007.[23]

The estimated volume of natural gas consumed by NGVs in the US has doubled since 2002, from 15 Bcf to 30 Bcf in 2008.[5] However, the EIA estimates that the total number of NGVs (CNG and LNG) in use has declined from a high of 123,500 in 2002 to 117,170 in 2007.[23] With the availability now of a dedicated natural-gas–fueled vehicle from a major auto manufacturer and the increasing public awareness of the benefits of NGVs in discussions about energy policy, the number of new and converted NGVs should begin to increase. There are 11.5 million NGVs in the world, with 5.7 million in the Asia-Pacific region (led by Pakistan with 2.5 million). Latin America accounts for another 4 million, led by Argentina with 1.8 million.[24]

Compressed natural gas (CNG) has distinct advantages over traditional transportation fuels. It burns more cleanly, requires less space for storage, and has a proven safety record.

On a Btu basis, the base cost of gasoline can be at least four times more expensive than natural gas as a transportation fuel.[25] Furthermore, the US Energy Policy Act of 2005 has made federal tax credits available to operators of manufactured NGVs and petroleum-fueled vehicles that have been retrofitted or repowered with EPA or CARB (California Air Resources Board) certified natural gas engines or conversion systems. *Liquefied natural gas* (LNG, but excluding LPG) is also is used in NGVs to a smaller extent, primarily in heavy-duty vehicles.

Natural gas is especially well suited as a transportation fuel for the US. There are 11.5 million government and corporate fleet vehicles, many of which return to a central location each night for refueling. Furthermore, because these sites, usually located within metropolitan areas, are already serviced by the local gas-distribution infrastructure, fuel delivery is of minor consequence. According to NGVAmerica, more than 60% of CNG vehicles presently are classified as light-duty. However, CNG applications more and more are directed toward medium- and heavy-duty vehicles, such as transit buses, school buses, utility vehicles, sanitation trucks, and off-road industrial, mining, and aviation vehicles, which typically have high fuel-consumption rates and, like fleets, can take advantage of centrally located refueling docks.[26]

Compared to the rest of the world, expanded use of NGVs in America has been constrained in large part by the limited, but growing, number of refueling stations, which number about 1,100, but only about one-half of which currently sell gas to the public. However, with the availability of home refueling units, which also qualify for federal and some state tax credits, a resurgence in car and light-duty NGV demand may be forthcoming. Chesapeake Energy is on track to convert its corporate fleet of 4,200 vehicles to CNG by 2014. Gazprom now controls the 130 CNG filling stations in the greater Moscow area.[15]

Important technical advancements continue to be made by the American Gas Association, Gas Technology Institute, DOE, and other groups in bringing more fuel-efficient NGVs into the marketplace.

9 The Role of Reserves and Potential Resources

As defined by the SPE/WPC/AAPG/SPEE Petroleum Resources Management System (2007) "Reserves are those quantities of petroleum anticipated to be commercially recoverable by application of development projects to known accumulations from a given date forward under defined conditions. Reserves must further satisfy four criteria: they must be discovered, recoverable, commercial, and remaining (as of the evaluation date) based on the development project(s) applied. Reserves are further categorized in accordance with the level of certainty associated with the estimates and may be sub-classified based on project maturity and/or characterized by development and production status." Reserves may be considered a working inventory, which is backed up by much larger volumes of technically recoverable and economically recoverable resources.

Estimates of world natural gas reserves in 2010 range between 6,261 Tcf[27] and 6,609 Tcf.[2] The distribution among countries with the highest gas reserves is shown in Tables 7 and 8, according to these two different sources. The reserves in different regions of the world and the top seven countries with the greatest gas reserves are shown in Table 7. Differences in absolute amounts are expected due to aggregation from different data sources by EIA and BP and varying vintages of the data themselves.

The top 10 countries according to Ref. 2 are shown in Table 8, along with their estimated reserves in 2000. Note that the total world reserves have increased by about 21% between 2000 and the end of 2010, a rather modest increase. The top 10 countries between them account for 77% of the world resources of natural gas.

Table 7:   World Natural Gas Reserves.

Region	Natural gas reserves (Tcf)	Country	Natural gas reserves (Tcf)
North America	319	Russia	1,680
Central and South America	267	Iran	1,046
Europe	161	Qatar	899
Eurasia	2,170	Saudi Arabia	263
Middle East	2,658	US	244.7
Africa	495	United Arab Emirates	210
Asia and Oceania	539	Nigeria	185
World Total	**6,609**		

Source: IEO (2010).

Table 8:   Natural Gas Reserves.[2]

Country	End of 2000 Tcf	End of 2010 Tcf	Share of total
Russian Federation	1,493.8	1,580.8	23.9%
Iran	918.2	1,045.7	15.8%
Qatar	508.5	894.2	13.5%
Turkmenistan	91.8	283.6	4.3%
Saudi Arabia	222.5	283.1	4.3%
US	176.6	272.5	4.1%
United Arab Emirates	211.9	213.0	3.2%
Venezuela	148.3	192.7	2.9%
Nigeria	144.8	186.9	2.8%
Algeria	158.9	159.1	2.4%
World Total	**5,449.1**	**6,608.9**	**100.0%**

Source: IEO (2011).[28]

The differences between the estimates in Tables 7 and 8 provide an indication of the uncertainties in these estimates.

The US domestic reserves situation suffered during the early to middle 1980s as a result of the so-called "gas bubble" — an abnormal surplus of production capacity rather than an excess of gas itself. This unusual circumstance resulted in dramatic declines in drilling activity, production, and reserves additions. The situation improved thereafter, largely through elimination of the federal government's disastrous attempts to control natural gas prices and restrict how gas was used. With disincentives removed, exploration rebounded. Although new discoveries continued to be made throughout this period, it was not until 1994 did the year-to-year change in proved reserves finally move into positive territory. The quantity of this *working inventory* continues to increase, particularly due to successes in using the unconventional gas resources, discussed earlier. Using gas from low-permeability shales as an example, the positive growth is illustrated in Fig. 2.

Proved reserves then become the basis for assessing a country's *future supply* of natural gas. To this value we add the assessment of potential resources — that is,

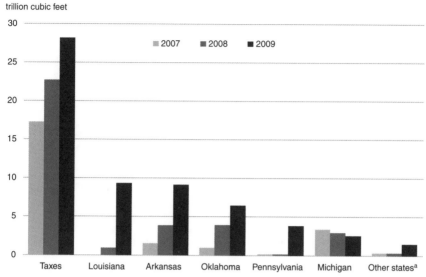

trillion cubic feet

aOther states include Indiana, Kentucky, Missouri, North Dakota, Tennessee, and West Virginia.
Source: U. S. Energy Information Administration.

Fig. 2.   US shale gas proven reserves by year.[29]

Table 9:   US Natural Gas Resources from Potential Gas Committee.[30]

Resources category	Mean values (Tcf) 2010
Traditional Gas Resources	
Probably Resources (current fields)	536.6
Possible resources (new fields)	687.7
Speculative resources (frontier)	518.3
Subtotal Traditional Resources	1739.2
Coalbed Natural Gas	
Probable Resources	13.4
Possible Resources	48.0
Speculative Resources	96.2
Subtotal Coalbed Gas Resources*	158.6
Total Potential Resources	**1897.8**
Proved Reserves (DOE/EIA)	272.5**
US Future Supply	**2,170.3**

*Mean values for Probable, Possible, and Speculative resources are *not* arithmetically additive in deriving the subtotal. Subtotal mean values *are* additive in deriving Total Potential Resources.
**Latest available figure is for year-end 2009.
Note: Totals are subject to rounding and differences due to statistical aggregation of distributions.

remaining *technically recoverable* natural gas resources. A view for the US is shown in Table 9 and Figs. 3 and 4.

Assessments of world resources (as compared to reserves) of natural gas are rare and dated. The most commonly quoted assessment is that published by the

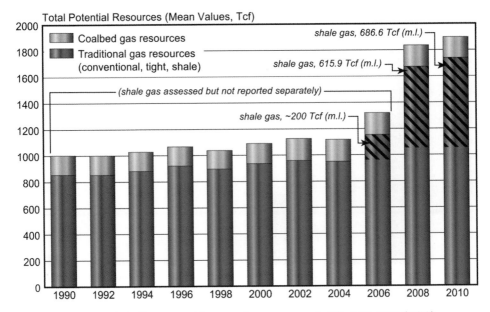

Fig. 3. Technically recoverable natural gas resources in US, 1990–2010 (PGC).

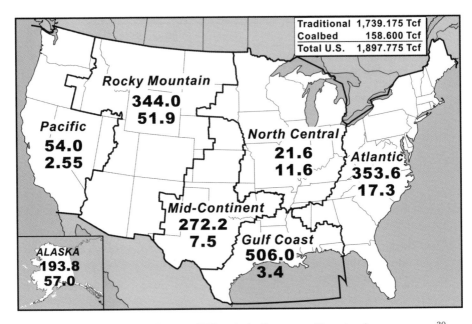

Fig. 4. Geographic distribution of US technically recoverable natural gas resources.[30]

US Geological Survey.[31] The total endowment (mean value) of 13,493 Tcf included 898 Tcf of Cumulative Production, 4,621 Tcf of Remaining Reserves, 3,305 Tcf of Reserve Growth (gas to be recovered from known fields, net of Reserves) and 4,669 Tcf of Undiscovered Conventional Resources. The cumulative production and reserves values were only for a subset of the world. The study did not address the Unconventional Gas Sources (shale gas, coalbed methane, and tight sands gas). Therefore the published numbers are quite conservative. For comparison, total world consumption in 2010 was 111.9 Tcf (BP, 2011).

Acknowledgments

Much of the material in this chapter was derived and updated from the *From Reservoir to Burner Tip — A Primer on Natural Gas* chapter of the 2009 Potential Gas Committee report, *Potential Supply of Natural Gas in the United States*. The assistance of my co-author of the Primer, Stephen D. Schwochow, and permission from my Committee to use the material is gratefully acknowledged.

References

1. M. W. H. Peebles, *Evolution of the Gas Industry* (New York University Press, New York, 1980), p. 235.
2. Statistical Review of World Energy, June 2011. Available at: bp.com/statisticalreview. Accessed 15 August 2011.
3. Energy Information Administration Office of Energy Markets and End Use, 2010, Annual Energy Review 2009. U.S. Dept. of Energy, Energy Information Administration, Rept. DOE/EIA-0384(2009), August, p. 446. Available at: http://www.eia.doe.gov/aer. Accessed 6 April 2011.
4. Energy Information Administration Office of Energy Markets and End Use, 2010, Natural Gas Annual Review 2008: U.S. Dept. of Energy, Energy Information Administration, Rept. DOE/EIA-0131(08) (2010), March, p. 193.
5. Energy Information Administration Office of Energy Markets and End Use, 2009, Annual Energy Review 2008: U.S. Dept. of Energy, Energy Information Administration, Rept. DOE/EIA-0384(2008), June, p. 407. Available at: http://www.eia.doe.gov/aer. Accessed 6 April 2011.
6. Energy Information Administration Office of Oil and Gas, 2009, Natural gas annual 2007: U.S. Dept. of Energy, Energy Information Administration, Office of Oil and Gas, Rept. DOE/EIA–0131(07), January, p. 185.
7. S. Bhattacharya, K. D. Newell, W. L. Watney and M. Sigel, Low-cost plant upgrades marginal gas fields. *Hart's E&P* **81**(8) (2008), pp. 102–103.
8. M. Golombok and D. Nikolic, Assessing contaminated gas, *Hart's E&P* **81**(6) (2008), pp. 73–75.
9. S. Bhattacharya, K. Newell, W. L. Watney and M. Sigel, Field tests prove microscale NRU to upgrade low-Btu gas, *Oil & Gas J.* **107**(40) (2009), pp. 44–53.
10. US Crude Oil, Natural Gas and NG liquids Proved Reserves, 2010. US Energy Information (August 2012).
11. Energy Information Administration Analysis & Projections, 2011, World Shale Gas Resources: An initial assessment of 14 regions outside the United States, (U.S. Dept.

of Energy, Energy Information Administration, 2011). Available at: www.eia.doe.gov/
analysis/studies/worldshalegas/. Accessed 15 August 2011.

12. U.S. Department of Energy, Shale Gas Production Subcommittee 90-Day Report,
    (2011), p. 42. Available at: http://www.shalegas.energy.gov/resources/081811_90_day_
    report_final.pdf. Accessed 15 August 2011.

13. Potential Gas Committee, Potential supply of natural gas in the United States, Report
    of the Potential Gas Committee (Potential Gas Agency, Colorado School of Mines,
    Golden, CO, USA, 2009), p. 400.

14. J. Tobin, P. Shambaugh and E. Mastrangelo, Natural gas processing — The crucial
    link between natural gas production and its transportation to market, (U.S. Dept. of
    Energy, Energy Information Administration, Office of Oil and Gas, 2006), p. 11.

15. Gazprom, Gazprom in questions and answers, (2011). Available at: http://www.
    iangv.org/tools-resources/statistics.html. Accessed 23 June 2011.

16. J. Tobin, Distribution of natural gas — The final step in the transmission process,
    (U.S. Dept. of Energy, Energy Information Administration, Office of Oil and Gas,
    2008), p. 14.

17. J. Tobin, U.S. underground natural gas storage developments — 1998–2005, (U.S.
    Dept. of Energy, Energy Information Administration, Office of Oil and Gas, 2006),
    p. 16.

18. Gazprom, Gazprom in questions and answers (2011). Available at: http://www.
    gazprom.com/production/transportation/underground-storage/. Accessed 23 June
    2011.

19. A. LaRose, Estimates of peak underground working gas storage capacity in the United
    States, 2009 update. (U.S. Dept. of Energy, Energy Information Administration, Office
    of Oil and Gas, 2009), p. 5.

20. J. Tobin, Natural gas market centers — A 2008 update. (U.S. Dept. of Energy, Energy
    Information Administration, Office of Oil and Gas, 2009), p. 14.

21. J. J. Conti et al., compilers, Annual Energy Outlook 2009, with projections to 2030,
    Rept. DOE/EIA-0383(2009) (U.S. Dept. of Energy, Energy Information Adminis-
    tration, Office of Integrated Analysis and Forecasting, 2009), p. 221 Available at
    http://www.eia.doe.gov/oiaf/aeo/. Accessed 15 August 2011.

22. P. L. Denholm and G. L. Kulcinski, Presented at the Energy Options for the Future,
    Meeting at the US Naval Research Laboratory, 11–12 March 2004. Available at:
    http://other.nrl.navy.mil/EnergyOptions/GGNR_Kulcinski/GGNR_Kulcinski.pdf.
    Accessed 15 August 2011.

23. Energy Information Administration, Alternatives to traditional transportation fuels,
    2007: Energy Information Administration, Data for 2003–07, 2009. Available at:
    http://www.eia.doe.gov/cneaf/alternate/page/atftables/afv_atf.html#supplied.
    Accessed 15 August 2011.

24. International Association for Natural Gas Vehicles April 2011 data through Dec 2010:
    Available at: http://www.iangv.org/tools-resources/statistics.html.

25. R. Rapier, How much natural gas do we have to replace gasoline? (2009). Available
    at: http://theoildrum.com/node/5615.

26. International Energy Outlook (2010) DOE/EIA/0484, p. 338.

27. U.S. Energy Information Administration, Natural gas prices report (2011). Available
    at: http://www.eia.gov/forecasts/ieo/table7.cfm. Accessed 16 January 2011.

28. S. Yborra Taking a second look at the natural gas vehicle: American Gas Magazine
    [American Gas Association], August–September 2006. Available at: http://www.aga.
    org / Content / ContentGroups / American _Gas _Magazine1/August _September _2006/
    Taking_a_Fresh_Look_at_NGVs.htm.

29. Energy Information Administration (EIA), Summary: U.S. crude oil, natural gas, and natural gas liquids reserves — 2009 Annual Report, (U.S. Dept. of Energy, Energy Information Administration, Office of Oil, Gas and Coal Supply Statistics, 2010).
30. Potential Gas Committee, Potential supply of natural gas in the United States, Report of the Potential Gas Committee (31 December 2010), Advance Summary, April, 2011, (Potential Gas Agency, Colorado School of Mines, Golden, CO, USA, 2011), p. 28.
31. T. Ahlbrandt, U.S. Geological Survey World Petroleum Assessment 2000: USGS World Energy Assessment Team, U.S. Geological Survey Digital Data Series 60 (CD-ROM) (2000).

Chapter 6

Nuclear Power

Bertrand Barré
AREVA
7, rue des Blanchisseurs, 92370 Chaville, France
bcbarre@wanadoo.fr

After two decades of quasi-stagnation, nuclear power is coming back, in spite of the accident at Fukushima, as a main source to generate reliable baseload electricity with very low greenhouse gas emissions. For this "renaissance" to be sustainable, there are prerequisites: no additional severe accidents, and an accepted way of disposing of long-lived waste. If nuclear power grows rapidly, new types of "generation IV" systems will have to be deployed.

1 Introduction

In 2009, 437 nuclear reactors operating in 30 countries supplied 5.9% of the world's primary energy and 14% of the world's electricity. Table 1 lists, by country, the reactors in operation and under construction as of 31 December 2011, by the number of units and capacity in GWe, as well as the amount of electricity generated during the year 2011, in billion kWh and in percentage of total electricity production.

Although nuclear fission was discovered in 1938 and the first nuclear "pile" went critical in December 1942, nuclear power production started only in the mid-1950s. The number and size of nuclear reactors grew rapidly in the 1970s and 1980s, but the world nuclear capacity grew very slowly from 1985 to 2005, and mostly in Asia, as shown in Fig. 1, from the PRIS database of IAEA.[1]

Since 2005, nuclear power is experiencing what the media call a "renaissance." Even though the percentage of the world's electricity generated by nuclear power decreased from 18% in 2000 to 14% in 2009, and the total amount of electricity generated by nuclear power decreased from 2,626 TWh in 2005 to 2,558 TWh in 2009, this is only the result of two decades of low construction rate and the progressive shutdown of the oldest plants. But an increasing number of countries are now planning to introduce nuclear in their energy mix and the number of reactors under construction, 27 in 2005, exceeded 60 in October 2010.

The accident which occurred in Fukushima on 11 March 2011 triggered a general re-evaluation of the robustness of nuclear plants against extreme natural disasters.

Table 1:   Nuclear Reactors in Operation and Under Construction, 31 Dec. 2011.

Country	In operation		Under construction		Production 2010	
	#	MWe	#	MWe	TWh	%
US	104	101,240	1	1,165	807.1	20
France	58	63,130	1	1,600	410.1	74
Japan	50	44,215	2	2,650	280.3	29
Russia	33	23,643	10	8,223	159.4	17
South Korea	21	18,700	5	5,560	141.9	32
Germany	9	12,068			133.0	28
Canada	18	12,624			85.5	15
Ukraine	15	13,107	2	1,900	84.0	48
China	16	11,688	26	26,660	71.0	2
Spain	8	7,567			59.3	20
United Kingdom	18	9,920			56.9	16
Sweden	10	9,298			55.7	38
Belgium	7	5,927			45.7	51
China-Taiwan	6	4,980	2	2,600	39.9	21
Switzerland	5	3,238			26.3	40
Czech Republic	6	3,678			26.4	33
Finland	4	2,736	1	1,600	21.9	28
India	20	4,391	5	2,708	20.5	3
Hungary	4	1,889			14.7	42
Bulgaria	2	1,906	2	1,906	14.2	33
Slovakia	4	1,816	2	782	13.5	52
Brazil	2	1,884	1	1,245	13.9	3
South Africa	2	1,800			12.9	5
Romania	2	1,300			10.7	20
Argentina	2	935	1	692	6.7	6
Mexico	2	1,300			5.6	4
Slovenia	1	666			5.4	37
Netherlands	1	487			3.8	3
Pakistan	3	725	1	315	2.6	3
Armenia	1	375			2.3	39
Iran	1	915				
TOTAL	**433**	**369,077**	**62**	**62,654**	**2,630**	**14**

Some reinforcement of safety procedures shall certainly be required by the relevant safety authorities: these requirements will likely increase the cost of nuclear power but make it even safer. Apart from Japan, very few countries reacted strongly to the accident, although Germany did shut down eight operating reactors and revived its 2001 policy of a complete phase-out of nuclear power by 2022. On a worldwide basis, however, one can expect that this accident will have delayed the "renaissance" by about two years, but without changing the trend.

There are at least three reasons for this continuing renaissance. From an economic point of view, nuclear power is immune from the erratic variations of oil and gas prices, since the price of uranium accounts for less than 5% of the total cost of a nuclear kWh. From a geopolitical point of view, uranium resources are well distributed across the continents in countries with very different regimes. Last but not

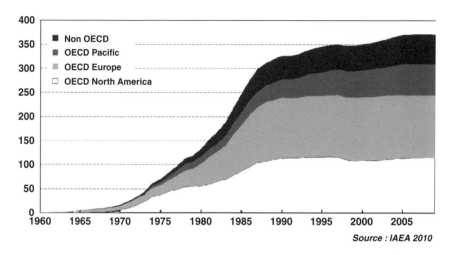

Fig. 1. World nuclear generating capacity (GWe) 1960–2009.

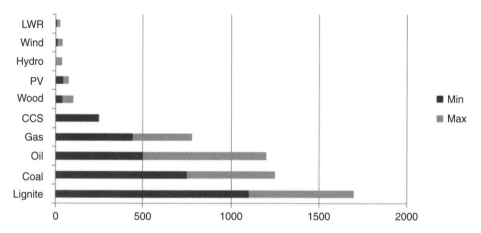

Fig. 2. Lifecycle GHG emissions, in gCO_{2eq} for 1 kWh of electricity.
Note: Ranges reflect differences in assessment technology, conversion efficiency, assessment boundary, etc.
Source: Weisser, IAEA (2006).

least, from an environmental perspective, nuclear power emits almost no greenhouse gases (GHGs) into the atmosphere over its complete lifecycle (Fig. 2).

2 Radioactivity, Fission, and Fusion

Whether solid, liquid, or gaseous, matter is made of atoms, alone or combined to form molecules or crystals. Nearly all the mass of the atom is concentrated in a positively charged nucleus, which is surrounded by "clouds" of negatively charged electrons. The nucleus itself is an assembly of positively charged protons plus neutrons which carry no electrical charge. Neutrons and protons have essentially the same mass, and their number A determines the atomic mass of a given atom. The number of protons, Z, identical to the number of electrons, determines the chemical nature of a given element, from hydrogen (Z = 1) to uranium (Z = 92) and beyond. Nuclides having the same Z but different A are *isotopes* of the same chemical element. There is a relationship between A and Z, which determines the stability of a nucleus. Stable elements are found in the so-called valley of stability in an (A, Z) diagram.

All nuclides found in nature are not stable: some do not have the right balance of neutrons and protons in their nucleus and, over time, they tend to reach the valley of stability by emitting radiation. This is called *radioactive decay*, or radioactivity. For instance, in a nucleus with too many neutrons, one neutron will turn into a proton and one electron will be expelled from the nucleus, which keeps the atom neutral. This is called β^- decay. The new nuclide, whose charge is now Z + 1, belongs to a different element. If the nucleus has too many neutrons and protons, it will expel an assembly made of two neutrons and two protons called an α particle, identical to the nucleus of a helium atom. If the nucleus has too much energy, it gets rid of its excess by emitting electromagnetic radiation, a γ-ray.

If a nucleus absorbs a particle, it will become unstable and undergo radioactive decay. The new nuclide will be "artificial," but the radiation it emits will still be "natural," α, β, or γ.

When a few heavy nuclei (^{233}U, ^{235}U, ^{239}Pu, ^{241}Pu) absorb a neutron, they become so unstable that they split into two *fragments*, releasing an enormous amount of energy and emitting two or three new neutrons with high velocity, around 20,000 km/sec (Fig. 3).

One such *fission* releases about i.e. 200 *million* electronvolts, i.e. 200 MeV (Table 2), while a typical chemical reaction releases only a few eV. The energy of the neutrinos emitted during the β^- decay cannot be recovered.

The new neutrons can, in turn, be absorbed by another *fissile* nucleus and propagate a *chain reaction* of fissions. The probability of a neutron being absorbed

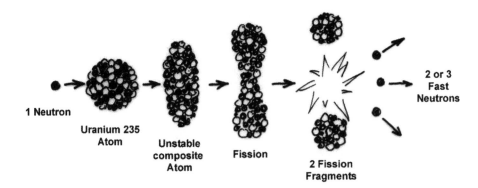

1 Neutron

Uranium 235 Atom

Unstable composite Atom

Fission

2 Fission Fragments

2 or 3 Fast Neutrons

Fig. 3.   Fission (schematic).

Table 2: Energy Released in the Fission of One ^{235}U
Nucleus by a Thermal Neutron.[2]

Fission fragments kinetic energy	166 MeV
Fission products β^- decay	7 MeV
Fission products γ decay	7 MeV
Prompt neutrons	5 MeV
Prompt γ	8 MeV
(neutrinos)	(10) MeV
Total *recoverable* energy	193 MeV

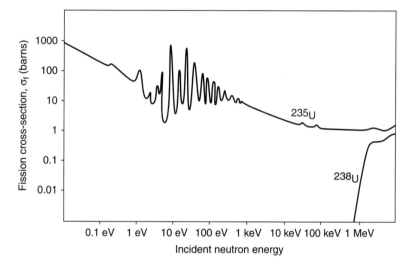

Fig. 4.   Neutron cross-sections for uranium to fission.[2]

increases when the velocity of the neutron decreases to the average velocity of
the surrounding nuclei, due to thermal agitation. The velocity of those "slow" or
"thermal" neutrons is close to 2.2 km/s.

Cross-sections (denoted σ) are a measure of the target surface that a nucleus
offers to a neutron for a given nuclear reaction i.e. a measure of the probability
that such a reaction occurs. They are expressed in *barns* (1 barn $= 10^{-24}$ cm^2).
Cross-sections are specific to a particular interaction between a given nucleus with
a neutron, and they are highly dependent upon the velocity of the neutron.

As shown in Figs. 4 and 5, there are three regions in the cross-section curves: At
low neutron energy, σ is proportional to the inverse of the neutron velocity. Then
there is a *resonance* region where σ may vary hugely for a small energy difference. At
high energy, above 1 MeV, some *threshold* reactions may suddenly appear, like the
"fast" fission of ^{238}U. Detailed cross-section plots are available from the Evaluated
Nuclear Data File at Brookhaven National Laboratory.[3] Table 3 summarizes the
main data useful in reactor design.

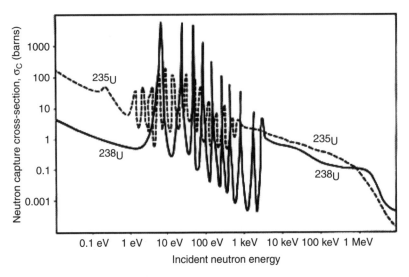

Fig. 5.   Cross-sections for uranium to capture a neutron (absorption without fission).[2]

Table 3:   Main Neutron Data for Fissile and Fertile Nuclei (Rounded Figures).

		^{233}U	^{235}U	^{239}Pu	^{232}Th	^{238}U
Fast Neutrons	Fission cross-section σ_f barns	2.8	2.0	1.9	0.01	0.05
	Capture cross-sections σ_c barns	0.3	0.5	0.6	0.35	0.30
	Neutrons per fission υ	2.5	2.5	2.9	2.3	2.75
	Neutrons per absorption η	2.2	2.0	2.2	—	—
Thermal Neutrons	σ_f barns	527	579	741	—	—
	σ_c barns	54	100	265	7.6	2.7
	υ	2.5	2.4	2.9	—	—
	η	2.2	2.0	2.1	—	—
	Delayed neutrons β_{eff} %	0.28	0.64	0.21	2.3	1.5

The fission of a nucleus of ^{235}U emits on average 2.4 *prompt* neutrons, and the figure ν reaches 2.9 for plutonium. But a few additional neutrons — their percentage, denoted $\beta_{\text{eff}} = 0.2\%$–$0.6\%$ — are emitted seconds or minutes later, during the radioactive decay of some fission fragments. These *delayed* neutrons are essential to control and stabilize the chain reaction in the core of a nuclear reactor: if there were only prompt neutrons, one could build nuclear bombs but one could not use nuclear power. The number $\eta = \nu.\sigma_f/(\sigma_f + \sigma_c)$ is the number of neutrons emitted per neutron absorbed by a fissile nucleus. For breeding purposes, η must be significantly larger than 2.0.

A few heavy nuclides (^{232}Th, ^{238}U) are called *fertile*. When they absorb a neutron, they undergo two successive β^- decays and turn into a fissile nuclide, respectively, ^{233}U and ^{239}Pu. The uranium found in the Earth's crust is mostly

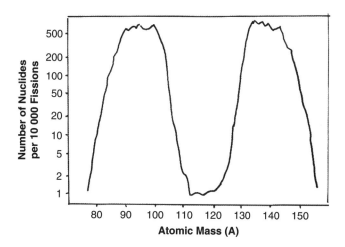

Fig. 6.   Relative yields of fission fragments (fission U235 by thermal neutron).

Table 4:   Half-Lives of Some Natural (N) and Artificial (A) Nuclides.

Nuclide	N/A	Half-life	Uses
Tritium ^{3}H	A	12.3 years	Nuclear fusion
Carbon 14	N	5,730 years	Dating old artifacts
Oxygen 15	A	2.02 min	Medical imaging
Cobalt 60	A	5.27 years	Industrial gammagraphy
Cesium 137	A	30.2 years	Cancer therapy
Radon 222	N	3.82 days	Source of natural irradiation
Radium 226	N	1,600 years	Used previously, now banned
Uranium 235	N	704 million years	Nuclear fission
Uranium 238	N	4.54 billion years	Nuclear fission
Plutonium 239	A	24,100 years	Nuclear fission

made of 99.3% fertile ^{238}U and 0.7% fissile ^{235}U, while thorium is 100% ^{232}Th and has no fissile nuclide.

Fission is a complex phenomenon that gives birth to a wide variety of pairs of fission fragments, the masses of which are distributed as shown in Fig. 6. Most of those fragments are radioactive and decay into a line of "daughter" nuclides. The fragments and their daughters are called *fission products*.

Each radioactive nucleus is characterized by its *half-life*, denoted $T_{1/2}$, which measures the time necessary for half of a given number of such nuclei to have decayed. Even though a *single* "short-lived" nucleus may survive almost indefinitely, statistics allows us to measure half-lives with incredible precision. Table 4 gives the value of the half-life of a number of nuclei of interest, and Table 5 gives the activities of some fission products as they decay with time.

Table 5:   Typical Fission Product Activities in 1 kg of Total Fission Products.[4]

Nuclide	Half-life	Activity (W) after			
		1 year	10 years	100 years	1,000 years
^{143}Ba	12 sec				
^{131}I	8 days				
^{103}Ru	40 days	20			
^{144}Ce	284 days	2.5×10^4	9		
^{106}Ru	1 year	1.9×10^3	2		
^{85}Kr	11 years	3×10^2	2×10^2	0.6	
^{90}Sr	29 years	3×10^3	2.5×10^3	2.9×10^2	
^{137}Cs	30 years	3.2×10^3	2.6×10^3	3.2×10^2	
^{151}Sm	90 years	4	4	2	0.002
^{99}Tc	200,000 years	0.5	0.5	0.5	0.5
129I	1.7 million years	0.001	0.001	0.001	0.001

3 How Does a Nuclear Reactor Operate?

A nuclear reactor, or nuclear power plant, is a machine within which a fission chain reaction is maintained to generate electricity. The chain reaction occurs in the *core* of the reactor, made from a number of *fuel assemblies* (or sub-assemblies) which contain the nuclear fuel. The heat released by fission is carried off by a *coolant* (gaseous or liquid) and used to boil water and produce high pressure steam. This steam activates a turbine coupled to an electrical generator.

In a pressurized water reactor, (PWR), the most widely used reactor type in the world, the coolant is ordinary water maintained at high pressure, 15.5 Mpa, so as to remain liquid at temperatures above 300°C. This water circulates within a closed *primary circuit*, which includes the core, located in a *pressure vessel* made of thick high strength steel, and several *loops*, 2 to 4 according to the size of the plant. Each loop comprises a *primary pump* and a *steam generator*. One of the loops is connected to a *pressurizer* in which a steam bubble maintains the required pressure within the primary circuit (Fig. 7).

Inside the steam generator, the water in the primary circuit transfers its heat to boil the water in the closed *secondary circuit*. The steam generated is dried and sent to the turbine before being condensed back to liquid water and recycled. This condensation is achieved in a large *condenser* cooled by water from the sea or a large river. Sometimes, the circuit cooling the condenser is a closed circuit, itself cooled inside a *cooling tower* by the evaporation of water from a river. There are also a number of auxiliary circuits, notably for the Emergency Core Cooling System (ECCS).

The power level of the reactor is controlled by adjusting the rate at which fissions occur within the core. This is achieved by inserting or removing *control rods* in and out of the core; these rods made of a material that strongly absorbs neutrons

PWR Schematic

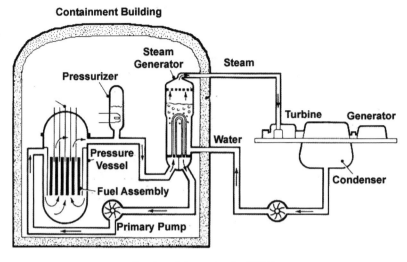

Fig. 7. Schematic of a PWR.

without undergoing fission. In addition, boric acid can be added to the primary water when the fuel is "fresh."

The PWR fuel is made of ceramic cylindrical *pellets* made of an oxide of enriched uranium (see later) or of a mixture of uranium and plutonium oxides. These pellets are inserted inside long thin hollow tubes made of an alloy of zirconium, a metal that does not absorb neutrons and is very resistant to corrosion by hot water. Those tubes, the *cladding*, are then welded closed to constitute the *fuel pins*.

Those pins are assembled around a "skeleton" of guide-tubes to make fuel assemblies with top and bottom end-pieces. A cluster of control rods can be inserted in the guide tubes, through the action of mechanisms located on top of the pressure vessel head (Fig. 8).

4 Reactor Types

A nuclear reactor needs a mixture of fissile and fertile materials to make the fuel, a fluid to act as coolant and, most of the time, a *moderator* to slow down the neutrons. We have already mentioned the possible choice of fissile and fertile materials. For the coolant, one can choose among liquids (ordinary water, heavy water,[a] molten sodium, or other metals, organic fluids, etc.) and gases (CO_2 or helium). The

---

[a]In a molecule of heavy water, the hydrogen atoms are "heavy": their nucleus consists of one proton and one neutron, instead of one single proton like the ordinary hydrogen nucleus. This heavy hydrogen is called deuterium and denoted D. Heavy water is therefore D_2O. Having already one neutron, D is not a neutron absorber. Ordinary water (H_2O) is sometimes called "light water" to emphasize the difference.

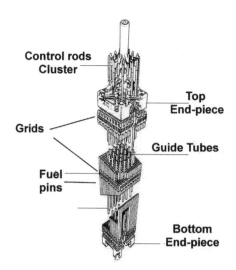

17 x 17 PWR Fuel Assembly

Fig. 8.   Typical PWR fuel assembly (*Courtesy*: AREVA).

Table 6:   Type and Net Electrical Power of Reactors Connected to the Grid and Under Construction.

Type of reactor	Operating #	31/12/2011 GWe	Under construction #	31/12/2011 GWe
PWR or VVR	270	248	52	51
BWR	84	78	4	5
GCR	17	9	—	—
PHWR (Candu)	47	23	4	2.5
LWGR (RBMK)	15	10	1	1
FBR	2	0.6	2	1.3
TOTAL	437	371	55	51

moderator must have light nuclei to efficiently slow down the neutrons by successive collisions and must not absorb too many neutrons in the process, which limits the practical choice to water (ordinary or heavy) and graphite. But fast neutron reactors can be designed with no moderator at all.

If, in addition, you take into account the various physical–chemical states of the fuel and its different possible shapes, the number of combinations is huge.

Between 100 and 200 different types of nuclear reactors have been designed, built, and operated in the 1950s and early 1960s, the "pioneer era." But today only a handful of nuclear plant types constitute the world nuclear fleet, as shown in Table 6. The oldest plants have a power in the range 300–500 MWe while the newest rate from 1,100 to 1,600 MWe.

Table 7:   Summary of Reactor Types.

Type	Fuel	Coolant	Moderator	Cycle
PWR	LEU or MOX	Ordinary water	Ordinary water	Indirect
BWR	LEU or MOX	Ordinary water	Ordinary water	Direct
GCR	Natural U or LEU	Carbon dioxide	Carbon dioxide	Indirect
Candu	Natural U	Heavy water	Heavy water	Indirect
RBMK	LEU	Ordinary water	Graphite	Direct
FBR	$(U, Pu)O_2$	Liquid sodium	—	Indirect

4.1 *"Generations" of Nuclear Reactors*

One usually describes the development of nuclear plants in terms of generations. Generation 1 (Gen1) plants were the early reactors of the pioneer era, a series of prototypes of increasing sizes and varied designs, with no standardization. Most of them are now shut down and decommissioned. Generation 2 (Gen2) consists of the reactors presently in operation, mostly LWRs. Generation 3 (Gen3) reactors are presently under construction: being modernized versions of LWRs, they embody significant improvements in safety and protection to prevent any severe accident. While Gen1 plants are being decommissioned, Gen2 plants are being operated, and Gen3 plants are under construction, international efforts are under way to develop Generation 4 nuclear systems, which are designed to fit the requirements anticipated to be in force around 2040–2050. A summary of reactor types is given in Table 7.

4.2 *Pressurized Water Reactors*

The pressurized water reactors (PWRs, or VVR for those designed in the Soviet Union or in Russia today) dominates the world market. This type was described in the previous section. In VVRs, the fuel pins are assembled with a triangular pitch instead of a square one, and the steam generators are horizontal rather than vertical.

4.3 *Boiling Water Reactors*

Like the PWR, the boiling water reactor (BWR) is cooled and moderated by ordinary water, but it uses a direct cycle: the water boils at the outlet of the core and the dried steam goes directly to the turbine while the liquid part is recycled within the pressure vessel. Instead of being assembled in a cluster, the control rods are cross-shaped and inserted in the core from below. The fuel assembly is similar to that of PWRs, but with an external wrapper which separates fuel channels inside the core. Together, PWR and BWR are called light water reactors (LWR).

4.4 *Gas-Cooled Reactors (Magnox, AGR, HTR)*

Gas-cooled reactors played a significant role in the beginning of nuclear power, notably in the UK and in France, but they survive only in the former. They use graphite as a moderator and CO_2 as coolant. The Magnox are fueled with rods of metallic natural uranium while the more recent AGRs are fueled with low enriched uranium dioxide.

A more advanced version of gas-cooled reactors, the High Temperature Reactor (HTR) has an all-refractory core cooled by very hot helium ($750°C$ or more). Several prototypes have been operated, in two families called "prismatic" or "pebble bed" according to the shape of their fuel element, but commercial maturity is expected only in Generation 4 reactors.

4.5 *Heavy Water Reactors (PHWR or Candu)*

The pressuized heavy water reactor (PHWR) is moderated by a huge tank of warm heavy water, which surrounds an array of horizontal pressure tubes inside which the fuel assemblies are cooled by hot pressurized heavy water. This "primary" heavy water goes to a collector and then to a steam generator very similar to that of a PWR, where ordinary water boils, the resulting steam being sent to the turbine.

As heavy water absorbs very few neutrons, the fuel pellets are made of natural uranium oxide. Therefore their fuel does not need enrichment. But heavy water is expensive to manufacture and the core is much larger than an LWR's, which means higher capital investment. Candu are mostly used in Canada and India.

4.6 *Light Water Graphite Reactors*

In a Reaktor Bolshoy Moshchnosti Kanalniy (RBMK) (a Russian high power channel-type reactor), arrays of vertical pressure tubes are installed throughout a huge graphite massif. Inside each tube, a fuel assembly is cooled by boiling ordinary water. The pressure tubes are connected to collectors where steam is separated and carried to the turbine. RBMKs were only located within the Soviet Union proper (Russia, Ukraine, and Lithuania) because their design enables them to produce weapon-grade plutonium. Unit 4 of Chernobyl, an RBMK, was at the origin of the worst ever nuclear accident in April 1986.

4.7 *Fast Breeder Reactors*

In a fast breeder reactor (FBR), neutrons are not slowed down by a moderator. The fissile materials inventory in their core must be larger than in a thermal reactor, but they can breed i.e. turn more fertile nuclei into fissile nuclei than they burn fissile nuclei in the process. By this means, through multiple recycling, they can make use of all the potential energy of the uranium resources. Reactors today use less than 1% of this energy. FBRs are fueled with a mixture of uranium and plutonium oxides and are cooled by molten sodium. Several prototypes have been (or are) in operation, but the technology is demanding and commercial maturity is expected only for Generation 4 reactors, when the threat of uranium scarcity makes them attractive.

5 Safety and Accident Prevention

Safety is a prime responsibility of the plant operator. In all countries using nuclear power, a national Safety Authority, fully independent of the operator, authorizes the operation, controls the compliance to safety regulations, and has the power to force the shutdown of facilities it deems no longer safe to operate.

5.1 *Barriers and Defense-in-Depth*

The nuclear industry extracts, produces, handles, conditions, and transports huge quantities of radioactive substances. The radioactive nuclei in those substances emit radiation, which can be harmful to living tissue if absorbed at high enough doses. The whole basis of nuclear safety is therefore to *contain* the radioactive substances behind suitable *"barriers"* able to stop the radiation so that it remains harmless. Containment is the name of the game everywhere: in reactors, fuel cycle facilities, transport casks and waste storage facilities or repositories. In the following, let us focus on reactors, taking the PWR as an example.

The radioactive nuclei resulting from fission within the fuel pellets are contained inside the leak-tight metallic cladding of the fuel pins, which constitutes the first barrier. If some pins start to leak, the radioactive elements will still be contained within the steel envelope of the closed primary circuit: a second barrier. In addition, the primary circuit is fully contained within the thick reinforced concrete walls of the reactor building, usually made of double enclosures. This containment building is the third (and ultimate) barrier preventing the radioactivity of the core from reaching the plant operators and the public-at-large. As long as at least one of these barriers remains intact, no radioactivity will escape to the environment.

Nuclear plants embody the concept of *defense in depth* to minimize the consequences of possible human errors or material failures:

- The plant must be designed safely, built accordingly, and correctly operated and maintained by well-trained specialists.
- In case of abnormal operation, redundant automatic *protection* systems will be activated to put the plant back on track.
- If these first two lines of defense fail, redundant automatic *safeguard* systems will be activated to bring the plant to safe shutdown, with at least one of the barriers still intact.
- If, and only if those three lines were to fail, will radioactivity escape: this is a *severe accident* and emergency measures must be taken to protect the population outside the plant.

To maintain the integrity of the first two barriers in a PWR, one must control the chain reaction to prevent reactivity accidents, and one must dissipate the heat produced within the fuel, *including* the residual *after heat* still generated by the radioactive decay of fission products once the chain reaction has been stopped.

Reactivity accidents are only minor contributors to the total risk of fuel melt-down in LWRs because of two negative feedback coefficients, intrinsic to the core design. If the core temperature increases, ^{238}U nuclei absorb more neutrons, while the moderation by water decreases: both phenomena create a negative *temperature coefficient* and the chain reaction stops by itself. Similarly, if the water density decreases by leakage or excessive boiling, the moderation is less efficient and this negative *void coefficient* stops the chain reaction. By contrast, in Chernobyl a strong **positive** void coefficient was one of the main triggers of the accident.

On the other hand, as much as 90% of the risk of fuel meltdown comes from the possible failure to properly cool it down. This is what happened in Three Mile Island 2 and Fukushima, even though automatic shutdown had operated perfectly. This explains why a series of diverse safeguard systems are devoted to the function of Emergency Core Cooling.

5.2 *The INES International Nuclear Events Scale*

Any abnormal event occurring in a nuclear facility is reported to the Safety Authority. Significant events are made public in the interest of full transparency.

Nuclear plants are complex facilities, and explaining the degree of significance of a given event to the public and the media is difficult. To convey the message, a logarithmic event scale, analogous to the well-known Richter scale, which describes the severity of earthquakes, has been defined internationally, ranging from 1 to 7 (see Fig. 1). Below level 1, events must still be reported but they have no safety significance. Levels 1 to 3 qualify as *incidents* of increasing significance. *Accidents* start at level 4 and reach 7 in case of a *major accident*. Chernobyl and Fukushima are to-date the only major nuclear accidents. TMI-2 was rated at level 5. In France, with 58 operating NPPs, more than 100, level 1, and two or three level 2 incidents are declared every year.

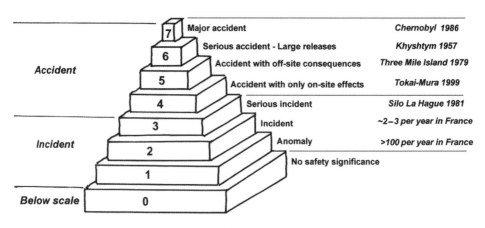

Fig. 9.   The INES scale.

5.3 *What Happened in Fukushima?*

The origin of the Three Mile Island accident was an ambiguous instrumentation signal which led to a severe operator error. The origin of the Chernobyl accident was an experiment carried out on a reactor which was unstable at low power, combined with several violations by the operators. The origin of the Fukushima accident was a very violent seismic event (magnitude 9 on the Richter scale) followed by a tsunami far exceeding the wave height against which the nuclear plant was protected.

On 11 March 2011, the Tohuku earthquake, the strongest ever recorded in Japanese history, did not significantly damage the 14 nuclear reactors located on the affected area, but destroyed the electric grid. The 11 nuclear reactors then in operation were automatically and successfully shut down, and the emergency diesel generators started, as expected, to supply the power necessary to cool down the reactor cores and remove the heat from decaying radioisotopes. Fifty minutes later, a huge tsunami destroyed a large strip of the north-east coast of Honshu Island, around the town of Sendai, destroying roads and houses and erasing whole villages as far as five miles inland. In the Fukushima Daiichi nuclear plant, the wave rendered five out of the six emergency power sources inoperable. One after the other, units 1 to 3, all BWRs, underwent similar scenarios.

Unable to properly remove the heat from decaying isotopes, the cores were overheated: the vapor pressure rose within the pressure vessels and the water levels dropped to the point where the top part of the fuel assemblies was no longer covered with liquid water. The fuel cladding reached the temperature where zirconium reacts with the water vapor, releasing hydrogen. The oxidized cladding released into the vapor all gaseous and volatile radioactive species it previously contained. The contaminated vapor reached the pressure suppression pool (wetwell) where condensation lowered the pressure for a time, and then reached the containment vessel itself (drywell). To save the integrity of the containment vessel, the operators released the vapor, together with the radioactive gases and particles as well as the hydrogen to the stack. But some of this vapor escaped to the top part of the reactor building, where the spent fuel storage pool is located. There, the hydrogen exploded in contact with the air, destroying the top of the buildings and releasing the radioactive vapor.

Some of the hydrogen from unit 3 flowed into unit 4, which was empty at the time of the earthquake, and exploded there. At the time of writing (May 2011) after weeks of heroic battle to regain control of the accident, the fuel degradation has stopped and the situation has stabilized, but the cooling of units 1 to 3 is still in open circuit, increasing day by day the amount of contaminated water to be stored, waiting for a closed circuit to be in place with water decontamination. A significant amount of very radioactive water from unit 2 has leaked into the Pacific Ocean. A 50-km-long and 10-km-wide band of land spreading north-west from the plant site is seriously contaminated. Addendum (July 2012): By December 2011, all damaged reactors were in safe cooling conditions and radioactive releases were reduced to below significant levels. Defueling the reactors and spent fuel pools should take 2 years, and then removing the core debris in reactors 1 to 3 may require 10 years.

Complete clean up and decontamination is expected to be accomplished in 30 years. A long period of clean-up and decommissioning will follow.

Very early, before any radioactive release, the Japanese authorities had evacuated the population within a 20-km radius from the damaged plant: as a result, no detectable health damage from radiation should affect the population. On the other hand, it is too early to assess the extent of land which will remain "off-limits" for housing and agriculture.

In view of the amount of radioactivity released, the Fukushima accident is classified as level 7 of the INES scale, but this amount is at least 10 times smaller than the radioactivity released during the Chernobyl accident, and the contaminated area is far more restricted.

6 The Nuclear Fuel Cycle

One cannot run a car on crude oil; in the same way, one cannot run a nuclear plant on uranium ore. The series of industrial steps necessary to prepare the nuclear fuel and manage the spent fuel after it is unloaded from the reactor core is called the nuclear fuel cycle (Fig. 10).

6.1 *Uranium Resources*

Uranium concentration in the continental crust is of the order of 3 ppm, intermediate between the usual and the precious metals. Some ore bodies in Canada have concentrations above 10% in uranium, but most deposits in the world rate 1% or less. The latest assessment of uranium resources is shown in Table 8, as compiled in the *Red Book* jointly issued in 2010 by IAEA and OECD/NEA.[5] Those figures, expressed in millions of tons, refer only to conventional uranium resources. Unconventional

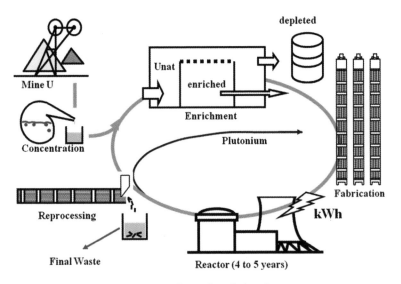

Fig. 10.   The nuclear fuel cycle.

Table 8:   Uranium Conventional Resources.

Mt U	Identified		To be discovered	
US$/kg U	Reasonably assured	Inferred	Prognosticated	Speculative
<40	0.57	0.23		
40–80	1.95	1.00	1.70	3.74
80–130	1.01	0.65	1.11	
130–260	0.48	0.42	0.09	0.16
>260	?	?	?	3.59
Sub-total	4.01	2.30	2.90	7.50
	6.31		**10.40**	

Table 9:   Identified Uranium Resources as a Function of the Cost (Thousands of Tons U).

Country	<US$40/kgU	<US$80/kgU	<US$130/kgU	<US$260/kgU
Australia		1,612	1,673	1,679
Kazakhstan	44	475	652	832
Russia	0	158	480	566
Canada	357	447	485	545
US	0	39	207	472
South Africa	153	233	296	296
Namibia	0	2	284	284
Niger	17	73	273	276
Ukraine	6	53	105	224
Uzbekistán	0	86	115	115
China	67	150	171	171
Total World	**796**	**3,742**	**5,404**	**6,306**

resources, notably as by-products in phosphate deposits, exceed 22 Mt U. There are over 4 billion tons of uranium in the oceans, but so diluted as to make its industrial exploitation unrealistic in the medium term.

For comparison, world uranium consumption amounted to 59,000 tU in 2008. Identified conventional resources would be able to feed the existing fleet of nuclear reactors for over a century, and we can expect to discover much more.

On the other hand, if a vigorous nuclear renaissance increases the world fleet from 440 to 2,000 reactors, lifetime uranium supply will become an issue, leading to the deployment of fast breeders.

Together, Australia, Kazakhstan, Russia, Canada, US, South Africa, Namibia, Niger, Ukraine, and China account for 85% of the world's identified reserves, out of which 27% are found in Australia alone (Table 9). In 2009, Kazakhstan, Canada, Australia, Namibia, Russia, Niger, Uzbekistan, and the US accounted together for 97% of the 51,000 tons of uranium produced in the world.

6.2 *Exploration, Mining, and Concentration*

Uranium itself is not very radioactive, but some of its daughter products emit strong γ-radiation. Once a potentially interesting region has been identified from geological maps, aerial surveys can detect this radiation, calling for closer local exploration to assess whether such anomalous uranium concentration qualifies as a commercially exploitable deposit.

According to its depth, uranium ore can be extracted by underground or open-air mining operations. Large and deep low-grade deposits in sparsely populated areas can also be exploited by in situ leaching (ISL). A number of wells are drilled into the ore body; an alkaline solution is injected into these wells to dissolve the uranium which is thereafter extracted by pumping up the solution through neighboring extraction wells.

As most deposits contain 1% uranium, or less, the ore must be concentrated before transportation. In the concentration mill near the mines, ore is crushed and ground, then dissolved to form a pulp. After clarification and purification using solvents or resins, uranium is precipitated in a concentrate called **yellowcake**. Yellowcake, which contains around 75% uranium, is packed in drums and shipped to the next facility in the cycle.

6.3 *Conversion and Isotopic Enrichment*

Natural uranium consists mostly of two isotopes, ^{238}U (99.28% and ^{235}U (0.72%). ^{238}U has a half-life of 4.47 billion years while ^{235}U has a half-life of 713 million years, which means that since the formation of Earth, half of the ^{238}U is still present while only 2% of the original ^{235}U still remains.

In an LWR, hydrogen nuclei absorb too many neutrons for a chain reaction to be sustained if the uranium contains only 0.72% fissile ^{235}U. One must *enrich* the natural uranium to reach 3%–5% ^{235}U before fabricating the fuel. As both isotopes have identical chemical properties, isotopic separation processes use the slight mass difference between the two isotopes.

First, the yellowcake is purified from any neutron-absorbing impurity and **converted** into uranium hexafluoride UF_6, a solid at room temperature but gaseous above 60°C. Natural fluorine has only one isotope, ^{19}F, so the UF_6 mass differences come only from the uranium isotopes. Early enrichment plants used a process called gaseous diffusion but all modern plants use the *centrifuge* process.

In a centrifuge (Fig. 11), a high narrow cylindrical bowl rotates at extremely high speed. The **feed** UF_6, noted F, is introduced at the center of the bowl. Heavier $^{238}UF_6$ molecules gather at the bottom and closer to the wall, where they are collected as **depleted** flow W, while lighter $^{235}UF_6$ molecules collect toward the top and farther away from the cylinder and are gathered as *enriched product* P. Each centrifuge increases the ^{235}U assay of the enriched flow but a *cascade of* several tens of centrifuges is needed to reach the target enrichment. As a single cascade contains very little material, enrichment plants are made of a series of many parallel mounted cascades.

The separative power of a cascade is expressed in a conventional unit called separative work unit (SWU). To produce 1 kg of 4% enriched uranium, 8 kg of

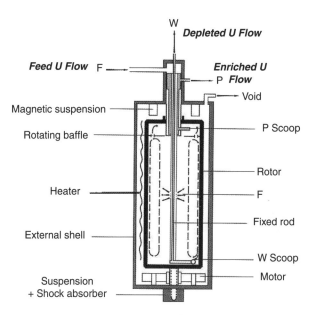

Fig. 11.   Centrifuge (schematic).

Table 10:   Enrichment Capacities (Million SWUs).

Production capacity	2006	2015
AREVA France	*10.8*	7.5
URENCO Europe	9.0	12.0
JNFL Japan	1.1	1.5
USEC USA	*11.3*	3.5
URENCO USA	—	3.0
AREVA USA	—	1.0
TENEX Russia	25.0	33.0
CNNC China	1.0	1.0
Total	**58.2**	**62.5**
Demand	**48.4**	**57–63**

natural uranium is required plus 6 SWUs. In addition, 7 kg of depleted uranium is stockpiled with an assay of 0.25% ^{235}U, which can be used later as a fuel for fast breeders. To enrich the uranium needed to supply a 1 MWe PWR, one uses around 120,000 SWUs per year.

Table 10 gives the world enrichment capacities in million SWU. Figures in italics refer to gaseous diffusion, the others, to centrifuge.

6.4 *Fuel Manufacture (PWR)*

Preliminary operations must be first carried out: zirconium alloy cladding tubes must be manufactured and assembly end pieces must be machined while UF$_6$ must be *defluorinated* into UO$_2$ powder. Spacing grids are assembled.

UO_2 is then pressed to make cylindrical pellets, which are sintered under reducing atmosphere to become ceramics. These pellets are machined and introduced into the cladding tubes, leaving an empty plenum at the top of the tube. The tube is pressurized with helium and a plug is welded on top, making it a completely leaktight fuel pin. The pins are then introduced within the "skeleton" constituted by the guide-tubes held in place by the grids. Top and bottom end pieces are fitted to complete the fuel assembly, ready to be shipped to the plants after many checks and inspections.

6.5 *Open Cycle or Closed Cycle?*

The "fresh" fuel assembly is loaded in the reactor core. It remains there for four or five years, being shuffled around the core during every reload. After this period of time, it can no longer produce power: it has lost many fissile nuclei (even though new nuclei were bred from the fertile material) and some of the fission products are neutron absorbers. In addition, its structural strength has been weakened by corrosion, creep, and radiation damage. It is therefore unloaded and put in temporary storage in the fuel pool of the nuclear plant. It is now *spent fuel*.

Some countries, like Finland and Sweden, consider this spent fuel as just radioactive waste, to be disposed of as such as we shall see in Sec. 7. They do not recycle anything, and one speaks of an *open cycle* — which means no cycle at all.

Other countries, like France and Japan, consider that spent fuel is a mixture of final waste and recyclable materials (Fig. 12): they operate a closed cycle by reprocessing the spent fuel to recover the leftover uranium and the bred plutonium and they store the properly conditioned final waste, metallic parts, fission products, and actinides, awaiting their disposal.

The recovered uranium has less than 1% ^{235}U and must be re-enriched to make new fuel. The recovered plutonium is mixed with depleted uranium to make mixed

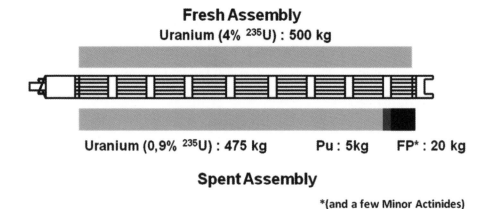

Fig. 12. Content of one spent PWR fuel assembly.

oxides (MOX) fuel. MOX pellets are loaded in fuel assemblies identical to uranium fuel assemblies. Plutonium having higher neutron absorption, the plutonium concentration in MOX pellets is twice the ^{235}U concentration in low enriched uranium (LEU) pellets.

6.6 *Reprocessing and Vitrification*

Reprocessing operations are aimed both at recovering recyclable fissile materials and at minimizing the long-term radiotoxicity of the final waste.

After one or two years in the spent fuel pool of the nuclear plant, spent fuel assemblies are shipped to the reprocessing plant in very massive casks, where they are stored a few more years under water for further cooling. The decrease in radioactivity, very significant in the first years after unloading from the reactor, eases the reprocessing operations.

At the request of the utility that owns the fuel, spent assemblies are removed from the pool to be reprocessed. *Shearing machines* are used to cut off the end pieces and then to cut up the fuel pins into segments, about 35 mm in length, which drop into a *dissolver*. In the dissolver, hot concentrated nitric acid completely dissolves the pellets, leaving only hollow segments of metallic cladding called "hulls".

End pieces and hulls are rinsed, dried and compacted under high pressure to make metallic pancakes, which will be piled up in cylindrical stainless steel containers.

After clarification, the solution that contains nitrates of uranium, plutonium, fission products and minor actinides is sent to the *chemical separation* facility. There, using solvent extraction processes, uranium and plutonium are separated from the waste products. In the PUREX process, the only industrial process, a second stage of extraction separates uranium and plutonium.

Plutonium is converted into oxide powder and temporarily stored before being shipped to the MOX fabrication plant. Uranium is either stored or sent to the conversion plant before re-enrichment.

The solution containing the waste products, fission products, and minor actinides is sent to the *vitrification* facility. There, the nitrates are heated in a rotating oven to become oxide *calcines*. Those calcines are mixed with glass frit in a furnace, where they are melted and refined to produce a homogeneous liquid glass. This liquid glass is poured into a cylindrical stainless steel container, where it solidifies into a *massive glass block* incorporating about 14% fission products.

Advanced reprocessing processes called partitioning and transmutation (P&T) are under study to further separate long-lived minor actinides from the fission products. Separated neptunium and americium can be fissioned, but only in future Generation 4 fast breeders. A modification of PUREX can prevent the complete separation of pure plutonium (COEX).

7 Radioactive Waste Management and Dismantling

7.1 *Waste Categories*

Several types of radioactive waste can be distinguished according to two criteria:

- The activity level i.e. the intensity of radiation emitted: this governs the level of shielding required to provide adequate protection against radioactivity.
- The radioactive half-life of the products contained; this determines how long they are potentially harmful and must be contained.

These two criteria are opposite for a given isotope: if a radioactive element is highly active, it decays very quickly and is therefore short-lived (and vice versa for a long-lived nuclide). What complicates matters is that a waste package very seldom contains a single radionuclide.

Radioactive waste is usually divided into three categories:

1. *Class A*: Low- and intermediate-level waste **LLW** with a short half-life (less than 31 years). Its radioactivity (β and γ) will be comparable to natural radioactivity in 300 years. It comes not only from nuclear power plants, but also from hospitals, laboratories, industry, etc.
2. *Class B*: Low- and intermediate-level (α) waste with a long half-life (several thousand years and more) **LL-ILW**. For example, the hulls remaining after spent fuel reprocessing.
3. *Class C*: High-level waste **HLW** giving off heat for several hundred years. This class emits α-, β-, and γ-radiation. It includes unprocessed spent fuel or glass containers from reprocessing.

7.2 *Radioactive Waste Disposal*

Class A waste constitutes the bulk of the waste volume but is only a small fraction of the radioactive waste in countries operating nuclear reactors. These wastes are disposed of in dedicated surface repositories, licensed in many countries.

The countries selecting an open cycle approach must dispose off their spent fuel as final waste. As spent fuel assemblies are not designed to contain radioactive products for millennia, they must be encapsulated in specially designed containers. Sweden and Finland have developed a high-integrity container made of pure copper. Loaded with several spent fuel assemblies gathered in a steel basket, and sealed with very careful friction welding, these copper waste packages will be disposed off in deep geological repositories dug in granite. In granite, "geological" water is reducing and will not corrode the copper containers.

Spent fuel reprocessing produces two waste packages: containers of compressed metallic waste of class B and containers of glass, class C. Both will be disposed off in deep geological repositories (clay, salt, tuff, and crystalline rocks are under

Table 11: Options for Deep Geological Disposal.

Country	Waste type	Geology	Underground laboratory	Repository construction	Repository operation
Belgium	Both	Clay	Yes	2025	2040
Finland	Spent Fuel	Granite	Yes	2012	2020
France	HLW	Clay	Yes	2016	2025
Germany	Both	Salt			2035?
Netherlands	HLW				>2110
Spain	Spent fuel				2050?
Sweden	Spent fuel	Granite	Yes	2015	2023
Switzerland	Both	Clay	Yes	∼2040	2050
UK	HLW				2040
US	Spent fuel	Tuff	Yes	Suspended	?

Note: "?" reflects uncertainty.

consideration by various countries). A number of in situ underground laboratories are used to characterize the selected strata, usually some 500 m deep. Table 11 describes the status of HLW disposal in various countries, as of 2010.

Many countries have not really chosen between open or delayed closed fuel cycles. They must keep their spent fuel in dry storage facilities, centralized or at the power plant site, waiting for a decision concerning their disposal. The Netherlands, for instance, has licensed such a facility for more than 100 years. Only one deep geological disposal for long-lived waste is already in operation: the WIPP waste isolation pilot plant near Carlsbad, New Mexico, was licensed in 1998, but it can receive waste only from the US defense program.

7.3 *Dismantling*

At the end of their operating life, nuclear facilities are shut down, dismantled, and then decommissioned. Decommissioning is an administrative step that releases the dismantled facility, totally or partially, from regulatory oversight. Dismantling consists of taking down production equipment, workshops, and structures where nuclear materials were present, and then disposing of the resulting radioactive waste.

The IAEA defines three stages of dismantling:

- Stage 1: Removal from the facility of all radioactive materials and heat transport fluids. The locked-up facility remains under surveillance, maintenance, and monitoring.
- Stage 2: The reactor is significantly decontaminated, and remaining areas with important residual radioactivity levels are sealed. Radioactive components that can be easily dismantled are removed. Some surveillance and monitoring are maintained.
- Stage 3: All radioactivity above acceptable levels is removed. No further inspection, surveillance, or monitoring is required.

B. Barré

8 Economics

Nuclear plants are expensive to build and inexpensive to operate. Capital costs
account for more than half of the levelized kWh cost which, therefore, is heavily
dependent upon the discount rate used in the cost calculations.

The International Energy Agency (IEA) and the Nuclear Energy Agency
(NEA), both part of OECD,[3] publish syntheses of future electricity cost studies
carried out by their member states on a regular basis. The latest issue refers to
plants being commissioned in 2015 and generating baseload electricity. A carbon
tax of $30 per ton of CO_2 is included in the costs of electricity from fossil-fuel–
powered plants, and decommissioning of nuclear plants is assumed to cost 15% of
their construction cost. Table 12 summarizes this study for real discount rates of 5%
and 10%. For each region surveyed, three figures are given in US$/MWh: median,
low and high.

With a low discount rate of 5%, nuclear power is very competitive, but less so
with 10%. Table 13 gives the breakdown of the median costs.

Of course, these calculations have assumptions on the future prices of coal, gas
and uranium. But Table 13 explains why nuclear costs are much less sensitive to
uranium prices than fossil costs are to coal prices, not to mention gas!

Table 12:   Projected Future Costs of Baseload Electricity (US$/MWh).

Region	Nuclear	Coal	Gas
DR = 5%			
U.S.	50 (48–51)	73 (68–15)	81 (77–91)
Pacific*	33 (29–50)	62 (54–88)	86 (67–105)
Europe	62 (50–80)	81 (68–120)	90 (85–119)
DR = 10%			
USA	75 (73–77)	90 (88–94)	88 (83–95)
Pacific*	49 (42–76)	75 (71–107)	89 (75–120)
Europe	108 (82–136)	100 (87–152)	97 (88–123)

*Pacific = Japan, South Korea, and Australia.

Table 13:   Breakdown of Projected Costs.

	Investment	O&M	Fuel	Carbon
DR = 5%				
Nuclear*	60%	24%	16%	—
Coal	28%	9%	28%	35%
Gas	12%	6%	70%	12%
DR = 10%				
Nuclear*	75%	15%	9%	—
Coal	42%	8%	23%	27%
Gas	16%	5%	67%	11%

*Nuclear costs include waste disposal and dismantling. Nuclear fuel
costs include uranium, conversion, enrichment, and fabrication.

9 Non-Proliferation

The term "proliferation" refers to the rise in the number of states in possession of nuclear weapons. The term "non-proliferation," on the other hand, refers to the political or technical means implemented to combat proliferation.

9.1 *Brief History*

Year	Country	Proliferation	Non-proliferation
1945	US	First A bomb	
1949	USSR	A bomb	
1952	United Kingdom	A bomb	
	US	First H bomb	
1953	USSR	H bomb	
	US		Atoms for Peace
1956	UN		Creation of the IAEA
1957	United Kingdom	H bomb	
1960	France	A bomb	
1963	US/USSR/UK		Moscow Treaty (to limit tests)
1964	China	A bomb	
1967	China	H bomb	
1968	France World	H bomb	Non-Proliferation Treaty NPT
1974	India	A "Test"	
	IAEA		"Trigger" list
	Exporters		London Suppliers Group NSG
1990	Iraq	Clandestine program	Gulf War
1991	South Africa		Dismantled weapons + joined NPT
1995	Former Soviet Union World		Weapons returned to Russia NPT extended indefinitely
1997	IAEA		Additional Protocol
1998	India	H bomb	
1999	Pakistan	A bomb	
2003	Pakistan, Libya, North Korea	A.Q. Khan "Bazaar"	
2006	North Korea, Iran	A bomb Enrichment crisis	Multilateral negotiations...

Note: Israel is credited with nuclear weapons but did not perform any nuclear test.

The US first tried to protect its military nuclear monopoly by preventing any transfer of nuclear knowledge. When proliferation occurred in the USSR anyway, President Eisenhower changed tack and allowed other countries access to reactor technology in exchange for their commitment to using the technology for peaceful applications only.

During the Cold War, the United Kingdom, then France, and lastly China, joined the Nuclear Weapon States (NWS). In 1968, the Non-Proliferation Treaty, (NPT), attempted to freeze the situation by recognizing five legitimate nuclear powers but no more. In exchange, the nuclear powers undertook to reduce their arsenal and give free rein to civil technology transfers. IAEA, set up by the UN in Vienna in 1956, was entrusted with the task of overseeing the peaceful use of nuclear materials. Many countries joined the NPT.

In 1974, India, which did not sign the NPT, broke the growing consensus by carrying out a "peaceful explosion," using plutonium produced in a heavy water reactor supplied by Canada. Exporting nations then agreed to regulate "sensitive" exports. In 1991, a similar shock was felt with the discovery of an extensive clandestine nuclear program in Iraq, a country that had signed the NPT. As a result, the powers and inspection capabilities of the IAEA were reinforced.

With the end of the Cold War and the disintegration of the USSR, the Russian Federation became the sole inheritor of the former nation's military nuclear power status. Significant disarmament programs were implemented in the nuclear weapon states (except for China). With the indefinite extension of the NPT in 1995, optimism prevailed.

But in 1999, Pakistan, India's rival since partition in 1948, crossed the "nuclear Rubicon." Then, in 2003, Libya revealed the existence of a black market of nuclear weapons technology managed by A.Q. Khan, the "Father of the Pakistani bomb." This, in turn, led to the discovery of a clandestine enrichment program carried out by Iran in violation of its international commitments as a party to the NPT. The most recent event was the test of a nuclear device by North Korea in 2006.

Despite the Iranian and North Korean crises, the NPT, backed up by IAEA inspections, now forms the universally acknowledged basis for all nuclear commerce. The only areas of resistance are Israel, India, and Pakistan and, now, Iran and North Korea.

9.2 *Proliferation and Civilian Nuclear Technologies*

So far, no country has ever proliferated by diverting nuclear materials or facilities under IAEA safeguards. Nuclear fission cannot be un-invented: any country ready to devote enough financial and technical efforts, and willing to pay the political price, can make weapons. North Korea is not the most advanced country in terms of industry or technology. Proliferation is much more a matter of political will than a matter of technology, but, to make weapons, highly concentrated fissile materials, highly enriched uranium (above 90% ^{235}U) or "weapons grade" plutonium are needed.

Not all areas of the nuclear industry are "sensitive" in terms of proliferation. Isotopic enrichment using centrifuges has become the most "proliferating" technology: it is easy to hide a clandestine facility and not too difficult to divert an existing facility from its purely civilian purpose. On the other hand, a reactor which can be refueled on-line has the capability to produce nearly pure ^{239}Pu in very low burnup fuel. Used in combination with spent fuel reprocessing, such a reactor would offer the most efficient way to build up an arsenal.

Table 14: Estimates of Total Electricity Generation and the Contribution of Nuclear Power.

Country group	2009			2030		
	Electricity TWh	Nuclear TWh	Nuclear %	Electricity TWh	Nuclear TWh	Nuclear %
North America	4,525	882	19.5	4,819	1,034	21.5
				5,249	1,339	25.5
Latin America	1,220	30	2.5	2,442	86	3.5
				2,725	178	6.5
Western Europe	2,967	781	26.3	3,466	666	19.2
				3,763	1,218	32.4
Eastern Europe	1,720	327	19.0	2,182	592	27.1
				2,448	778	31.8
Africa	628	12	1.8	1,734	50	2.9
				3,338	126	3.8
Middle East + South Asia	1,573	17	1.1	4,037	205	5.1
				6,122	363	5.9
SE Asia + Pacific	729			1,547	6	0.4
				1,797	39	2.2
Far East	5,195	510	9.8	9,027	1,399	15.5
				10,412	1,898	18.2
World Total	**18,558**	**2,558**	**13.8**	**29,254**	**4,040**	**13.8**
				35,654	**5,938**	**16.6**

It is very unlikely that a would-be proliferator could make bombs using plutonium recovered from spent LWR fuel. As noted by Pellaud,[4] nobody has ever done so. Conversely, surplus weapon-grade plutonium from the Cold War inventories can be "demilitarized" by burning it as MOX fuel in civilian power plants.

10 Prospects

Despite the Fukushima accident, most international studies still forecast a "growth" of nuclear power, for the reasons listed in the introduction to this chapter. Table 14 gives the forecast of the IAEA[8] for the short term (for nuclear power, 2030 is tomorrow), with low and high estimates. The difference between both sets of figures is mostly due to hypotheses about the projected lifetimes of existing plants. In summary, nuclear power will grow everywhere, with the possible exception of Western Europe.

Since its reference "business-as-usual" scenario would lead to unacceptable world CO_2 atmospheric emissions, the IEA[9] has developed a so-called "450" scenario designed to decrease those emissions by 50% by 2050 and limit the GHG atmospheric concentration to 450 ppm. Such a target can only be reached by a number of voluntary policies which include the strong development of nuclear power, as shown on Fig. 13.

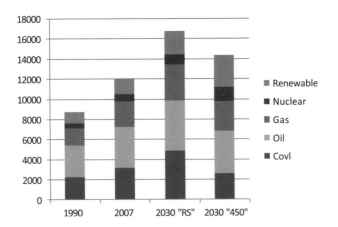

Fig. 13.   World primary energy demand by source (Mtoe).

Between the reference scenario and the 450 ppm scenario, nuclear power generation increases by 49%. After Fukushima, the IEA developed a "low nuclear" scenario, but this would make it impossible to attain the 450 ppm target.

References

1. IAEA Reactor database PRIS. Available at: http://www.iaea.org/programmes/a2/index.html. Accessed May 2011.
2. W. Marshall., *Nuclear Power Technology* (Clarendon Press, 1986).
3. ENDF-B files. Available at: http://nndc.bnl.gov/exfor/endf00.jsp. Accessed May 2011.
4. R. M. E. Diamant, *Atomic Energy* (Ann Harbor Science, MI, USA, 1982).
5. OECD, *Uranium 2009: Resources, Production and Demand* (OECD, 2010).
6. OECD IEA/NEA, *Projected Costs of Generating Electricity* (OECD IEA/NEA, 2010).
7. B. Pellaud, Proliferation aspects of plutonium recycling, *J. Nucl. Mater. Management* **Fall** (2002), pp. 30–38. www.inmm.org.
8. IAEA, Reference Data Series 1, 2010.
9. OECD/IEA, *World Energy Outlook* 2009.

Further Suggested Readings

All About Nuclear Energy, From Atom to Zirconium (AREVACOM, France, 2008).
B. Barré and P. R. Bauquis. *Nuclear Power, Understanding The Future* (Hirlé editions, France, 2010).
D. Cacuci *et al.*, *Handbook of Nuclear Engineering* (Springer, Berlin, 2010).
I. Hore-Lacy. *Nuclear Energy in the 21st Century* (WNU Press, UK, 2006).
OECD, *Nuclear Energy Today* (2003).

Chapter 7

Magnetic Fusion Energy

R.J. Goldston* and M.C. Zarnstorff[†]
Princeton Plasma Physics Laboratory
PO Box 451, Princeton, NJ 08540, USA
*rgoldston@pppl.gov
†zarnstorff@pppl.gov

Fusion of light nuclei confined by magnetic fields is being developed as an effectively inexhaustible energy source without the production of greenhouse gases or long-lived radioactive waste and without the risk of runaway accidents. This chapter presents the characteristics of Magnetic Fusion Energy and methods to harness it as an energy source, including an overview in Section 1 of fuel resources, waste, safety, and proliferation risks. Section 2 discusses the physics and technology of magnetic fusion systems. Section 3 describes current major experiments, fusion energy achievements, and the role of the ITER burning-plasma experiment. Section 4 summarizes the international activities to develop fusion as an attractive energy production system.

1 Overview

The fusion of light nuclei powers the stars and our sun and is the primary origin of almost all terrestrial energy. Controlled use of fusion as a direct human energy source has been pursued for more than 50 years because it offers an effectively inexhaustible energy supply that would be available to all nations without the production of greenhouse gases or long-lived radioactive waste, without the possibility of catastrophic runaway accidents, and with low weapons proliferation risks. In addition, fusion offers a steady source of baseload power that does not require extensive land use, long-distance energy transmission, or large-scale energy storage. The challenge of fusion is to confine and control extremely hot, ionized gas called plasma, and this has driven an increased understanding of plasma physics, the study of ionized gasses that make up almost all the visible universe. The understanding and development of fusion-related technology is resulting in rapid progress through worldwide collaboration toward providing abundant fusion energy. The next major step will be taken by ITER, an international experiment under construction near Marseille, France, to study burning-plasma conditions, in which the plasma is dominantly heated by its own fusion reactions, at approximately power-plant scale. Beyond

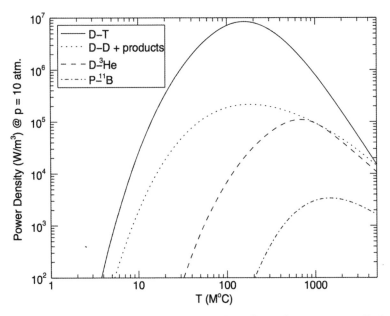

Fig. 1.   Fusion power density versus fuel temperature for different fusion reactions. Fuel pressure is fixed at 10 atm.

ITER, the challenges will be to make fusion energy production practical, reliable, and cost-effective.

The easiest fusion reaction for use in energy production is between deuterium and tritium (heavy isotopes of hydrogen) creating helium and neutrons. Tritium is radioactive, decaying with a half-life of 12.3 years and thus must be manufactured. The helium ion produced from the fusion of deuterium and tritium carries 20% of the released energy and heats the plasma to sustain the temperature $T \sim 160\,\text{M}^\circ\text{C}$ needed for the peak reaction rate at a given pressure (see Fig. 1). A blanket-structure surrounds the plasma and absorbs the neutrons, capturing the neutron's energy thermally and producing tritium fuel through a neutron reaction with lithium in the blanket. A fusion power plant will convert thermal power to electrical power using turbine generators, similar to other power generation systems, with a conversion efficiency expected to be in the range from 1/3 to 1/2 depending on blanket temperature. The production of 1 GW-year (3.2×10^{16} J) of fusion electrical energy with a thermal conversion efficiency of 1/3 requires 88 kg of deuterium and 263 kg of lithium-6 (^{6}Li), producing 350 kg of helium. Higher thermal conversion efficiencies, due to blankets being operated at a higher temperature, reduce the required fuel quantities.

At the very high temperature needed for fusion, the plasma cannot be in contact with solids or liquids, or it would immediately cool. Two general strategies have been developed to isolate and hold a hot plasma and produce fusion energy: (1) Magnetic Fusion Energy (MFE), discussed in this chapter, where the particles in the ionized gas travel along strong magnetic fields (like beads on a string) and are confined in a magnetic torus and (2) Inertial Fusion Energy (IFE), where the plasma is rapidly compressed and heated to produce a burst of fusion reactions, confined only by inertia before it flies apart, which is the subject of Chapter 8.

Magnetic Fusion Energy has been pursued as a large-scale international collaboration since it was declassified in 1958, as part of the second "Atoms for Peace" conference. Due to its attractive characteristics and the need for long-term energy supply without greenhouse gas emission, there are large MFE programs in China, Europe, India, Japan, South Korea, Russia, and the US, with smaller programs in other countries. Both Europe and the US developed experiments which produced significant fusion power in the late 1990s, demonstrating that the needed conditions can be achieved and the fundamental scientific concepts behind MFE are sound. These included producing 10.7 MW peak in a 0.4 sec pulse (US)[1] and 16.1 MW peak (EU) in a 0.8 sec pulse, where the durations are at half-peak value, with the latter plasma achieving an energy multiplication factor of $P_{fusion}/P_{input} \approx 0.62$.[2] Of course, the heating power injected into the plasma also comes out again. So, in the EU experiment, 42 MW of power came out for 26 MW injected. Heating pulses of 5 sec duration produced 20 MJ of fusion energy. A Japanese experiment demonstrated plasma confinement equivalent to an energy multiplier of 0.8,[3] but was not able to use tritium fuel. The ITER experiment[4] under construction will continue the development of MFE understanding and technology to approximately the scale of a power plant. It is designed to produce up to 500 MW of fusion power with an energy multiplication factor ≥ 10, and multiplication factors > 5 for pulses of 3,000 sec. It will prototype the tritium producing blanket modules, the tritium handling system, and other technologies needed for a power plant. ITER is a partnership of the seven large international fusion programs (China, Europe, India, Japan, South Korea, Russia, and the US), representing more than half the world's population.

The available fuel supply for fusion on earth is enormous and very widely distributed. Deuterium atoms make up 156 ppm of all hydrogen atoms on earth. Deuterium is present in all water as HDO molecules, at a level of one per 3,200 light-water molecules. Heavy water is separated in large industrial quantities via chemical reactions in several countries as well as by distillation and electrolysis. Lithium is widely distributed in the earth's crust, with reserves at mines estimated at 13 million tons,[5] and total world resources estimated at 34 million tons. Lithium is present in seawater at a concentration of approximately 0.2 ppm and a total estimated quantity of 230 billion tons. The South Korean company POSCO has started a program to commercialize extraction of lithium from seawater.[6,7] Even if such a process resulted in lithium production at 20 times the current commercial price, this would add less than 0.1 US cent per kWh to the cost of electricity from fusion. (The energy density of Li in seawater, fueling fusion, is 2,500 times that of

^{235}U in seawater fueling light-water fission reactors.) ^{6}Li is the preferred isotope for producing tritium, since it releases energy in the process and can use neutrons of any energy. It has an average natural abundance of $\sim 7.5\%$ of all lithium and has been separated using chemical reactions in several countries. The more abundant ^{7}Li (92.5%) can be retained for other uses, such as batteries. Thus, even if fusion supplies all of the world's electrical energy, the available ^{6}Li should last for millions of years. For both deuterium and lithium, prepared using current technology, the cost of fusion fuel is negligible when compared with the value of its energy content.

In addition to the helium produced by the fusion reaction, radioactive waste is produced due to neutron activation of the material structures near the plasma, including the blanket structure for capturing the neutron's energy and producing tritium from lithium. Studies of fusion power plant conceptual designs have found materials for these components that produce only short-lived radioactivity and could have a radiological hazard potential more than 10,000 times lower than fission waste[8] if impurities are controlled. Such waste would not require long-term isolation and could qualify for shallow burial disposal.

Fusion is much safer than fission-based energy systems as the amount of fuel available in the plasma is small, typically only sufficient for a few minutes of operation without additional supply injection. Any substantial increase in energy release will cause influx of the wall materials, cooling the plasma and stopping the fusion reactions. Thus no runaway accident is possible. Neutrons activate the structure surrounding the plasma, but with proper choice of low-activation materials, the fractional density of radioactive atoms can be low, and the systems designed to ensure that radioactive materials cannot be released by after-heat from radioactive decay.

The proliferation risks from magnetic fusion are much less than those from fission.[9] Since no fertile materials, ^{238}U or ^{232}Th, need be near a fusion plasma, it is relatively straightforward for inspectors to detect if the operator attempts to expose such materials to fusion neutrons in order to generate fissile materials that can be used in nuclear explosives. It also appears impractical to operate a clandestine, undeclared fusion system to create such materials, since a large amount of power would be required and the environmental signatures could be easily detected. In the case of a "breakout" scenario, in which safeguard inspectors are expelled, no fissile material should be available at the time of their expulsion, unlike the case with fission systems. It is possible to disable a fusion power plant remotely, so that it cannot be used to produce fissile material, by damaging power supplies, cooling systems or cryogenic systems, with little risk of dispersing radioactive material. Since the tritium that is produced in fusion systems can be used to boost the yield of nuclear weapons, safeguards for fusion power plants should include not only assurance of the absence of fertile materials but also careful accounting for tritium. There is no overlap between the physics of magnetic fusion and nuclear weapons, so there is no proliferation risk in sharing MFE knowledge and systems.

In magnetically confined fusion power plants, cryogenic superconducting coils will produce the magnetic field to avoid large power-losses from resistive coils. A number of medium- to large-scale fusion experiments around the world have been built using low-temperature superconducting coils and liquid helium cooling, and these will be used for ITER. Liquid helium is also frequently used for cryogenic pumping of the plasma exhaust and it may be used as a thermal-transfer fluid in power plants for high-temperature, high-efficiency operation. The estimated total available world helium resources are approximately 8.6 million tons,[5] which is much more than that needed for fusion power plants to supply the world's electricity demand. However, since helium is a byproduct of natural gas production, there is a question of whether sufficient helium will be in production when fusion energy systems are commercialized. This has been analyzed recently[10] using a model of the natural gas market and future production. The authors project that the helium production rate will decrease in the second half of the 21st century. They conclude that there will be sufficient helium to deploy and sustain up to 3.5 TW of fusion power by ~2110, provided that the helium loss-rate is substantially reduced, helium is efficiently recycled on plant shutdown, and methods to substitute for some helium usage are developed. This magnitude of fusion energy could provide the projected total requirement for nuclear energy (fission + fusion) by 2100 to mitigate global warming.[11] The new superconducting tokamak in South Korea has already achieved a routine operating helium loss rate of 1%/year (M. Kwon, private communication, 2012), which the study projected as a best-case goal for the year 2100, and the Korean team is working to reduce accidental releases. Helium substitution opportunities include the use of high-temperature superconductors with other cryogenic fluids (e.g., Ne), other high-temperature heat-transfer fluids (e.g., CO_2), and mechanical exhaust pumps.

2 MFE Physics and Technology

2.1 *Breakeven, Gain, and Ignition*

Energy can be generated both by splitting the heaviest atoms (fission) and by joining together the lightest ones (fusion). The fuel for almost all fission reactors is the lighter isotope of uranium, ^{235}U. The most favorable fuel for fusion comprises two isotopes of hydrogen, deuterium (D) (one proton + one neutron) and tritium (T) (one proton + two neutrons). Their reaction is

$$D + T \rightarrow n + {}^4He + 17.6\,\text{MeV}, \tag{1}$$

where 1 MeV (million electron volts) $= 1.602 \times 10^{-13}$ J. About 80% of the reaction energy emerges as kinetic energy of the neutron, while about 20% is invested in the helium nucleus, or α particle. Since T does not exist in significant quantities in nature, it is necessary to use the neutrons from fusion to "breed" the required fuel, though the reaction

$$n + {}^6Li \rightarrow T + {}^4He + 4.8\,MeV. \tag{2}$$

Some neutrons are absorbed in structural materials, and it is necessary to produce slightly more T than is burned in order to provide T to startup future power plants. Thus the primary fusion neutrons are collided with materials such as lead or beryllium, which undergo (n, 2n) reactions: one neutron colliding with a Pb or Be nucleus produces two lower-energy neutrons, each capable of the reaction described in Eq. (2).

The DT fusion reaction rate, in reactions per second per m^3, is given by $n_D n_T \langle \sigma v \rangle_{DT}$, where n_D is the number density of deuterium ($\#/m^3$), n_T is the number density of tritium, and $\langle \sigma v \rangle_{DT}$ is the average over a thermal distribution of particles of the relative particle velocities, multiplied by the cross-section for the DT reaction. Thus the power density in a fusion system is given by

$$p_{fus}(watts/m^3) = n_D n_T \langle \sigma v \rangle_{DT} E_{fus}. \tag{3}$$

where

$$E_{fus} = 17.6 \times 1.602 \times 10^{-13}\,J.$$

Figure 1 shows the fusion power density that can be produced as a function of fuel temperature, for a fixed fuel pressure of 10 atm. Clearly the DT reaction can provide the most power density for a given pressure, and very hot fuel indeed is required to obtain interesting values of power density. At these high temperatures, the fuel is fully ionized: electrons are dissociated from atomic nuclei. The resulting cloud of charged particles is called plasma. Since the positive charge of the fuel nuclei is balanced by electrons, keeping the plasma nearly electrically neutral, the total plasma pressure is given by the sum of the electron and ion pressures.

For a fusion plasma to come to a steady temperature, the input power to the plasma must equal its loss of power:

$$P_{aux} + 0.2\,P_{fus} = P_{loss}, \tag{4}$$

where P indicates total power in watts, not power density, and we have assumed the presence of some "auxiliary" power injected from outside to sustain the plasma, plus that part of the fusion power invested in the electrically charged α-particles, which can be contained by the magnetic field. The power in the neutrons escapes from the fusion plasma and is absorbed in the "blanket" that surrounds it.

The gain of a steady fusion system is defined as

$$Q = P_{fus}/P_{aux}. \tag{5}$$

Magnetic fusion power plants require $Q > 25$ to have a favorable power balance because practical heating systems can have efficiencies in the range of 40%, as can power conversion systems. (As a first approximation we assume that the plasma heating system dominates the power requirements for plant operation.) In this case, four times more fusion power is produced than is required for heating.

$$P_{\text{e,gross}} = 0.4\,P_{\text{fus}}$$
$$P_{\text{e,aux}} = P_{\text{aux}}/0.4 \tag{6}$$
$$P_{\text{e,gross}}/P_{\text{e,aux}} = 0.4 \times 0.4\,P_{\text{fus}}/P_{\text{aux}} = 0.16 \times Q = 4.$$

Ignition in magnetic fusion is defined as $Q = \infty$, which is not required for fusion power plants.

The ITER experiment is designed to bridge the gap in Q from present experiments by operating at $Q = 10$ for pulses of 300–500 sec, at $P_{\text{fus}} = 500$ MW, for a total thermal energy production per pulse of ~200,000 MJ. ITER will address not only the physics of burning plasmas, plasmas heated by their own fusion reactions, but also many of the technological issues for fusion energy development, as required to handle this high-energy throughput.

2.2 *Magnetic Confinement*

At fusion temperatures, the fuel of a fusion system is fully ionized and composed of charged particles, so it responds strongly to the presence of magnetic fields (Fig. 2). In particular, both ions (atomic nuclei) and electrons spiral along magnetic fields. Thus a strong toroidally directed magnetic field can "levitate" an ionized gas, or plasma, and confine its heat. Indeed, experiments around the world have achieved central plasma temperatures of up to 500,000,000°C, well above what is required for fusion (see Fig. 1). Nevertheless, confinement of a fusion-grade plasma entails scientific and technological challenges.[12]

2.2.1 *Transport and turbulence*

The logical primary challenge is to show that it is possible to produce net power from fusion. This means $Q > {\sim}6$ [see Eq. (6)], which can be traced back through Eq. (5) to a requirement for low $P_{\text{aux}}/P_{\text{fus}}$, and through Eq. (4) to low $P_{\text{loss}}/P_{\text{fus}}$. However, the core of a fusion-grade plasma must be very hot, as we have seen, and of course the temperature of the material surfaces that surround it must be much cooler, forcing a temperature gradient of order 10^{8}°C/m. Strong temperature gradients cause turbulence in fluids, and plasmas are no exception. The turbulence causes particles to travel across the magnetic field and carry heat out of the plasma.

Fortunately, detailed experimental measurements and advanced numerical simulations of plasma turbulence both show that the turbulence intensity varies proportionally to the radius of the particle spirals — the gyro-radius — divided by the radius of the plasma. This ratio is denoted as ρ^{*}. In other words, the tight spirals of the particle motion serve to constrain not only the particle motion but also the turbulence amplitude. The net effect is that the time that energy remains in the

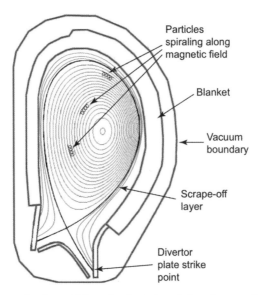

Fig. 2.   Cross section at fixed toroidal angle of a tokamak-based magnetic confinement system
for fusion energy, identifying the components near the plasma. The nested contours represent the
surfaces traced out by the magnetic field lines inside the plasma.

plasma increases very rapidly with the size of the plasma and with its magnetic
field, indeed as the cube of ρ^*, as shown in Fig. 3.

The "energy confinement time," $\tau_{E\,exp}$, is defined in steady conditions as the
total energy stored in the plasma divided by the total heating power supplied either
through auxiliary heating or through fusion reactions. Modern high-power experiments achieve stored energies of $\sim$10 MJ, with heating powers of $\sim$20 MW, for
confinement times above 0.5 sec at 3 T[a] magnetic field. Figure 3 also illustrates that
confinement times measured on many fusion systems worldwide form a consistent
data set projecting to successful operation of ITER with about 4 sec confinement
time at 5.3 T. This performance should provide a wide operational window to explore
the physics of high-gain plasmas.

2.2.2 Stability

The next scientific challenge is associated with the global stability of the plasma,
which is governed by the plasma pressure. For fixed optimum temperature, the
fusion power density goes as the square of the fuel density [Eq. (3)] and so, again for
fixed temperature, as the square of the pressure. For an economically practical fusion
power system, pressures in the range of 10 atm are required. The key parameter
governing the underlying plasma physics is denoted as β, the pressure in the plasma

---

[a]Tesla (T) is the unit of magnetic flux density (**B**) in the International System (SI) of units.

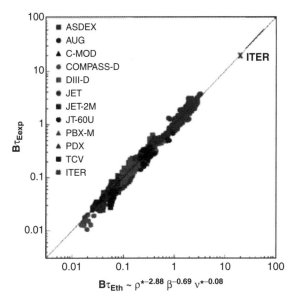

$$B\tau_{Eth} \sim \rho^{*-2.88} \beta^{-0.69} \nu^{*-0.08}$$

Fig. 3. Experimental energy confinement time, $\tau_{E\,exp}$, multiplied by magnetic field strength B versus a multi-experiment confinement time fit expressed in key dimensionless plasma parameters: gyro-radius divided by torus radius, ρ^*, normalized plasma pressure, β (see text), and collision rate between particles divided by time to circumnavigate the torus, ν^*. Projection to ITER is favorable.

divided by the pressure in the magnetic field $(B^2/2\,\mu_0)$.[b] The magnetic field pressure is limited by magnet technology, so it is desirable to raise β to as high a value as practical. ITER, for example, will normally operate at $\beta = 2.5\%$–3%, while a fusion power plant may require $\sim 6\%$, a value already exceeded for short times in current experiments. Experiments in ITER and in superconducting devices operated in parallel with ITER will explore this higher range for long pulses.

2.2.3 *Sustainment*

It is desirable to sustain the operation of fusion plasmas for long pulses, in order to minimize cyclic thermal fatigue and mechanical stresses and the need for energy storage. Tokamak plasmas such as ITER require an electrical current to flow around the torus in the long direction, so that the magnetic field lines circumnavigate the toroidal plasma in both the long and short directions, a key requirement for magnetic confinement. This current can be maintained by a combination of transformer action, external injection of power that pushes the electrons relative to the ions, thereby driving a net electrical current, and pressure-driven currents analogous to piezoelectricity. The distribution of current in a tokamak affects the plasma confinement and stability, determining the β value. Instabilities involving the current can

---

[b] μ_0 is the permeability of free space and $= 4\pi \times 10^{-7}$ V sec/A m.

produce significant transient heat fluxes and magnetic forces on the wall. ITER is planned to operate for 300–500 sec without significant external current drive and for up to 3,000 sec with a combination of strong external current drive and pressure-driven current. The challenge of limiting the required current drive power, to maximize Q in this operating mode, will be investigated in ITER and the other superconducting experiments in China, South Korea, Japan, and Germany.

2.2.4 *Plasma–material interaction*

With success comes the challenge of handling the 20% of the fusion power that is confined by the magnetic fields, heating the plasma. This heat, along with P_{aux}, flows out of the plasma in a thin "scrape-off-layer" into a divertor region away from the main plasma, ultimately striking divertor plates (Fig. 2).

This configuration has the advantages that it concentrates the He "ash" from the fusion reaction into a region where it can be pumped away, and any particles that are ejected from the material surface are far from the main plasma, reducing contamination of the DT plasma. However, the point of interaction between the thin layer of heat flux and the divertor plate (the divertor "strike point" in Fig. 2) presents a scientific and technological challenge. The leading material candidate for this surface is tungsten, a high-atomic-mass metal. But if the material erodes into the plasma, it can strongly reduce the temperature of the system through atomic excitation of its many electrons and subsequent ultraviolet radiation. One alternative is to use a low-atomic-mass liquid metal, such as lithium, at the strike point, which is far less damaging to plasma operation and can be replenished as it erodes. The heat flux from tokamak plasmas is not always steady but can have high transients. An advantage of a liquid metal surface is that it can withstand a sudden transient heat flux through evaporation and can subsequently be replenished. A melted tungsten surface would likely need to be replaced.

2.2.5 *Neutron–material interaction (including tritium breeding)*

Another consequence of success will be the production of copious amounts of 14.1 MeV neutrons from the fusion reaction. These high-energy neutrons both displace atoms in surrounding materials and drive (n, α) reactions in which an energetic neutron causes the production of α-particles, helium nuclei, inside the material. These effects can cause swelling, creep, and embrittlement of structural materials. One can confidently predict based on experiment and calculation that modern low-activation ferritic-martensitic steels developed for fusion applications will operate effectively in the temperature range of 400–600°C, but only at displacements per atom (dpa) up to about 10^{13} which corresponds to about 1 MW-yr/m² of 14.1 MeV neutrons. Beyond 10 dpa, the effects of the helium produced in the material are uncertain. These steels can be improved through the inclusion of nanosized oxides dispersed at very small scales, with the promise that the nanodispersed particles both strengthen the steel at high temperature and also scavenge He, keeping it from forming bubbles of significant size, and so avoiding embrittlement. Silicon carbide composite materials are also being developed for fusion structures, as they show strong resilience to neutron bombardment and extremely low long-term activation.

A DT fusion plasma must be surrounded with material that contains lithium, so that the reaction described in Eq. (2) can proceed. ITER is designed with special ports for "test blanket modules" to study this process. If one takes into account the various ports that are required for plasma heating, and the divertor region that will shield some neutrons from the blanket, it is challenging to achieve a tritium breeding ratio much in excess of unity, and this is an area of continuing R&D.

2.2.6 *Magnets*

The magnets being constructed for ITER will be the largest and most powerful superconducting magnets in the world. They will produce 5.3 T magnet field in a torus of major radius 6.2 m. The maximum field at the surface of the magnet is designed to be 11.8 T and the stored energy in the magnet is an impressive 41 GJ. This system is based on Ni-Sn superconducting cables surrounded by thick load-bearing conduits, cooled to 4 K. Future fusion magnet systems may strive for higher magnetic field, since fusion power density varies as B^4 at fixed β and plasma temperature. That said, the ITER magnet system is very much in the range of the magnets that will be required for fusion power plants.

2.2.7 *Magnetic field configurations*

Many configurations of the magnetic field have been studied during the fusion program. The highest absolute performance levels have been obtained in tokamaks (Fig. 4) for times of order 1 sec.

Two other configurations that have demonstrated similar performance characteristics are the spherical tokamak and the stellarator. The spherical tokamak is a tokamak with a minor radius that is nearly as large as its major radius — thus a torus with a very small hole in the center. This configuration can run at very high β but its magnetic field is restricted by the small region available for the inner leg of the magnet. It may be a route to lower-cost fusion systems, since the magnets can be relatively inexpensive water-cooled copper systems and can be disassembled through mechanical joints, allowing easier access for maintenance. However, significant power is required to operate such magnets, reducing Q. Experimental facilities in the US and the UK are investigating the physics properties of the spherical tokamak.

The stellarator is an important complement to the tokamak. In stellarators, the magnetic field lines are coaxed around the torus in the short direction by three-dimensional, non-axisymmetric shaping of the magnets (Fig. 5).

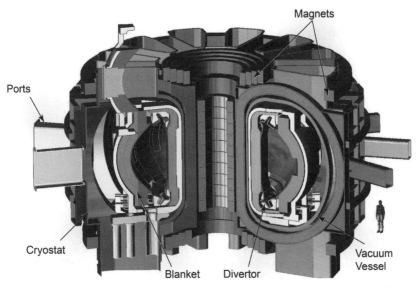

Fig. 4.   Drawing of a tokamak fusion pilot-plant design. The horizontal ports are used for external current-drive and heating systems, and for diagnostics. The large top ports are used for maintenance access (T. Brown, PPPL).

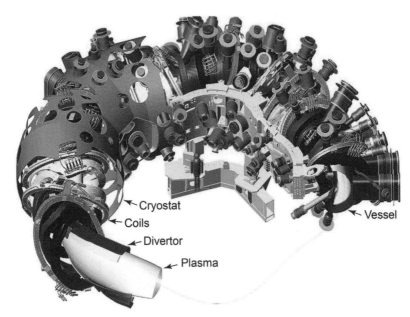

Fig. 5.   Cut-away drawing of the superconducting stellarator experiment Wendelstein 7-X, under construction in Germany (T. Klinger, Max-Planck/IPP). The faint curve at the bottom of the figure represents the trajectory of the plasma center across the cut-out.

This means that current drive power is not required in stellarators, and it is easier to foresee steady-state high Q operation. Furthermore, stellarator plasmas generally do not create large transient heat fluxes and forces, a significant advantage for the integrity of divertor plates and blankets. Major superconducting stellarator experiments in Japan and Germany will investigate the physics of steady-state, high-β operation in parallel with ITER.

Both the spherical torus and the stellarator configurations are close to the tokamak and share a common theoretical basis, so the improved understanding from ITER results should directly apply to all three configurations and inform their use in fusion energy systems.

An ideal magnetic fusion configuration would have such low turbulence that it could be much smaller than ITER and have lower magnetic field; it would support very high β, for high power density at this low magnetic field; it would operate in steady state with little recirculating power; and it would allow exhaust heat to be expelled from the magnet system along diverging magnetic field lines. It would be topologically simply connected, allowing easier access and maintenance. Its performance would be so high that it could even burn fuels such as p-^{11}B (Fig. 1) that generate no neutrons. Such systems would be challenging for fundamental reasons, but ideas continue to be pursued, and should be, that may have the prospect of exceeding the performance of tokamaks, spherical tokamaks, or stellarators.

3 Progress Toward Fusion Energy

3.1 *National and International Research Facilities*

The magnetic fusion energy research program is a broad international effort, based on peer-reviewed scientific analysis, with no barriers to communication and replication of results due to classification. Every two years, the International Atomic Energy Agency sponsors a Fusion Energy Conference.[14] In the most recent conference, there were papers presented by experimental and theoretical groups from Australia, Brazil, Canada, China, England, France, Germany, Holland, Italy, Japan, Kazakhstan, Russia, South Korea, Spain, Sweden, Switzerland, the US, and Ukraine. The leading research facilities in the world are currently located in China, England, France, Germany, Japan, South Korea, and the US (Table 1). Particularly notable are major experiments with superconducting magnetic coils operating in China, Japan, and South Korea, and new large superconducting experiments under construction in Germany and Japan. The largest fusion experiment currently operating is managed by the European Commission in England. This device has about one-half of the linear dimensions of ITER, about two-thirds of the magnetic field strength, and pulse length $\sim$20 sec versus 300–3,000 sec in ITER.

The progress in energy production in magnetic fusion experiments was exponential from 1970 to 2000, as shown in Fig. 6. Since that time, no magnetic fusion

Table 1: Facility Parameters for the Major MFE Experiments, Worldwide, Including Experiments under Construction and being Upgraded.

Location	Type	Major radius (m)	Minor radius (m)	Magnetic field (T)	Max heating power (MW)	Max pulse length (sec)	Comments
China	Tokamak	1.85	0.45	3	26	1,000	Superconducting
England	Tokamak	3.1	1	4	51	20	EU project DT capable
England	Spherical Tokamak	0.85	0.65	0.75	7.5	5	Stage 1 of upgrade project
France	Tokamak	2.38	0.72	3.8	3	1,000	Superconducting No divertor
Germany	Tokamak	1.65	0.5	3.3	34	10	
Germany	Stellarator	5.5	0.5	3	20	1,800	Superconducting Under construction
Japan	Stellarator	3.7	0.6	3	36	3,600	Superconducting
Japan	Tokamak	2.97	1.2	2.25	39	100	Superconducting Under construction
South Korea	Tokamak	1.8	0.5	3.5	24	300	Superconducting
US	Tokamak	0.68	0.22	5.4	8.5	5	$B = 8T$ at reduced pulse length
US	Tokamak	1.67	0.67	2.2	35	10	Includes proposed power and pulse length upgrades
US	Spherical Tokamak	0.9	0.6	1	16	5	Under upgrade $P = 21$ MW for shorter pulse
ITER	Tokamak	6.2	2	5.3	75	300–3000	Superconducting Under construction

Note: The pulse lengths and heating powers listed are the estimated maximum expected after currently planned upgrades.

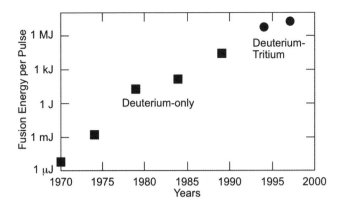

Fig. 6.   Energy produced per pulse in magnetic fusion experiments worldwide, 1970–2000.

energy experiments have operated with DT plasmas, as the world awaits the completion of ITER. First plasma in ITER is anticipated in about 2019; its 200,000 MJ point (500 MW for 400 sec) is anticipated within the decade of the 2020s.

3.2 *ITER: Role and Characteristics*

ITER will advance the science and technology of MFE along all of the dimensions described in Sec. 2. The large jump in size and the high magnetic field of ITER pushes the parameter ρ^* into new territory, where turbulence should be further suppressed, and we will, for the first time, be able to explore the physics of plasmas heated dominantly by fusion power production, so-called "burning plasmas". At $Q = P_{\text{fus}}/P_{\text{aux}} = 10$, the power from the fusion-produced helium nuclei will be double that from auxiliary heating, $P_\alpha = 0.2P_{\text{fus}} = 2P_{\text{aux}}$.

ITER will also allow studies of the stability of plasmas in this new parameter range. The highest βs, however, will be explored for long pulses in the new superconducting tokamak experiments in South Korea, China, and Japan. ITER will also explore the sustainment of tokamak plasmas by current drive, and again will be complemented by the large superconducting tokamaks. The sustainment properties of the complementary stellarator configuration will be examined in superconducting facilities in Japan and Germany, while this topic will be examined for the spherical tokamak in the US and England.

ITER will cross a major threshold in the area of plasma–material interactions. Both the steady and transient heat fluxes anticipated in ITER are capable of damaging the components that face the plasma, potentially requiring frequent replacement. Thus a major goal of ITER, and experiments in parallel with ITER on smaller facilities, will be to master the plasma–material interface and clear the path for even higher-power and longer-pulse interactions in a demonstration power plant.

ITER will make a major contribution to the study of the tritium-breeding blankets required for fusion, because ports on ITER have been designated to expose Test Blanket Modules to the ITER plasma. These modules will be supplied by the ITER partners. The total fluence (flux × time) of neutrons from ITER is too low to challenge materials properties, however, so parallel efforts will be required in this area.

ITER's magnets will stand, literally, as a testament to the fact that magnetic fields of the class required for commercial fusion can be constructed and operated reliably.

3.3 *Theory and Modeling*

It is important to recognize that advances in fusion energy production have been accompanied by similar advances in theoretical understanding and computational modeling of magnetic fusion plasmas. These advances were required to extrapolate with confidence from present experimental results to ITER. Theoretical codes now predict the observed properties of turbulence in the hot core of tokamak plasmas, including the absolute rate at which energy is lost. Stability limits are routinely calculated with high precision, and the current drive techniques required for sustainment are understood to high accuracy. The physics of the plasma edge and scrape-off region may be the most mysterious, but recently, theoretical models have been able to capture both the magnitude and scaling of the scrape-off width for the first time. Furthermore, advanced computation has been critical for the design of new materials to operate in the 14 MeV neutron environment and for interpreting experimental results from neutron exposure. Computational tools are used extensively in the design and optimization of powerful superconducting magnets and the supporting structure.

4 Development Plans and Design Studies

ITER will substantially advance our knowledge of burning plasmas and technical capabilities toward practical fusion energy. However, ITER is designed as an experiment, not a prototype power plant. With the launch of ITER construction and preparations for its operation, several of the ITER partner countries and country-groups have established programs to resolve the remaining challenges, and to design the experiments and facilities needed to advance to a demonstration magnetic-fusion power plant (Demo). The remaining challenges have been identified and extensively documented (e.g. see Ref. 13), and fall into three broad groups:

- *Reliable, predictable, steady-state operation at high performance.* In order to produce substantial net electricity, the fusion power multiplier must be significantly higher than the steady-state value of 5 expected for ITER. This requires that the plasma pressure (β) be higher for Demo. It also requires that less external current drive power be used and that more current be self-generated by the plasma pressure, which requires increased β. Higher β increases instabilities, is more challenging to control, and has not been demonstrated in tokamaks with steady-state current operation. This is a research focus for superconducting tokamaks worldwide.

In principle, this is not a challenge for stellarators, which do not require external current drive and have demonstrated steady high-β operation. However, stellarators have not demonstrated adequate confinement simultaneously with high-β. This is a research focus for the superconducting stellarator in Japan and for the new superconducting stellarator under construction in Germany.

- *Reliable exhaust of plasma heat and helium ash.* The steady heat flux in the exhaust structures will be much higher for Demo than for ITER. In addition, fusion neutron bombardment and transmutation will degrade the properties of solid materials, including their heat conductivity. Plasma instabilities can generate large transient increases in the exhaust flux. Either the instabilities must be eliminated or the transient fluxes absorbed. The high heat flux in Demo must be handled without excessive erosion or infiltration of the plasma by the wall material. Research directions include developing new materials, methods to spread the edge magnetic field to spread the exhaust heat flux, methods to cool the edge plasma to reduce erosion, and possible use of liquid surfaces to self-heal damage and handle higher power fluxes. Development of these strategies may benefit from a dedicated medium-scale experiment.

- *Capturing neutrons and self-sufficient production of tritium.* The energy from the fusion reactions must be captured and converted to electricity, including both the 14.1 MeV neutrons and the α particles (helium nuclei). The neutrons must also interact with lithium in the blanket to produce tritium. The tritium must be gathered, purified, stored, and injected into the plasma for continued fusion burning. The 14.1 MeV neutrons bombard the structural materials around the plasma and blanket, causing displacement damage and transmuting the constituent atoms, including generating H and He gas internally. Materials and designs must be validated as being able to survive and function safely in this environment. In addition, the blankets must shield the superconducting coils from the neutrons, to prevent conductor and insulator damage. These issues will be studied on ITER, including its Test-Blanket Module program (TBM). Studies with higher neutron fluence may require a dedicated high-energy neutron sources, such as the proposed International Fusion Materials Irradiation Facility (IFMIF), and may benefit from staged studies on DT fusion facilities approaching Demo.

The solutions to these challenges must be integrated and made reliable and maintainable with high availability to have a practical, economical energy system.

Possible solutions to these integration challenges have been explored in many preconceptual design studies for possible fusion power plants, based on tokamaks, stellarators, and other magnetic configurations. Recent examples include the EU PPCS studies,[15] the US ARIES studies,[16,17] and the Japanese SlimCS study,[18] which explore different design choices and development risks.

Several national and international programs have been launched to address the post-ITER challenges and prepare for a Demo power plant, expected to be the last step before commercialization. Differing needs for new energy sources drive the various national programs, producing differences in plan details. Nations with the most time-urgent needs favor larger steps and higher technical risk per step. However, all programs conceptually aim to provide fusion power by approximately the middle of this century. The goals and plans of the various programs were reviewed and compared in a recent workshop.[19] The major Demo-related programs include:

- The Broader Approach[20] is a large joint program by the EU and Japan to carry out R&D to prepare for Demo and complement ITER, established by signed agreement in 2007. The Broader Approach has three major parts: (1) a new superconducting long-pulse tokamak, to explore steady-state high β operation and optimization of the plasma configuration for Demo. (2) Design, R&D, and prototype development for an accelerator-based intense 14.1 MeV neutron source for testing materials and sub-components for future fusion power systems. (3) The International Fusion Energy Research Center, which includes Demo design studies, Demo R&D coordination, a shared supercomputer center, and remote operation facilities for ITER.
- Europe[21] is undertaking a broad program of activities whose construction would start in approximately 2030. The aim is to have a preconceptual design by 2014, and a conceptual design by 2020.
- China is engaged in a multi-institutional preconceptual study of various approaches for developing fusion and advancing toward Demo, driven by an urgent national need for new energy supplies. The goal is to prepare for a possible Engineering Design Phase decision in 2014, and first operation in 2025. The facility may have multiple phases, starting with sub-system qualification and demonstration of integrated operating scenarios.
- South Korea has started a Demo preconceptual design study with a goal to begin construction in the mid-2020s and operation in the mid-2030s. The first phase would develop and qualify technological solutions. A second phase would demonstrate full-scale electricity generation by fusion.

Other countries, including India, Russia, and the US, are evaluating facilities to conduct specific research to prepare for an eventual Demo. These include neutron sources, integrated test facilities, and upgradation of existing facilities. In addition, individual laboratories and collaborations are engaged in design studies and R&D program planning.

5 Summary

Controlled fusion offers an effectively inexhaustible energy source without production of greenhouse gases or long-lived radioactive waste, without the possibility of catastrophic runaway accidents, and with low proliferation risk. It can be made available to all nations. There has been dramatic progress in understanding the physics and technology needed for magnetically confined fusion (MFE), and producing the conditions needed for substantial fusion power production. The next major step will be ITER, providing industrial levels of fusion power for long pulses, and enabling the study of burning plasmas and fusion power technology. Beyond ITER, the challenge will be to make fusion practical and economical. Several countries are already engaged in R&D and design programs for post-ITER fusion power systems, preparing the steps needed to achieve fusion's potential as an attractive new energy source.

References

1. R. J. Hawryluk, Results from deuterium-tritium tokamak confinement experiments, *Rev. Mod. Phys.* **70** (1998), p. 537.
2. J. Jacquinot, Deuterium-tritium operation in magnetic confinement experiments: results and underlying physics, *Plasma Phys. Control Fusion* **41** (1999), p. A13.
3. T. Fujita and the JT-60 Team, Overview of JT-60U results leading to high integrated performance in reactor-relevant regimes, *Nucl. Fusion* **43** (2003), p. 1527.
4. Available at: http://www.iter.org/. Accessed 19 September 2012.
5. USGS Commodity Summaries, 2012. Available at: http://minerals.usgs.gov/minerals/pubs/commodity/. Accessed 19 September 2012.
6. POSCO news release, 5 February 2010. Available at: http://www.posco.com/homepage/docs/eng/jsp/prcenter/news/s91c1010025v.jsp?mode=view&idx=1272. Accessed 19 Sept. 2012.
7. *The Korea Times*, 23 February 2012. Available at: http://www.koreatimes.co.kr/www/news/nation/2012/02/182_105554.html. Accessed 19 September 2012.
8. M. Kikuchi and N. Innoue, Role of fusion energy for the 21 century energy market and development strategy with International Thermonuclear Experimental Reactor in *WEC-18*, World Energy Council, 2001.
9. A. Glaser and R. J. Goldston, Proliferation risks of magnetic fusion energy: clandestine production, covert production and breakout, *Nucl. Fusion* **52** (2012), p. 043004.
10. R. H. Clark and Z. Cai, "Helium and Fusion Energy" in *The Future of Helium as a Natural Resource*, eds. W. H. Nuttall, R. H. Clarke, and B. A. Glowacki, (Routledge, London and New York, 2012), p. 235.
11. R. J. Goldston, Climate Change, Nuclear Power, and Nuclear Proliferation: Magnitude Matters, *Science & Global Security* **19** (2011), p. 130.
12. Research needs for magnetic fusion energy sciences. (US DOE 2009). Available at: http://science.energy.gov/~/media/fes/pdf/workshop-reports/Res_needs_mag_fusion_report_june_2009.pdf. Accessed 19 September 2012.
13. Fusion Energy Sciences Advisory Committee Report on Opportunities for Fusion Materials Science and Technology Research Now and During the ITER Era, February (2012), p. 68. Available at: http://science.energy.gov/~/media/fes/pdf/workshop-reports/20120309/FESAC-Materials-Science-final-report.pdf. Accessed 19 September 2012.
14. Available at: http://www-pub.iaea.org/iaeameetings/38091/23rd-IAEA-Fusion-Energy-Conference. Accessed 19 September 2012.
15. D. Maisonnier *et al.*, Power plant conceptual studies in Europe, *Nucl. Fusion* **47** (2007), p. 1524.
16. F. Najmabadi *et al.*, The ARIES-AT advanced tokamak, Advanced technology fusion power plant, *Fusion Eng. Design* **80** (2006), p. 3.
17. F. Najmabadi and A. R. Rafray, *The ARIES-CS compact stellarator fusion power plant, Fusion Sci. Technol.* **54** (2008), p. 655.

18. K. Tobita *et al.*, Compact DEMO, SlimCS: Design progress and issues, *Nucl. Fusion* **49** (2009), p. 075029.

19. G. H. Neilson, G. Federici, J. Li, D. Maisonnier, and R. Wolf, Summary of the International Workshop on Magnetic Fusion Energy (MFE) Roadmapping in the ITER Era; 7–10 September 2011, Princeton, NJ, USA, *Nucl. Fusion* **52** (2012), p. 047001. Available at: http://advprojects.pppl.gov/ROADMAPPING/presentations.asp. Accessed 19 September 2012.

20. Available at: http://www.broaderapproach.org/. Accessed 19 September 2012.

21. Available at: http://www.efda.org/collaborators/efda-departments/pppt-department/. Accessed 19 September 2012.

Chapter 8

Progress Toward Inertial Fusion Energy

Erik Storm

Lawrence Livermore National Laboratory
P.O. Box 808, Livermore, CA 94551, USA
Storm1@llnl.gov

Inertial Confinement Fusion (ICF), a thermonuclear reaction in a millimeter-sized fuel capsule filled with deuterium and tritium (DT), has been the subject of theoretical and experimental studies since the early 1970s. Recent results at the National Ignition Facility (NIF) in the US provide confidence that net fusion energy gains of 10–20 (fusion yield/laser input energy) will be demonstrated with indirectly driven ICF targets by 2013 and that gains of 60–70 from a 2–2.2 MJ driver is achievable. Similar experiments are planned for the Laser Megajoule (LMJ) in France a few years later. The European HiPER and Japanese FIREX-II programs are considering a direct drive ignition approach for the early 2020s. Successful demonstration of ignition and gain will be a transforming event for ICF and is likely to focus the world's attention on laser-driven Inertial Fusion Energy (IFE) as an option for commercial power plants.

1 Introduction

Fusion is an attractive energy option for the future. Inertial Confinement Fusion (ICF) uses lasers or heavy ion beams to rapidly compress a capsule containing a mixture of deuterium and tritium (DT). As the capsule radius decreases and the DT density and temperature increase, DT fusion reactions are initiated in a small spot in the center of the compressed capsule. These DT fusion reactions generate both α-particles and 14.1-MeV neutrons, and a fusion burn front propagates, generating significant energy gain.[a] In contrast, magnetic fusion energy (MFE) uses powerful magnetic fields to confine a low-density DT plasma and to generate the conditions required to sustain the burning plasma for a sufficiently long time to generate energy gain.

The roadmap for transforming fusion energy into a source of electricity can be thought of as a five-step process: (1) demonstrate understanding of the underlying physics principles, (2) achieve net energy gain, (3) execute an R&D program to develop the required technologies (4) build and demonstrate a prototype power

---

[a]For ICF, the gain G is defined as $G = $ *fusion yield* laser input energy.

plant, and (5) build commercial power plants. The progress made over the last 50 years means that Step 1 has been achieved both for MFE and ICF. Step 2 for ICF will likely be achieved within the next few years.

The experiments will use a central hot spot ignition (HSI) target in an indirect drive configuration. HSI relies on simultaneous compression and ignition of the spherical DT-filled capsule. In the indirect-drive configuration, the capsule is placed inside a cylindrical cavity of a high-Z metal (a hohlraum), and the implosion pressure is provided by focusing the laser energy onto the interior walls of the hohlraum and converting it to X-rays. The small (a few percent of the total DT fuel mass), high-temperature central part of the imploded fuel provides the "spark," which ignites the cold, high-density portion of the fuel. The scientific basis for HSI targets has been intensively developed over the last 40 years, and ignition and gains of 10–20 with HSI targets is expected by 2012. Target gains > 60 that are desirable for efficient, cost-effective IFE with HSI targets could be achieved a few years later with laser energies around 2 MJ.

After nearly 40 years of R&D, IFE research is thus at a key juncture. The demonstration of fusion ignition and energy gain in an experimental setting will provide the basis for the transition from scientific research to considering Inertial Fusion Energy (IFE) as an option for commercial power plants.

2 Review of Basic ICF Physics

The basic idea of ICF is for a driver to deliver sufficient energy, power, and intensity to a capsule containing DT to make the capsule implode. The implosion is designed to compress and heat the DT mixture to a temperature at which thermonuclear burn is ignited. The self-sustaining thermonuclear reaction releases the energy of the DT mixture, and if the energy output is greater than the driver energy supplied, the gain is >1. To cause the $D + T \rightarrow ^3He + n + 17.6\,MeV$ reaction to take place to a reasonable degree, the fuel must be raised to a temperature of 5–10 keV or greater,[b] and for the reaction to be self-sustaining, there must be a sufficiently high particle density. Densities of interest to ICF are in the 200–1200 g/cm^3 range, depending on the target and performance required.

The sequence of events leading to DT burn is as follows (Fig. 1). The driver deposits its energy either by electron conduction (with the driver — lasers or particle beams) directly focused on the capsule or indirectly in the outer layers of the fuel capsule (via X-rays generated by converting the driver energy to X-rays in a high-Z cavity surrounding the capsule). This energy deposition in turn ablates the surface of the fuel capsule, and the ablation acts as a rocket exhaust to drive the implosion of the capsule.

---

[b]Energy and temperature are related by Boltzmann's constant, $k = 1.3807 \times 10^{-23}$ J/K. Therefore 10 keV is equivalent to 1.16×10^8 K i.e. about 116 million Kelvin.

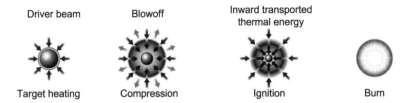

Fig. 1.   Basic sequence of events for high-gain ICF.

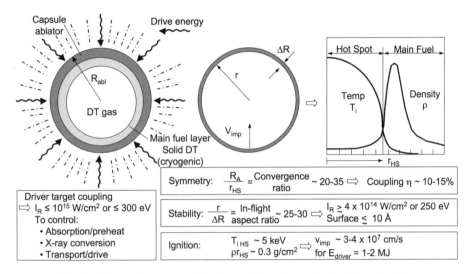

Fig. 2.   Target physics specifications and requirements for high-gain ICF.

The ablation generates pressures from 1 to about 100 Mbar over a period of about 15×10^{-9} sec and accelerates the capsule inward to velocities of $3-4 \times 10^7$ cm/sec. The acceleration continues until the internal pressures exceed the ablation pressure. At that time, the rapidly converging shell begins to decelerate and compresses the central portion of the fuel. In its final configuration, the fuel is nearly isobaric at pressures of up to ~200 Gbar but consists of two very different regions — a central hot spot containing ~2% of the fuel and a dense main fuel region with a peak density of ~1,000 g/cm³. Fusion initiates in the central hot spot, and a thermonuclear burn front propagates through the dense, main fuel producing high-energy gain. To achieve these conditions, high-gain ICF targets have features and requirements similar to those shown in Fig. 2.

This rather straightforward sequence of events is complicated by the impracticality of building drivers of arbitrarily high energy, power, and focusability, and by some basic facts of nature, which are: (1) the implosion must be carried out in a reasonably isentropic manner until a late stage of the implosion, or else the continual heating of the fuel will make a high compression too demanding on driver energy and (2) potential hydrodynamic instabilities demand a specific type of ablative implosion to avoid mixing of the outer shell with the fuel and subsequent degradation of the burn efficiency.

The underlying questions with achieving high-gain ICF have always been whether an appropriately configured fuel capsule can be made; a sufficiently uniform pressure pulse can be produced to cause the 30- to 40-fold radial convergence of the capsule required to achieve the necessary high compression; and we fully understand the driver/target interaction physics, the hydrodynamics, and the physics of DT burn on which our models are based? A few simple arguments will indicate the design constraints for a successful high-gain target.

2.1 *DT Burn Physics*

The thermonuclear burn rate for a DT plasma can be written

$$dn/dt = n_D \, n_T \langle \sigma v \rangle, \qquad (1)$$

where $\langle \sigma v \rangle$ is the Maxwellian averaged cross-section and $n_D(t)$ and $n_T(t)$ are the instantaneous deuteron and triton number densities. We define the fuel burn-up fraction as $\phi = n/n_o$, where n_o is the initial total fuel number density. If we assume a nearly constant burn rate, then for an equimolar mixture of D and T, we can integrate (1) to find

$$\phi/(1 - \phi) = n_o \tau / 2 \langle \sigma v \rangle. \qquad (2)$$

For DT at about 20 keV, we have approximately

$$\phi = n_o \tau / (n_o \tau + 5 \times 10^{15}) = \rho r / (\rho r + 6), \qquad (3)$$

where n has units of particles per cubic centimeter and τ has units of seconds. In the equality on the right, ρr has units of g/cm^2 and is obtained from the formula containing $n\tau$ by setting $n = 2.4 \times 10^{23} \rho$ for DT and $\tau = r/3v_s$, where r is the radius of the compressed fuel and v_s is the sound speed.

This formula for ϕ for ICF agrees well with detailed burnup simulations of most high-gain ICF targets. One also notes that in order to achieve reasonable burnup, say between 1/4 and 1/3, the fuel ρr should be between 2 and 3 g/cm^2.

Ignition of the DT fuel occurs when the α-particle energy deposition into the fuel plasma during one energy confinement time equals the energy required to heat the plasma. If the electron and ion temperatures are equal, then the energy required for heating the fuel with 100% efficiency equals

$$E_{\text{DT heating}} = 0.115 \times 10^9 \, T \, \text{(J/g)} = 2.3 \times 10^9 \, \text{(J/g)}, \quad \text{at } 20 \, \text{keV} \quad \text{with } T_{\text{ion}} = T_e. \qquad (4)$$

A DT plasma produces 3.34×10^{11} J/g if it burns completely. The specific thermonuclear energy produced is given by the product of the burn efficiency and this specific energy. Of this, about 20% is in α-particles. The neutrons have too large a

range to heat either an MFE plasma or an ICF pellet, so the specific thermonuclear
energy available to heat the plasma is given by

$$E_{\text{thermonuclear } \alpha\text{-particle}} = 6.68 \times 10^{10} n\tau / (n\tau + 5 \times 10^{15}) \, (\text{J/g}) \quad \text{at } 20 \, \text{keV.} \quad (5)$$

Setting these two equal in this simple model, the requirement for ignition is $n\tau >$
$1.7 \times 10^{14} \, \text{s/cm}^3$. This is known as the Lawson Criterion, and corresponds to $\phi =$
0.034, which is adequate for MFE.

2.2 *Compression and Central Ignition*

For ICF, this is equivalent to a $\rho r \sim 0.3 \, \text{g/cm}^2$ at $20 \, \text{keV}$. However, if the fuel is
surrounded by a denser region of DT of $\rho r \sim 2\text{--}3 \, \text{g/cm}^2$, the ideal ignition condition
can be relaxed to $\rho r \sim 0.3 \, \text{g/cm}^2$ at $\sim 5 \, \text{keV}$. The increased confinement time then
allows the α-deposition to bring the fuel to ignition temperature ($20 \, \text{keV}$) before
decompression. For ICF, this is also called marginal ignition. However, ignition
is inadequate for ICF, which must first overcome a factor of 10–20 in implosion
inefficiency. For IFE power generation, we must also account for another factor of
5–10 in driver efficiency. ICF achieves this combined approximately two-order-of-
magnitude improvement by utilizing two effects: compression and central, or hot
spot ignition.

The first order of magnitude comes from **compression**. If a ρr of $2.5 \, \text{g/cm}^2$
can be achieved, the burn efficiency rises to 30%. This ρr is more than sufficient
to absorb the 3.5-MeV α-particles, while the 14.1-MeV neutrons escape without
contributing their energy to the fuel. However, if the DT is uncompressed, a ρr
of $2.5 \, \text{g/cm}^2$ requires a sphere of solid DT almost $25 \, \text{cm}$ in diameter with a mass
of $\sim 1.5 \, \text{kg}$. Not only would such a large fuel mass produce an unacceptably large
yield ($\sim 1.5 \times 10^{14} \, \text{J}$, or $\sim 35 \, \text{kT}$) but it would take about $10^{13} \, \text{J}$ to heat it to ignition.
Since mass m $\sim (\rho r)^3 / r^2$, the fuel mass and hence the driver energy required (which
is proportional to the fuel mass) for a given ρr can be reduced by a large factor by
compression.

The kinetic energy of the imploding fuel shell must compress the fuel to the
required density and then heat it to ignition temperature ($5\text{--}10 \, \text{keV}$). If the fuel
is compressed to a density of $500 \, \text{g/cm}^3$, the mass required for a ρr of $2.5 \, \text{g/cm}^2$
drops to $\sim 2 \, \text{mg}$ for a spherical DT shell of radius R and shell thickness $\Delta R = R/2$
(a simple approximation to the radial density distribution shown in Fig. 2). The
fusion yield now drops to a manageable $200 \, \text{MJ}$.

Compression is attractive from an energetic point of view, however, only if it
is "cheap" compared to the energy required for heating the DT. If the compression
can be done with the fuel in a near Fermi-degenerate state (which is possible if the
driver energy is delivered with a precise time history), this is indeed the case. The
specific energy for compression of Fermi-degenerate DT is

$$\varepsilon_{\text{FD}} \, (\text{J/g}) = 3 \times 10^5 \rho^{2/3}, \quad (\text{with } \rho \text{ in g/cm}^3) \text{ or } \varepsilon_{\text{FD}} \sim 1\text{--}3 \times 10^7 \, \text{J/g}, \quad (6)$$

for DT in the $200\,\mathrm{g/cm^3}$ to $1,000\,\mathrm{g/cm^3}$ range. This is only a few percent of the energy needed to heat DT to the 5–10 keV required for ignition. (The specific heat for DT per keV is $\sim1.15\times10^8\,\mathrm{J/gm}$). While only about 40 kJ is needed for the compression of 2 mg to an average density of $500\,\mathrm{g/cm^3}$ and $\rho r = 2.5\,\mathrm{g/cm^2}$, an implosion energy of $\sim$1–2 MJ is required to heat the 2 mg to 5–10 keV.

If we assume a driver-to-implosion efficiency of 5%, the driver would have to deliver 20–40 MJ. While the yield of $\sim$200 MJ would be sufficient for IFE, the gain of 5–10 is inadequate and the cost of the driver would be prohibitive.

To overcome this difficulty and provide the second order of magnitude, ICF capsules use **central ignition** and propagation of the burn into compressed DT to achieve high gain. In a typical high-gain capsule, only a few percent ($\sim$2%) of the fuel has to be heated to ignition conditions. The energy invested to heat this part of the 2 mg of fuel, would only be $\sim$40 kJ, and α-particles produced in this region ignite the rest of the surrounding highly compressed, cooler fuel. If this is done, the energy used to ignite the fuel becomes comparable to the compression energy. If we account for the fact that the main fuel is also heated to about 100 eV, the total kinetic energy required would be about 100 kJ.

An implosion velocity of $\sim3.3\times10^7$ cm/sec is required to give the 2 mg of fuel this kinetic energy. This results in a laser of about 2 MJ, not an unreasonable requirement. With the fusion yield of 200 MJ, this would produce a gain of about 100, more than adequate for IFE.

2.3 *Fluid instabilities, Mix, and Low-Entropy Implosions*

Although energetically attractive, the physics of compression and hot spot ignition is very demanding. At ignition, the radius of the hot spot is smaller than the initial radius of the capsule by a factor of 30–40. This reduced radius is achieved by forming the fuel into a spherical shell, and it increases the effective volume over which PdV work can be done on the fuel and lowers the pressure required. With peak pressures of about 100 Mbar, convergence ratios of about 40 can be achieved by this technique. To maintain a near-spherical implosion with this convergence ratio, requires a drive flux uniform to about 1%–2% over the surface of the capsule.

A second hurdle for successful ICF implosions is the effect of hydrodynamic instabilities. A description of the detailed analysis of the growth of Raleigh–Taylor instabilities (caused by the low-density ablation region pushing on the high-density imploding shell) is outside the scope of this note. Both analytical models and detailed numerical simulations show that the growth of instabilities is proportional to the ratio of the shell radius (R) to the shell thickness ΔR. Since the shell is compressed during the implosion, we must use the smallest value of the shell thickness during the implosion to get the so-called in-flight aspect ratio. To keep the imploding shell from breaking up and the resulting mix of cold ablator from mixing into the hot spot and quenching the ignition, this in-flight aspect ratio should be kept to a maximum of about 25–30. The surface of the ablator must also be smooth to about 1 nm and the DT ice layers to about $0.5\,\mu\mathrm{m}$.

A third constraint is the requirement that the entropy generated by the ablative process be limited to about that generated by a 1-Mbar shock passing through the initially uncompressed layer of solid or liquid DT. This allows the compression process to keep the fuel as near as possible to a Fermi-degenerate state; i.e. as low an entropy generation and heating as possible. Keeping the first shock to about 1 Mbar and each succeeding shock to not more than four times the pressure of the preceding shock satisfies the requirement for low entropy generation.

The ablation pressures (P_{abl}) required to generate the implosion velocities and the compression for high-gain ICF implosions are around 100 Mbar.

The development of the technology to fabricate targets with the precision mentioned and the development of the experimental and computational tools to accomplish implosions with this kind of accuracy has been one of the major achievements of the worldwide ICF programs over the past 40 years.

2.4 *Indirect- and Direct-Drive Approaches to ICF*

The two fundamental approaches to ICF are differentiated by the mechanism chosen for the energy transport to drive the ablation and implosion (see Fig. 3).

In the **direct-drive** approach, laser (or ion) beams are arrayed around the target in a near-uniform pattern and focused directly on the fuel capsule. The energy is transported from the absorption region to the ablation surface by electron conduction. The absorption occurs at a particle density equal to or less than the critical plasma density $n_c(\text{cm}^{-3}) = 10^{21}/\lambda^2$, where λ is the laser wavelength in μm. Typically, particle densities at the ablation front are $2-3 \times 10^{23}\,\text{cm}^{-3}$, and electron conduction must bridge the gap between the two.

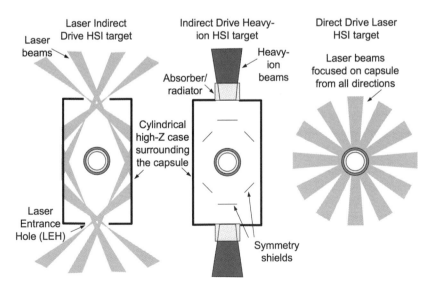

Fig. 3.   Schematic laser and heavy-ion indirect-drive and laser direct-drive HSI targets.

Two fundamental requirements for direct drive are efficient absorption of the laser light in the underdense plasma region by collisional processes (to avoid excessive generation of hot electrons) and a pressure distribution at the ablation surface uniform to <1%–2%. To avoid high levels of laser-plasma instabilities and generation of energetic electrons, direct-drive targets typically employ laser wavelengths <0.35 μm to reduce the scale length of the underdense plasma region. This, however, reduces the region between the absorption and ablation front, which reduces the potential smoothing by electron conduction. Uniformity must therefore be obtained by overlapping a large number of nearly ideal beams and the use of beam smoothing techniques. Direct drive is a more efficient approach, and simulations project about 2X higher gain than indirect drive HSI, but the very stringent illumination uniformity requirements tend to offset some of this potential advantage.

In **indirect drive**, the laser energy is first converted into X-rays inside a high-Z enclosure (a hohlraum). The fuel capsule is contained inside the hohlraum, and the X-rays are used to drive the ablation of the surface of the capsule, decoupling the absorption and transport process. The hohlraum provides capsule drive uniformity, allowing for a wide variety of possible laser geometries and greatly relaxing the requirements on laser beam uniformity. In general, laser-driven hohlraums are cylinders with two laser entrance holes (LEHs).

Target physics specifications for laser-driven indirect-drive targets are shown in Fig. 2. Driver-target coupling issues limit the laser intensities inside the hohlraums to $<10^{15}$ W/cm^2, resulting in a radiation temperature limit of about 300 eV. This limitation is primarily the result of laser-driven plasma instabilities that result in the scattering of laser light and the generation of high-energy electrons. Light scattering degrades symmetry, and high-energy electrons cause capsule and fuel preheat, which reduces the compression. With the convergence requirement of ~30, the X-ray flux on the capsule must be uniform to 1%–2%. To achieve this level of uniformity, the hohlraum areas must typically be 15–25 times larger than the capsule surface area. This large case-to-capsule area limits the coupling efficiency of the driver energy to the capsule to 10%–15%, although it should ultimately be possible to increase this to about 20%–25%, by optimizing hohlraum and LEH geometries with respect to capsule surface area.

Relative to direct drive, radiation drive targets are also less sensitive to the effects of hydrodynamic instability during the implosion. However, since it is a two-step process (absorbed laser light is first converted to X-rays and the capsule only absorbs a fraction of these X-rays), indirect-drive targets are less efficient.

Heavy-ion indirect drive. Capsule performance is essentially independent of the source of the X-rays as long as they have the appropriate spectral, temporal, and spatial characteristics. This means that the X-rays to drive the capsule can also be produced by heavy-ion (HI) beams. The hohlraum would still be a cylindrically shaped high-z case, but the ion beams would be focused on absorbers placed in the

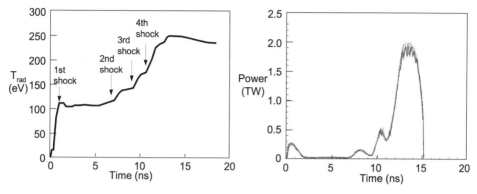

Fig. 4. X-ray drive pulse for the capsule for indirectly driven HIS target (left) and the required laser power as a function of time (right). Laser data from one National Ignition Facility beamline from 16 shots.

same location as the LEH for a laser-driven target. The ion beam energy is then reradiated in the form of X-rays. A typical HI beam indirect-drive target is shown in Fig. 3.

Pulse shaping. Reaching a peak drive pressure of some 100 Mbar coupled with the condition for low entropy generation means that four or more shocks, each one not more than a factor of 4 greater than the preceding one, must be used to reach the peak pressure. The scheme is to send the shocks through the solid (or liquid) DT fuel in such a way that they all coalesce near the center and there produce the hot spot required for ignition. Figure 4 shows the X-ray drive pulse for an indirectly driven target. The laser pulse required to produce this X-ray drive pulse is also shown. A laser pulse with a shape similar to the X-ray drive pulse of Fig. 4 would be required for a direct drive HSI target.

Wavelength dependence. Nearly all of the experimental data for ICF targets (both directly and indirectly driven) over the past 40 years have been obtained with frequency converted solid-state lasers operating at 0.53, 0.35, or 0.25 μm, (Table 1) although some work has also been done with KrF at 0.25 μm and iodine-based gas lasers at longer wavelengths, primarily in Russia. The coupling of laser energy to the target, both for the indirect drive and the direct-drive approach, is greatly enhanced at shorter wavelengths since, the shorter the laser wavelength, the higher the absorption of the laser energy and the smaller the conversion of laser energy to deleterious high-energy electrons (which would preheat the shell and DT fuel and cause performance degradation).

High-gain direct-drive HSI target designs generally assume 0.25-μm laser light, although 0.35 μm may be acceptable. Indirectly driven HSI targets with v_{imp}3.5−4× 10^7 cm/sec generated by hohlraum temperatures greater than 250 eV require laser wavelengths of 0.35 μm. Higher yield indirect-drive HSI targets that can operate at temperatures lesser than 250 eV may tolerate 0.53 μm.

Table 1:   Main Institutions Engaged in ICF R&D.[a]

Institute-location	Facility; laser type	Laser parameters	Main R&D efforts
ILE, Osaka Univ, Japan	Gekko XII; FIREX-I; Nd:glass	2–5 kJ/2ns; 500 J/1 ps	DD; DD FI; HFP
LLE, Univ Rochester, US	Omega/Omega EP; Nd:glass	40 kJ/1–2 ns; 2.5 kJ/10 ps	DD; DD FI; HFP
Rutherford Labs, UK	Vulcan; Nd:glass	2.6 kJ/1 ns; 500 J/0.5 ps	DD-FI; FIP; HFP
Ecole Polytech., France	LULI-200; Pico2000; Nd:glass	2 kJ/1.5 ns; 200J/1ps	DD; FIP; HFP
NRL, Wash. D.C. US	Nike, KrF	5 kJ/4 ns	DD
LLNL, Livermore, US	NIF; Nd:glass	1.3–1.8 MJ/10–20 ns	Ignition; ID; HEDP
CEA/CESTA, France	LIL; Nd:glass	20–30 kJ/4–8 ns	ID; HEDP

[a]DD = Direct-Drive HIS; ID = Indirect-Drive HIS; DD-FI = Direct-Drive Fast Ignition; FIP = Fast Ignition Physisc; HFP = High Field Physics ($I \geq 10^{20}$ W/cm^2); HEPD = Basic High Energy Density Physics.

2.5 *Alternative Ignition Concepts*

For conventional HSI targets, the driver pulse (laser or heavy ion) assembles the fuel at high density and imparts a sufficiently high velocity ($\sim 3.5 \times 10^7$ cm/s) to the imploding shell so that its PdV work create both the high, main fuel density, and the central ignition hot spot on stagnation.

In contrast, there are target options (such as fast ignition and shock ignition) that decouple the fuel assembly and ignition phases. In both the options, the cryogenic DT shell is first compressed to a $\rho r \sim 2$–3 g/cm^2 at ~ 300–500 g/cm^3 at a low velocity ($\sim 2 \times 10^7$ cm/sec) using a driver of lower peak power and lower total energy. For a shock ignition target, the assembled fuel is then separately ignited from a central hot spot heated by a strong, spherically convergent shock driven by a high-intensity spike at the end of the compression pulse. For the fast ignition target, an extremely high peak intensity ($\sim 10^{20}$ W/cm^2) short pulse (10–20 ps) laser focused to a small (~ 20–30 μm diameter) spot at the edge of the high-density compressed fuel provides MeV ions to initiate ignition and propagating burn which spreads through the remainder of the fuel (Fig. 5).

The majority of the laser energy is contained in the main, compression portion of the pulse, with $\sim 20\%$ for the separate ignition pulse. Because the implosion velocity

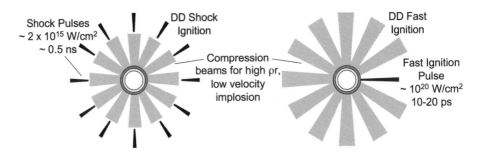

Fig. 5.   Schematic of directly driven shock ignition and fast ignition targets.

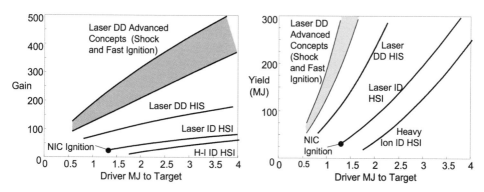

Fig. 6. Target gain and fusion yield for a variety of ICF targets and drivers.

is less than that required for conventional HSI, more fuel mass can be assembled for the same kinetic energy in the shell, offering higher fusion gains/yields for the same laser energy. For the fast ignition scheme, the compression could be achieved either by direct or indirect drive, although the overall gain will be slightly lower with indirect drive. For shock ignition, it appears that only direct-drive is realistically possible.

Summary of ICF target performance. Figure 6 shows the performance for HSI targets for both indirect and direct drive with lasers and HI, and direct-drive shock and fast ignitions. In contrast to directly driven HSI targets, which will likely require lasers with wavelengths of $\sim$0.25 μm, simulations show that directly driven shock and fast ignition targets can achieve adequate symmetry and convergence with 0.35-μm laser wavelength, available with frequency converted solid-state lasers. The performance of indirectly driven fast ignition targets is not shown but would be between the performance band of the direct-drive advanced concepts and the performance for direct-drive HSI targets.

From Fig. 6, it would appear that directly driven ICF targets, in particular, fast ignition or shock ignition targets, would have an advantage over indirect-drive targets for IFE, and IFE efforts in Europe and Japan are primarily focusing their R&D programs on these targets. However, as will be discussed in Sec. 4, the gain of an ICF target is not an adequate metric to choose an IFE system.

We also note that the projected gain for various IFE target concepts is based on widely varying levels of knowledge and assessment.

Direct-drive HSI targets are more sensitive to hydrodynamic instabilities and remain higher risk than indirect-drive HSI. If, in addition, the calculated higher gain can only be realized by using a KrF laser whose "wall-plug" efficiency" is about half that of high average power solid state lasers, the higher gain may not translate into an advantage for IFE (see Sec. 4).

Fast ignition and shock ignition targets are at an early stage of investigation, with complex physics and engineering issues still to solve, and it is not clear what

Table 2:   ICF facilities Under Construction.

Institute-location	Facility; laser type; date	Laser parameters	Main R&D efforts
CEA/CESTA, France	LMJ; Nd:glass (∼2016)	1.3–1.8 MJ/10–20 ns	Ignition; ID; HEDP
CEA/CESTA, France	PETAL; Nd:glass (∼2015)	3.5 kJ/5 ps	FIP; HFP

level of the postulated high-gain performance, if any, will remain after these issues have been resolved, nor how long such an R&D effort would take.

3 Progress Toward Ignition and High-Gain ICF

The physics of ICF has been the subject of extensive theoretical and experimental studies around the world for more than 40 years. The principal institutions currently doing experiments are listed in Table 1. Table 2 shows principal future facilities under construction with anticipated completion dates.

Indirect-drive ICF was also extensively studied with the frequency converted glass lasers Nova at LLNL, and Phébus at the CEA laboratory in France through the 1990s. By the early 1990s, the progress in the understanding of high energy density and laser plasma interaction physics and the development of experimental and computational tools of ICF was sufficient to lead the Department of Energy (DOE) in the US and The Commissariat à l'énergie atomique et aux énergies alternatives (CEA) in France to construct National Ignition Facility (NIF) and Laser Megajoule (LMJ). The goal of these facilities is to demonstrate ignition and high energy gain with indirectly driven HSI targets.

The main effort for direct-drive HSI over the past decade has been conducted at LLE and their Omega facility. There has been impressive progress in cryogenic target technology and in improving the beam smoothing required for these targets. Although the 40 kJ of 0.35-μm laser energy available on Omega is not sufficient for ignition, their goals were to demonstrate the capability to implode and model a capsule implosion on an ignition-type adiabat and also achieve the required implosion velocity. This has proven to be a challenging task. With sufficient shell thickness to avoid debilitating hydroinstabilities, the velocities obtained were only ∼3×10^7 cm/s, and the intensities required to achieve ignition level velocities generate too many high energy electrons. Although the achieved ρr values of 0.3 g/cm^2 are close to calculated values (implying low adiabat implosions), the neutron yields are just a few percent of the calculated values. This has led LLE to consider joining the Japanese and Europeans in moving from direct-drive HSI targets to direct-drive Fast or Shock Ignition options for IFE.

The potentially higher gain from Fast Ignition (FI) was first proposed in the early 1990s. The shift from HSI to FI for IFE and, in particular, the proposals for the FIREX II project in Japan, and the European HiPER project to demonstrate ignition and IFE technology with FI by 2020, were primarily due to a series of seminal, low-energy FI experiments at the GEKKO XII laser at Osaka University in Japan in 2001. The results implied that 25%–30% of the energy of the 10–20 ps short-pulse ignitor could be coupled to the compressed core. This suggested that a short-pulse energy of 80–100 kJ would be sufficient to ignite a compressed target. Several facilities are currently studying FI physics, including Omega EP at LLE, FIREX-I at ILE, Vulcan at Rutherford, and LULI/PICO2000 at LULI.

Recent experimental results, and the more realistic simulations made possible by the advances in high-performance computers, have shown that the generation and transport/focusing of high-energy electrons is significantly less efficient than what the Osaka results implied. It appears that short-pulse ignitor energies in excess of 300 kJ may be required. Since more than 600 kJ of conventional long-pulse laser energy would be required to generate this ignitor pulse, and the basic physics of fast ignition is still largely unproven, the HiPER project has recently made Direct-Drive Shock Ignition their mainline approach for IFE, and extended their IFE ignition demonstration goal to the early-mid 2020s.

The first demonstration of ignition and net energy gain for IFE will therefore be done with indirect drive HIS targets on NIF and LMJ. Figure 7 shows the indirect-drive HIS point design and a cut-away of NIF with the two laser bays, each with 96 beams and two switchyards, which rearrange the 192 beams into two sets of 24 groups of 4 beams entering into the target chamber from the top and bottom. NIF has demonstrated that it can meet the ignition energy performance goals simultaneously with the requirements for temporal pulse shaping, focal-spot conditioning, and peak power of 500 TW for representative ignition pulses.[c] The goal of the ignition campaign is to demonstrate a repeatable ignition platform with indirect-drive HIS targets by the end of 2013.

The campaign to demonstrate ignition is proceeding in four phases. In the first or "drive" phase, the empty hohlraum is tuned to produce the necessary radiation

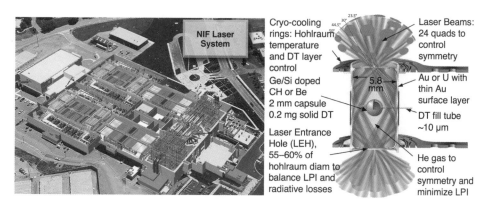

Fig. 7.   The National ignition facility and the indirect-drive ignition point design.

[c]On July 5, 2012, NIF performed an implosion experiment with 1.855 MJ in a 22 ns ignition-shaped pulse at a peak power of 522 TW.

drive on the capsule as a function of time. In the second "capsule tuning" phase, non-cryogenic and cryogenic capsules are used to adjust the hohlraum symmetry, shock timing, velocity, and mass ablated to produce the conditions in the imploding capsule required for ignition when a cryogenic fuel layer is incorporated. The third phase uses layered cryogenic implosions with a mixture of T, H, and D in a ratio 74:24:2 to make the THD capsules hydrodynamic analogues to DT implosions. The low D content leaves them unaffected by thermonuclear energy production, yet their radiation and hydrodynamic transport mirror that of DT implosions up to the point where DT implosions become perturbed by α-particle production and deposition. The reduced yields allow the full diagnostic suite to be used and the required pre-burn temperature and fuel ρr to be verified. The final phase is ignition with DT implosions.

The first two phases measured drive temperatures of 300 eV with backscatter of 10% to 15% and radiation symmetry control was demonstrated to compressed cores to <10%. In these hohlraums, beam propagation in underdense plasma and underdense plasma production from hohlraum blow off was more important than in previous Nova and Omega experiments and have brought new understanding to hohlraum performance. Energy transfer between crossing laser beams is an important effect and can be controlled and used to modify the hohlraum environment. New models have been developed for the radiating plasma and are being used to refine the ignition target design.

The third phase began in September 2010, and THD capsule implosion results with 1.3 MJ at 0.35 μm have demonstrated the hohlraum temperature of 300 eV and the symmetric X-ray environment required for high convergence. Hot spot ion temperatures of 3.7 keV and a main fuel ρr of 1 g/cm^2 were obtained (Fig. 8). The DT ignition campaign began in fall of 2011, and DT yields of 6–8 × 10^{14}, main full Pr of 1.3 g/cm^2, peak fuel densities of 850 g/cm^3 and implosion velocities of

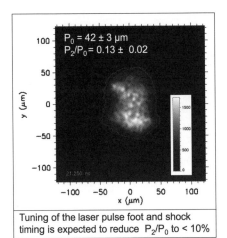
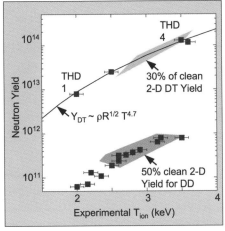

Fig. 8.   THD implosions at 1.3 MJ approach ignition symmetry specifications, demonstrated ion temperature of 3.7 keV, a main fuel ρr of 1 g/cm^2 and low mix (30% of clean 2-D calculated yield).

350 km/sec have been achieved. By moving to a lower adiabat implosion (modifying the detailed pulse shape and shock timing) and operating at 1.9 MJ and 500 TW, it is anticipated that the implosion velocity and the main fuel pr will increase to the ignition levels of 2–$2.5\,g/cm^2$. The DT ignition campaign will begin in the fall of 2011, and fusion yields of 10 to 25 MJ are expected by the end of 2012.

Extending NIF ignition design to IFE. Although the fusion performance for NIF-ignition style targets would be adequate for IFE, the Au or U hohlraums are not suitable both from the perspective of cost and half-lives of activation levels. Additionally, the technique used to form the solid DT layers require cool-down time of 10–14 h and place severe constraints on high repetition rate (~15 Hz) target production. Using Pb for the hohlraum material solves the cost and activation issues, and wicking liquid DT into a low density (~20 mg/cc) nanoporous carbon foam shell allows rapid fill and freeze options. However, these design choices reduce target performance. Use of Pb for the hohlraum reduces conversion of laser light to X-rays by 5%–10%, and use of the low-density foam may reduce the yield by ~10%.

However, by changing the cylindrical shape of the hohlraum to a "rugby-like" shape while maintaining the overall length and the diameter at the central point and placing small shields between the capsule and the laser entrance hole to reduce X-ray losses and changing the "X-ray environment" in the hohlraum (see Fig. 9), we can not only compensate for these performance losses but also actually improve the overall target performance. The benefit of the rugby shape has been verified in recent joint LLNL/CEA experiments at the Omega facility, which measured an

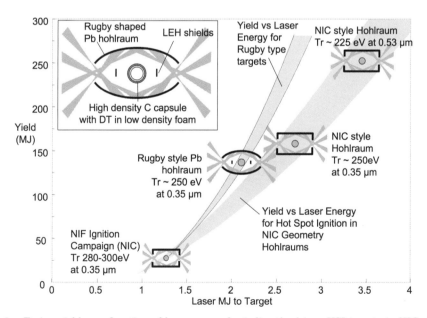

Fig. 9. Fusion yield as a function of laser energy for indirectly driven HSI targets in NIC-style and rugby-style hohlraums optimized for IFE. The laser wavelength used to drive the hohlraums is shown. The width of the band indicates uncertainty in hohlraum performance.

increase of 20% in the coupling efficiency. This target option is also the baseline
design for the ignition program planned for the LMJ facility in France in about
2016. Overall coupling efficiency increases of ~15% are calculated for LEH shields.
Experiments to verify these results at ignition scales and energy levels are planned
for 2012. The overall potential net improvement in target performance is also shown
in Fig. 9.

Once gain with NIC-based targets have been demonstrated, variants such as
the Rugby-type targets optimized for IFE systems are also likely to achieve ignition
and gain with high confidence and will be evaluated. By optimizing target and
laser performance, fusion yields of 130–150 MJ with laser energies of 2–2.2 MJ and
net energy gains of 60–70, adequate for inertial fusion energy systems, could be
demonstrated a few years after ignition.

4 IFE Systems

4.1 *Review of IFE Basics*

Once high-gain ICF target performance is available, the extension to an IFE power
plant is — in principle — relatively straightforward. (In this introductory discussion,
we assume laser IFE with indirect-drive HSI targets.) Referring to Fig. 10, the target
is injected into a chamber at a repetition rate ω (Hz). The laser with energy E_d
(also at ω) is focused on the target producing the fusion yield E_F and a fusion power
of $E_F \omega$. (The chamber is shown with just one opening. In reality, the target would
be injected into a separate port, and there would be 48 laser ports arranged in a
NIF-like configuration.) The target gain G is E_F/E_d. With an indirect-drive target,
~20% of the fusion output comes in the form of ions and X-rays (~10% each). The

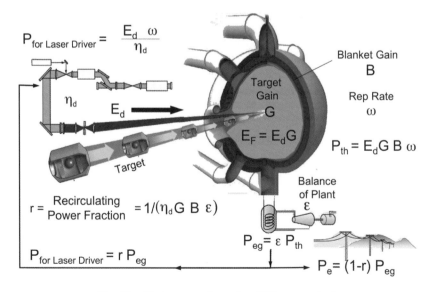

Fig. 10.   Basic parameters of an IFE power plant.

chamber is filled with Xe at a density of $\sim 6\,\mu\text{g/cc}$. This gas fill will stop 100% of the ions within a few tens of centimeters and approximately 90% of the X-rays but is low enough to have no impact on the frozen DT layer of the capsule, since it is protected by the thermally robust hohlraum and thin IR reflecting membranes (Fig. 9). The remaining X-rays are absorbed in the first wall.

After passing through the first wall and the first wall coolant, the remaining neutrons are stopped in a flowing liquid blanket. The coolant in the first wall as well as the blanket can be liquid lithium, lithium-bearing liquid metals, or lithium-bearing molten salt. The neutron–lithium interactions produce the T needed for the DT targets. There is also a net chamber energy gain (often called the Blanket Gain, B). It is defined as the ratio of the sum of the nuclear heating (neutrons and neutron-induced gamma-rays), X-ray heating, and ion heating to the initial energy of 17.6 MeV that is released from every fusion reaction. The thermal power generated is therefore $P_{\text{th}} = E_d\, G\, B\omega$.

By running the first wall coolant and the blanket through a heat exchanger and using the heat to drive a generator with an overall thermal to electric conversion efficiency ε, the gross electric power generated is $\varepsilon P_{\text{th}}$. A fraction of the electric power generated is required to run the driver. If the overall "wall-plug" efficiency of the driver is η_d, the recirculating power fraction r is

$$r = (E_d\,\omega/\eta_d)/(\varepsilon P_{\text{th}}) = (E_d\,\omega/\eta_d)/(\varepsilon E_d\,GB\,\omega) = 1/(\eta_d\,GB\,\varepsilon) = 1/G_{\text{eff}}, \quad (7)$$

where $G_{\text{eff}} = \eta_d\,GB\,\varepsilon$ is the effective system gain, and the net electric power available

$$P_e = (1-r)\varepsilon P_{\text{th}} = (1-r)E_F B\,\varepsilon\,\omega. \quad (8)$$

(The small fraction ($\sim 2\%$) of gross electric power required to run auxiliary "Balance of Plant" equipment is — for simplicity — not included above.)

IFE metrics. Common IFE wisdom has been that the target gain G should be > 100, and that $\eta_d\,G$ is the relevant metric for judging IFE systems, should be >10. Since for optimally designed systems $B\,\varepsilon$ is ~ 0.5, this means that $\eta_d\,GB\,\varepsilon \geq 5$.

Figure 11 shows $(1-r)$, or the fraction of electric power that can be sold, as a function of $\eta_d\,GB\,\varepsilon$ or G_{eff}. For this example, we chose $B = 1.25$, $\varepsilon = 0.47$ (values expected for properly optimized IFE systems) and assumed $\eta_d = 0.15$, expected for a diode-pumped solid-state laser (DPSSL).

A value of $G_{\text{eff}} < 3$ (and certainly < 2.5) is not economically interesting and is also too sensitive to small system changes. Similarly, once G_{eff} values of 5 (and certainly 6) are available, further increases in G_{eff} is not necessarily optimum from an overall systems perspective, (e.g. pushing the driver technology and/or operating parameters to a point where the risk of damage is increased or maintainability decreased just to gain a small increase in η_d).

Clearly, for given values of η_d, B, and ε, the higher the gain, the smaller the recirculating power fraction and the more of the gross electric power can be sold.

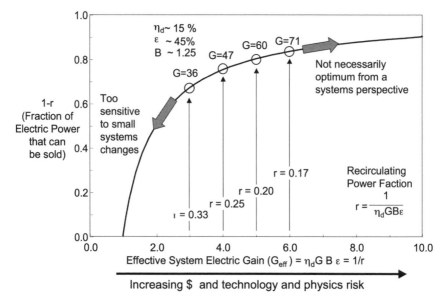

Fig. 11.   A target gain $G > 100$ is obviously useful, but not required.

However, we note that for this example (indirect-drive HSI targets and DPSSL), gains of 50 to 60 are more than adequate. Additionally, if any extra gain (and hence reduction in r) is "bought" at the expense of a larger driver (higher cost) or a target concept that has higher risk, or is incompatible with practical chamber options, a full systems evaluation is required to see if the higher G is "worth it."

There is a considerable variation in target performance for the different target options. There is also a 2X difference in the efficiency between KrF and DPSSL and another 2X difference between DPSSL and HI drivers. However, the thermal-to-electric conversion and net chamber/blanket gain will essentially be the same for all reasonably optimized IFE systems. This allows a comparison of the possible operating space for IFE driver/target options that provide $1{,}000\,\mathrm{MW_e}$.

Referring back to Fig. 10, we can show that:

$$E_F = P_e/(\varepsilon\, B\, \omega) + E_d/(\varepsilon\, B\, \eta_d). \tag{9}$$

Figure 12 shows the result for $P_e = 1{,}000\,\mathrm{MW_e}$, for $\omega = 7.5$, 15 and $30\,\mathrm{Hz}$, and for three typical driver efficiencies ($\eta_d = 0.075$ representative of Krf, $\eta_d = 0.15$ representative of DPSSL and $\eta_d = 0.30$ representative of HI accelerators), where we have assumed that all systems have:

- Thermal-to-electric conversion efficiency $\varepsilon = 0.47$
- Net chamber/blanket gain $B = 1.25$.

(The conclusions will not change if more conservative values are assumed.)

Also shown are the E_F versus E_d curves for the various targets discussed in Secs. 2 and 3. The intersections of the target performance curves with Eq. (11) show

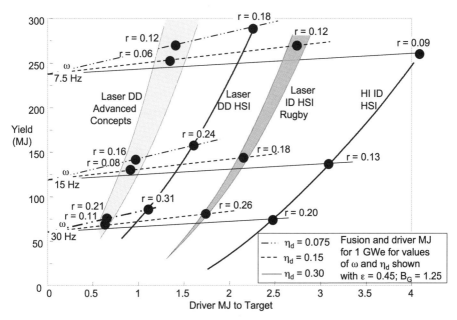

Fig. 12. Recirculating power fraction for 1,000-MW$_e$ IFE plants for various target and driver options for driver efficiencies typical for KrF, DPSSL and HI accelerators for blanket gains and thermal-to-electric conversion efficiencies expected for optimized IFE plants.

operating points for 1,000 MW$_e$ with the specific target and driver combinations. The "circles" show the recirculating power fraction r for the various target and driver combinations. We show direct-drive advanced concepts for both KrF ($\eta_d = 0.075$) and DPSSL ($\eta_d = 0.16$), since the less stringent implosion requirements for these targets would likely allow implosions driven with both 0.25-μm and 0.35-μm lasers, whereas only 0.25-μm KrF laser are assumed for the direct-drive HSI targets.

Although certain combinations have quite low recirculating power fractions, all the systems have $G_{\text{eff}} > 4$ (except for the 30-Hz direct-drive HSI system with KrF, which still has an acceptable $G_{\text{eff}} = 3.2$).

The 30-Hz systems were included on Fig. 12 to make a point. Proponents of shock and fast ignition targets for IFE, argue that they have acceptable gains (>100) at low driver energies. However, an IFE system needs a minimum of 2,000 MW of fusion power for 1,000 MW$_e$. (It is beyond the scope of this chapter to go into the details, but IFE systems do not scale economically favorably to small, $\sim$500 MW$_e$ plant sizes). Even with a gain 100 target at 0.5 MJ (at the upper range of predictions) a 40-Hz operation would be required. An examination of the constraints between target injection speed, the impact of acceleration on the integrity of the DT ice layer, target "fratricide," chamber clearing, it is difficult to find a self-consistent solution for repetition rates >20 Hz. Thus, even if the postulated gains from shock or fast ignition targets could be realized, they would still need to operate in the 10–15 Hz regime, or with drivers >1 MJ.

Clearly, neither G by itself, nor $\eta_d G$ or $G_{\text{eff}} = \eta_d \, GB \, \varepsilon$ is a sufficient metric to choose between alternate IFE systems. Rather it is self-consistent target-chamber-driver physics and technology choices and combinations that matter. This is the subject of the next subsections.

4.2 *Review of IFE Subsystems — Targets, Driver, Chamber, Balance of Plant*

An IFE plant has essentially four separate "sub-systems": a target system (fabrication, injection, tracking), the driver, a chamber where the fusion reactions occur, and the "Balance of Plant" to generate electricity.

IFE Targets. The two types of HSI targets, indirectly and directly driven, as well as the advanced target concepts, such as fast ignition and shock ignition (primarily directly driven) were covered in detail in Secs. 2 and 3.

The **IFE target system** must deliver of the order of 10^6 targets per day to the center of the target chamber. The maximum allowable target cost will be a function of the cost of electricity. With today's prices, this results in a maximum allowable cost of US$0.50 per target for a 1,000 MW$_e$ IFE plant. By using low-cost materials (e.g. Pb for high-Z materials and CH or high-density C for the capsule) and adopting conventional low-cost, high-throughput, fully automated techniques several studies have estimated ICF target cost to be from about US$0.17 (simple direct drive capsules) to US$0.45 (HI-driven indirect-drive targets).

A combination of high-speed (200–300 m/sec), high-precision injection and tracking must be capable of determining where the target is and deliver the focused laser energy to an accuracy of a few tens of micrometers, or a few microradians with the anticipated stand-off distance for the final optics. The target system must also include recovery and recycling of the imploded target material residue. Significant progress has been made in recent years with the construction of prototypical gas-gun and induction injector and tracking systems at General Atomics. Experiments with room temperature surrogate achieved injection velocities of $\sim$250 m/sec at up to 20 Hz (batch mode operation), with repeatable accuracy of delivery to chamber center of $\sim$100 μm.

The **IFE driver** must be able to deliver several MJ in a 10–20 ns precision-shaped pulse with peak powers of 400–500 TW and intensities up to 10^{15} W/cm^2 to millimeter-sized ICF targets. For certain target types, ignition pulses with intensities of 10^{20} W/cm^2 in $10-20 \times 10^{-12}$ sec (ps) are also required.

Lasers are attractive because of their demonstrated ability to execute ICF experiments. The flash-lamp–pumped, frequency-converted glass lasers used in most ICF facilities around the world have been most intensely studied and developed. However, efficiency and repetition rate limitations make them unevitable for IFE. To overcome these limitations, research is being carried out on diode-pumped solid-state lasers (DPSSL) and krypton–fluoride (KrF) gas lasers.

The laser medium in a **KrF laser** is a gas that can be circulated for heat removal, making it possible to achieve repetition rates of 5–10 Hz. High rep-rate operations have been tested using the Electra facility at NRL.

KrF lasers operate at a shorter wavelength (0.248 μm) than the typical frequency-tripled wavelength (0.35 μm) of the DPSSLs and is the preferred laser for direct-drive HIS targets. Since direct-drive targets also have a higher coupling efficiency than indirectly driven targets, direct drive with KrF lasers have calculated target gains ~2X higher than indirect-drive HSI targets (Fig. 6). This advantage is offset by lower driver efficiency of 7%–7.5%. R&D is ongoing to develop a high beam uniformity required for optimum laser target physics, the required intensity on target, a modular architecture, and a pulsed power-based industrial technology that scales to a power plant–sized system.

DPSSL is built using the ICF glass laser technology but use diodes instead of flashlamps to pump a solid-state medium, dramatically reducing the cool-down time needed between laser pulses. Further improvement in repetition rate has been achieved using new, compact laser architectures and helium gas cooling. The Mercury laser at LLNL is a 10-Hz prototype DPSSL operating at fluence and surface heat-loads (W/cm^2) and amplifier thermal stress loads relevant for IFE. A current DPSSL laser point design for IFE applications (Sec. 5) has a predicted efficiency of ~18% (electrical-to-frequency converted optical output).

Final optics presents a durability issue for both KrF lasers and DPSSLs. In both systems, these optics must survive the high-intensity ultraviolet laser beams and the debris, neutrons, and X-ray radiation from the fusion target. One approach uses grazing incidence metal mirrors such as aluminum-coated silicon carbide. Multilayer dielectric mirrors offer another potential option, and thin, transmissive fused silica gratings are the preferred option for DPSSLs.

HI drivers for fusion share the same basic technology as existing accelerators used for a range of scientific and engineering pursuits. This experience indicates that an HI driver could meet the IFE efficiency, repetition rate, and durability requirements. (Efficiencies of ~30% are projected for HI drivers for IFE applications.)

Magnetic lenses outside the target chamber would focus the HI beams on the target. The penetration of heavy ions into dense matter is greater than ultraviolet or X-ray photons. This feature yields more efficient energy coupling to the target. It also allows more efficient penetration through the higher vapor pressures if liquid thick-walled chambers were used (see target chamber discussion below).

A major challenge for the HI approach is to demonstrate the ability to focus 10 kJ level beams onto a target with the required short-pulse length (few nanoseconds) and small spot size (of few millimeters radius). To manage the significant excess electric charge of the ions, US conceptual designs use induction accelerators with many beams in parallel and that of European plan to accumulate charge gradually in a series of storage rings. Both approaches also require the beam duration to be severely reduced from its initial value, by about three orders of magnitude for induction machines and six for radio-frequency accelerators, and R&D is primarily directed at meeting these stringent requirements.

The basic functions for all **IFE chamber** concepts are similar. The structure must maintain a very low density in the central cavity where the target is injected and ignited. The fusion target releases energy in the form of high-energy neutrons, X-rays, and energetic ions. Some chamber designs include mechanisms to attenuate the ions and the X-rays, but the first structural wall, must (for safety and economic considerations) be designed to have a "life-time" measured in years. Various **first wall** designs have been proposed:

- Dry wall, (with or without gas fill) where the first wall is a solid structure designed to handle the fusion power. This is the simplest concept, but in the absence of a moderate density high-Z gas fill $(5–10 \, \mu g/cm^3)$ ion spall damage to the first wall is a serious concern and may severely limit the first wall lifetime, or require very large radius chambers.
- Wetted wall, where a thin liquid layer coats the first wall and absorbs the short-range X-rays and ions before they can damage the wall. However, the repeated cyclical stress of the shock-wave transmitted by the ion and X-ray–heated liquid is a serious issue.
- Thick ($> 50 \, cm$ lithium-bearing) liquid wall that flows between the target and first structural wall and provides protection from X-rays, ions, and neutrons. A drawback is the need for low-incidence angle driver beam illumination (half angle $< 20°$), eliminating this chamber concept for direct-drive targets, as well as laser indirect-drive targets, and the relatively low repetition rate ($< 7 \, Hz$) to restore the flow of the thick liquid wall between shots.

Not only must the chamber/first wall "manage" the fusion output, it must also effectively manage the intrashot recovery — the conditions inside the chamber (such as any vapor and droplet density) that must be recovered between each shot so that the next target can be injected and the laser or HI beams can propagate through the chamber to the target. The chamber and all of its sub-systems need to be designed using materials that avoid the necessity of high-level waste disposal.

A **blanket**, at or just behind the first wall, converts the energy pulses into a steady flow of high-grade heat and breeds the T to continue to fuel the IFE plant. To accomplish these two functions, the blanket has to be thick enough to slow down and absorb the neutrons to extract their energy, and it must contain lithium to react with those slowed neutrons to create the tritium. The X-rays and ions from the target would also heat the blanket (either directly or indirectly by heat transfer from any gas or liquid that attenuates/absorbs them). When lithium bearing liquids are used for tritium breeding, the liquid is generally circulated as the primary coolant for the fusion chamber. When solid breeders such as lithium oxide are used, high-pressure helium serves as the chamber coolant.

The blanket must operate at temperatures $>500°$ C to achieve high efficiency in the **power conversion system**. With 600°C and current super critical steam technology, this translates into thermal to electric conversion efficiencies of 45%–47%, and 52%–55% at temperatures of 700–720°C are projected within the next 10 years. More advanced technologies such as supercritical CO_2 or closed He Brayton cycles offer efficiencies close to 60%.

4.3 *Self-Consistent IFE Systems*

From the description of the various target, driver, and chamber options, one could construct the matrix of possible IFE systems shown in Table 3.

Upon more detailed examination of the target–driver–chamber options, it is clear that neither are all combinations physically realistic nor are they equally attractive or practical. An IFE system must not only be composed of self-consistent sub-systems but also respond to realistic physical constraints. If we also want an IFE system that could respond to the need for low carbon energy sources for baseload power plants (GW_e power levels) within a few decades, we require technologies that can credibly be ready at the required scale within this time frame. The impact of consequences imposed by target–driver–chamber self-consistency constraints are examined below.

Target constraints imposed by self-consistency considerations. High-gain ICF requires that the DT fuel be a cryogenic solid shell inside the capsule ablator. For direct-drive targets, the chamber pressure must be < 10 mTorr, for the solid DT layer to survive the transit. (With a thin IR reflective coating, pressures up to 50 mTorr may be possible). About 10 Torr-m of Xe is required to range out 3.5 MeV α-particles, so at these low pressure the ions from ~2,200 MW of fusion (required for 1,000 MW_e) would result in an erosion (spall) of ~1 cm/year for the first wall for a 15-m-radius chamber, and a chamber radius >100 m would be required to sufficiently attenuate the αs and maintain the integrity of the ice layer. Although physically possible, this is not a practical option.

With a density of 5 $\mu g/cm^3$ of Xe (such as could be used for indirect-drive targets), the αs would range out in 10–15 m, but the capsule would now need to be protected. Proposals have been made to inject the capsule in a protective sabot, which would separate and release the capsule shortly before the laser beams hit the target. However, the Xe gas would heat up to >1,000 K, and even with a sabot releasing a few centimeters away, the DT layer would not survive.

Nanoengineered first wall materials to mitigate the erosion or magnetic deflection schemes to keep the ions away from the first wall have been postulated and,

Table 3: Potential IFE System Target, Driver, and Chamber Options.

ICF target type	Driver	Chamber
DD HSI	KrF	Dry Wall w/gas fill
DD Fast/Shock Ignition	DPSSL	Wetted Wall
ID HSI	Heavy Ion	Thick Liquid Wall
ID Fast Ignition		

if achievable, could allow a vacuum dry wall chamber for direct-drive targets. But since the ~2X higher calculated gain direct drive has over indirect drive targets would be offset by the 2X lower efficiency for the 0.25-μm KrF laser required, it is not clear whether the R&D investment is warranted.

With the significantly higher gain calculated for advanced direct drive options (fast or shock ignition), coupled with the fact that higher efficiency DPSSLs could probably be used, it could be argued that an R&D effort to develop advanced first wall options might be justified. However, these target concepts are still at a very early stage of investigation, with complex physics and engineering issues still unsolved and are yet to go through the complex set of tradeoffs and technology developments the driver and targets need to actually achieve ignition and gain. It is not clear what level of the postulated high-gain performance, if any, will remain after those tradeoffs have been made for these concepts, or how long such an R&D effort would take. (In fact, as mentioned in Sec. 3, the HiPER program has recently abandoned the fast ignition concept in favor of shock ignition for just these reasons.)

When the direct-drive HSI targets as well as the advanced ignition options are viewed in the context of self-consistent and realistic target and target/chamber systems, we judge only indirectly driven HSI targets meet the requirements for a practical fusion system.

We now consider **driver constraints** for indirectly driven targets. DPSSLs are obviously acceptable being an extension of the Nd:glass laser technology used for the majority of ICF experiments over the past 40 years. With diodes replacing the flashlamps and He gas cooling, DPSSLs are capable of operating at 10–20 Hz, and projected "wall-plug" efficiencies (allowing for the cooling) of ~16% are more than adequate for IFE.

Even though the relatively low efficiency of KrF would make it less attractive for IFE than DPSSL systems with indirect-drive HSI targets, the intrinsic limit of ~7.5% would not, in and of itself, rule out this combination. However, the relatively low saturation intensity of KrF lasers limits the intensity out of the KrF amplifiers to about 10 MWcm^2 at pulselengths of interest to ICF (this is to be compared with output intensities in the multi GW/cm^2 range for DPSSLs). Higher output intensities are possible for KrF, but only at the cost of reduced overall efficiency. Intensities ~10^{15} W/cm^2 are required both to focus the laser beams through the LEH of the indirect-drive hohlraum and to obtain the required radiation temperatures. Additionally for IFE, these intensities must be provided by final focusing optics that are more than 20 m away to survive the target neutrons and X-rays. The combination of low efficiency and the constraints imposed by the intrinsic brightness of KrF lasers make them (in our judgment) less attractive than DPSSLs for indirect-drive HSI targets for IFE.

HI accelerators are projected to meet the efficiency, repetition rate, and durability performance required for IFE applications, and they are quite well suited for indirect-drive targets. However, even after about 40 years of R&D, there are still

Table 4: Not All Combinations are Physically Realistic or Self–Consistent.

ICF target type	Driver	Chamber
DD HSI	KrF	Dry Wall w/gas fill
DD Fast/Shock Ignition	DPSSL	Wetted Wall
ID HSI	Heavy Ion	Thick Liquid Wall
ID Fast Ignition		

Note: Dark gray: not self-consistent IFE options. Light gray: unlikely, but cannot be definitely ruled out as long-term options.

no ICF experiments that have been performed at any relevant ICF energies and intensities. Thus, heavy ion accelerators may be an attractive driver for indirect-drive HSI targets some time in the future, but technology is many decades behind lasers for IFE applications and, in particular, DPSSLs.

Chamber constraints. Thick ($> \sim 50$ cm) liquid wall chambers would allow actual structural first wall to survive the ~ 50 years of full plant-life time. A drawback, however, is the requirement of low-incidence angle driver beam illumination (half angles ideally less than $\sim 20°$). This eliminates this chamber concept for rugby-style laser indirect-drive targets.

Wetted wall chambers, where a thin liquid layer coats the first wall and absorbs the short-range X-rays and ions before they can damage the wall, are an interesting concept, but the repeated cyclical stress of the shock-wave transmitted by the ion and X-ray–heated liquid is a potential issue for first wall lifetime.

With these constraints and practical limitations, as well as the goal of IFE systems that could respond to the need for low carbon options for baseload power plants within a few decades, the possible combinations shown previously in Table 3 are reduced to the subset shown in Table 4.

In summary, indirectly driven advanced ignition targets (yet to be defined) or indirect-drive HSI targets with heavy ion drivers may be attractive options in the future. However, we judge that indirect-drive central hotspot target driven by a DPSSL at ~ 2.2 MJ at ~ 15 Hz and a dry wall chamber filled with $\sim 6\,\mu$g/cc of Xe, has adequate IFE system performance for a 1000-MW$_e$ IFE system. Section 5 describes such a Laser IFE system.

5 Progress Toward Laser IFE Technologies

A fusion power plant must meet a number of top-level requirements consistent with commercial operation. These include standardized proven technology, maintainability, and constructability; a high level of quality assurance; competitive economics; and environmental sustainability. The complex set of interrelated performance requirements presents major challenges to demonstrating IFE as a commercially

Fig. 13.   A 1,000-MW$_e$ laser fusion plant. The vacuum chamber with the fusion chamber is in the center of the drawing. The two other vacuum chambers show how they would be removed for rapid replacement of the first wall. The laser bays are the circular areas above and below the chamber. The heat exchange systems are to the left. The target "factory" is behind the chamber.

attractive energy source. Fortunately, great advantage can be taken from the inherent "separability" of the IFE subsystems and plant components. Once ignition and gain has been achieved, the laser IFE system described below would build on demonstrated ICF physics and credible extensions of current driver and materials technologies, and would allow a path forward for IFE.

The laser IFE power plant system (Fig. 13) comprises a 16-Hz DPSSL with a "wall-plug" efficiency of approximately 16%, a target factory, a target chamber surrounded by a lithium blanket to convert the fusion power to thermal power and also breed the T needed, and the balance of the plant (heat exchange and thermal to electric conversion systems). The system is designed for ICF gains of about 60 and fusion yields of about 135 MJ to provide 2,100 MW of fusion power. With a blanket gain of 1.25 and a supercritical steam cycle thermal to electric conversion efficiency of ~47%, this laser IFE plant delivers a net electric output of 1,000 MW$_e$.

Laser-driven inertial fusion energy target. The indirect-drive HSI targets and fusion yield as a function of laser energy as well as the rugby Pb-hohlraum adaptation that will be used for this laser IFE system were described in Secs. 1 and 2 and will not be further discussed here.

Target chamber and thermal-to-electric conversion system. Figure 14 shows a model of the vacuum vessel with the first wall, blanket and support structure (these combine to form "the chamber") sitting inside. The chamber consists of eight identical sections, which would be factory built and shipped to the power plant site. Liquid lithium is the primary coolant for both the first wall and blanket. The two systems are independently plumbed to allow greater flexibility in optimizing

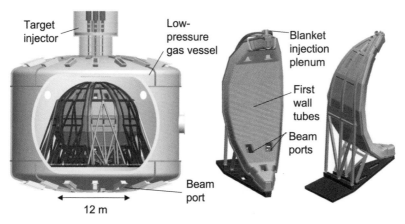

Fig. 14. The chamber has eight identical modules assembled into $1/4$ -sections for transport to the target bay. Modules are made of steel tubes mounted to coolant plena on the sides of the blanket.

flow rates and coolant temperatures. Lithium is a low-activation coolant that offers excellent tritium breeding capability.

The chamber structural "first wall" would be made of 10-cm-diameter, 1-cm-thick tubes of ferritic martensitic steel such as 12YWT or oxide-dispersion strengthened ferritic steel (ODS-FS) (Fig. 14). A pipe-based first wall was chosen, in part because of the high strength-to-weight ratio and ease of fabrication. The void swelling lifetime of ferritic-martensitic steels is likely to be more than 100 dpa or > four full power years for a 1,000-MW$_e$ system, at which point the first wall sections would be replaced. The modular construction and assembly makes it possible to replace first wall/chamber sections in less than a month, and treat the replacement of first wall/chamber sections as a maintenance process. The deciding metrics are that the cost of the replacement and impact on the plant availability be economically acceptable. The first wall/chamber concept lends itself to this "exchange philosophy" and has a negligible impact on the cost of electricity. The target chamber and beam path will be filled with xenon at about 6 μg/cm^3 to absorb ~90% of the X-ray energy and range out the ions emitted by the target. The hot gas will then cool via radiation over hundreds of microseconds, sufficiently long to prevent damage to the first wall tubes. For a 135-MJ, 16 Hz, Rugby target and a 6-m-radius chamber, the first wall tubes will only experience a 250°C temperature spike. The gas will be pumped out between shots through the laser beam entrance ports. A clearing ratio of just 1% per shot is sufficient to remove target debris for disposal or possible recycling.

An advantage of the inherent "separability" of LIFE subsystems is the flexibility it allows in the choice of design and materials of the "first wall." The concept for a commercial LIFE chamber described above would allow a 4- to 5-year replacement time for the first wall and blanket sections for a 1000 MWe LIFE system and eliminates the need for geologic storage of end-of-use chamber materials. By careful attention to the levels of impurities in the ODS-FS (e.g. terbium and cobalt), the

Table 5:   Laser and Target Illumination Parameters for NIF and a Laser IFE System.

Parameter	NIF	IFE Laser
Laser Pulse Energy/Peak Laser Power	1.8 MJ/500 TW	2.2 MJ/500 TW
# of Beamlines/# Laser Ports	192/48	384/48
Port-to-port energy variation	<4% rms	<4% rms
Beam pointing error	<50 μm rms	<100 μm rms
Repetition Rate	> three shots per day	16 Hz
Electrical-to-3ω-Optical Efficiency (w/o cooling)	—	>18%
Laser-System Availability	>98.9% of shots	>99% of time
Lifetime	>3 × 10^4 shots	>3.0 × 10^{10} shots

first wall and blanket sections would immediately qualify as low-level waste (shallow land burial) by the end of its life. Unshielded waste containers could be disposed of via remote handling. If contact handling was desired, this could be achieved either by using shielded waste containers or allowing for an interim storage time of about 20–50 years. The final configurations (choice of levels of impurities, chamber radius, replacement time, design of waste drums to contain first-wall materials, choice of manual or remote handling of drums) and cool-down time to allow manual handling would be an operations/cost tradeoff.

This chamber design produces sufficient tritium without the use of beryllium or lithium isotopic enrichment. The point design has a T breeding ratio (TBR) of 1.59 and a chamber blanket gain (B) of 1.10. Excess TBR can be traded for additional blanket gain, and a gain (including penetrations for beamports, target injection and pumping), as high as 1.25 still provides adequate TBR.

The chamber and energy conversion system blanket operates at 600°C and, with current super-critical steam technology, results in a thermal-to-electric conversion efficiency of 47%. Second generation systems with more advanced materials (such as ODS) could operate at 700–720°C and supercritical steam systems are projected to achieve electric conversion efficiencies of 52%–55%. More advanced technologies such as supercritical CO_2 or He Brayton cycles offer the possibility of thermal efficiencies close to 60%.

Laser systems for a 1000 MW$_e$ plant. In many ways, lasers such as NIF and LMJ are prototypes for an IFE laser. Table 5 compares key laser parameters.

The laser consists of a large number of independent beamlines, each using a multipass architecture to optimize performance. This choice would allow reuse of much of the technology and manufacturing base. However, the laser design differs in several respects. High average power operation is enabled by replacing passive cooling system with high-speed helium gas to remove heat from active components. To achieve baseline laser efficiency from electrical power to the frequency converted third harmonic laser light at 0.35 μm of ~18%, the flash lamps are replaced by laser-diodes. An increase in repetition rate by nearly five orders of magnitude results in average output power of order 100 kW per beamline. Many of the other technologies required for high-average-power DPSSLs, such as the thermal management of the

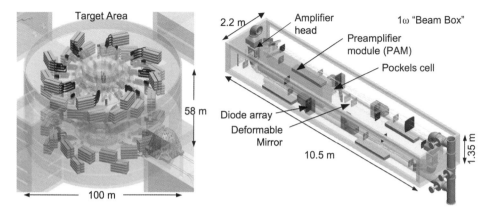

Fig. 15.   The "1ω beam boxes" and target chamber building.

optics, use of adaptive optics to correct thermal wavefront distortions and methods for harmonically converting high-average-power laser, have been demonstrated with LLNL's Mercury laser.

The main building block is the "1ω Beam Box," which generates 8 kJ of 1.05-μm laser energy with a "wall-plug" efficiency of 25% (Fig. 15). The 1ω beam box is the largest line-replaceable unit (LRU) within the laser system; however, it is small enough to be transported from factories to the power plant with trailers and to be handled gracefully by installation and removal equipment in the laser bay. Figure 15 also shows how 384 of these boxes are arrayed in a compact laser bay, and stacked in 4-high × 2-wide arrays arranged on circles centered on the chamber. The 1ω beam boxes and other LRUs can be inserted and removed without disturbing neighboring beamlines. Using this approach, the laser system can achieve high system availability (>99%) with moderate beamline lifetime (MTBF > 2000 h) and replacement time (4 h).

Final Transport Optics. The final transport optics represents a fundamental change relative to current ICF lasers. A set of pinholes in thick lead sheets transport the laser beam while filtering neutrons from the target chamber to levels acceptable for workers (~0.04 rem/y). The final optic, which is directly exposed to the output of the target, experiences the greatest threats. The ions and X-rays are absorbed by the xenon in the chamber and the beam tubes, so the main challenge is the 14-MeV neutrons with an average exposure of 1.5×10^{17} n/m^2 sec. The final optic must efficiently transmit the 0.35-μm laser light and allow high reliability operation, rapid replacement, and adequate mean-time-between-failure. The architecture that meets these requirements is a thin focusing grating (5-mm fused silica) that focuses and deflects the beam to target. Irradiation studies of fused silica indicate that the neutron-induced absorption saturates to acceptable levels and allows efficient transmission of the beam to the target. The relatively small size and weight of the optic enable its automated extraction and replacement.

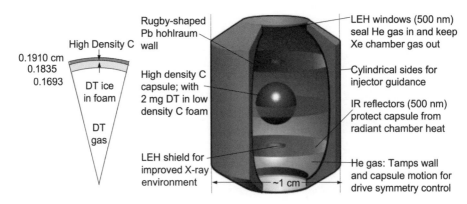

Fig. 16.   Details of the baseline indirect-drive IFE target.

Fusion Target Systems for Laser IFE. The primary challenge is to develop low cost fabrication techniques for the high-precision target shown in Fig. 16.

The manufacturing strategy is to use conventional low-cost, high-throughput, fully automated techniques and to increase the batch size of chemical processes to reduce cost and increase throughput. An example is the use of die-cast hohlraum parts and an ultra-large batch chemical vapor deposition (CVD) diamond coating process for the capsule ablator and membranes. The choice of materials involves the consideration of several factors: the implosion physics; material cost and availability; potential for developing a low-cost fabrication technique; thermal, mechanical and optical properties to withstand the forces of manufacturing and injection into the hot chamber; ability to recover the post-explosion debris; limiting laser propagation interference and enable recycling or disposal of waste products. By limiting the number of different low-cost materials used, the set of processes that must be developed can be reduced. The bulk material set can be reduced to lead and carbon-based (CVD diamond, and graphene oxide, DCPD) materials. DT is added to the capsule and small amounts of metals for dopants and reflectors.

Costs for the targets of Fig. 16 were estimated for each of the manufacturing processes. The numbers of machines for each process along with the floor space needed, the work for each process and the associated capital, consumables, and personnel cost have been estimated in consultation with industry. The contribution of each process to the total target cost was evaluated for a repetition rate (PRF) of 15 Hz, resulting in a total target cost of US$0.23. The total contribution of the materials to the overall target cost is small, at ~13%.

The integrated process of **target injection, in-flight tracking, and beam engagement** is a key component. The injector must accelerate the target to 250–300 m/sec and deliver it to the center of the chamber to ±500 μm, with a tilt of < 40 mrad in order for the laser to "hit" the target to the required ±100 μm. The target tracking and engagement system must track the position of the target to ±50 μm to allow the laser to point to the target to this ±100 μm requirement.

Electromagnetic steering would likely be used to improve the accuracy of the target trajectory. An optical tracking system would measure trajectory, velocity and tilt. The output of each laser beam to be focused on the target would be stabilized with a high-bandwidth pointing loop. The loop would be closed on an alignment beam co-aligned with the 15-Hz pulsed laser target beam. Rotating shutters would be used to shield the injector and steering equipment and especially the targets and the DT ice layer from radiation damage and heating from the fusion output from the previous target.

6 Conclusion

After nearly 40 years of R&D, IFE is at a key juncture. Demonstration of fusion ignition and gain will provide the basis for the transition from scientific research for developing and demonstrating the technology required for commercial IFE.

The science and technology for an integrated demonstration of an IFE system can be developed at the modular level in appropriately scaled facilities. The demonstration of optimized target gain will be done independently on NIF and other facilities such as the LMJ. Demonstration of mass production techniques for the targets at required precision and cost scalability can be done "off-line." Target delivery, tracking, and target engagement, as well as chamber clearing can be demonstrated with surrogate targets and low power lasers in separate facilities. The DPSSL technology will be demonstrated with a single 100-kW beamline. Such independent demonstrations could allow a sub-scale integrated performance of many of the Laser IFE systems by about 2020, a fully integrated, pre-commercial system by the mid-2020s and construction of commercial Laser IFE plants to begin by mid-2030s.

Further Suggested Readings

Fusion as Part of a Global Energy Strategy

A. Gibson and the JET Team, Deuterium-tritium plasmas in the Joint European Torus (JET): Behavior and implications, *Phys. Plasma* **5** (1998), pp. 1839–1847.

R. J. Hawrykuk, S. Batha, W. Blanchard *et al.*, Fusion plasma experiments on TFTR: A 20 year retrospective, *Phys. Plasma* **5** (1998), pp. 1577–1589.

Physics of Ignition and High Energy Gain

P. Chang, R. Betti, B. Spears *et al.*, Generalized measurable ignition criterion for inertial confinement fusion, *Phys. Rev. Lett.* **104** (2010), p. 135002.

M. J. Edwards, J. D. Lindl, B. K. Spears *et al.*, The experimental plan for cryogenic layered target implosions on the National Ignition Facility — The inertial confinement approach to fusion, *Phys. Plasma* **18** (2011), p. 051003.

S. Glenzer, B. MacGowan *et al.*, Symmetric inertial confinement fusion implosions at ultra-high laser energies, *Science* **327** (2010), p. 1228–1231.

S. W. Haan, J. D. Lindl, D. A. Callahan *et al.*, Point design targets, specifications, and requirements for the 2010 ignition campaign on the National Ignition Facility, *Phys. Plasma* **18** (2011), p. 051001.

O. L. Landen, J. Edwards, S. W. Haan *et al.*, Capsule implosion optimization during the indirect-drive National Ignition Campaign, *Phys. Plasma* **18** (2011), p. 051002.

J. D. Lindl, *Inertial Confinement Fusion* (AIP Press, Springer Verlag, 1998). (Has a wealth of references for all aspects of ICF target physics.)

J. D. Lindl and E. I. Moses, Special topic: Plans for the National Ignition Campaign (NIC) on the NIF: On the threshold of initiating ignition experiments, *Phys. Plasma* **18** (2011), p. 050901.

N. Meezan, L. Atherton, D. Callahan *et al.*, National Ignition Campaign Hohlraum Energetics, *Phys. Plasma* **17** (2010), p. 056304.

T. C. Sangster, V. N. Goncharov, R. Betti *et al.*, Shock-tuned cryogenic-deuterium-tritium implosion performance on Omega, *Phys. Plasma* **17** (2010), p. 056312.

E. Storm, J. D. Lindl, E. M. Campbell *et al.*, Progress in laboratory high gain ICF and prospects for the future, LLNL/UCRL-9942 (1988).

Key Requirements for an IFE Power Plant

R. Linford, R. Betti, J. Dahlburg *et al.*, A review of the US Department of Energy's Inertial Fusion Energy Program, *J. Fusion Energ.* **22** (2003), pp. 93–126.

Drivers

R. Betti, D. Hammer, G. Logan, D. Meyerhofer *et al.*, Advancing the science of high energy density laboratory plasmas, Fusion Energy Science Advisory Committee report (United States Department of Energy, January 2009).

KrF

P. M. Burns, M. Myers *et al.*, Electra: An electron beam pumped KrF Rep-Rate Laser System for IFE, *Fusion Sci. Tech.* **56** (2009), p. 349.

S. Obenschain, D. Colombant *et al.*, Pathway to a lower cost high repetition rate ignition facility, *Phys. Plasma* **13** (2006), p. 056320.

J. Sethian, M. Friedman and R. Lehmberg, Fusion energy with lasers, direct drive targets, and dry wall chambers, *Nucl. Fusion* **43** (2003), pp. 1693–1709.

DPSSL

A. Bayramian *et al.*, Compact, efficient laser systems for laser inertial fusion energy, *Fusion Sci. Technol.* **60** (2011), p. 28.

Y. Kozaki, Way to ICF reactor, *Fusion Eng. Des.* **51–52** (2000), pp. 1087–1093.

C. Orth, S. Payne and W. Krupke, A diode pumped solid state laser driver for inertial fusion energy, *Nucl. Fusion* **36** (1996), pp. 75–116.

HI Accelerators

B. Logan, L. Perkins and J. Barnard, Direct drive heavy-ion-beam inertial fusion at high coupling efficiency, *Phys. Plasma* **15** (2008), p. 072701.

W. Meier, A. Raffray, S. Abdel-Khalik, G. Kulcinski, J. Latkowski, F. Najmabadi, C. Olson, P. Peterson, A. Ying and M. Yoda, IFE chamber technology — Status and future challenges, *Fusion Sci. Technol.* **44** (2003), pp. 27–33.

P. Roy, S. Yu, E. Henestroza, A. Anders *et al.*, Drift compression of an intense neutralized ion beam, *Phys. Rev. Lett.* **95** (2005), p. 234801.

S. Yu, W. Meier, R. Abbott, J. Barnard *et al.*, An updated point design for heavy ion fusion, *Fusion Sci. Technol.* **44** (2003), pp. 266–273.

Fast Ignition

S. Atzeni, A. Schiavi and C. Bellei, Targets for direct-drive fast ignition at total laser energy of 200–400 kJ, *Phys. Plasma* **14** (2007), p. 052702.

N. Basov, S. Yu and L. Feokistov, Thermonuclear gain of ICF targets with direct heating of ignitor, *J. Sov. Laser Res.* **13** (1992), pp. 396–399.

M. H. Key, Status and prospects for fast ignition inertial fusion concept, *Phys. Plasma* **14** (2007), p. 055502.

R. Kodama, P. A. Noreys, K. Mima *et al.*, Fast heating of ultrahigh-density plasma as a step towards laser fusion ignition, *Nature* **412** (2001), p. 798.

M. Tabak, J. Hammer, M. Glinsky *et al.*, Ignition and high gain with ultrapowerful lasers, *Phys. Plasma* **1** (1994), pp. 1626–1634.

Shock Ignition

L. J. Perkins, R. Betti, W. H. LaFortune *et al.*, Shock Ignition: A new approach to high gain inertial confinement fusion on NIF, *Phys. Rev. Lett.* **103** (2009), p. 045004.

W. Theobald, R. Betti *et al.*, Initial experiments on the shock-ignition inertial confinement fusion concept, *Phys. Plasma* **15** (2008), p. 056306.

IFE Power Plant Components

W. Meier, A. Raffray, S. Abdel-Khalik, G. Kulcinski *et al.*, IFE chamber technology — Status and future challenges, *Fusion Sci. Technol.* **44** (2003), pp. 27–33.

A. R. Raffray *et al.*, Conceptual study of integrated chamber core for laser fusion with magnetic intervention, *Proc. 22nd IEEE/NPSS Symposium on Fusion Engineering* (2007), pp. 106–110.

S. Reyes, R. Schmitt, J. Latkowski and J. Sanz, Liquid wall options for tritium-lean fast ignition inertial fusion energy power plants, *Fusion Eng. Des.* **63–64** (2002), pp. 635–640.

S. Sharafat *et al.*, Micro-engineered first wall tungsten armor for high average power laser fusion energy systems, *J. Nucl. Mater.* **347** (2005), pp. 217–243.

Chapter 9

Energy from Photovoltaics

Ignacio Rey-Stolle

Instituto de Energía Solar, Universidad Politécnica de Madrid
(Solar Energy Institute, Technical University of Madrid)
ETSI de Telecomunicación, Avda. Complutense 30, 28040 Madrid, Spain
ignacio.reystolle@upm.es

This chapter summarizes the current state and perspectives of photovoltaic solar energy. The fundamentals of solar radiation and photovoltaic (solar) cells are covered; the basic technology of photovoltaic modules and systems is briefly discussed; and a general outlook is provided for the uses, market, and environmental impact of photovoltaic solar energy.

1 Introduction

The Earth receives annually around 1.5×10^{18} kWh of solar energy, which is by far the most abundant energy resource available for mankind. If adequately harnessed, only a minuscule fraction of this energy ($\sim 0.01\%$) would suffice to supply the world primary energy demand, which in 2008 was 1.4×10^{14} kWh.[1] The primary energy is processed by the energetic system into different forms of readily consumable energy, among which electricity is considered the key technology for the next decades. Accordingly, the direct generation of electricity — the preferred consumable form of energy — from solar radiation, which is the richest resource, is a topic of the highest relevance and is the essence of photovoltaics (PV). There was a gap of more than a century between the discovery of the photovoltaic effect in 1839 by French physicist Alexandre-Edmond Becquerel and the first successful application of photovoltaic panels to power the Vanguard I satellite launched in 1958.[2] However, in the last 50 years, the PV industry has evolved from Watt-ranged applications to the GW systems planned today. In the first decade of the 21st century, having installed more than 16 GW worldwide, PV technology has demonstrated the maturity to become a major source of power for the world. This robust and continuous growth is expected to continue in the decades ahead in order to turn PV into one of the key players in the pool of technologies involved in generating electricity for the 21st century.

2 Solar Radiation

2.1 *Fundamentals*

Solar radiation is a general term that refers to the electromagnetic energy flux emitted by the sun's surface (i.e. the photosphere). This emission of electromagnetic waves does not take place at a single wavelength but spans a continuum of different wavelengths from X-rays to deep infrared photons. Accordingly, the **solar spectrum** is the distribution of the electromagnetic power emitted by the sun (per unit wavelength and unit area) as a function of wavelength. The solar spectrum reaching the earth outside the atmosphere may be well approximated by that of a blackbody at 5778 K,[3] as shown by Fig. 1.

The integral of the solar spectrum in Fig. 1 yields the solar power per unit area reaching the outer surface of Earth's atmosphere. This magnitude is known as the solar constant (B_0) and its most accepted average value is 1,367 W/m². Throughout the year (and from year to year) the solar constant varies slightly ($\sim$7%) as a result of the change in the sun–earth distance along the Earth's orbit and variations in the sun's activity (solar cycles). The actual value of B_0 can be simply calculated just by multiplying the solar constant by a correction factor (ε_0), taking into account the eccentricity of the Earth's orbit as a function of the ordinal day in the year (d_n; for January 1, $d_n = 1$; while for December 31, $d_n = 365$):

$$\varepsilon_0 = 1 + 0.033 \cdot \cos\left(2\pi \cdot d_n / 365\right). \tag{1}$$

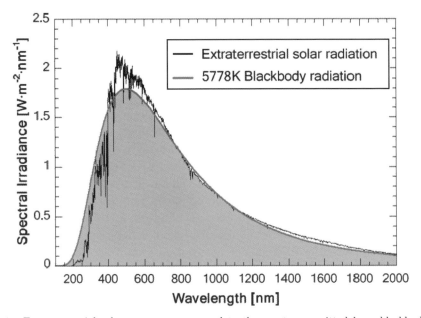

Fig. 1. Extraterrestrial solar spectrum compared to the spectrum emitted by a blackbody at 5778 K.

2.2 *Basic PV Terminology and Notation for Solar Radiation*

The irradiance at a given surface is the total energy per unit area per second (i.e. power per unit area) reaching that surface in the form of solar radiation. Units used for irradiance are W/m^2 and is typically noted using the letter G. The irradiation at a given location is the total energy per unit area reaching that location in the form of solar radiation in a certain period. Typical units used for irradiance are Wh/m^2 or J/m^2 and is noted using the symbol G_p, where the subscript p denotes the period over which the irradiation is calculated as

$$G_p = \int_p G \cdot dt. \tag{2}$$

Typical periods for calculating irradiation are hours (G_h), days (G_d), or months (G_m). It is also common to use averages of daily or hourly irradiations taken over different periods. In this case, no uniform notation is found in the literature but in general, subscripts are added to indicate the period for averaging (G_{dm} = average daily irradiation taken over a month).

2.3 *Components of Solar Radiation*

When the sun's radiation traverses the atmosphere it is partially reflected by the clouds, partially absorbed, and partially scattered by particles and gases. Thereby the radiation reaching the Earth's surface is largely attenuated, filtered, and randomized in terms of direction. Accordingly, the irradiance reaching a surface on Earth may be expressed in terms of three main components, namely, beam or direct irradiance (B), diffuse irradiance (D), and albedo or reflected irradiance (R):

$$G = B + D + R. \tag{3}$$

The beam component refers to the irradiance coming directly from the solar disc and the circumsolar ring, whilst the diffuse component refers to the irradiance coming from the rest of the sky. The albedo or reflected component accounts for radiation that reaches the target surface after reflection from the ground, buildings, snowy hills, or any other reflecting surface.

On clear days, when attenuation is the main effect of the atmosphere, the global irradiance on a horizontal plane on the Earth surface can be approximated with a very simple empirical expression:

$$G = B_0 \cdot \varepsilon_0 \cdot 0.74^{AM \times 0.678}. \tag{4}$$

where B_0 and ε_0 are the solar constant and eccentricity factor as defined in Sec. 2.1; whilst AM is the air mass, which is defined as the ratio of the length of the beam irradiance path through the atmosphere to the vertical length of the atmosphere. Accordingly, in PV terminology AM0 refers to the extraterrestrial irradiance; AM1.5 to an irradiance traversing an atmosphere length 1.5 times its vertical length, and analogous definitions can be given for other AM values. AM varies through the day and season (as apparent sun movement does) and depends on location (latitude). Simple geometrical considerations lead to[3]:

$$AM = 1/\cos\theta_{ZS}, \tag{5}$$

where θ_{ZS} is the solar zenith angle, which is itself a function of latitude, hour of the day, and day of the year.

Following Eq. (3), irradiation can also be expressed in terms of the different components of irradiance integrated over a period (p) of time:

$$G_p = B_p + D_p = \int_p B \cdot dt + \int_p D \cdot dt, \tag{6}$$

where the albedo component has been neglected since it is site dependent and thus it is typically measured.

2.4 *World Distribution of Solar Radiation*

Among energy resources, solar radiation is the one more evenly distributed on a world scale. Virtually any location with latitude below 55° is suitable for installing PV systems. Figure 2 visually summarizes the overall availability of solar resource in the world.

2.5 *Solar Radiation Collected by PV Systems*

In PV engineering, the key question regarding solar radiation is how much irradiance will reach the surface of a PV system i.e. the so-called in-plane irradiance. Following Eq. (3) to calculate the in-plane irradiance, it is necessary to model or to measure the beam and diffuse irradiances (and albedo when applicable) on the target surface for the desired location during the course of the day and throughout the seasons in a year.

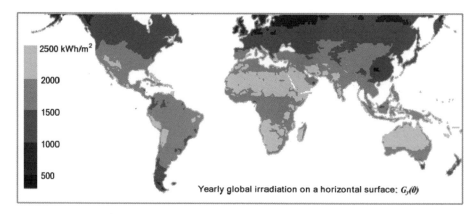

Fig. 2.   Yearly global irradiation (kWh/m^2) on a horizontal surface.

Many approaches exist for calculating the in-plane irradiance in PV systems[3-5] that provide the average irradiance and irradiation of arbitrarily oriented surfaces for diverse periods of time (hours, days, months, year). Another approach, with increasing use in PV engineering, is to obtain these values from reference databases that integrate complex solar radiation models with extensive data from ground stations or satellites. Examples of these databases are PVGIS[6] for Europe and Africa; and the NREL US Dynamic Solar Atlas.[7]

3 Solar Cells

3.1 *Definition*

A **solar cell** or **photovoltaic cell** is a device that directly transforms solar radiation into electrical energy by means of the photovoltaic effect without any thermal cycles, mechanical cycles, or chemical reactions.[8-10]

3.2 *Basic Solar Cell Equations and Equivalent Circuits*

3.2.1 *Simple equivalent circuit for a solar cell and I-V characteristic*

The simplest circuit model associated with a p/n junction solar cell is the one shown in Fig. 3. This circuit consists in the parallel connection of: (1) a current source modeling the photogenerated current (I_L); (2) a diode modeling the dark current through the p/n junction (I_D); (3) a parallel resistor (R_P) modeling possible shunts across the junction; and these three elements are connected in series with an additional resistor (R_S) modeling ohmic losses in the device (contacts, substrate, etc.).

According to this equivalent circuit, the I-V curve of a solar cell has the following expression:

$$I = I_L - I_0 \left[\exp\left(\frac{V + I \cdot R_S}{n \cdot V_t} \right) - 1 \right] - \frac{V + I \cdot R_S}{R_P}, \tag{7}$$

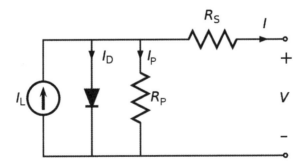

Fig. 3. Equivalent circuit of a solar cell with a single diode.

where I_L is the photogenerated current in Amperes; I_0 is the diode dark current in Amperes; n is the diode ideality factor; R_S is the series resistance in Ohms; R_P is the parallel or shunt resistance in Ohms; and V_t is the thermal voltage in Volts given by the well-known expression:

$$V_t = kT/q \Rightarrow V_t \cong 0.026\,\text{V for T} = 300\,\text{K}, \tag{8}$$

where k is the Boltzmann constant ($k = 1.3806 \times 10^{-23}\,\text{J/K}$); q is the elementary charge ($q = 1.6022 \cdot \times 10^{-19}\,\text{C}$); and T is the absolute temperature in Kelvin.

3.2.2 *General equivalent circuit for a solar cell and I-V characteristic*

For some solar cells, a single conduction mechanism across the p/n junction is not enough to model the dark losses. In this case, a circuit with multiple diodes in parallel is considered, where each diode accounts for a different mechanism (recombination in neutral regions, recombination at the space charge region, recombination at the junction perimeter, tunneling, etc). In such case, the new expression for the I-V curve is:

$$I = I_L - \sum_{i=1}^{n} I_{0i}\left[\exp\left(\frac{V + I \cdot R_S}{n_i \cdot V_t}\right) - 1\right] - \frac{V + I \cdot R_S}{R_P}, \tag{9}$$

where I_{0i} is the diode dark current for conduction mechanism i in Amperes; and n_i is the diode ideality factor for the ith conduction mechanism.

3.3 *The I-V Curve of a Solar Cell*

3.3.1 *General look and key parameters*

The I-V curve of a solar cell under illumination is schematically depicted in Fig. 4. This curve has three characteristic points, which are: (1) the **short circuit current** (I_{SC}) which is the current produced by the solar cell at zero voltage; (2) the **open circuit voltage** (V_{OC}), which is the voltage produced by the solar cell at zero current; and (3) the **maximum power point** or **MPP** (V_m, I_m), which is the point at which the power delivered by the solar cell is maximum (i.e. the product $V \cdot I$ reaches its maximum).

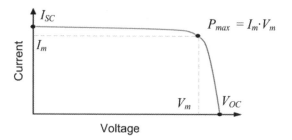

Fig. 4.   Solar cell I-V curve.

Table 1:   Key Parameters of a Solar Cell with Symbol and Units.

Parameter	Symbol	Units
Short circuit current	I_{SC}	A
Open circuit voltage	V_{OC}	V
Current at maximum power point	I_m	A
Voltage at maximum power point	V_m	V
Fill factor	FF	%
Efficiency	η	%

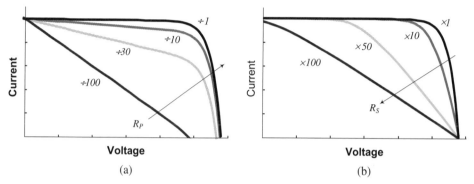

Fig. 5.   (a) Effect of decreasing the parallel resistance and (b) increasing the series resistance on a solar cell I-V curve.

Based on the curve in Fig. 4, it is possible to define a key figure of merit for a photovoltaic cell known as the fill factor (FF) where:

$$FF = \frac{I_m \cdot V_m}{I_{SC} \cdot V_{OC}}. \tag{10}$$

The efficiency of a solar cell is given by the ratio of the maximum power delivered to the load over the optical power received on the cell surface:

$$\eta = \frac{I_m \cdot V_m}{G \cdot A} = \frac{FF \cdot I_{SC} \cdot V_{OC}}{G \cdot A}, \tag{11}$$

where G is the irradiance on the solar cell in W/m^2; and A is the area of the solar cell in m^2.

Table 1 summarizes the key parameters of a solar cell with symbol and units.

3.3.2 *Effect of variations in series and parallel resistance on the I-V curve*

The effect of decreasing the magnitude of the parallel resistance or increasing the magnitude of the series resistance can be seen in Figs. 5(a) and 5(b), respectively. When the parallel resistance decreases, the solar cell becomes effectively shunted, and V_{OC} and FF of the cell decrease and the slope of the I-V curve changes markedly in the vicinity of I_{SC}, which remains unchanged. When the series

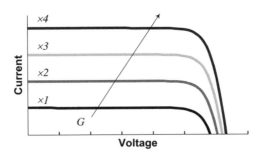

Fig. 6.   Effect of increasing the irradiance on a solar cell I-V curve.

resistance increases, the internal load in the solar cell increases, and the I_{SC} and FF of the cell decrease and the slope of the I-V curve changes markedly in the vicinity of V_{OC}, which remains unchanged.

3.3.3 *Effect of variations in irradiance on the I-V curve*

The effect of changing the irradiance impinging on a photovoltaic cell can be visualized as a vertical shift of the I-V curve (Fig. 6).

Let G_{STC} be a reference irradiance and G any other irradiance value. We can then define the concentration factor (X) (sometimes referred to simply as concentration) as the ratio between G and G_{STC}:

$$X = G/G_{STC}. \tag{12}$$

The evolution of I_{SC} and V_{OC} of a solar cell with the concentration factor (i.e. with irradiance) are rather simple:

$$I_{SC}(G) = I_{SC}(G_{STC}) \cdot X, \tag{13}$$

$$V_{OC}(G) = V_{OC}(G_{STC}) + V_t \cdot \ln X. \tag{14}$$

On the other hand, the variation of FF (and thus maximum power point and efficiency) with irradiance is more complicated. At moderate irradiance levels, when the effects of series resistance are negligible, the fill factor increases with irradiance, while at high irradiance level, series resistance losses counterbalance this effect and the fill factor starts to decrease with concentration.

3.3.4 *Effect of variations in temperature*

When the operating temperature of a solar cell increases (Fig. 7), I_{SC} increases slightly, V_{OC} decreases more significantly, and P_{max} also decreases. All these variations are quantified using three linear coefficients associated with I_{SC}, V_{OC}, and P_{max}. For silicon solar cells the I_{SC} temperature coefficient (α) is around

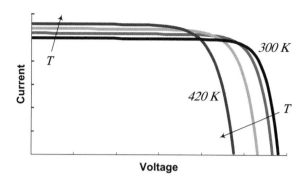

Fig. 7.   Effect of changing the operating temperature on a solar cell I-V curve.

$\alpha = 0.1\%/^\circ$C; the V_{OC} temperature coefficient (β) is around $\beta = -2\,\text{mV}/^\circ$C; and the $P_{\max}$ temperature coefficient (γ) is around $\gamma = -0.5\%/^\circ$C.

3.3.5 *Standard test conditions for solar cells*

In order to be able to compare the performance of different solar cells (and PV modules) in a consistent manner, efficiencies are reported using standard test conditions (STCs). These conditions are a device temperature of 25°C and an irradiance set to the value and spectral distribution as defined in reference spectra, included in ASTM standard G173-03.[11] The most frequently used spectra in PV are AM0 for space solar cells, AM1.5 global for conventional non-concentrating solar cells, and AM1.5 direct for concentrator solar cells.

3.4 *Overview of Solar Cell Technologies*

The core of any solar cell is a light-absorbing material, where the photons are absorbed and generate charge carriers (free electrons and holes) via the photovoltaic effect.[8–10] In the history of PV, many materials have been investigated — most of them crystalline semiconductors — in the quest for an optimum, low-cost, high-performance solution. Today the portfolio of materials used in PV is dominated by crystalline silicon, either monocrystalline or multicrystalline, though the use of the so-called thin film materials, which include hydrogenated amorphous silicon (a-Si:H), cadmium telluride (CdTe), and copper indium gallium diselenide (CIGS), is starting to gather momentum.[12–14] The III–V crystalline semiconductors are today limited to space applications though they might be a key player in concentrator pholtovoltaics (CPV).[15] Other PV materials, which are at present limited to laboratory research or marginal production, include dye-sensitized[16] and organic polymers.[17] Table 2 gives an overview of laboratory and commercial results for different solar cell technologies.[18, 19]

Table 2:   Overview Laboratory and Commercial Results for Different Solar Cell Technologies.

Technology Material	Silicon		Thin film				CPV
	Mono-c	Multi-c	a-Si:H	a-Si/ μc:Si	CdTe	CIGS	III–V
Record cell efficiency	25.0%	20.4%	10.4%	13.2%	16.7%	20.3%	43.5%
Commercial cell efficiency	16%–22%	14%–18%	4%–8%	7%–9%	10%–11%	7%–12%	37%–39%
Commercial module efficiency	13%–19%	11%–15%					25%–30%
Module area needed per kWp[a]	$\sim 7\,m^2$	$\sim 8\,m^2$	$\sim 15\,m^2$	$\sim 12\,m^2$	$\sim 10\,m^2$	$\sim 10\,m^2$	

[a]For a definition of this unit (W_p = Watt peak) see Sec. 4.3.1.

4 PV Modules

4.1 *Fundamental Principles*

4.1.1 *Concept and mission*

A **PV module** or **solar panel** is an interconnected assembly of solar cells packaged in a robust, weather-proof casing. Solar cells are brittle devices and provide voltage and current levels inadequate for almost any electrical load. Thereby, the role of a solar panel is: (1) to provide adequate levels for electrical current and voltage; (2) to preserve solar cells from ambient degradation and mechanical damage; (3) to maintain electrical insulation; and (4) to facilitate the convenient handling of solar cells and their mounting outdoors.

4.1.2 *Construction*

Figure 8 schematically shows the construction of PV modules made of wafer-based silicon solar cells. These are constructed as a laminate where the assembly of solar cells is embedded into a transparent encapsulant (EVA); which is then sandwiched between a highly transparent (low iron) tempered glass and a tedlar backsheet. The laminate is then framed with anodized aluminum inserting a sealant to prevent moisture penetration. Electrical terminals are made accessible in a plastic weather-proof (IP65) connection box fixed at the rear of the module.

Figure 9 shows a schematic representation of a solar panel based on thin film technology. In this case, the module is not based on the connection of discrete solar cells previously fabricated. On the contrary, in this technology, the solar cell material is deposited on the module backsheet (glass, metal, plastic) or superstrate (glass). Individual cells are manufactured by scribing the solar cell material (either with a laser beam or mechanically) and the cell interconnections are made during the deposition. Typically, each single solar cell is in the form of a long narrow strip and is connected in series with the adjacent strip (i.e. cell). Accordingly, being a

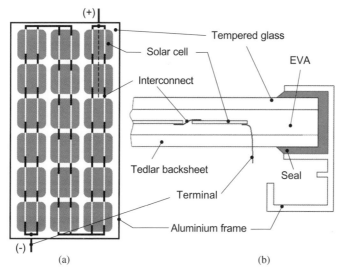

Fig. 8.   (a) Schematic representation of the front-view of a solar panel based on silicon solar cells (b) Cross section of the panel at approximately the location marked as a dashed line.

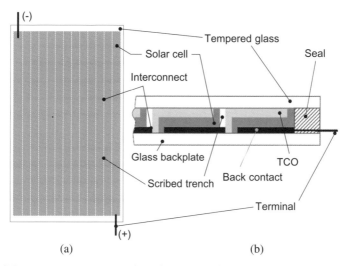

Fig. 9.   (a) Schematic representation of the front-view of a solar panel based on thin film technology. (b) Cross-section of the panel at approximately the location marked as a dashed line.

series connection of solar cells, the current of the module is determined by the area of each strip and the voltage by the total number of strips (cells). This connection scheme minimizes the inter-cell area, giving the module a very uniform appearance, which is even enhanced by the fact that thin film modules are typically frameless — just insulated with edge seal. The aesthetic impact of these two facts has allowed a great penetration of thin film technology in the building integration market.

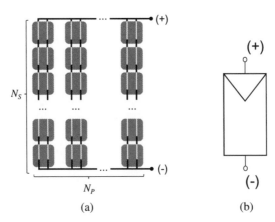

Fig. 10. (a) Generic electrical configuration of a PV module with N_P branches in parallel and N_S cells in series per branch. (b) Electrical symbol for a PV module.

4.2 Characteristic Equation and I-V Curve

4.2.1 I-V characteristic of a PV module

Figure 10(a) sketches a generic configuration of a module with an arbitrary number (N_P) of cell strings in parallel, each string consisting of N_S cells in series. Considering that the solar cells in the module are identical and that they operate at the same conditions (irradiance and temperature), the I-V characteristic of such a connection can be deduced from the I-V characteristic of the individual solar cells presented in Sec. 3:

$$I = N_P I_L - N_P I_0 \left[\exp\left(\frac{V + IR_S N_S}{N_S \cdot nV_t}\right) - 1\right] - \frac{V + IR_S N_S}{N_S R_P}, \qquad (15)$$

where I_L is the photogenerated current per each individual cell in Amperes; I_0 is the diode dark current of each cell in Amperes; n is the diode ideality factor of each cell; R_S is the series resistance of each cell in Ohms; R_P is the parallel or shunt resistance of each cell in Ohms; and V_t is the thermal voltage in Volts. Equation (15) can be presented in a more compact solar-cell-like form as:

$$I = I_{Lm} - I_{0m} \left[\exp\left(\frac{V + IR_{Sm}}{n_m V_t}\right) - 1\right] - \frac{V + IR_{Sm}}{R_{Pm}}, \qquad (16)$$

where all parameters used are analogous as those of the cell as defined in Table 1.

Accordingly, the equivalent circuit of Fig. 2 can also be used for modules, with the parameters calculated as in Table 3.

As can be deduced from Eq. (16), the I-V curve of a solar panel under illumination (Fig. 11) has the same shape as that of the individual cells, except that it is scaled according to the relations expressed in Table 3. If the parallel resistance is high enough to be neglected, the characteristic equation of a solar panel can be

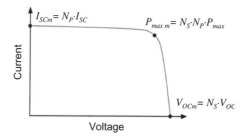

Fig. 11.   I-V curve of the photovoltaic module of Fig. 10(a).

Table 3:   Relation Between Cell and Module I-V Parameters.

Symbol	Parameter	Eq.
I_{Lm}	Module short circuit current	$I_{Lm} = N_P \cdot I_L$
I_{0m}	Module equivalent dark current	$I_{0m} = N_P \cdot I_0$
n_m	Module equivalent ideality factor	$n_m = N_S \cdot n$
R_{Sm}	Module equivalent series resistance	$R_{Sm} = N_S \cdot R_S$
R_{pm}	Module equivalent parallel resistance	$R_{Pm} = N_P \cdot R_P$

also expressed in terms of the parameters of the I-V curve as:

$$I = I_{SCm} \left[1 - \exp \left(\frac{V - V_{OCm} + IR_{Sm}}{n_m V_t} \right) \right] \qquad (17)$$

4.3 *Electrical Performance*

4.3.1 *Standard test conditions*

As in solar cells, the electrical performance of PV modules is generally rated under a set of predefined conditions referred to as **Standard Test Conditions** (STCs). These conditions are: irradiance of $1{,}000\,\mathrm{W/m^2}$ impinging perpendicularly on the tested module, a solar spectrum of AM1.5 global[11] and module temperature at 25°C. The maximum electrical power supplied by the module at STC is called the peak power and is given in **Watts peak** (W_p). As STCs are very favorable, the peak power of a solar panel may be seen as the maximum electrical power it would deliver in real operation more than its typical or average output power. STCs are very difficult to achieve outdoors in virtually any location, therefore **solar simulators** are used to rate PV modules at these conditions.

4.3.2 *Factors affecting the electrical power of solar panels under real operation*

The operating conditions in the field deviate considerably from STC (negatively), so any solar panel operating outdoors will typically exhibit a lower power output

than its peak power at STC. The main factors affecting solar panel power output
are:

Irradiance: As shown in Sec. 3.3.3 for solar cells, low irradiance levels degrade the
open circuit voltage and make the effects of the shunt resistance more important.

Angle of incidence of light: For solar modules mounted on fixed structures, the angle
of incidence of the beam radiation (and thus the reflection losses) depends on the
position of the sun; in addition, diffuse radiation reaches the solar panels from all
angles.

Spectrum of light: It never reproduces exactly the standard (AM1.5g) since it varies
annually with the season; daily with the AM value; and locally with altitude,
weather, and pollution.

Module temperature: As show in Sec. 3.3.4, an increase in temperature degrades the
open circuit voltage.

Soiling: Dust, dirt debris, or other soiling hinder the absorption of light by the solar
cells causing an effective drop of irradiance.

Ageing: As for any other industrial product, PV modules show signs of wear with
time. Typically, manufacturers warrant that the power output will degrade less
than 10% during the first 10 years of operation and less that 20% during the first
25 years.

To avoid the deleterious effects that partial shadowing might have on module performance or integrity (i.e. hot spots[20]) sometimes bypass diodes are connected in
parallel but with opposite polarity to solar cells in a module. The mission of these
diodes is to short-circuit shadowed cells. The number of bypass diodes per module
varies from one per module up to three or four.

4.3.3 *Modeling the equilibrium cell temperature in a PV module*

The first step of most models describing the performance of PV modules out of
STC is modeling the equivalent cell temperature. The most common method used
for this purpose is based on an empirical parameter (supplied by the manufacturer)
named *NOCT* (**nominal operating cell temperature**). *NOCT* is defined as the
cell temperature when the ambient temperature is $20°C$ and the irradiance on the
module is $800 \, W/m^2$. Typical values of *NOCT* are around $45°C$. Using the *NOCT*,
the cell temperature (T_{cell}) for any irradiance and ambient temperature (T_{amb}) can
be calculated as:

$$T_{cell} = T_{amb} + \frac{G}{800 \, W/m^2}(NOCT - 20°C). \tag{18}$$

4.3.4 *Electrical power of solar panels at any irradiance and temperature*

Given certain operating conditions (in-plane irradiance and cell temperature), the most precise way to calculate the electrical output power from a PV module is to calculate its I-V curve for those conditions (Sec. 4.2.1 and Secs. 3.3.3 and 3.3.4), and then determine its maximum power point.[4,9]

Despite being quite accurate, this model is costly in terms of calculations. Therefore simpler approaches are often used that implement first-order (linear) corrections to the module power (P_M) as a function of cell temperature (T_{cell}) and in-plane irradiance (G):[21]

$$P_M = P_{M,STC}\, G/G_{STC}\, [1 + \gamma(T_{cell} - T_{STC})], \tag{19}$$

where $P_{M,STC}$ is the module power rating at STC in W_p; and G_{STC} is the irradiance at STC ($= 1000\,W/m^2$); T_{STC} is the temperature at STC ($= 25°C$); γ is a parameter that defines the relative change in module efficiency per $°C$ (γ is typically negative, so power decreases as temperature increases).

4.3.5 *Electrical energy from a PV module*

Once the instant power from a PV module is known, its electrical energy output $(E_{M,p})$ over a given period of time (p) can be calculated just by integrating:

$$E_{M,p} = \int_p P_M(G, T_{cell})dt. \tag{20}$$

Typically, the continuous evolution of the irradiance and cell temperature over the period p will be not known. On the other hand, a most likely situation is to take a set of discrete values (or averages) of G and T_{cell} at regular intervals (Δt) along period p. Thus, a more convenient formulation of Eq. (20) is:

$$E_{M,p} = \sum_i P_M(\langle G \rangle_i, \langle T_{cell} \rangle_i)\Delta t \tag{21}$$

where $\langle G \rangle_i$ is the average for the irradiance over interval i in kW/m^2; and $\langle T_{cell} \rangle_i$ is the mean value for the cell temperature over interval i in $°C$.

5 PV Arrays and Systems

5.1 *Basic Definitions*

A **PV array** is the combination of a certain number of PV modules to constitute an electrical power generator providing a desired electrical output. A **PV system** is a system including a PV array and ancillary elements designed to provide electrical power demonstrating the desired performance, safety and reliability.

In PV terminology, all the elements in a PV system other than the PV array are altogether referred to as the balance of system or BOS. BOS may include diverse elements such as mounting structures, wiring, fuses, maximum power point trackers, charge controllers, batteries, inverters, electrical safety elements, and meters.

5.2 *Balance of System Components of PV Systems*

5.2.1 *Power conditioning*

Solar panels produce DC electricity while most electrical applications demand AC electricity. The **inverter** is the equipment that transforms the DC electricity from the PV array into AC electricity to power AC-loads or inject to the grid. In addition to DC-to-AC electronics, modern inverters typically integrate circuits intended to force the PV array to operate at optimum conditions by interfacing them with the adequate impedance: these are the **maximum power point trackers**. Other ancillary functions also performed by modern inverters are synchronization with the grid, monitoring and logging the production of the PV system, and implementing some protections (such as anti-island operation).

5.2.2 *Storage*

In PV systems, as in any other system based on a renewable resource, energy storage is needed to marry production with demand. In PV, this energy reservoir typically consists of a set of **batteries**. When using electrochemical energy storage, special care has to be taken to preserve the life and adequate operation of the batteries. Therefore dedicated electronics, namely the **charge controllers**, are inserted between the PV array, the load and the batteries to avoid overcharging or over-discharging these elements. Many different battery technologies are available today, though the classic lead-acid electrochemical cell remains the dominant choice in the PV market.

5.2.3 *Electric components*

PV installations are typically constituted by many elements and may cover large areas and therefore need many meters of wiring and a large number of connectors. Weatherproof **cables** of the correct cross-section have to be used to ensure safety and reliability and minimize risks, voltage drops, and energy losses. To preserve critical elements (such as batteries and inverters) **fuses** and **circuit breakers** are used. Since electricity generation in a PV array cannot be suppressed as long as there is sunshine, extra **insulation switches** are included in several parts of the system. Finally, in grid-connected systems **meters** are used to record PV electricity injection.

5.2.4 *Mounting structures*

PV arrays have to be firmly and securely mounted on rigid structures guaranteeing the maximum exposure to sunlight, as well as ease of installation, access to connectors, ventilation, durability, and cleaning. Depending on the location (such as roof, wall, pole, ground) a variety of fixtures exist. For some applications (PV power plants, PV pumping, satellites, etc.), sometimes PV arrays are mounted on **solar trackers**. These are moving structures that follow the sun during the day, maintaining the angle of incidence between the incoming light and a PV array close to the normal at all times and thus minimizing reflection losses and providing around 30%–40% more power than a fixed array.

5.3 *Types of PV Systems*

Depending on their final use, PV systems combine the elements mentioned in the latter section in different topologies. Figures 12 to 15 show schematic representations of PV systems for grid-connected operation without (Fig. 12) and with storage

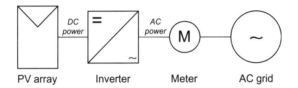

Fig. 12.   Grid-connected PV system without storage.

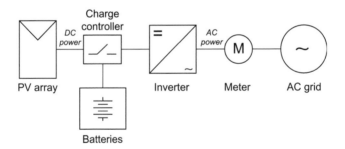

Fig. 13.   Grid-connected PV system with storage.

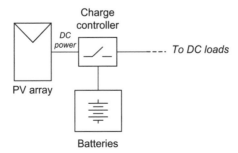

Fig. 14.   Off-grid DC PV system.

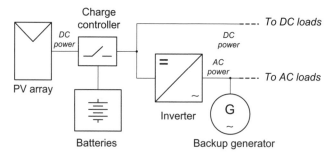

Fig. 15. Off-grid DC/AC hybrid PV system.

(Fig. 13), an off-grid system with only DC output (Fig. 14) and an off-grid hybrid system with both DC and AC supply and a backup generator (Fig. 15). For the sake of simplicity, fuses, insulation switches, and circuit breakers have not been included in the figures.

5.4 *Designing a PV System*

5.4.1 *Location*

The location, with a certain latitude and climate, determines the available solar resource and thus is a first-order impact factor on the performance of PV systems. The availability of long-term records of meteorological data is a key factor in assessing the suitability of a given location for PV use. PV systems should be located far from obstacles that might produce shading, mostly in the central hours of the day and throughout the year. Snow, dirt, or other soiling are location dependent and are also sources of shading. Shading small parts of a PV array may cause severe losses in its energy output. If shading is unavoidable, a good design including adequate module interconnection, protective elements such as blocking diodes and special inverters, may minimize the losses.

In many situations, there is little flexibility in choosing the location of a PV array. The system should be close to the demand in off-grid applications, mounted on the available structures in building integrated PV, or near a high-voltage transmission line for power plants. In all these cases, a careful design is the tool to overcome restrictions associated with the location.

5.4.2 *Orientation and tilt*

The **orientation** of a PV array, also referred to as **azimuth angle** (α), is defined as the angle measured clockwise between true south (or sometimes due north depending on the definition used) and the array plane. The **tilt** of a PV array, also referred to as **elevation angle** (β), is defined as the angle formed by the array plane and the horizontal plane.

The optimum orientation of a PV array in the northern hemisphere is true south ($\alpha = 0$), whilst on the southern hemisphere it is due north. For moderate latitudes ($\Phi < 30$), it is generally accepted that the tilt angle that maximizes the annual energy collection is approximately equal to the latitude ($\beta = \Phi$). In higher latitudes, the tilt angles used are approximately equal to the latitude minus 10° ($\beta = \Phi - 10°$). In some applications, it is important to maximize the energy collection during the worst conditions. This is the case in off-grid systems, where the PV installation has to meet the user's needs through the year, including the months with less insolation. To maximize the collection in winter the tilt angles used are approximately equal to the latitude plus 10° ($\beta = \Phi + 10°$).

For PV arrays with tilt roughly equal to the latitude, even deviations in orientation as high as 30° from the optimum azimuth only cause minor losses (below 10%) in the irradiation collected. Accordingly, installing cumbersome structures to modify the orientation of a preexisting support (such as roof, façade, wall) is not justified in most cases.

5.4.3 *Sizing*

In the most general case, sizing a PV system essentially implies determining the PV array size and configuration (number of modules and interconnection), the inverter size (if needed), and the battery bank size (if needed) as a function of location (available resource and statistical variability), load distribution and target loss-of-load probability of the system (the latter being the probability that the power demand will exceed the generating capacity of a system during a certain period). The problem of sizing acquires a critical engineering dimension in stand-alone (off-grid) PV systems, since a failure of the system to meet the desired operation may cause an interruption in the electric supply. Accordingly, many sizing methods exist for stand-alone systems.[22–25] Conversely, the design of grid-connected systems is often an exercise more determined by economics (calculating return of investment, as a function of feed-in-tariffs, tax rebates, green certificates, and interest rates) than by PV engineering itself.

5.5 *PV System Performance*

5.5.1 *Output power of PV systems*

The maximum output power of an array formed by the connection of an arbitrary number (N_M) of identical PV modules can be calculated as:

$$P_A = N_M \times P_M(G, T_{\text{Cell}}) \times \eta_A, \qquad (22)$$

where P_M is the actual power supplied by each module as calculated using Eq. (19) and η_A is a correction factor that accounts for mismatch losses between the modules in the array and additional losses in the cables and wiring of the system. The array power is sometimes referred to as the DC power of the PV system.

Most PV systems, as shown in Figs. 12 to 15, will include ancillary elements that may drain some of the energy produced before it is transmitted to the load (i.e. inverters, storage, charge controllers, etc.). Such losses should be accounted for depending on the configuration of the system, including as many efficiencies or correction factors as needed. For instance, in a grid-connected system without storage (Fig. 12) the overall system output power (P_S) would be:

$$P_S = P_A \times \eta_i\,(P_A), \tag{23}$$

where η_i is the efficiency of the inverter, which is typically a function of the input DC power. In systems with inverters, the overall system output power is sometimes referred to as the AC power of the PV system.

5.5.2 *Energy rating of PV systems*

Analogously as described for PV modules in Sec. 4.3.5, the electrical energy output of a PV system (E_p) over a given period of time (p) can be calculated as:

$$E_p = \sum_i P_{S,i} \times \Delta t, \tag{24}$$

where $P_{S,i}$ is the average system output power during the ith interval in p.

5.5.3 *Alternative (simpler) energy rating of PV systems*

The use of Eq. (24) needs the determination of the average in-plane irradiance and mean array temperature for each time interval i, as well as many efficiencies and correction factors for other elements in the PV system. These data are not always available or, simply, such a detailed modeling of the system is not needed. For these situations, IEC standard 61724[26] defines a series of parameters for a much simpler determination of the energy output of a PV system. Their definition is as follows:

The PV system final yield (Y_f) is defined as the net energy output of the system over a given period of time (E_p) — typically over a year — divided by the nominal power ($P_{A,\mathrm{STC}}$) of the PV array:

$$Y_f = E_p / P_{A,\mathrm{STC}}. \tag{25}$$

The PV system final yield is measured in $\mathrm{kWh/kW_p}$ or simply "hours" and thus it is sometimes referred to as the "equivalent hours" of the system since it represents the number of hours that the PV array would need to operate at STC to provide the same energy. Typical values for the (yearly) final yield vary from location to location being $Y_f = 700\,\mathrm{h}$ for Germany and the Netherlands, $Y_f = 830\,\mathrm{h}$ for Switzerland, $Y_f = 1{,}400\,\mathrm{h}$ for Spain or California, and up to $Y_f = 1{,}600\,\mathrm{h}$ for Israel.

The PV system reference yield (Y_r) is a magnitude associated with the solar resource available at the system location and is defined as total in-plane irradiation on the system (G_p) over a certain period of time (typically a year) divided by the reference irradiance at STC ($G_{\text{STC}} = 1,000\,\text{W/m}^2$):

$$Y_r = G_p/G_{\text{STC}}. \tag{26}$$

PV system reference yield is measured in hours and represents the number of equivalent hours at the reference irradiance. Sometimes the reference yield is also referred to as the Peak Solar Hours of the system.

The performance ratio (PR) of a PV system is defined as the system final yield (Y_f) divided by the system reference yield (Y_r). In other words, the PR is the ratio of the actual and theoretically possible energy outputs of a PV system.

$$PR = Y_f/Y_r. \tag{27}$$

Accordingly, PR accounts for the overall effect of losses in the system compared to the rated (nominal) output power. In addition to the factors affecting the efficiency of PV modules (discussed in Sec. 4.3.2), in a complete PV system, the PR may be affected by mismatch between the modules forming the PV array, inverter inefficiency, losses in the wiring or connections and other BOS component failures. PR values can be calculated on a daily, weekly, monthly, or yearly basis, the latter being the most frequent case. For most systems, yearly PR values span a range from 0.6 to 0.8. Because losses in a PV system may change during the year (due to factors such as temperature, soilings, shadows), PR values fluctuate in a 10%–20% range, being greater in the winter than in the summer.[27]

Combining Eqs. (25) to (27), it is straightforward to obtain a reasonably accurate estimate of the annual energy output of a PV system as a function of its performance ratio, nominal power and irradiation data:

$$E_p = P_{A,\text{STC}} \times Y_r \times PR. \tag{28}$$

6 Uses and World Market of PV Solar Energy

6.1 *Overview on the Uses of PV Energy*

Over its first decades of development, PV technology encountered many niche applications where the conventional electric grid did not exist or access to it was uneconomical, impractical or even impossible. Therefore, the initial uses of PV systems were off-grid applications (supplying electrical power to professional equipment, rural electrification or consumer products). In the first decade of the 21st century, this situation has changed and now grid-connected applications (including utility owned or de-centralized rooftop-mounted power plants) account for more than 80% of new systems installed in the field.[28] The driving force for this change has been the implementation of various incentives (in the form of special feed-in-tariffs, tax rebates, green certificates, etc.) in many countries to favor the penetration of PV technology into the electric market. Figure 16 summarizes the main application areas and uses of PV systems.

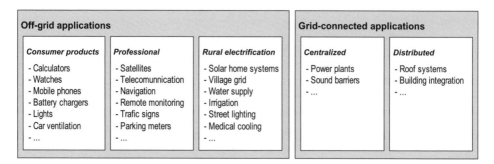

Fig. 16.   Application areas and uses of PV systems.

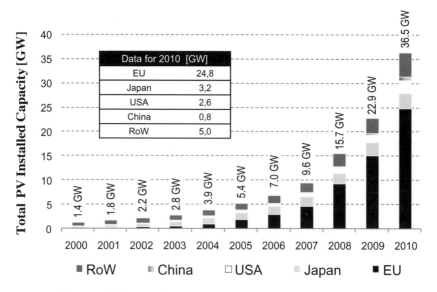

Fig. 17.   Evolution of the global PV installed capacity worldwide.

6.2 *World PV Market*

6.2.1 *Size and historic evolution of the world PV market*

Figure 17 shows the recent historic evolution of the global PV installed capacity per world region.[28-30] As of end of 2010, the European Union accounts for about two-thirds of the global PV installed capacity.

In terms of the distribution of PV installations by country, Table 4 lists the top five countries as of the end of 2010, which constitute more than 75% of the worldwide PV installed capacity.[30] On the production side, Table 5 summarizes the PV cell production by region/country.[31] Tables 4 and 5 show that while market concentration is in Europe, production is in Asia (mostly in China).

Table 4:   Top Five Countries in PV Installations as of December 2010.

	Country	Installed capacity	
		(MW)	(%)
1	Germany	17.193	43%
2	Spain	3.784	10%
3	Japan	3.622	9%
4	Italy	3.494	9%
5	US	2.528	6%
	Total	30.621	77%

Table 5:   PV Cell Production by Region/Country as of December 2010.

	Region	(MW)	(%)
1	China/Taiwan	14.193	59%
2	RoW	3.280	14%
3	Europe	3.127	13%
4	Japan	2.182	9%
5	North America	1.116	5%
	Total	23.889	100%

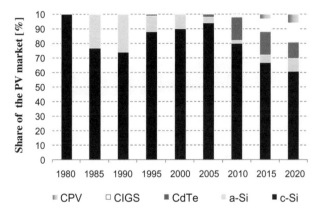

Data for 2010	
c-Si	80%
a-Si	2%
CdTe	16%
CIGS	2%
CPV	0%

Fig. 18.   Historic evolution of the share of solar cell production per technology and expected trends until 2020. c-Si includes mono- and multi-crystalline silicon; CPV stands for concentrator PV.

6.2.2 *PV cell production by technology*

Figure 18 shows the evolution and prediction of solar cell production by technology from 1980 to 2020.

6.2.3 *Evolution of PV module costs and PV electricity*

Over the past decades, the PV industry has achieved remarkable price decreases. The price of PV modules has reduced by around 20% each time the cumulative installed capacity has doubled (see Fig. 19).

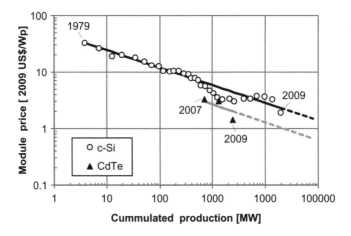

Fig. 19.   PV module price experience curve for crystalline silicon and CdTe technology.

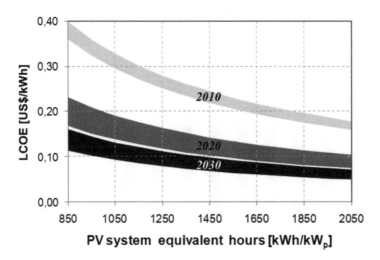

Fig. 20.   Evolution of the levelized cost of electricity for PV systems.

Figure 20 shows current and future Levelized Cost of Electricity (LCOE) i.e. electricity generation costs, for large ground-mounted PV systems with PR = 85%. The data is based on the price reductions summarized in Fig. 19 and forecasts costs of PV kWh below $0.20 for almost any sunny place by 2020.[30]

7 Material Usage and Environmental Impact of PV Solar Energy

7.1 The Value Chain of PV Technology

Different PV technologies include different processes and services in moving from raw materials to turn-key systems, as shown by Fig. 21. These differences

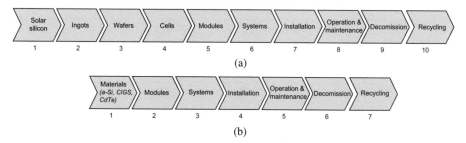

(a)

(b)

Fig. 21. (a) Value chain for wafer-based PV technology; (b) value chain for thin film PV technology.

Table 6:   Material Inventory for Different PV Technologies.

	Material inventory (kg/m^2 and %)			
Category	Ribbon-Si	Multi-Si	Mono-Si	CdTe
Cell material	0.9 (*6%*)	1.6 (*10%*)	1.5 (*10%*)	0.065 (*0%*)
Glass	9.1 (*58%*)	9.1 (*55%*)	9.1 (*55%*)	19.2 (*88%*)
EVA	1.0 (*6%*)	1.0 (*6%*)	1.0 (*6%*)	0.6 (*3%*)
Frame	3.0 (18%)	3.0 (*18%*)	3.0 (*18%*)	0.0 (*0%*)
Connections and packaging	1.8 (*12%*)	1.8 (*11%*)	1.8 (*11%*)	2.0 (*9%*)
Total	**15.8 (*100%*)**	**16.5 (*100%*)**	**16.4 (*100%*)**	**21.8 (*100%*)**

Note: Numbers in brackets represent the share of each material in the total weight per unit area of the PV module.

concentrate at the first stages of the manufacturing process. For wafer-based technology (either mono- or multi-crystalline silicon) modules are manufactured as assemblies of discrete solar cells, previously fabricated. In thin film technology, the whole solar cell material in a module is deposited at once on a substrate and subsequently separated into cells as described in Section 4.1.2.

7.2 *Material Usage of PV Technology*

While the solar cell is the essence of a PV system, in terms of weight, it accounts for only a small fraction of the total materials required to produce a solar panel. As shown in Table 6, the largest share of the total mass of a PV module is glass followed by the aluminum frame.[32–34]

Some materials used in thin film PV technology are rare (In, Cd, Se, Mo, Te) and this could become an important problem for its large-scale deployment.[35] Table 7 summarizes some of the key characteristics, in relation with their scarcity, of these elements.[36]

7.3 *Energy Payback Time of PV Systems*

The **energy payback time** (EPBT) of a PV system is the time needed for the system to produce the energy invested in its manufacturing and installation.[37–40]

Table 7: Some Characteristics of Rare Metals used in PV.

Material	Scarcity (relative to Si)	2009 spot price (US$/Kg)	Max. spot price 2000–2010 (US$/Kg)	2010 World primary production (in Tons)	2010 World reserves (in Tons)	Origin
Cadmium	10^{-7}	4	12	22,000	6.6×10^5	Zn mining
Indium	5×10^{-8}	350	900	574	?	Zn mining
Molybdenum	10^{-6}	65	70	234,000	9.8×10^6	Porphyry
Selenium	5×10^{-8}	75	100	2,260	8.8×10^4	Cu and Ni mining
Tellurium	4×10^{-9}	200	220	< 200	2.2×10^4	Cu mining

Table 8: Breakdown of the Energy Cost for Manufacturing a Multi-Crystalline Silicon Module and an Amorphous Silicon Module using State-of-the-art Technology.

Process	Embedded energy (MJ/m^2)	
	Silicon	Thin film
Cell material	3,200	50
Cell/module processing	300	400
Module encapsulation material	200	350
Overhead operations and equipment manufacture	500	400
Total frameless module	**4,200**	**1,200**
Aluminum module frame	400	400
Total framed module	**4,600**	**1,600**

Accordingly, when calculating the EPBT, all energy invested in the value chain (Fig. 20) up to the system level has to be accounted for. A breakdown of these costs for silicon and thin film PV technology is included in Table 8.

Obviously, the energy output of a PV system depends on a variety of factors (location, orientation, tilt, design decisions, etc.) and thus all those issues will yield different energy payback times for the same PV technology. As shown in Fig. 22, EPBT of 0.8–2.1 years are found for South European or South US locations (irradiation around 1,700 kWh/m^2/yr).

7.4 Greenhouse Gas Emissions of PV Systems

During operation, PV systems do not produce greenhouse gas (GHG) emissions. It is only at the manufacturing, installation, and recycling stages of PV systems when GHG are produced and are mostly (90%) caused by energy use.[41] Fig. 23 summarizes the GHG emissions for several PV technologies.[38, 42] As shown by Fig. 23, lifecycle GHG emissions are in the 21–54 g/kWh range.

7.5 Operational Hazards of PV Systems

PV systems (as any power station) have the obvious hazards associated with electricity. However, under normal operation, PV systems do not produce any emissions, nor is it possible to be exposed to the small amount of toxic substances (some

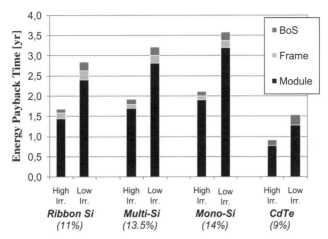

Fig. 22. Energy payback time for different PV technologies installed at two different locations (high irradiation: $1{,}700\,\mathrm{kWh/m^2/year}$; low irradiation: $1{,}000\,\mathrm{kWh/m^2/year}$). Percentages included in brackets are average efficiencies assumed for each technology.

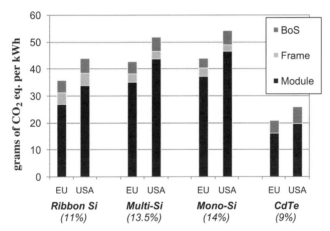

Fig. 23. GHG emissions for several PV technologies for a ground-mounted system, receiving an irradiation of $1{,}700\,\mathrm{kWh/m^2/yr}$, with a $PR = 0.8$, and lifetime of 30 years Two electrical grids are considered, the European (UCTE) grid mixture and the US grid mixture. Percentages included in brackets are average efficiencies assumed for each technology.

semiconductors or solder pastes) embedded into them.[43] The only possible exposure route is via inhalation if the modules are consumed in the event of a fire,[44, 45] which would itself be a more severe hazard.

7.6 *PV Module Decommissioning and Recycling*

After their useful life — estimated to be in the 20- to 30-year range — PV modules will need decommissioning and recycling to both ensure that potentially harmful materials are not released into the environment and to reduce the need for new raw materials. Recent studies show that recycling the useful materials (typically glass and metals) is technically and economically feasible[46, 47] and will help to preserve the environmentally friendly nature of PV technology. Accordingly, PV manufacturers are implementing voluntary take-back and recycling programs for end-of-life PV modules.[34, 48]

References

1. International Energy Agency, *Key World Energy Statistics 2010* (IEA Press, Paris, 2010).
2. J. Perlin, *From Space to Earth: The Story of Solar Electricity* (AATEC Publications, Ann Arbor, MI, 1999).
3. M. Iqbal, *An Introduction to Solar Radiation* (Academic Press, New York-London, 1983).
4. E. Lorenzo, Energy collected and delivered by PV modules, in *Handbook of Photovoltaic Science and Engineering*, 2nd edn., eds. A. Luque and S. Hegedus (Wiley, Chichester, 2011).
5. J. Page, The role of solar radiation climatology in the design of PV systems, in *Practical Handbook of Photovoltaics*, eds. T. Markvart and L. Castañer (Elsevier, Oxford, 2003).
6. Available at: http://re.jrc.ec.europa.eu/pvgis/. Accessed 1 September 2012.
7. Available at: http://www.nrel.gov/gis/solar.html. Accessed 1 September 2012.
8. S. J. Fonash, *Solar Cell Device Physics*, 2nd edn. (Academic Press, Oxford, 2009).
9. M. A. Green, *Solar Cells: Operating Principles, Technology and System Application* (Prentice-Hall, New Jersey, 1982).
10. J. Nelson, *The Physics of Solar Cells* (Imperial College Press, London, 2003).
11. ASTM G173-03(2008) Standard Tables for Reference Solar Spectral Irradiances: Direct Normal and Hemispherical on 37° Tilted Surface.
12. R. H. Bube, *Photovoltaic Materials*, (Imperial College Press, London, 1998).
13. R. W. Miles, K. M. Hynes, and I. Forbe, Photovoltaic solar cells: An overview of state-of-the-art cell development and environmental issues, *Prog. Cryst. Growth Charact. Mater.* **51** (2005), pp. 1–42.
14. V. Avrutin, N. Izyumskaya and H. Morkoç, Semiconductor solar cells: Recent progress in terrestrial applications, *Superlattices Microstruct.* **49** (2011), pp. 337–364.
15. D. J. Friedman, J. M. Olson and S. Kurtz, High efficiency III-V multijunction Solar cells, in *Handbook of Photovoltaic Science and Engineering*, 2nd edn., eds. A. Luque and S. Hegedus (Wiley, Chichester, 2011).
16. Md. K. Nazeeruddin, E. Baranoff and M. Graetzel, Dye-sensitized solar cells: A brief overview, *Solar Energy* **85**(6) (2011), pp. 1172–1178.
17. H. Hoppe and N. S. Sariciftci, Organic solar cells: An overview, *J. Mater. Res.* **19**(7) (2004), pp. 1924–1945.
18. M. A. Green, K. Emery, Y. Hishikawa and W. Warta, Solar cell efficiency tables (version 37), *Prog. Photovol. Res. Appl.* **19**(1) (2011), pp. 84–92.
19. Solar Generation VI. A report by the European Photovoltaic Industry Association and Greenpeace Brussels-Amsterdam (2011).
20. W. Herrmann, W. Wiesner and W. Vaaßen, Hot spot investigations on PV modules — new concepts for a test standard and consequences for module design with respect to bypass diodes, in *26th IEEE PV Specialist Conference*, Anaheim, 1997, pp. 1129–1132.
21. C. R. Osterwald, Translation of device performance measurements to reference conditions, *Solar Cells* **18** (1986), pp. 269–279.

22. S. Silvestre, Review of system design and sizing tools, in *Practical Handbook of Photovoltaics*, eds. T. Markvart and L. Castañer (Elsevier, Oxford, 2003).

23. J. M. Gordon, Optimal sizing of stand-alone photovoltaic solar power systems, *Solar Cells* **20**(4) (1987), pp. 295–313.

24. M. A. Egido and E. Lorenzo, The sizing of stand-alone PV-systems. A review and a proposed new method, *Sol. Energy Mater. Sol. Cells* **26** (1992), pp. 51–69.

25. T. Markvart, A. Fragaki and J. N. Ross, PV system sizing using observed time series of solar radiation, *Sol. Energy* **80**(1) (2006), pp. 46–50.

26. Photovoltaic System Performance Monitoring — Guidelines for Measurement, Data Exchange, and Analysis, IEC Standard 61724 (1998).

27. J. A. del Cueto, Comparison of energy production and performance from flat-plate photovoltaic module technologies deployed at fixed tilt, in *29th IEEE PV Specialists Conf.* New Orleans, 2002.

28. Trends in Photovoltaic Applications: Survey report of selected IEA countries between 1992 and 2009, International Energy Association (IEA) (2010).

29. 2010 Market Outlook, European Photovoltaic Industry Association -EPIA Brussels (2010).

30. US Solar Market Insight. 2010 Year in Review, Solar Energy Industries Association (SEIA) Washington DC (2011).

31. PV News, May 2011 issue, Greentech Media Research, Boston (2011).

32. M. J. de Wild-Scholten and E. Alsema, Environmental life cycle inventory of crystalline silicon photovoltaic module production, Presented at MRS Fall 2005 Meeting.

33. V. M. Fthenakis and H. C. Kim, Energy use and greenhouse gas emissions in the life cycle of CdTe photovoltaics, *Presented at MRS Fall 2005 Meeting*. Boston (2005).

34. K. Sander, Study on the development of a take back and recovery system for photovoltaic products, PV cycle (2007). Available at: http://www.oekopol.de/en/Archiv/Stoffstrom/pv-cycle/Report%20PVCycle%20en.pdf. Accessed 1 September 2012.

35. V. M. Fthenakis, W. Wanga, H. C. Kima, Life cycle inventory analysis of the production of metals used in photovoltaics, *Renew. Sustain. Energy Rev.* **13**(3) (2009), pp. 493–517.

36. Mineral Commodity Summaries 2001. United States Geological Survey — USGS, Reston VA (2011).

37. W. Palz and H. Zibetta, Energy pay-back time of photovoltaic modules, *Int. J. Sol. Energ.* **10**(3–4) (1991), pp. 211–216.

38. E. Alsema, Energy payback time and CO_2 emissions of PV systems, in *Practical Handbook of Photovoltaics*, eds. T. Markvart and L. Castañer (Elsevier, Oxford, 2003).

39. V. M. Fthenakis and E. Alsema, Photovoltaics energy payback times, greenhouse gas emissions and external costs: 2004–early 2005 status, *Prog. Photovolt. Res. Appl.* **14** (2006), pp. 275–280.

40. E. Alsema, M. J. De Wild and V. M. Fthenakis, Environmental impacts of PV electricity generation: a critical comparison of energy supply options, *21st European PV Energy Conference*, Dresden, 2006.

41. R. Dones and R. Frischknecht, Life cycle assessment of photovoltaic systems: Results of swiss studies on energy chains, *Prog. Photovolt. Res. Appl.* **6**(2) (1998), pp. 117–125.

42. V. M. Fthenakis, H. C. Kim and E. Alsema, Emissions from photovoltaic life cycles, *Environ. Sci. Technol.* **42** (2008), pp. 2168–2174.

43. V. M. Fthenakis, Overview of potential hazards, in *Practical Handbook of Photovoltaics*, eds. T. Markvart and L. Castañer (Elsevier, Oxford, 2003).

44. P. D. Moskowitz and V. M. Fthenakis, Toxic materials released from photovoltaic modules during fires: Health risks, *Sol. Cells* **29** (1990), pp. 63–71.

45. V. M. Fthenakis, M. Fuhrmann, J. Heiser, A. Lanzirotti, J. Fitts and W. Wang, Emissions and redistribution of elements in CdTe PV modules during fires, *Prog. Photovolt. Res. Appl.* **13** (2005), pp. 713–723.
46. V. M. Fthenakis, End-of-life management and recycling of PV modules, *Energy Policy* **28** (2000), pp. 1051–1058.
47. J. K. Choi and V. M. Fthenakis, Design and optimization of photovoltaics recycling infrastructure, *Environ. Sci. Technol.* **44**(22) (2010), pp. 8678–8683.
48. Available at: http://www.pvcycle.org/. Accessed 1 September 2012.

Chapter 10

Concentrating Solar Thermal Power

Wes Stein

Division of Energy Technology, CSIRO
P.O. Box 330, Newcastle, Australia
wes.stein@csiro.au

In harnessing solar power, the power of the sun is concentrated to provide a source of heat to generate a hot fluid for a downstream energy conversion process. It can be used to generate electricity, produce solar fuels, or process heat for industry. The fundamental part of the technology is the solar collector, the most common types being parabolic trough, linear Fresnel, central receiver, and paraboloidal dishes. Approximately 1,800 MW of plants are presently operating commercially, and 10,000 MW of plants are presently at various stages of planning and approval. The majority of existing plants are parabolic troughs, using oil to generate steam for a steam turbine. However, there is presently a strong emerging interest in central receivers (power towers) due to their ability to produce much higher temperatures to suit higher efficiency power cycles, and ultimately lower cost. The key for concentrating solar power lies in its ability to incorporate thermal storage and thus provide dispatchable solar power, which will also benefit the penetration of other renewables such as photovoltaics and wind. It is likely that the future for concentrating solar power will involve high efficiency power cycles and high temperature storage for electricity generation, and solar fuels for transport and industry.

1 Introduction

More than 80% of the world's electricity is derived from a fuel source that drives a thermal process (Fig. 1). The single largest source is coal/peat, used for generating steam for steam turbines; nuclear power uses heat from the fission of uranium to generate steam for steam turbines; natural gas is used both for steam generation for steam turbines and as a combustion fuel for gas turbines; and oil is used for steam raising and direct combustion in gas turbines and engines.

Concentrating solar power (CSP) provides a solar alternative for generating the hot fluids needed to power these cycles. This has the attraction that much of the significant wealth of knowledge, experience, and technology in the power industry today is also directly beneficial and transferable to CSP. Indeed, the advances

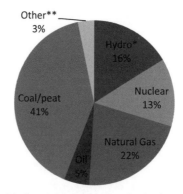

World electricity generation by fuel type
Total global electricity 2009 - 20,055 TWh

Fig. 1.   Fuel use for world electricity supply in 2009 (data from Ref. 1).
*Excludes pumped hydro.
**Other includes geothermal, solar, wind, biofuels and waste, and heat.

that continue to be made in gas and steam turbomachinery and power technology generally will also be beneficial to CSP.

Solar energy arrives at the surface of the earth as a relatively diffuse source of energy. For example, the solar energy falling on one square meter of land over an average day is similar to the energy contained in one kilogram of coal. Concentrating the solar radiation provides a more thermodynamically useful source of heat for use in energy applications based on thermal energy. This is known as *concentrating solar power.*

This chapter provides a background to methods of concentrating solar radiation, the basic relationships of concentrating optics and energy collection, typical applications and thermodynamic cycles for CSP, and an overview of the current status of, and future projections for, the CSP industry.

2 Solar Radiation and Concentration

The solar irradiance at the outer edge of earth's atmosphere is $1,367\,\mathrm{W/m^2}$, known as the solar constant, G. This radiation is scattered and absorbed by atmospheric constituents prior to reaching earth's surface, and arrives as a combination of diffuse and direct radiation. The direct radiation is emitted from the disc of the sun, whereas diffuse radiation is emitted from all angles. When concentrating radiation from the sun, only the direct radiation is useful.

The sun itself is a disc of finite size with a half angle of $0.267°$ ($4.653\,\mathrm{mrad}$). All but the very highest concentrators can also accept *circumsolar radiation,* or light from a narrow area outside of the solar disc. The finite size of the disc means that when concentrating the radiation from the sun, it cannot be assumed that the radiation consists of parallel rays. Figure 2 shows the principle of concentration.

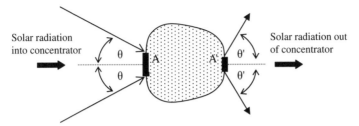

Fig. 2. Geometric quantities of a solar concentrator, where the outlet area A' (with angular spread θ') is less than the inlet area A (which has angular spread θ). An outcome of the second law of thermodynamics is that radiation can only be concentrated by increasing its angular spread.

Assuming that radiant flux is conserved as it passes through the concentrator, then it can be shown[2] that the concentration ratio (CR) can be given as:

$$CR = A/A' = \sin\theta'/\sin\theta \qquad \text{(linear concentrators)}, \qquad (1)$$

$$CR = A/A' = \sin^2\theta'/\sin^2\theta \quad \text{(point concentrators)}, \qquad (2)$$

assuming the entrance and exit media have a similar refractive index.

With $\theta = 0.267°$ (corresponding to the half-angle subtended by the solar disc), the maximum concentration ratios of point and linear focus concentrators can be found by setting θ' to $90°$, giving:

(a) linear concentrator $\qquad$ $CR_{max} = 215$
(b) point focus concentrator $\quad$ $CR_{max} = 46,200$

In practice, the image generated by any concentrator is imperfect due to reflective surface irregularities and off-axis optics, and thus is usually represented by a Gaussian distribution of relative radiation density.

3 Receiving and Absorbing Solar Radiation

The *receiver* is located at the focus of the concentrator and is the device that absorbs the incoming concentrated solar radiation as heat into the fluid. Once the radiation has entered the receiver aperture, the aim is to maximize thermal absorption and minimize thermal losses back to the colder environment.

3.1 *Energy Balance*

The performance of a receiver is determined by the net energy balance of solar radiation entering (gain) against thermal energy leaving (loss). Thermal losses are primarily in the form of re-radiation and convection (due to wind and natural circulation), with some loss due to conduction also occurring. The re-radiation loss is particularly important in high temperature systems due to the effect of this component increasing in proportion to $\{T^4_{\text{receiver}} - T^4_{\text{surrounds}}\}$. Note that in some receiver

W. Stein

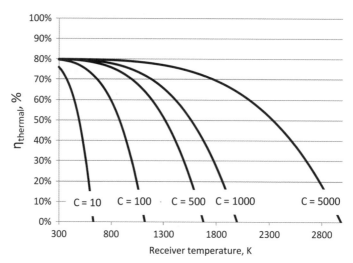

Fig. 3. Receiver thermal efficiency as a function of temperature and concentration ratio (assuming solar irradiance $= 900\,\mathrm{W/m^2}$, $T_{\mathrm{ambient}} = 300\,\mathrm{K}$, $\mathrm{F}\cdot\alpha = 0.8$, $\mathrm{F}\cdot\varepsilon = 0.8$, and zero convective losses.[3]

designs the radiation entering the receiver will tend to be attenuated if a glass cover is used in order to reduce convective loss.

Figure 3 shows that the receiver's thermal efficiency increases with concentration ratio but decreases with temperature. Concentration ratio has an effect because thermal losses are proportional to the surface area, meaning a system with higher concentration has less area from which losses emanate. Higher temperatures reduce efficiency because they result in greater re-radiation of energy to the environment.

In high concentration systems, it is important to absorb as much energy as possible. A selective surface (Sec. 3.2) can greatly improve the performance of lower concentration systems such as linear troughs but has less relative effect in a point focus system where the concentration ratio is, in practice, one to two orders of magnitude higher.

3.2 *Selective Surface Theory*

A *selective surface* is a coating which minimizes thermal losses by exploiting the fact that the spectrum of incoming solar radiation is different to that of thermally re-radiated energy.

The performance of a selective surface material depends on its absorptivity and emissivity properties at the operating temperature. A surface can be characterized by Kirchoff's Law for opaque materials, which states that for a body in thermodynamic equilibrium the absorptivity α is equal to the emissivity ε at a given wavelength and angle of incidence,

$$\alpha(\lambda, \theta) = \varepsilon(\lambda, \theta). \tag{3}$$

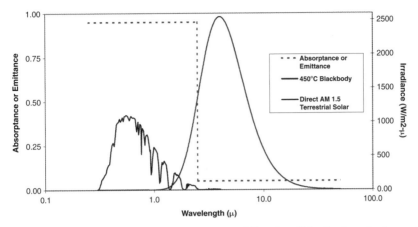

Fig. 4. Spectra for solar irradiance, a blackbody at 450°C and an "ideal" solar selective surface.[4] The solar spectrum can be approximated by application of Planck's spectral distribution corresponding to a black body at 5777 K.

In addition, reflectance ρ is related via

$$\alpha(\lambda, \theta) = 1 - \rho(\lambda, \theta). \tag{4}$$

The majority of the energy of solar radiation for collector purposes arrives within a wavelength range of approximately 0.3 to $3\,\mu$m. The absorbing surface, however, can be treated as a blackbody emitter at its surface temperature, which generally results in longer wavelengths for the re-radiated energy. Application of a selective surface treatment to the absorber allows it to have low reflectance and hence high absorptivity [Eq. (4)] at wavelengths less than a critical value — thus allowing maximum absorption of incoming solar radiation — while having high reflectance and hence low absorptivity and emissivity [Eq. (3)] for wavelengths greater than this value, hence minimizing re-radiation. An ideal selective surface would exhibit a vertical crossover as shown in Fig. 4.

Figure 4 shows that at low absorber temperatures, there is little overlap between the solar and blackbody spectra. As the blackbody temperature of the absorbing surface increases, the blackbody curve moves to the left. This increases the degree of overlap with the solar spectrum and the spectrally weighted ratio of α/ε approaches 1, such that the effectiveness of a selective surface eventually becomes negligible. Selective surfaces therefore offer more significant performance improvement potential for lower-temperature concentrating solar thermal systems than those at higher temperature.

4 Types of Solar Collectors for Power and Fuels

The term *solar collector* refers to a device that collects solar radiation into a fluid in the form of heat. This chapter will generally deal with concentrating solar applications, where the solar collector system is a combination of the solar concentrator (e.g. the mirrors) and the receiver (the device that absorbs the solar radiation and converts it to heat). Refractive concentrators such as lenses are not covered — they have applications for concentrating PV systems, but generally they become too large and unwieldy for larger capacity thermal systems. However, two types of solar thermal technologies — solar ponds and the solar chimney — do not involve concentration, but are included here due to the fact they have been used for solar thermal power generation.

4.1 Solar Pond

The most common form of *solar pond* is based on salt water. It is relatively shallow — usually no more than a few meters deep — and has a salt density gradient that increases with depth. At the bottom, the salinity is near-saturation, and it is here in this 1–2 m thick zone that maximum temperatures are reached. The solar radiation penetrates through the upper zones to the bottom, while the salinity gradient prevents upward natural convection. Because the salt tends to naturally diffuse upward over time, the salt gradient needs regular maintenance or refreshing.

One of the major requirements for a salt pond is in fact fresh or low salinity water for the upper levels due to evaporation. The salt can be common salt (NaCl), able to provide salt water densities of up to $1,200\,\text{kg/m}^3$, or other salts such as magnesium chloride which can achieve higher densities (over $1,300\,\text{kg/m}^3$) and better performance. The maintenance cost of a salt pond is a critical issue, with salinity gradient, excursion events, clarity, and chemical balance all requiring careful control.

Temperatures above 90°C have been demonstrated in solar ponds, though the potential thermal to electric efficiency is very low — less than 5% in practice. Their application may be most useful for low temperature thermal processes in salinity-affected areas such as desalination or production of some chemicals. An additional attraction of the salt pond is that it provides thermal storage.

A number of small plants have demonstrated the production of electricity from solar ponds, including at the Dead Sea, Israel (5 MWe); Alice Springs, Australia (15 kWe); El Paso, USA (120 kWe), and Pyramid Hill, Australia (60 kWth).[5]

4.2 Solar Chimney

The *solar chimney*, also referred to as the *updraught solar tower*, works on the principle that a large "greenhouse" solar collector heats air such that its density is less than the surrounding ambient air outside the chimney. The resulting pressure difference causes air to move up the chimney, passing through one or more wind turbines installed at the base of the chimney and generating electricity (see Fig. 5).

The power generated is a balance between the driving force provided by the pressure drop in the chimney (including frictional effects) and the pressure drop

Fig. 5.   Illustration of a 5 MW solar chimney (*Courtesy*: Schlaich Bergermann and Partner).

across the turbine. It can be shown that electrical power is optimized when the pressure difference across the turbine is two-thirds of the total available, not considering the transient coupling effects between collector and chimney, allowing maximum power to be estimated for a system with collector efficiency $\eta_{\text{collector}}$, chimney height H_{chimney}, and collector area $A_{\text{collector}}$, for given conditions solar irradiance I, ambient temperature T_0 and specific heat capacity of air c_p[6]:

$$P_{\max} = 2/3 \, \eta_{\text{collector}} \left[g/(c_p T_0) \right] H_{\text{chimney}} \cdot A_{\text{collector}} \cdot I. \tag{5}$$

A typical large-scale solar chimney may have air temperature at the base of the chimney about 18°C above ambient, air velocity through the turbine of approximately 11 m/s,[7] collector thermal efficiency based on global horizontal radiation in the range 40%–60%, and overall solar to electric efficiency of the order of 1%–2% depending on tower height.

An experimental solar chimney was constructed and operated successfully from 1982–1988 at Manzanares, Spain.[8] It had a 50 kW$_e$ rating, 195 m high chimney, 46,000 m^2 collector area, 10.2 m chimney internal diameter (at base), and 10 m turbine rotor diameter. The plant was constructed with an intended lifespan of just three years but ended up operating for eight. The tower collapsed in a storm in 1989 due to a corroded guy wire, though the structure survived beyond its design life.

The technology has been developed and advanced largely through the German engineering firm Schlaich Bergermann and Partner. Commercial projects from 30 to 200 MW have been proposed for Jordan, India, Australia, and the Southwestern US. A 200 MW plant would typically require a 1 km high chimney (with 120 m diameter at the base) and 7 km diameter collector, with annual solar radiation of 2,300 kWh/m^2/yr such a plant could generate 680 GWh/yr of electricity.[8]

The attraction of the technology is its simplicity, with capital costs able to be minimized in countries with low labor costs. However, operation and maintenance is likely to be an area of concern. If plastic is used as the collector, the lifetime will likely be considerably less than for glass; for either plastic membrane or glass, cleaning cost could be significant given the essentially flat arrangement.

4.3 *Parabolic Trough*

Parabolic trough technology, often referred to simply as a *solar trough*, uses linear parabolic reflectors to concentrate solar radiation to a focal line. A fluid flows through a receiver tube positioned along the focus and is heated for use in a downstream process.

The concentration ratio of a parabolic trough collector (PTC) is given by the ratio of collector aperture area (planar) and the circumferential area of the absorber tube, with typical values ranging from 10 to 100.[9]

Parabolic troughs generally have a single axis of rotation that is aligned in either an east–west or north–south orientation. A north–south axis provides greater annual solar energy collection because the angle of incidence of the sun's rays to the normal of the trough aperture (θ) is less on an annual average basis' leading to $\cos\theta$ being closer to the ideal value of one. Additionally, during the summer months when some three to four times more solar energy falls than in winter (depending on latitude and seasonal climate), average daily incidence angle is lowest meaning collection is greater. An east–west system, by contrast, provides a lower seasonal variation in energy delivery.

For a solar trough at latitude ϕ on day of year n and at time of day $t(0 \leq t < 24)$, the angle of incidence between the incoming beam and the normal of the aperture of the reflector θ can be found by the following relationships.

For a parabolic trough with east–west horizontal continuous tracking,

$$\cos\theta_{\text{ew·hor}} = (1 - \cos^2\delta \cdot \sin^2\varphi)^{1/2}, \tag{6}$$

and for north–south horizontal continuous tracking,

$$\cos\theta_{\text{ns·hor}} = [(\sin\phi \cdot \sin\delta + \cos\phi \cdot \cos\delta \cdot \cos\varphi)^2 + \cos^2\delta \cdot \sin^2\varphi]^{1/2}, \tag{7}$$

where the quantities δ and φ are defined by

$$\sin\delta = -\sin(23.45°) \cdot \cos[360°(n + 10)/365.25], \tag{8}$$

and

$$\varphi = 360° \cdot (t - 12)/24. \tag{9}$$

Polar axis continuous tracking

$$\cos\theta_{\text{ns·polar}} = \cos\delta. \tag{10}$$

For comparison, a dish concentrator with ideal two axis tracking gives

$$\text{Cos}\,\theta_{\text{two-axis}} = 1. \tag{11}$$

In large PTCs for solar power stations, a number of trough modules are connected in series such that the overall length might be up to 150 m with apertures over 6 m wide. In these cases, the collector unit is rotated by hydraulic motors, due to the large forces and high precision required.

The reflectors of PTCs are usually made from rear-silvered low iron glass (92%–94% spectral reflectance) which has been thermally sagged to give the correct shape. Alternative reflectors currently in use are polished aluminium sandwich construction sheets and polymer film reflectors on a substrate. It is noted that rear-silvered glass reflectors have now been in operation in a variety of applications and environments for over 20 years and have also been subjected to accelerated life tests that indicate negligible drop in reflective performance with time.

The heart of the PTC is the receiver tube. In PTC power stations, these tubes comprise a steel pipe (to contain the fluid pressure) with an annular glass cover. To minimize convection losses, the space between the glass and the steel pipe is evacuated, requiring a metal bellows and special glass-to-metal weld to seal the glass to the metal pipe and allow for differential expansion. The vacuum is maintained through the use of "getters" — chemical sponges that maintain and indicate vacuum status. The evacuated tube is especially valuable when the operating temperature is above temperatures of the order of 250°C, where the reduction in thermal losses outweighs the extra cost. The other mechanism for reducing thermal loss is the selective surface, which has been discussed in Sec. 3.2.

Because the receiver tube rotates with the trough reflector, a suitable means of transporting the fluid to the ground is required. Both flexible hoses and ball joints have been used, with ball joints being used more commonly in modern troughs that use oil as the heat transfer fluid (HTF) due to lower maintenance requirements and reduced pressure drop.

Table 1 provides characteristics of the nine SEGS (Solar Energy Generating System) parabolic trough plants built by Luz Industries that have been in commercial operation for more than 20 years. Figure 6 shows a typical parabolic trough plant layout.

4.3.1 *Heat transfer fluid*

The fluid used in the PTC power plants being installed today is most commonly a thermal oil, in particular VP-1, a eutectic mixture of 73.5% diphenyl oxide and 26.5% diphenyl. The advantage of this oil is that it can reach reasonably high temperatures of 395°C before decomposition becomes significant. However, it has a relatively high melting point (12°C) requiring auxiliary heating, and at atmospheric pressure it boils at 257°C, requiring an inert gas blanket.

Table 1:   Characteristics of the SEGS I–IX Plants.[10]

SEGS plant	1st year of operation	Net output MW$_e$	Solar field outlet temp. (°C/°F)	Solar field area m^2	Solar turbine eff. (%)	Fossil turbine eff. (%)	Annual output (MWh)
I	1985	13.8	307/585	82,960	31.5	—	30,100
II	1986	30	316/601	190,338	29.4	37.3	80,500
III & IV	1987	30	349/660	230,300	30.6	37.4	92,780
V	1988	30	349/660	230,500	30.6	37.4	91,820
VI	1989	30	390/734	188,000	37.5	39.5	90,850
VII	1989	80	390/734	194,280	37.5	39.5	92,646
VIII	1990	80	390/734	464,340	37.6	37.6	252,750
IX	1991	80	390/734	483,960	37.6	37.6	256,125

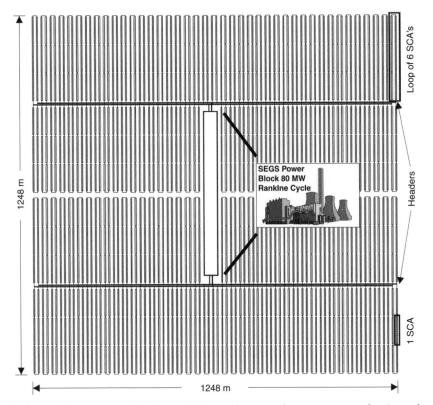

Fig. 6.   Typical layout of an 80 MW parabolic trough power plant, no storage, showing a density of mirror aperture to ground covered of approximately 30% and approximately 0.5 MW$_e$/ha.[10]

Other heat transfer oils are available that extend this operating temperature range slightly but they are expensive. Two other heat transfer fluids are under active investigation and demonstration — steam (known as *direct steam generation* — DSG) and *molten salt*.

The attraction of steam based on demineralized water is that it can be used directly in the Rankine cycle with no need for an intermediate heat exchanger;

it is cheaper than oil; the environmental hazards are less; higher temperatures are possible. The difficulties are that the pressure associated with useful steam temperatures becomes high, and thus collector field piping and flexible joint fittings become expensive. In addition, the two-phase flow in the evaporation portion of the receiver field creates hydrodynamic flow instabilities and temperature control issues over long distances under transient solar radiation. DSG has been researched since the 1980s in the US. LUZ built an LS4 DSG prototype in Israel to follow their successful LS2 and LS3 PTCs, which used oil as the HTF, and a considerable research program was conducted by CIEMAT, DLR, ZSW, and Siemens in the 1990s. A European R&D project named DISS (Direct Integrated Solar Steam) test facility was built at the PSA in Spain in 1998 and has been extended and used for considerable testing since.[11] The current configuration is a single row of length 750 m and 3,822 m^2 aperture area, providing a nominal 1.8 MW$_{th}$ with 1 kg/s of superheated steam at 400°C and 10 MPa pressure. The experimental results have indicated that no additional heat transfer enhancement devices are required to keep within allowable temperature gradients. The preferred arrangement to ensure reliable control is recirculation rather than once-through, with two-phase stratification in the evaporator section prevented (or minimized) by water injectors and mass flux control.

The other fluid of great interest is molten salt. This has been well developed as the working fluid for central receivers and as the bulk storage medium for central receivers and trough plants. The most common salt used in CSP plants today is a binary nitrate salt of 60% by weight sodium nitrate and 40% by weight potassium nitrate. The commercial trough projects deployed today with salt storage are unable to make use of the full benefit of salt, because though the salt can operate to 560°C, the use of oil as the HTF within the trough limits the storage temperature to less than 390°C. A 5 MW$_e$ demonstration (Archimedes) has recently been commissioned to show molten salt used for the first time as the HTF in a PTC.[12] Though the temperature from a PTC will still not reach the levels possible from towers, it still enables higher operating temperatures, higher cycle efficiencies and potentially more cost-effective storage.

4.4 *Linear Fresnel*

Large reflectors broken down into multiple Fresnel sub-elements can help to improve the manageability of moving components. The concept was first advanced by Baum *et al.* in 1957, and later important development work in both linear and point focus Fresnel systems was undertaken by Francia in 1968. Essentially, the linear Fresnel emulates the trough, and the central receiver (a point-focus Fresnel system) emulates the dish.

The linear Fresnel concept was revived for its potential commercial application in 2000.[13] The attractions of the technology include that the moving structural components are at ground level with reduced wind loading and cost compared to a trough, and the receiver is an inverted cavity fixed in space (see Fig. 7). A fixed receiver can be designed without the limitation of flexible joints, making direct

Fig. 7.   Illustration of a compact linear Fresnel reflector in which reflectors can target one or another receiver depending on time of day and solar incidence angle.[15]

Fig. 8.   Novatec Solars PE1: a 1.4 MW$_e$ demonstration plant at Murcia, Spain. PE2 (30 MW$_e$) is presently under commissioning on an adjacent site (*Courtesy*: Novatec Solar).

steam generation quite feasible, as well as allowing the possibility of very long lengths, reducing end losses. Some receiver designs have proposed a simple boiler tube arrangement, and others are using evacuated tubes from trough technology with a secondary reflector.

The disadvantage of the linear Fresnel is that annual optical efficiency is around 43% (depending on latitude) compared to a PTC field of approximately 57%. The aperture-specific investment cost of a linear Fresnel field in order to break even with a parabolic trough field has been calculated to between 29% and 78% of the PTC cost, depending on optical and thermal performance assumptions, the definition of aperture, as well as O&M costs.[14]

Over the last 10 years, there has been a rapid rise in linear Fresnel demonstrations and projects. In Australia, a $1.3\,MW_{th}$ demonstration loop was integrated with the Liddell coal-fired power station by Solar Heat and Power (now AREVA) in 2005, and is presently being upgraded to a $9\,MW_{th}$ system by Transfield and Novatec Solar.[16] Further north in Queensland, AREVA are building a $44\,MW_e$ linear Fresnel as a solar boost to the dry-cooled supercritical Kogan Creek coal-fired power station.[17] More recently, an AU\$1.2 billion, $250\,MW$ linear Fresnel and natural gas hybrid project called Solar Dawn has been announced for Australia.[18]

Other linear Fresnel projects include the AUSRA's $5\,MW_e$ Kimberlina demonstration project in California,[19] the Man Ferrostaal $1\,MW_{th}$ loop in Spain (2007), and the plants PE1 ($1.3\,MW_e$) and PE2 ($30\,MW_e$) built by Novatec Solar in Murcia, Spain see (Fig. 8).[20]

4.5 *Central Receiver (Power Tower)*

Central receiver systems comprise many ground-mounted reflecting panels (*heliostats*) that individually track the sun in two axes and reflect it to a receiver mounted on a tower (Fig. 9).

4.5.1 *Heliostats and field layout*

Heliostats are typically based on rear-silvered glass, though some stretched membrane designs have been trialed. Sizes range from $1\,m^2$ to $150\,m^2$ each, with prototypes of up to $200\,m^2$ having also been developed. They are driven around each axis by a motor and gearbox, linear actuators, or a hydraulic system. Control

Fig. 9. Illustration of the central receiver concept using multiple heliostats to reflect solar radiation to a central fixed receiver (*Courtesy*: CSIRO).

is most often accomplished with open-loop controllers using well-known algorithms for the solar position; however, closed-loop heliostat positioning systems are also used for some designs.

The function of the control and actuation system is primarily to maintain the normal of the reflective surface such that it bisects the angle between the sun, heliostat, and receiver, with refined systems enabling target-alignment strategies on the receiver. The most common actuation mechanism in use involves rotation about the azimuth axis with adjustment for the altitude angle (i.e. elevation); however, others such as pitch/roll and target-aligned also offer particular benefits.

4.5.2 *Receivers*

There are generally three types of central receivers in use.

In *external receivers*, the radiation directly heats the tubes that contain the heat transfer fluid with tubes exposed to the atmosphere. They are most often used for medium-temperature applications such as steam generation.

Cavity receivers use a tube bank as the absorber surface within a cavity, but the aperture where the concentrated radiation enters is much smaller than the absorbing surface and is useful for minimizing re-radiation and convection losses in higher temperature applications. Note that partial cavities are possible.

Volumetric receivers are those where, instead of using tubes to contain the fluid (which may fatigue under a high number of high temperature cycles and which limits the allowable incident flux due to heat transfer limitations), the receiver comprises a porous arrangement of material arranged to absorb very high concentration flux. This also provides a high surface area for transfer of heat to the fluid. Volumetric receivers may be at atmospheric pressure [see Fig. 10(a)] or, if pressurized, require a transparent window with the absorber in a cavity behind [see Fig. 10(b)].

Although point-focus concentrators are capable of delivering very high concentration ratios to the receiver, the ability to absorb the energy is dependent on the heat transfer characteristics of the interface between surface and fluid. Thus in many cases, the peak allowable flux on the receiver is heat transfer limited rather than limited by the solar concentrator. For example, pressurized air or superheated steam in a tubular receiver might need to be limited to an irradiance of under $200\,kW/m^2$ because of the poor heat transfer properties of a gas, whereas the evaporator section of a solar boiler might handle radiation of the order of $500\,kW/m^2$, molten salts up to $1,000\,kW/m^2$ or more (Fig. 11), and particle receivers even higher.

4.6 *Dish Concentrators*

From an optical perspective, the ideal solar concentrator is the *paraboloidal dish*. It tracks the sun continuously around two axes, and because the receiver moves with the dish it is always collinear with the sun, resulting in no cosine losses. The dish provides the highest levels of concentration and is useful for applications requiring high solar flux.

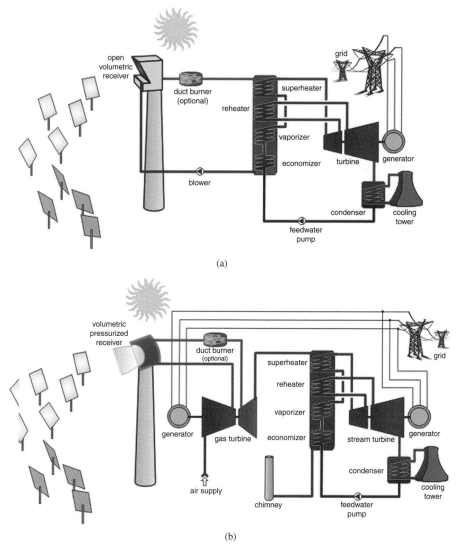

(a)

(b)

Fig. 10. Two concepts for the central receiver power cycle: (a) a volumetric receiver at atmospheric pressure that absorbs energy to heat air for subsequent steam generation for a Rankine cycle; (b) a cavity receiver heating air at pressure to very high temperatures for subsequent use in a Brayton cycle, with the option of a combined cycle system for higher efficiency.[21]

Source: http://www.volker-quasching.de/publis/klimaschutz_e/index.php

As with heliostats, a wide range of solar dish sizes has been developed. Some very small applications (under 1 kW) have been proposed, but generally sizes have ranged from approximately 40 m^2 up to the largest built of nearly 500 m^2.

Dish systems have been classified as distributed or centralized. *Distributed dish systems* have generally been based around the Stirling engine, though microturbines (based on the Brayton cycle) are also under investigation. In each case, electricity is

Fig. 11. Solar Two in California — a $10\,MW_e$ demonstration of a receiver using molten salt as the heat transfer fluid. Operating from 1996 to 1999, it provided the technology basis for new commercial projects currently operating or under development that promise cost-effective long-term thermal storage (*Courtesy*: SANDIA).

generated directly from the engine/generator which is mounted at the focus of the dish. This allows relatively simple collection of energy from many point sources.

In *centralized dish systems*, an array of dishes is used to generate a high-temperature fluid. This provides high solar to thermal efficiency at the receiver but requires the aggregation of fluid from multiple point sources to the central facility. Each dish also requires flexible joints to allow the fluid to travel from the moving receiver to the ground.

The largest centralized dish demonstration to date was the *Solarplant 1* system built in Southern California by the Lajet consortium in 1983–1984. It comprised 700 dishes, with a total collecting area of $30,590\,m^2$. The dishes generated steam in their cavity receivers, with 600 of the dishes producing saturated steam and others taking the saturated steam and superheating it to 460°C. The project experienced problems with the reflective film used on the dish facets and long start-up times.[22]

More recently, Wizard Power have proposed a project based on large dish technology generating high temperature steam for a $40\,MW_e$ steam turbine[23] using dish collector technology developed at the Australian National University (see Fig. 12).[24]

A variety of dish reflector concepts have been developed. Some early designs used a fiberglass mould to which reflective tiles were glued. An interesting stretched membrane concept was used for a number of the dish Stirling demonstration units. In this design, the dish is shaped into a parabola by plastically yielding the membrane using a combination of uniform and non-uniform loading, with mirrored tiles or reflective film providing the reflection. Membranes for the Schlaich Bergermann and Partner dish Stirling units in Spain ($44.2\,m^2$, $9\,kW_e$) were made from 0.23 mm

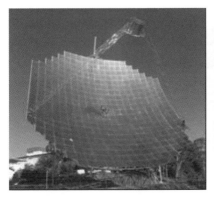 

Fig. 12. Left: One of the largest solar dishes in the world at the Australian National University, Canberra, shown here with a steam receiver. It has a 494 m^2 aperture, 13.4 m focal length, 50.1° rim angle and total mass of 19.1 tonnes. The reflector uses square panels with 1 mm rear-silvered glass adhered to a pre-formed sandwich substrate (*Image Courtesy*: ANU). Right: 10 kW$_e$ dish Stirling units installed at PSA, Spain (*Image Courtesy*: DLR).

thick stainless steel sheet. Shape is maintained in practice by drawing a slight vacuum behind the membrane. If a temperature excursion occurred at the receiver, the vacuum could be released quickly to disrupt the flux.

5 Thermal Storage

Solar and wind power are generally disadvantaged by the intermittent nature of the source. Not only does this lead to uncertainty for the grid operator but also invested capital such as the turbine and the balance of plant equipment remains unused for much of the year. For CSP, the solar field requires about one-third to half of the total capital cost, and of course this cannot be used when the sun is not shining. However, if thermal storage is employed, the balance of plant equipment, comprising the electricity generating equipment such as the turbine, can be used to continue generating electricity regardless of solar conditions.

The basic concept is that solar energy during the day is used to charge a thermal store. Then, on demand, the stored heat is used to generate steam for a steam turbine or otherwise heat a working fluid or feedstock in a controlled manner for a downstream energy conversion process (see Fig. 13).

The US Department of Energy has established a target cost for thermal storage of under US$15/kWh$_{th}$ based on round trip efficiency.[25]

5.1 *Two-Tank Molten Salt*

The thermal storage technology of choice in today's commercial CSP systems is the two-tank molten salt system. A number of 50 MW$_e$ trough projects, with oil as the heat transfer fluid in the PTC field interfacing with a salt storage loop, are now operating with this approach, though this limits temperatures to that of the heat transfer oil. (A new PTC receiver tube designed for molten salt is under early operation — refer to Sec. 5.5.)

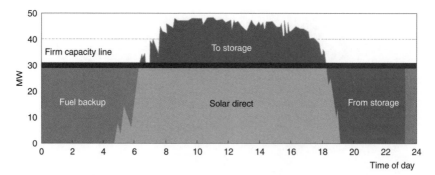

Fig. 13.   Schematic of the basic principle of storage in a CSP system. The solar field, storage, and turbine may be sized for any desired outcome — from simply charging the storage during solar hours for later turbine operation, through to simultaneously running the turbine and charging the store during solar hours with the turbine then continuing to operate into the night.

Source: Geyer (2007). SolarPACES Annual Report.

A 20 MW$_e$ tower developed by Gemasolar in Spain began operation in 2011 using molten salt as both the receiver heat transfer fluid and storage medium.[26] Based on the pilot scale concept originally developed at Solar Two in California, it uses salt as the HTF in the receiver (565°C) and a two-tank storage system. Designed for 15-h storage (6,500 h annually), it has demonstrated 24-h operation on solar energy. Many new CSP projects are incorporating two-tank molten salt storage, and several large projects, both tower and PTC, are under construction.

5.2 *Single Tank Molten Salt*

One way to reduce storage costs may be to use a single tank system with a thermo-cline, where the hot and cold molten salt is contained in a single vessel, separated vertically by a small high-temperature gradient region. The single-tank system has a much lower volume than the total volume of a two-tank system. The two isother-mal regions remain vertically stratified through thermal buoyancy forces. A low-cost filler material (such as quartzite rock) with high specific heat and low void fraction (< 0.25) is used in the tank to further reduce the inventory of the relatively expen-sive molten salt. Comparative studies have shown that the thermocline approach may offer a 35% cost reduction over the two-tank system.[27] The issues that have been raised are thermal ratcheting (when a filler material is used) and also the risk related to disturbance of the stratified regions when charging or discharging and the consequent loss of temperature difference. This latter effect is managed through well-placed inlet and outlet points, distributor regions, and suitable flow rates.

5.3 *Alternative Thermal Storage Options*

Molten salt offers a number of attractions such as relatively low cost and good specific heat and density, but poses problems both at the low temperature end (due to elevated freezing temperatures) and the upper end (due to instability at temperatures that would be even more attractive for thermal power cycles). Given that point-focus solar collectors can reach temperatures in excess of salt upper temperatures, alternative materials are constantly under development to enable thermal storage at temperatures well in excess of the present $\approx 600°C$.

Solids such as graphite, silicon carbide, alumina, high-temperature concrete, and fluidized sand have been investigated as storage media. Generally, the weight of these materials for any significant storage capacity precludes them from being located at the receiver in the case of central receivers or dishes, although they may be useful for overcoming operational transients. Thus a high-temperature heat transfer medium is usually needed to move the heat from the receiver into the solid material. The heat transfer fluids considered include supercritical fluids such as steam and CO_2, lower pressure gases such as helium and air, and even solid particles used in a "falling curtain" arrangement in the receiver. Given the solid material will not have freezing point issues and will generally have a much higher upper temperature limit than today's solar salt formulations, the HTF/bulk solid material combination may be a cost-effective form of storage, particularly for higher-temperature CSP applications.

The Jülich Solar Tower project has reported results for temperatures up to 680°C with success, using alumina porcelain as the packed bed material in a regenerative design.[28] High-temperature concrete has been investigated and has shown long-term stability for temperatures up to approximately 500°C, suitable for parabolic trough power plants.[29] As an indication of thermal capacity, approximately $50,000 m^3$ of concrete would be needed for a $50 MW_e$ with design storage capacity of $1,100 MWh_{th}$ plant similar to that shown in Fig. 14.

The above thermal storage options rely on sensible heat, and thus, as heat is drawn from the storage media, the temperature begins to drop. Phase-change materials (PCMs) provide a constant discharge temperature over much of the inventory by operating within the latent heat region. This is particularly valuable for the evaporation section of steam generators. However, this requires the development of PCMs for the desired operating temperature. Many PCMs have been formulated for low temperatures; however, there are few low-cost synthetic options available for CSP at present. The phase-change temperature of KNO_3 (333°C) and $NaNO_3$ (308°C) may be suitable for low-temperature application, and NaCl (802°C), Na_2CO_3 (851°C) and K_2CO_3 (891°C) for high-temperature application but present management difficulties such as corrosion. Liquid metals are a form of PCM though they are relatively expensive. PCMs often exhibit poor thermal conductivity — one option is to encapsulate the PCM material within a matrix of higher thermal conductivity to improve surface area available for heat transfer.

5.4 *Thermochemical Storage*

An alternative to sensible heat storage is to store the solar energy through a thermochemical process. A closed-loop thermochemical cycle provides storage only, whereas an open-loop thermochemical cycle provides a means to store solar energy prior to

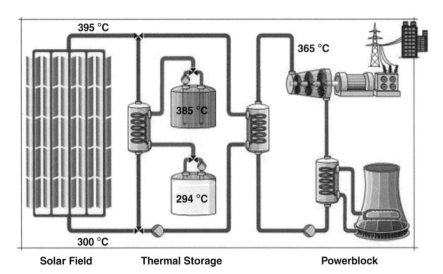

Turbine Capacity (gross)	50 MW$_e$
Turbine Efficiency (Full Power)	37%
Storage Thermal Rating	1010 MWh, 7.4 hrs
Storage Tank Size, H × D	2 tanks, 14 m × 38.5 m each
Storage Fluid	Solar salt (60% NaNO$_3$, 40% KNO$_3$)
Storage Heat Exchanger	16000 m^2
Solar Collector Field Aperture Area	510,000 m^2
Solar Collector Assembly, W × L	5.77 m × 148 m
Solar Collector Row Spacing	16.2 m
Solar Collector Fluid (HTF)	Therminol VP-1

Fig. 14.   Typical arrangement and operating data for a 50 MW$_e$ trough plant with 7.4 h storage. *Source*: SolarPACES.

using it as a fuel. The latter concept is presented in Solar Fuels in Sec. 7 of this chapter.

The principle of closed-loop thermochemical storage is that solar energy is used to drive dissociation or reforming reaction, which is preferably highly endothermic. The products are then stored, and later recombined in an exothermic reaction as and when the heat is needed. This reverses the endothermic reaction and closes the loop.

The Australian National University has developed the ammonia-based system shown in Fig. 15 over the last few decades. Production of ammonia is one of the world's largest chemical process industries, and so it essentially introduces a solar step using a point focus concentrator at around 700°C into a well-known process.[30]

Research and pilot-scale demonstration has also been carried out on a thermochemical cycle based on a methane-reforming process. In this cycle, methane and steam and/or carbon dioxide enter the solar receiver, whereupon they are converted to a carbon monoxide/hydrogen mixture in an exothermic reaction. The cycle is completed with an exothermic methanation reaction releasing heat.[31,32]

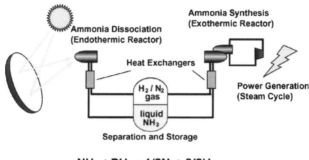

$$NH_3 + DH \rightleftharpoons 1/2N_2 + 3/2H_2$$

Fig. 15. Schematic showing the concept of using solar-dissociated ammonia in a closed loop energy storage system (*Courtesy*: ANU).

5.5 *Cost Reduction of Thermal Storage Through Higher Temperatures*

One method of significantly reducing thermal storage cost is simply utilizing higher temperature storage. Modern parabolic trough plants use oil as the HTF, which limits the maximum salt storage temperature to about 385°C even though conventional nitrate salt formulations are capable of working up to 560°C. In this low-temperature configuration for PTCs, the temperature difference is low ($\Delta T \sim 100$°C). This is because cold storage is kept at approximately 290°C in order to keep conditions well away from the salt melting point of 220°C. This low-temperature difference results in a much larger storage volume (and hence salt and tank cost) being required for a given storage capacity. A PTC has recently been demonstrated with a receiver tube that uses salt as the HTF, as noted in Sec. 4.3.1.

Central receivers can reach the upper operating temperature of today's salt formulations with relative ease, providing a working temperature difference that is two to three times higher than possible with PTC using oil as the HTF (though the ΔT that is thermodynamically useful will depend on the cycle). Thus the inventory of salt and size of tanks required is less, though the temperatures are higher. This is expected to lead to lower costs of thermal storage.

A number of research programs are under way to develop salt formulations that both decrease the melting point (Sandia have demonstrated a salt with melting temperature below 100°C) and increase the upper operating temperature to make full use of the capabilities of point-focus concentrators.

6 Concentrating Solar Power Systems

A CSP power plant is an optimization of the type of collector, its operating parameters, and the thermodynamic cycle employed. In theory, the heat engine is the defining component as concentrating solar collectors can provide heated fluids for most temperatures. For commercial systems, the cycles of choice are Rankine,

Brayton, and Stirling cycles. Each presents specific features such as preferred fluids, temperatures, efficiencies, and importantly, part load response curves.

An important point is that each of these cycles are able to operate independently of a solar input, which affords the ability for CSP to be hybridized with fossil fuel in such a way as to improve performance and reduce the cost of the solar delivered.

6.1 *Rankine Cycle*

All of the large-scale commercial CSP projects in the world at present are based on the *Rankine cycle*, the most common form of which is the steam turbine (though it is noted that a 1 MW parabolic trough power plant using an organic Rankine cycle has been built in Arizona, US). The majority use the PTC with oil as the HTF — the only real variations are that different developers have developed different trough structures. The majority use receiver tubes available commercially from a small number of suppliers.

The parabolic trough power plants typically operate with steam temperatures of about 380°C at sub-critical pressures and steam cycle efficiencies of around 37%–38%. Even lower temperature operation has been considered for linear Fresnel systems using saturated steam turbine designs from the nuclear industry. However most interest at present is in higher temperature turbines (due to higher efficiency) and thus the development of receiver tubes using molten salt or direct steam generation.

Central receivers can easily achieve the temperatures of steam turbines used in many coal-fired power stations today (∼540°C), and even the conditions needed for supercritical steam turbines of around 25 MPa and over 600°C, where cycle efficiencies are of the order of 43%.

One way in which CSP may be introduced most readily to the market is through solar steam supplementation of existing or new coal or gas-fired steam plants — this allows the solar steam to benefit from the higher efficiencies of larger steam turbines and utilizes existing power block capital; the downside is that the solar plant is then constrained to the solar irradiance of that site and the solar contribution is generally low. Overall however, such an option provides excellent transitional opportunities for CSP. In addition to integration with a conventional coal- or gas-fired Rankine cycle, the solar steam can also be integrated with the steam cycle (bottoming cycle) of a combined cycle plant (see Fig. 16).

6.2 *Brayton Cycle*

Central receivers and dishes can be used to provide hot air for *Brayton cycles* i.e. gas turbines. In the industrial context, gas turbines can be hundreds of megawatts capacity and operate with turbine inlet temperatures exceeding 1,300°C. Point-focus collectors have demonstrated they can reach these temperatures, though the materials and fluid handling for such temperatures needs to be carefully considered — transporting such high-temperature gas streams to a turbine

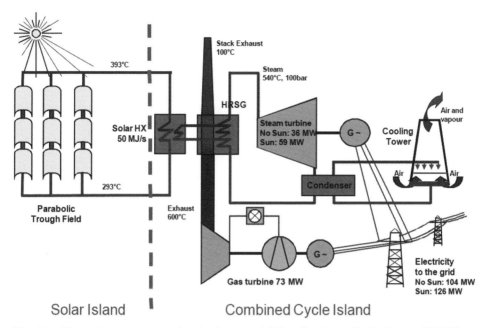

Solar Island | **Combined Cycle Island**

Fig. 16. The cycle arrangement for the Integrated Solar Combined Cycle System (ISCCS) at Kuraymat, Egypt (*Courtesy*: Fichtner Solar GMBH).

at ground level incurs either a high temperature loss or expensive insulation; alternatively, mounting a large gas turbine next to the receiver can be structurally expensive. It is noted that, at present, the allowable gas turbine combustor inlet temperature is generally much lower than the solar receiver can generate, so if solar/gas hybrid operation is desired, some significant gas turbine redesign is required or the solar air temperature would need to be reduced, which would yield a low solar contribution. Though an open-cycle gas turbine requires no water for cooling, it is often less efficient than a large steam turbine. Thus the advantage for the Brayton cycle lies in the modularity and lower cost of gas turbines for small capacity systems, or in the very high efficiency (~60%) of gas turbine combined cycles at larger capacities [see Fig. 8(b)]. A number of research and demonstration programs for the solar tower Brayton cycle are under way or been previously undertaken.[33,34]

A difficulty with such high-temperature cycles is the ability to integrate competitive thermal storage. Whilst electrochemical storage is possible after the generator (as with wind and PV), this denies the opportunity to use the inherent advantages of thermal storage of lower cost and "spinning reserve." A cycle under development for CSP is the closed-loop Brayton using *supercritical CO_2* as the working fluid. The particular properties of this fluid mean the power used to drive the compressor is significantly less than in the conventional air Brayton cycle. In an optimized configuration, this yields very high cycle efficiencies of over 50% even with dry cooling at high ambient temperature. A bottoming cycle can improve this further. Most importantly, as shown in Fig. 17, this efficiency can be

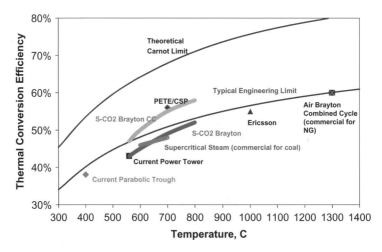

Fig. 17.   Plot of comparative cycle efficiencies versus capacity (*Courtesy*: N. Siegel).

achieved at temperatures of the order of 650–700°C, which is a relatively comfortable operating range for central receivers and is within the range of thermal storage developments.

6.3 *Stirling Cycle*

The *Stirling cycle* offers one of the highest efficiencies of the thermal cycles and has been under development and demonstration for many years. It is a compact device, and a solar dish can provide ideal concentrator characteristics. It is based on a combination of constant temperature and adiabatic processes, with the receiver being externally heated. The working fluid is usually hydrogen or helium. Demonstrated capacities have been in the range of tens of kilowatts, and solar-to-electric efficiencies of approximately 30% have been reliably reproduced — one of the highest demonstrated by solar energy in the field to date.[35]

7 CSP and Solar Fuels

Solar technologies are generally known for their generation of electricity. However, biofuels have demonstrated that a renewable resource may have great potential in the non-electricity market. Given that many of the processes for generating fuels are thermal in nature, CSP is an ideal candidate for production of fuels in addition to electricity. The attraction of solar energy in the form of a fuel rather than electricity is the versatility provided — the fuel can be used as a form of storage (it can be stored at low temperature because the energy is chemical rather than thermal), it can be used in high efficiency electricity cycles such as combined cycles or fuel cells, it can be used as a transport fuel, and it can be used in the minerals and chemical industries.

How it works – thermochemical routes for the production of solar fuels (H₂, syngas)

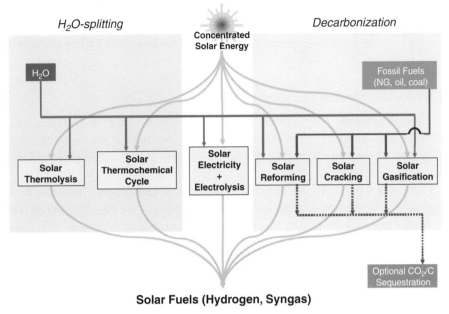

Fig. 18. Schematic of processes using CSP for fuel production.[36]

A variety of potential processes are available as illustrated in Fig. 18, including:

- steam reforming of natural gas for hydrogen production, using readily achievable temperatures of around $800°C$[37];
- redox cycles based on metal oxides for pure solar hydrogen production, but requiring quite high temperatures of over $1200°C$[38];
- solar thermal gasification of carbonaceous feedstocks such as biomass or low rank coal[38]; and
- multi-step water splitting for production of hydrogen.

No solar fuel processes have been commercialized yet. However, in the same way that CSP electricity generation has built on the foundation of industrial turbine technology, solar fuels could build on the significant technology that exists in the petrochemical industry today.

The commercial pathway often lies through appropriate transitional technologies. In the case of solar fuels, an attractive option is to use mixed reforming (a combination of steam and CO_2 as the reforming agents) to produce solar syngas (H_2/CO mix). This is the feedstock for the Fischer–Tropsch process, which can be used to produce liquid fuels such as diesel. Being a synthetic diesel derived from natural gas and solar energy, it has very clean combustion and a low CO_2 footprint. Most importantly it can be used as a "drop-in" fuel in today's vehicle fleet and is easily stored and transported from the sunbelt regions to population centers around the world. The solar syngas can also be used to produce other valuable liquid fuels such as methanol, DME, and gasoline.

Solar liquid fuels are attractive from a commercial perspective because, using a common US$/GJ basis, the market price of energy for transport fuels is significantly greater than that for electricity.

8 CSP in the Market

The first commercial CSP plants were those built by LUZ in California from 1985–1991 (see Table 1). Over 2,000,000 m^2 of trough collectors were installed in nine projects, providing 354 MW$_e$ in total, which are still in commercial operation today.

There was little commercial activity until 2006, since when there has been a strong growth in deployment of commercial CSP plants (see Fig. 19). These plants have been supported financially through various mechanisms such as up-front grants and loans, through feed-in-tariffs, or mandatory targets, reducing over time.

The CSP industry momentum has been fostered largely through the incentives offered in Spain under Royal Decree; however, growth is again underway in the US, and a range of other countries and regions are now accelerating their own programs for commercial deployment — Middle East and North Africa, South Africa, Australia, Israel, China, India, Brazil, Mexico, and Chile. An exciting concept known as Desertec aims to deploy plants in the sunbelt region of north Africa and export much of the electricity to Europe.

In Spain, legislation was such that plants needed to be restricted to 50 MW$_e$, with the aim of fostering competition and preventing monopolization. This strategy

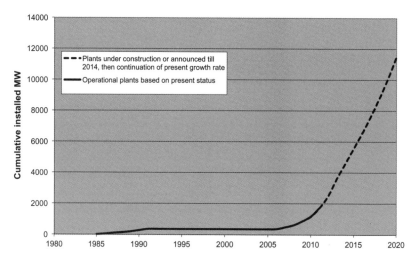

Fig. 19.   Installed global CSP capacity as at 2011, plants under construction, and predicted to be installed by 2020.

has borne fruit, with numerous companies entering the market through both "bottom-up" growth and technology acquisition. Many of these companies now have a global presence. Now, around the rest of the world, plants are typically sized at over 100 MW each, when no regulated "ceilings" are in place, with a number greater than 250 MWe. Thermal storage of upto several hours is now regularly incorporated to improve grid integration and financial rate of return. There is presently approximately 1,800 MW of CSP plants in commercial operation and some 10,000 MW at various stages of planning and approval, predominantly in the US.

There are several strategic issues presently in play for the CSP industry, including consideration of placement in relation to grid capacity strength (where dispatchability is an advantage), a move toward dry cooling (which is well understood by the power industry), and the possibility of large "solar zones" to centralize infrastructure or, conversely, the development of more modular/compact CSP technologies to expand the range of sites. However, the single most important issue is cost reduction, to be achieved through parallel paths of:

- technology improvement from R&D;
- "learning by doing" from plant deployment and operation;
- component cost reduction through mass production; and
- reduction of cost of financing through the risk reduction resulting from the above.

The US Department of Energy Sunshot initiative and various CSP roadmaps have identified the key technology areas that most effectively impact cost and have established targets that would lead to CSP being a low-cost source of dispatchable solar electricity (see Fig. 20). These generally relate to higher temperature receivers for associated advanced cycles (particularly supercritical cycles), lower cost collectors, and higher temperature storage.

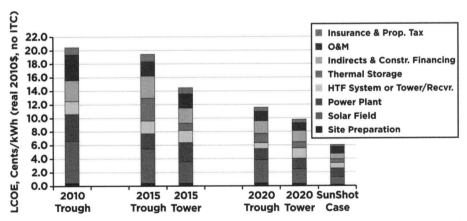

Fig. 20. The targets established by the US DoE Sunshot initiative and corresponding road-maps.[39]

9 Conclusions

Concentrating solar power is in a strong commercial growth phase at present, building on many years of strong R&D activity. The present trend continues the theme of larger capacity plants, most of which are based on parabolic trough technology. However, central receivers using molten salt as the heat transfer fluid and storage medium are now being increasingly deployed as a technology of choice.

There is a general move toward higher temperature systems for improved efficiency, as this leads directly to a reduction in collector area, which is the single most significant capital cost component, and lower storage cost. The industry has developed rapidly since 2006, when it consisted of relatively few companies with in-house technology, to now, where a diverse range of large multi-national companies has been established, providing strong competition in components, construction and project management, and finance. Risk still exists as the projects are large and strongly dependent on capital that needs to be underwritten by power purchase agreements and, for now, other incentives.

New and advanced technology opportunities are also emerging. These include very high efficiency supercritical CO_2 Brayton cycles, and the opportunity to apply all of the collector and receiver knowhow developed for high-temperature CSP electricity generation as the basis for developing a new market in commercial solar fuel technology.

The key for the future of CSP lies in its ability to deploy dispatchable solar energy in large quantities, building on much of the knowhow and technology available in the energy industry today.

References

1. International Energy Agency, *Key World Energy Statistics 2011* (IEA Press, Paris, 2010).
2. C. J. Winter, R. L. Sizmann and L. Vant-Hull, *Solar Power Plants: Fundamentals, Technology, Systems, Economics* (Springer-Verlag, Berlin, 1991).
3. R. Pitz-Paal, High temperature solar concentrators, in *Solar Energy Conversion and Photoenergy Systems*, eds. J. B. Galvez and S. M. Rodriguez (*Encyclopedia of Life Support Systems* (EOLSS). Developed under the Auspices of the UNESCO, EOLSS Publishers, Oxford, UK, 2007). Available at: http://www.eolss.net. Accessed 21 February 2012.
4. C. E. Kennedy, Review of mid to high temperature solar selective surfaces, Technical Report NREL/TP-520-31267 (2002).
5. A. Akbarzadeh, J. Andrews and P. Golding, Solar pond technologies: A review and future directions, Chapter 7 in *Advances in Solar Energy*, Vol. 16, ed. D. Y. Goswami (Earthscan, USA, 2005).
6. D. R. Mills, Solar thermal electricity, in *Solar Energy—The State of The Art — ISES Position Papers*, ed. J. M. Gordon (James and James, 2001), p. 635.
7. Available at: http://www.solar-updraft-tower.com/de#commercial_sut/scale_up. Accessed 1 June 2012.
8. Available at: http://www.solar-updraft-tower.com/de#commercial_sut/costs. Accessed 1 June 2012.

9. P. DeLaquil, D. Kearney, M. Geyer and R. Diver, Solar thermal electric technology, in *Renewable Energy—Sources for Fuel and Electricity*, eds. T. B. Johansson, H. Kelly, A. K. N. Reddy and R. H. Williams, exec. ed. L. Burnham (Earthscan Publications Ltd and Island Press, 2001), p. 219.

10. Available at: http://www.nrel.gov/csp/troughnet/power_plant_data.html. Accessed 1 June 2012.

11. R. Pitz-Paal, J. Dersch, B. Milow, F. Téllez, A. Ferriere, U. Langnickel, A. Steinfeld, J. Karni, E. Zarza and O. Popel, Development steps for parabolic trough solar power technologies with maximum impact on cost reduction, *J. Solar Energy Eng.* **129** (2007), p. 371.

12. Available at: http://www.nrel.gov/csp/solarpaces/project_detail.cfm/projectID=19. Accessed 1 June 2012.

13. D. R. Mills and G. L. Morrison compact linear Fresnel reflector solar thermal power-plants, *Solar Energy* **68**(3) (2000), pp. 263–283.

14. G. Morin, J. Dersch, W. Platzer, M. Eck and A. Haberle, Comparison of linear fresnel and parabolic trough collector power plants, *Solar Energy* **86**(1) (2012), pp. 1–12.

15. Available at: http://www.physics.usyd.edu.au/app/research/solar/clfr.html. Accessed 1 June 2012.

16. Available at: http://www.novatecsolar.com/79-1-Australias-Largest-Solar-Thermal-Plant-to-Expand.html. Accessed 3 May 2012.

17. Available at: http://www.areva.com/EN/operations-415/concentrated-solar-power-technology.html. Accessed 12 March 2012.

18. Available at: http://minister.ret.gov.au/MediaCentre/Speeches/Pages/Round1Solar FlagshipsProjects.aspx. Accessed 12 March 2012.

19. Available at: http://www.nrel.gov/csp/solarpaces/project_detail.cfm/projectID=37. Accessed 15 May 2012.

20. Available at: http://www.novatecsolar.com/8-1-Projects.html. Accessed 3 May 2012.

21. Available at: http://www.volker-quaschning.de/articles/fundamentals2/index_e.php. Accessed 12 February 2012.

22. P. DeLaquil, D. Kearney, M. Geyer and R. Diver, Solar thermal electric technology, in *Renewable Energy — Sources for Fuel and Electricity*, eds. T. B. Johansson, H. Kelly, A. K. N. Reddy and R. H. Williams, exec. ed. L. Burnham (Earthscan Publications Ltd and Island Press, 2001), p. 268.

23. Available at: http://www.wizardpower.com.au/index.php?option=com_content&view= article& id=36&Itemid=35. Accessed 12 February 2012.

24. K. Lovegrove, G. Burgess and J. Pye, A new $500\,\mathrm{m}^2$ paraboloidal dish solar concentrator, *Solar Energy* **85**(4) (2011), pp. 620–626.

25. Available at: http://www1.eere.energy.gov/solar/sunshot/vision_study. Accessed 2 July 2012.

26. Available at: http://www.nrel.gov/csp/solarpaces/project_detail.cfm/projectID=40. Accessed 15 May 2012.

27. D. A. Brosseau, P. F. Hlava and M. J. Kelly, Testing thermocline filler materials and molten salt heat transfer fluids for thermal energy storage systems used in parabolic trough solar power plants, Sandia Report, SAND2004–3207 (2004).

28. S. Zunft, M. Hanel, M. Kruger, V. Dreißigacker, F. Gohring and W. Wahl, Julich solar power tower—Experimental evaluation of the storage subsystem and performance calculation, *J. Solar Energy Eng.* **133**(3) (2011), p. 031019.

29. D. Laing, High-temperature solid-media thermal energy storage for solar thermal power plants, *Proc. IEEE* **100**(2) (2012), pp. 516–524.

30. R. Dunn, K. Lovegrove, G. Burgess, A review of ammonia-based thermochemical energy storage for concentrating solar power, *Proc. IEEE*, **100**(2) (2012), pp. 391–400.
31. M. Epstein and I. Spiewak, Design and operation of the Weizmann Institute 480 kW solar reformer in an energy storage cycle, *Proceedings of the 7th International Symposium on Solar Thermal Concentrating Technologies*, **4**, Moscow, Russia (1994).
32. J. H. Edwards, G. J. Duffy, R. Benito, K. T. Do, N. Dave, R. McNaughton and S. P. S. Badwal, CSIRO's advanced power generation technology using solar thermal — Fossil energy hybrid systems, *Proceedings of the Fifth International Conference on Greenhouse Gas Technologies*, Cairns, Australia (2000), pp. 863–868.
33. Available at: http://ec.europa.eu/research/energy/pdf/solgate_en.pdf. Accessed 29 November 2011.
34. Available at: http://sfera.sollab.eu/downloads/4th_presentation.pdf. Accessed 18 February 2012.
35. T. Mancini, P. Heller, B. Butler *et al.*, Dish-stirling systems: An overview of development and status, *J. Solar Energy Eng.* **125** (2003), pp. 135–151.
36. Available at: http://www.solarpaces.org/Library/docs/Solar_Fuels.pdf. Accessed 1 June 2012.
37. W. Stein, J. Edwards, J. Hinkley and C. Sattler, Fuels–hydrogen production — Natural gas: Solar thermal steam reforming, in *Encyclopedia of Electrochemical Power Sources*, Jürgen Garche, Ed. in Chief, (Elsevier, Amsterdam, 2009), pp. 300–312.
38. Available at: http://www.pre.ethz.ch/publications/0_pdf/books/Encyclopedia_of_Energy_Solar_Fuels_and_Materials.pdf. Accessed 10 April 2012.
39. *SunShot Vision Study*. Available at: http://www1.eere.energy.gov/solar/sunshot/vision_study. Accessed 2 July 2012.

Chapter 11

Biomass

Mark Downing* and Anthony F. Turhollow Jr.[†]

Oak Ridge National Laboratory
Oak Ridge, TN, USA
* *downingme@ornl.gov*
[†] *turhollowaf@ornl.gov*

This chapter summarizes some of the major biomass resources available and conversion technologies in use today. In addition to understanding what biomass is, it is important to understand the constituents of biomass, which are cellulose, hemi-cellulose, and lignin. We summarize a diverse array of biomass feedstock resources such as woody crops, herbaceous perennials, forest resources, and agricultural crop residues. For several of the significantly available and used resources, we attempt to summarize relative productivity as well as costs of production. We summarize an array of conversion technologies to include those used for biomass to electricity, liquid fuels, and biochemicals. For each case, we identify specific countries where these biomass resources are concentrated and utilized, as well as the products of conversion (e.g. sugarcane resources to ethanol in Brazil). We conclude that there is a great deal of biomass in the world, although the useable quantities may be more efficiently used locally, and that specific biomass resources have certain qualities that are better utilized in some conversion technologies than others. Finally, it is important to note that the estimates of biomass availability do not reflect specific cost of production strategies, and the demand for end-products from conversion does not imply availability of these commodities in every part of the world.

1 Introduction

Biomass is organic matter that can be transformed into energy. Included in biomass is organic material of agricultural origin (grains, crop residues, vegetable oils, sugar crops, manure, and purpose-grown cellulosic crops) and woody material (residues and thinnings, pulping liquors, fuelwood, and purpose-grown short rotation woody crops). These resources are quite substantial, with more than 1 billion metric tons estimated to be available for energy use in the US alone.[1,2] According to Macqueen and Korhaliller,[3] biomass makes up 10% of the world's primary energy consumption.

There are multiple methods of converting biomass into other forms of energy. At present, most biomass used for energy is combusted, either in traditional fireplaces or stoves. Biomass is made up of carbohydrates (sugars and starch), cellulose, hemi-cellulose, lignin, lipids (vegetable oil), protein, and ash (inorganics). It can also contain significant amounts of water, up to 80% (wet-weight basis) in fresh material. Starch in maize (corn) and other grains and sucrose from sugarcane are converted into ethanol by fermentation. The oils from oilseed crops and fats are converted into biodiesel by transesterification.

2 Ethanol

The primary liquid biofuel produced is ethanol, 87×10^9 billion liters (L) in 2010, with the US (58%) and Brazil (30%) accounting for the majority of the production.[4] Ethanol production and consumption have been increasing since 2001. The top dozen countries producing ethanol are shown in Table 1. The world production from 2001 to 2009 shows a fourfold increase during that period.

Presently, ethanol is produced primarily from first-generation feedstocks — corn and sugarcane. Ethanol yield per hectare (ha)[a] from maize in the US and sugarcane in Brazil are 4,000 and 6,000 L per ha, with yields expected to increase to 5,500 and 9,000 L per ha in the US and São Paulo state (the biggest ethanol-producing region in Brazil) by 2015–2017, respectively (Fig. 1). In 2004, Brazil's ethanol yield was the 6,000 L per ha, by 2010, ethanol yield had increased to about 7200 L per ha.[7]

3 Ethanol Production from Sugarcane in Brazil

Brazil is the second largest producer of ethanol in the world, producing 31 billion L in 2010, utilizing sugar from sugarcane. The production and use of sugarcane-derived ethanol for fuel began in 1975, when Proálcool was started in response to high oil prices and a crisis in the sugar market. Use of ethanol as a fuel was stimulated by feedstock availability, a supportive government policy, and improvements in sugarcane production and ethanol conversion processes.[7]

Table 1: World Ethanol Production 2001–2009 (in Billions of Liters per Year).[5]

	2001	2003	2005	2007	2009
Canada	0.23	0.23	0.26	0.80	1.09
US	6.68	10.62	14.78	24.68	41.40
Brazil	11.47	14.47	16.04	22.56	26.10
Colombia	0.00	0.00	0.03	0.27	0.30
Jamaica	NA	0.15	0.13	0.28	0.40
France	0.12	0.10	0.15	0.54	1.25
Germany	0.00	0.00	0.16	0.39	0.75
Spain	0.00	0.20	0.30	0.36	0.46
Australia	0.00	0.00	0.02	0.08	0.20
China	0.00	0.80	1.20	1.67	2.15
India	0.17	0.19	0.21	0.26	0.34
Thailand	0.00	0.00	0.07	0.17	0.40
World	**18.76**	**27.02**	**33.91**	**53.63**	**77.03**

---

[a]1 hectare (ha) $= 10{,}000\,\mathrm{m}^2 = 2.471$ acres.

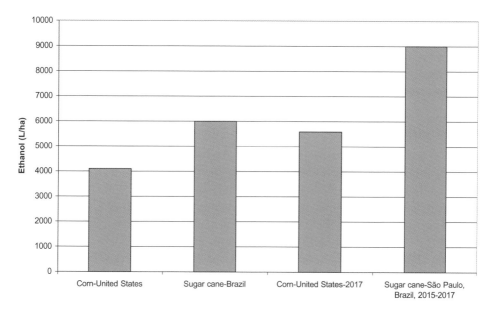

Fig. 1. Per ha ethanol production from corn and sugarcane.
Source: Sugarcane: Brazil (in 2004) and São Paulo — Goldemberg[6]; Corn: US, based on average 2008–2010 corn yield 9.86 Mg per ha[8] and assuming 417 L per Mg, Corn: US, 2017 based on USDOE[9] corn yield 12.92 Mg per ha and assuming 432 L per Mg.

Based on Fig. 5 in Valdes,[7] sugarcane yields have increased from about 50 tons per ha in 1970–1975 to 90 tons per ha in 2010, and ethanol yield has increased from about 40 to 80 L per ton in the same timeframe. (Note that the Valdes publication lists tons[b] and it is not clear whether this is metric tons or short tons). Ethanol yield has increased from 2000 to 7200 L per ha from 1970–1975 to 2010. In São Paulo province, where the majority of ethanol is produced, yields are even higher.

In the early 1980s, vehicles that could use hydrous ethanol (using neat-ethanol engines) became attractive to consumers because the government ensured that the price of hydrous ethanol was 64.5% that of gasoline. Sales of neat-ethanol vehicles and their market share increased to over 90% of vehicle sales. However, supply of hydrous ethanol was not guaranteed, and a shortage of ethanol in the early 1990s led to a crisis and gradual abandonment of neat-ethanol vehicles. In 2003, flex-fueled vehicles were introduced, which are capable of utilizing ethanol–gasoline blends ranging from 0% to 100% ethanol.[6]

The area planted to sugarcane has been increasing, with 4.273, 4.805, and 9.191 million ha in 1990, 2000 and 2010, respectively, and is expected to reach

---

[b]1 tonne = 1 metric ton = 1,000 kg = 10^6 g = 1 megagram = 1 Mg = 2,205 lbs.
1 long ton = 2,240 lbs = 1.016 tonne; This long ton was the pre-metric British system ton.
1 short ton = 2,000 lbs = 0.907 tonne.

10.045 million ha in 2020. About 60% of the sugarcane produced is used for ethanol production, and this is expected to remain about the same. Yields have been increasing over time and investments are being made in higher-yielding varieties and practices. Brazil's ethanol industry is operating at about 75% of installed capacity and capacity is expected to reach 46.4 million L by 2018.[7]

4 Biodiesel

Biodiesel is produced via a process known as transesterification, utilizing vegetable oils and animal fats with an alcohol (typically methanol) to produce an ester (typically methyl ester). Biodiesel production increased rapidly from 2001 to 2009, increasing by almost 15 fold. Biodiesel production was 18 billion L in 2009, with the majority produced in the European Union (56%), and followed by Asia (12%), the US (11%), and Brazil (9.0%). The countries with the highest production[10] are Germany, France, the US, Brazil, and Argentina (Table 2). In the US, soybean oil and animal fats each account for about half the feedstock, while in the EU, rapeseed oil is the most widely used feedstock.

Biodiesel can be substituted for petroleum-derived diesel. However, there can be issues in cold weather with biodiesel. But the biggest drawback with using vegetable oils is its cost. A liter of biodiesel requires 0.90 kg of vegetable oil. With the recent increase in agricultural commodity prices, soybean oil, typically the least expensive vegetable oil in the US, is expected to be about $1.20 per kg in 2011. Just the feedstock costs alone amount to $1.07 per L.

5 Thermal Processes

There are a number of thermal processes that can be used to convert biomass into biofuels and electricity: pyrolysis, gasification, and combustion. While a biological fermentation process might get up to 460 L of ethanol per dry Mg of biomass,

Table 2: World Biodiesel Production 2001–2009 (in Billions of Liters per Year).[10]

Country	2001	2003	2005	2007	2009
Germany	0.39	0.91	2.26	4.54	2.97
France	0.34	0.42	0.49	1.09	2.38
US	0.03	0.05	0.34	1.85	1.91
Brazil	0.00	0.00	0.00	0.40	1.61
Argentina	0.01	0.01	0.01	0.44	1.34
Italy	0.16	0.31	0.45	0.53	0.76
Spain	0.09	0.12	0.19	0.19	0.64
Thailand	0.00	0.00	0.02	0.07	0.61
Belgium	0.00	0.00	0.00	0.19	0.47
China	0.01	0.01	0.05	0.35	0.46
Poland	0.00	0.00	0.07	0.05	0.39
Austria	0.02	0.03	0.09	0.30	0.35
World	1.22	2.08	4.48	11.77	17.88

Table 3: Pyrolysis Products Depending on Conditions.[11]

Mode	Conditions	Liquid	Solid (char)	Gas
Fast	500°C, 1 s vapor residence	75%	12%	13%
Intermediate	500°C, 10–20 s vapor residence	50%	20%	30%
Slow	400°C, long solids residence	30%	35%	30%
Gasification	800°C, long vapor residence	5%	10%	85%

a thermal process, depending on the feedstock could get up to 530 L of ethanol per dry Mg.

5.1 Pyrolysis

Pyrolysis is heating in the absence of oxygen, while controlling temperature, pressure, and residence time to determine the formation of gaseous, liquid, and solid products (Table 3). Fast pyrolysis sets conditions [short residence time, lower temperature process (350–500°C), rapid quenching] such that mostly liquid fuel (referred to as pyrolysis- or bio-oil) (70 wt %+) is produced. Pyrolysis takes place at atmospheric pressure and requires a low moisture ($< 10\%$) content feedstock (to minimize moisture in the bio-oil) and small particle size ($\sim$2–3 mm).[11,12]

There are issues associated with the acidity and stability of the bio-oil. Bio-oil has a low pH (2.5), high water content (15%–30%), and a heating value of 16–19 MJ per kg (higher heating value).[13,14] Using a pyrolysis process, wood and herbaceous materials have the potential of producing 570 and 530 L per dry Mg (148 and 126 gallons per dry ton) of ethanol, respectively.[15]

Wood has a higher output of product because of its higher carbon (energy) content compared to herbaceous material. Wood is higher in lignin than herbaceous materials and thermal processes take advantage of this property.

Bio-oil can be utilized for a number of products — liquid boiler fuel; petroleum-like fuels by direct utilization, upgrading via hydrotreating and hydrocracking (Fig. 2), gasification, fermentation, reform to hydrogen; and phenolics replacement. The solid produced (char) may be utilized as soil amendment, activated carbon, and process heat or for energy for the pyrolysis process. The gaseous product can also be used for providing energy for the pyrolysis process.

5.2 Gasification

Gasification converts the carbon fraction in biomass into synthesis gas (carbon monoxide and hydrogen). The synthesis gas has wide potential for a variety of uses: combustion, in gas engines, in gas turbines, in fuel cells, upgraded to methane and methanol, fermented to ethanol, or via the Fischer–Tropsch process into various liquid fuels, including diesel. Two of the major research challenges for gasification are developing improved catalysts for liquid fuel production and developing efficient gas

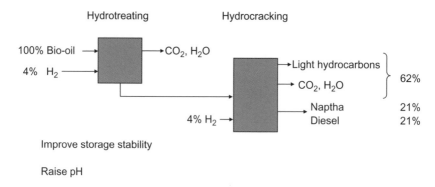

Fig. 2.   Upgrading pyrolysis oil.[15]

Table 4:   Some Possible Gasification to Product Pathways.

Step 1 (Biomass conversion)	Step 2 (Syngas conversion)	Step 3 (Product generation)
Gasification to syngas	Synthesis into mixed alcohols	Distill to pure ethanol
	Catalytic conversion into hydrocarbon fuels	Refining to diesel and naptha (for jet fuel) fractions

cleaning technologies.[16] Two potential gasification conversion pathways are shown in Table 4.

5.3 Combustion

In thermal processes, all the fractions of the biomass are converted into energy, excluding the ash fraction. In the case of wood, this allows for up to 99% of the biomass to potentially be utilized to make fuel products. Combustion of biomass is generally not energy efficient, with conversions in the 20%–25% range for electricity generation (i.e. 20%–25% of the energy in the biomass is converted into electricity). There are two primary reasons for this: (1) the scale of the electricity generation facility is usually small, less than $25\,\mathrm{MW}_e$, which makes investments uneconomic that can make the process more energy efficient and (2) the moisture content of the biomass is usually high, which makes the conversion less efficient (Fig. 3). First gasifying the biomass and then generating electricity would be more energy efficient.

6 Biological Processes

6.1 Anaerobic Digestion

In anaerobic digestion, microorganisms break down biodegradable organic material in the absence of oxygen and produce a mixture of gases, primarily methane and carbon dioxide. The biogas produced has a medium (as compared to natural gas) heat value that ranges from 19,000 to 29,000 kJ per m^3. (Natural gas has a higher

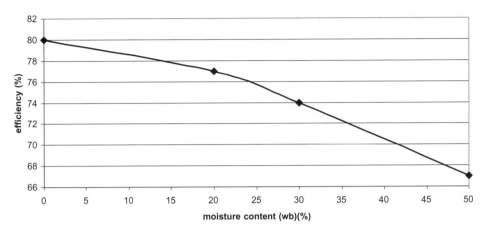

Fig. 3.   The effect of moisture content on conversion efficiency.

heating value of around 38,200 kJ per m^3.) The biogas can be upgraded to natural gas quality. Anaerobic digestion is used commercially at sewage treatment facilities and industrial facilities to treat organic wastes and on farms to treat animal wastes. India and China make widespread use of anaerobic digesters in rural areas to provide fuel. Kangmin and Ho[17] report that, at the end of 2005, there were 17 million anaerobic digesters in China.

On-farm digesters are used to treat livestock manure, although other organic feedstocks such as crop residues or off-farm food processing wastes can also be processed. For manure, after processing, the remaining organic matter now has reduced odor pollution, reduced pathogens, and reduced biological oxygen demand, while retaining most of its nutrient value for use as a fertilizer. In the US, most anaerobic digesters supply biogas to an engine that produces electricity.

The three main on-farm digester designs are: (1) covered lagoons; (2) plug-flow; and (3) mixed or stirred.[18] Crenshaw[19] reports costs for two covered lagoons as, \$778,586 for 496 cows and \$1,265,194 for 1,600 cows, and Lazarus[18] reports regression equations for costs based on Crenshaw[19] (see Table 5).

6.2 *Fermentation*

Fermentation is the main process currently used to produce liquid transportation fuel in the form of ethanol from maize in the US and sugarcane in Brazil. Fermentation converts carbohydrates into ethanol, although other products can be produced as well. The lignin fraction is not converted in fermentation processes and is a byproduct.

Cellulosic feedstocks can utilize many pathways to be converted into ethanol. Both grasses and hardwoods have xylose as their main hemi-cellulosic sugar, whereas softwoods have mannose as their main hemi-cellulosic sugar. The discussion that follows is geared toward xylose as the main hemi-celluslosic sugar.

Table 5: Anaerobic Digester Costs (US$).[16]

	Digester type	
	Plug-flow	Mixed
Digester costs	$678,064 + $563 per cow	$354,866 + $61 per cow
Ancillary items (grid connection, hydrogen sulfide removal	13%	
Cost for 700 cow herd	$1.2 × 10⁶	
Cost for 2,800 cow herd	$2.5 × 10⁶	
Operating and maintenance for digester with electrical generation	5%	
Output	0.71 to 3.82 m³ per cow-day	
Electrical output based on 1.98 m³ cow-day^{-1}, 60% methane in biogas, 27% thermal conversion to electricity, 90% engine run time	1,000 kWh per cow-day	
Breakeven electricity cost ($ per kWh)		
700 cows	$0.22	
2,800 cows	$0.12	

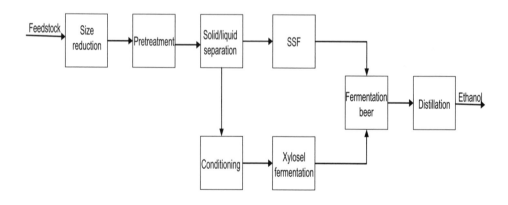

Fig. 4. One pathway for biological ethanol production.

The main steps are: (1) size reduction, (2) pretreatment to release xylose and cellulose, (3) solid (to simultaneous saccharification and fermentation)/liquid (liquor to conditioning) separation, (4) conditioning prior to xylose fermentation, (5) xylose fermentation, (6) simultaneous saccharification and fermentation of glucose (SSF), and (7) distillation of the ethanol product form the two fermentation processes (see Fig. 4). There are many possible variations on this process, such as using the stream (or part of it) from xylose fermentation to dilute the solids going into the SSF reactor.

There are a number of choices for the pretreatment step [e.g. dilute acid, concentrated acid, steam explosion, ammonia fiber expansion (AFEX), organosolv].

The National Renewable Energy Laboratory has published a number of design and economic studies on cellulosic ethanol,[20−22] with an update to be released.[23]

6.3 *Algae*

Algae have the potential of producing large quantities of biomass-derived fuels on small areas of land and to utilize saline waters. However, algae have harvest and handling challenges. The two methods of cultivating microalgae and cyanobacteria are open ponds and closed photobioreactors. Macroalgae (seaweed) has different requirements and are grown in coastal facilities or offshore.

Research on algae for fuels started in the 1950s with work on utilizing algae to produce methane via anaerobic digestion. Interest was revived after the oil crisis in the 1970s, focusing on methane and hydrogen. The US Government–sponsored research on microalgae was centered at the Solar Energy Research Institute (later renamed the National Renewable Energy Laboratory) with the Aquatic Species Program (ASP), and the focus turned to lipid oils and diesel replacement. There was still work on hydrogen. The ASP started in 1978 and continued through 1996. Over 3,000 species were screened and 300 promising species selected primarily green algae (Chlorophyceae) and diatoms (Bacillariophyceae). Choosing the right starting species is important. While a foundation was laid, much work remains to be done. There is a need to understand biosynthetic pathways and regulatory mechanisms and develop genetic strategies. Dewatering and lipid extraction are significant technical and cost challenges.[24,25]

Like lignocellulosic crops, microalgae can be converted into numerous products, depending on the conversion process chosen.

7 Dedicated Energy Crops

Dedicated energy crops include both herbaceous (e.g. switchgrass, sorghum *Miscanthus*) and short rotation woody crops (e.g. hybrid poplar, willow, Eucalyptus). The main advantages dedicated energy crops have over conventional food crops is the potential to produce much higher yields per unit land area and to utilize lands that food crops do not use.

7.1 *Switchgrass*

Switchgrass (*Panicum virgatum*) is a perennial, thin-stemmed grass native to North America. It is known as a C4 plant. The two main photosynthetic pathways in plants are C3 and C4. The C3 plants fix carbon dioxide through photorespiration and require stomatal opening to acquire CO_2. The C4 plants acquire CO_2 from malate and do not require open stomata, thus providing higher water use efficiency and producing more biomass in hotter, drier climates than C3 plants. Under conditions of moderate temperatures and available soil water, C3 plants typically have an advantage in CO_2 fixation and thus overall growth.[26]

Parrish and Fike[27] and Mitchell *et al.*[28] provide overviews of switchgrass. It is well adapted to low rainfall areas and is efficient in its use of nutrients. It has two major ploidity levels (tetraploid and octaploid) and two ecotypes — upland and lowland. The upland variety has finer stems but has a lower potential yield than lowland varieties, where lowland varieties are adapted. Yield is linked to temperature and the timing of reproductive growth and is connected to photoperiod. By moving lower latitude cultivars to higher latitudes, the length of the growing season can be extended and hence the potential yield increased. However, frost tolerance can be an issue. Planting is done with seed. Weed control is necessary for successful stand establishment. Full yield is not expected until the third year of growth. Nitrogen fertilization is needed for sustained optimum yield. If harvesting is carried out after a killing frost, then 13 to 16 kg N per Mg of harvested yield is needed.[28] One harvest per year after senescence and translocation of nutrients allows stands to persist for an extended period. Two harvests per year can increase the yield in some areas, but additional nutrients will need to be applied to compensate for nutrients removed in the midseason harvest, and stand longevity may suffer. In addition to its potential use as a biofuel, switchgrass can also be utilized for soil erosion control and as a filter strip along streams to intercept soil, nutrients, and pesticides.

Most of the switchgrass currently grown is from varieties developed for forage. In the past two years varieties specifically bred for bioenergy have been released. F_1 hybrids of Kanlow (lowland variety) and Sumner (upland variety) produced higher yields than either parent at Mead, Nebraska (Table 6).

7.2 *Sorghum*

Sorghum (*Sorghum bicolor*) is a morphologically diverse species with grain, forage, sweet, and energy varieties. It is a C4 grass. Grain sorghum is short stemmed and produces as much residue as grain. Forage sorghum is designed for use as livestock feed. Sweet and energy sorghums have potential application as bioenergy feedstocks.[30] As an annual, as opposed to a perennial crop, sorghum can be readily rotated with traditional food crops, allowing added flexibility to production systems.

Table 6:   Hybrid versus Parent Switchgrass Yield.[29]

Strain	Yield (Mg per ha) (relative to Kanlow)
Kanlow	14.7 (1.00)
Sumner	12.4 (0.84)
Kanlow × Sumner F_1 hybrid	20.9 (1.42)

Sweet sorghum has a sweet stalk (contains sugar) but also produces a large amount of lignocellulosic biomass in the stalk and may also produce grain. The International Crops Research Institute for the Semi-Arid Tropics (ICRISAT) is developing sweet sorghums adapted to Asia, Africa, and the Americas. In areas with longer growing seasons, it may be possible to grow two crops in a single year. ICRISAT is developing sweet sorghum that not only produce sugar but also relatively large amounts of grains.[31,32] Sweet sorghum also has the potential to complement sugarcane by being harvested during the period of the year when sugarcane is not being harvested. While sweet sorghum can give sugar yields comparable to sugarcane, its sugar is not used for refined sugar production because its high starch content interferes with sucrose crystallization and hastens the conversion of sucrose to glucose and fructose.[33]

Energy sorghums are specifically bred for energy purposes (producing high biomass yields). Energy sorghums will be photoperiod sensitive and do not flower in temperate climates, thus utilizing the entire possible growing season to accumulate primarily lignocellulose. Texas A&M University is conducting a breeding program for energy sorghum, with an annual yield goal of 34 to 45 dry Mg per ha.[34]

7.3 *Miscanthus*

Giant Miscantus (*Miscanthus x gignateus*) (hereafter referred to as *Miscanthus*) is a tall, thick-stemmed grass related to sugarcane. It is a sterile triploid and is a C4 grass. It can be highly productive and has the advantage over other herbaceous crops of being able to be harvested in the spring after overwintering. The yield will be about one-third lower in the winter but the harvested plant matter will be drier than in the fall. A major disadvantage of *Miscanthus* is that it is vegetatively propagated (using rhizomes) and requires a specialized planter for mechanical planting.

Aravindhakshan *et al.*,[35] Jain *et al.*,[36] and Khanna *et al.*[37] estimated the cost to establish Miscanthus at $705, $2,957, and $613 per ha, respectively. The cost in Jain *et al.* is so much higher because they assume a cost of $0.25 per rhizome and 10,000 rhizomes per ha, for a cost of $2,500 per ha. Aravindhakshan *et al.* base their rhizome cost on Khanna *et al.*, which is a planting density of 10,000 rhizomes per ha at a cost of $335 per ha. Two papers[35,37] have compared the economics of switchgrass and *Miscanthus* production and, because of differing assumptions, come to different conclusions on the relative economics of the two crops. If harvested in the spring, *Miscanthus* has very low nutrient requirements because most of its nutrients have been translocated back into its roots. Christian *et al.*[38] reported no response to nitrogen in a 14-year study in the United Kingdom, while Ercoli *et al.*[39] did report response to nitrogen in Italy. Rather than looking at switchgrass and *Miscanthus* as competing crops, because of their differing harvest windows, they can be seen as complementary crops by providing fresh supplies of biomass at different times of the year and by diversifying biomass supply sources.

7.4 *Sugarcane and Energy Cane*

Sugarcane (*Saccharum* spp.) is a thick-stemmed tall grass that is cultivated as a perennial crop, primarily for its storage of sucrose (a disaccharide consisting of glucose and fructose) in its stem. It is a C4 grass and is a genetically complex crop from interspecific hybridization efforts, primarily from *Saccharum officinarum* and *Saccharum spontaneum*. In the US, sugarcane is grown as a monoculture and replanted every four to five years. Sugarcane is vegetatively propagated, with one acre of seedcane being able to plant 6–10 acres of sugarcane. Vegetative planting is expensive both in materials and labor.

Energy cane is bred for higher biomass yields (and has lower stalk sugar) than sugarcane. Stalk numbers and heights are higher than sugarcane, and as a result, one acre of energy cane can provide 13 acres of seed cane. Energy cane is also expected to have at least two more harvests than the four or five one gets from sugarcane. Thus annualized planting costs are expected to be about $143 per ha, versus $309 per ha for sugarcane. In Louisiana, US, energy cane varieties have yielded 12 5green Mg per ha, with 18 to 31 dry Mg per ha of fiber and 9 to 13 dry Mg per ha of sugar.

Because of its thick stalks with a waxy coating and the need to get energy cane off the field before the next growing season, energy cane, like sugarcane, will need to be harvested green and dewatered or, for energy cane, ensiled if it is to be stored for later use. Sugarcane breeding begins with hybridization with the introgression of desirable traits from the wild relative of sugarcane, *Saccharum spontaneum*. Early generation progeny from these crosses exhibit high levels of hybrid vigor, which includes increased cold tolerance, better rationing ability (re-sprout from the roots), increased tolerance of moisture extremes, increased disease and insect tolerance, and more efficient nutrient utilization.[40] The early-generation hybrids are ideal candidates for energy cane for biomass.

7.5 *Hybrid Poplar*

The genus *Populus* includes almost 30 species. Hybrids within and among these species are referred to as "hybrid poplars." For bioenergy purposes, poplars are planted at relatively dense spacing (e.g. 1680 trees per ha) using cuttings. After planting, it is necessary to control weeds using herbicides and/or mechanical cultivation. Attention must also be paid to insects and diseases. Harvest is done utilizing standard pulpwood harvest systems or purpose designed machines that combine felling and chipping or bundling in one machine.

Commercial yields of poplar for pulp are not available. Yields of newly selected poplar genotypes in small plot experiments have exceeded 15 dry Mg per ha on good agricultural soils in southern Wisconsin and Iowa.[41,42] Sustainable yields of 10, 13, and 20 dry Mg per ha can be expected in the Midwestern, Southern, and Northwestern US, possibly increasing to twice these levels with appropriate research and development (USDOE, 2011). Production costs (to the forest landing) in the US are estimated at between $28 and $66 per dry Mg.[2]

7.6 *Willow*

Interest in shrub willow (*Salix* spp.) as an energy crop is focused in Europe and North America, particularly in Sweden, the US, and Canada. Trials began in Sweden in the mid 1970s and in the US in 1986. Characteristics that make willow a good candidate for biomass include (1) ease of propagation from dormant hardwood cuttings; (2) a broad underutilized genetic base; (3) ease of breeding for multiple characteristics; and (4) ability to re-sprout after multiple cuttings.[2]

To produce biomass from willow, genetically improved varieties are planted in prepared fields where weeds have been controlled. Weed control utilizes mechanical and chemical means and should begin in the fall before planting if perennial weeds are present. Planting of willows takes place as early in spring as feasible at 15,000 unrooted, dormant cuttings per ha using mechanized planters attached to tractors, capable of planting 0.8 ha per hour. After the first year of growth during the dormant season, the willows are cut back to just above ground level to encourage coppice regrowth of multiple stems. After three to four years of growth, the willows are mechanically harvested during the dormant season after the leaves have been dropped, utilizing forage harvesters with cutting heads specifically designed for willow coppice material. The forage harvester produces uniform, consistently sized chips that can be collected and delivered with no additional processing and then additional size reduction and/or drying can take place at an end-user facility.

Willow needs to be fertilized at about 100 kg per ha in the spring after each harvest. They are allowed to grow another three or four years before harvest, and it is projected that seven harvests are possible before replanting is required because the willow stools have expanded too much to allow harvesting.

Fertilized, irrigated yields of unimproved willow varieties have exceeded 27 dry Mg per ha.[43,44] On non-irrigated research-scale trials in central New York (in the US), yields ranged from 8.5 to 11.6 dry Mg per ha.[43,45,46] Second rotation yields of the five best producing unimproved varieties showed yield 18% to 62% higher than first rotation.[47] Breeding and selection of willow in the US began in the mid-1990s. New varieties from the initial rounds of breeding in the late 1990s have produced yields up to 40% and 70% greater than first and second rotations using standard varieties used in the early yield trials.

In an economic model of willow production[48] (EcoWillow ver. 1.4 (Beta)), the major costs for production are harvesting, establishment, and land rent. If the frequency of harvesting is reduced by increasing the rotation length from three to four years, harvesting costs are reduced by $14 to $16 per dry Mg. About two-thirds of establishment costs are for planting stock. If cost per cutting can be reduced from $0.12 to $0.10 per cutting, then establishment costs are reduced by $262 per ha. Costs can also be significantly reduced by increasing yields.

7.7 Eucalyptus

Eucalyptus spp. is the world's most widely planted hardwoods. There are over 700 species of Eucalyptus, but less than 15 are commercially significant in the world. Eucalypts are widely grown in the tropics and subtropics. They have been bred for fast, uniform growth, self-pruning, and coppicing (regrowing from the stump). Eucalypts are a major source of fiber. Brazil and other South American countries are major producers of Eucalypts. In Brazil, in addition to roundwood and pulp, Eucalypts are extensively utilized for charcoal. Eucalypts have great potential for use as an energy feedstock. Research on Eucalyptus is ongoing in the Southeast US.

For conventional pulpwood production stands are typically planted at 1500–2500 trees per ha and harvested every 6 to 10 years. For biomass production, Eucalypts in Florida, US, may be planted at 8400 trees per ha, harvested every 3 to 4 years.[49] *Eucalyptus* spp. yielded 17–32 dry Mg per ha-year after 3 to 5 years of growth in central Florida on a clay settling area, which is similar to 20–31 dry Mg per ha-year estimated for Florida.[50]

7.8 Oilseeds

Oilseeds can provide feedstock for biodiesel production. Different feedstocks give very different per ha biodiesel yields (Fig. 5). There are some feedstocks that may potentially provide high biodiesel yields, including jatropha and algae (Table 7). Dar[51] reports that in India, the National Biofuel Centre of the Petroleum

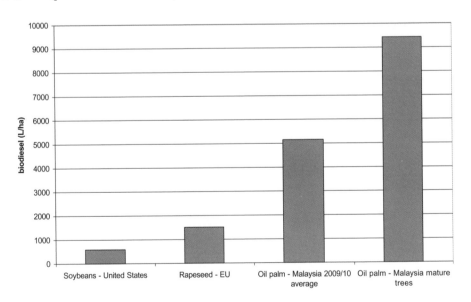

Fig. 5.   Per ha biodiesel production.
Source: Soybeans US estimates based on soybean yield of 1.15 Mg per ha[8] and 19.0% oil; Rapeseed EU rapeseed yield of 3.3 Mg per ha and 41.5% oil[52]; Oil palm: Malaysia 2009–2010 average and mature trees based on palm oil production of 17,765,000 Mg (100% oil) and palm kernel production of 4,519,000 Mg (46.75% oil) and 4,300,000 ha and 2,346,000 mature [tree] ha equivalents[53]; and assuming 0.899 kg oil per L of biodiesel.

Table 7:   Biodiesel yIelds of Current and Potential Sources.

Crop	Yield L per ha	Data source
Soybeans–US current	603	
Rapeseed–EU current	1522	
Jatropha–India estimate	540–3400	Dar[51]
Oil palm–Malaysia mature current	9,427	
Microalgae–projected	9350–60,770	USDOE/EERE[54]

Conservation Research Association estimates jatropha seed yield of 1.5 Mg per ha, or an oil yield of about 540 L per ha, while the Centre of Excellence for Jatropha Biodiesel Promotion in Rajasthan estimates for intensively managed plantations a seed yield of 10 Mg per ha or an oil yield of about 3400 L per ha.

8 Future Use of Biofuels

The World Energy Assessment of the United Nations[56] reported that in 2001, traditional biomass accounted for 39 exajoules[c] (EJ) of primary energy and modern biomass for 6 EJ of primary energy. This corresponds to 9.3% of total world primary energy use from traditional biomas and 1.5% of primary energy use from modern biomas, respectively.

There have been many studies that have estimated the future biomas use, 17 of which are reviewed by Berndas *et al.*[57] In addition, Hoogwijk *et al.*[58] estimate that in 2050, biomas (both traditional and modern) could contribute as little as 33 EJ and as much as 1,135 EJ of primary energy. The reason for the large difference in these estimates is because of the large uncertainty in biomas production on agricultural lands.

References

1. R. D. Perlack, L. L. Wright, A. F. Turhollow, R. L. Graham, B. J. Stokes and D. C. Erbach. Biomass as feedstock for a bioenergy and bioproducts industry: The technical feasibility of a billion-ton annual supply, ORNL/TM-2005/66 (Oak Ridge National Laboratory, Oak Ridge, TN, USA, 2005).
2. USDOE, Biomass as feedstock for a bioenergy and bioproducts industry: An update to the billion-ton annual supply, R. D. Perlack and B. J. Stokes (Leads), ORNL/TM-2010/224 (Oak Ridge National Laboratory, Oak Ridge, TN, USA, 2011).
3. D. Macqueen and S. Korhaliller, Bundles of energy: The case for renewable biomass energy, Natural Resource Issues No. 24, (International Institute for Environment and Development (IIED), London, United Kingdom, 2011).
4. Renewable Fuels Association (RFA). (2011). Statistics. Available at: http://www. ethanolrfa.org/pages/statistics. Accessed 22 April 2011.
5. USDOE/EIA. (2011). Fuel ethanol production, U.S. Department of Energy, Energy Information Administration. Available at: http://www.eia.doe.gov/cfapps/

---

[c]1 exajoule $= 10^{18}$ joule.

ipdbproject/iedindex3.cfm?tid=79&pid=81&aid=1&cid=regions&syid=2000&eyid=
2009&unit=TBPD. Accessed 28 April 2011.

6. J. Goldemberg, The Brazilian biofuels industry, *Biotechnol. Biofuels* **1**(2008), p. 6.

7. C. Valdes, Brazil's ethanol industry: Looking forward, BIO-02. U.S. Department of
Agriculture. Economic Research Service (2011).

8. USDA/NASS, Crop production 2010 summary. Cr Pr 2-1 (11)a (U.S. Department of
Agriculture, National Agricultural Statistics Service, 2011).

9. L. P. Ovard, F. H. Ulrich, D. J. Muth Jr, J. R. Hess S Thomas and B. J. Stokes,
(eds.). Office of Biomass (2010) billion-ton update: "High-yield scenario," Workshop
Summary Report. USDOE, DOE-EERE, INL/EXT-10-18930 (2010).

10. USDOE/EIA, Biodiesel production, (U.S. Department of Energy, Energy Information Administration, 2011). Available at: http://www.eia.doe.gov/cfapps/ipdbproject/
iedindex3.cfm?tid=79&pid=81&aid=1&cid=regions&syid=2000&eyid=2009&unit=
TBPD. Accessed 28 April 2011.

11. T. Bridgwater, Biomass pyrolysis, in IEA Bioenergy Annual Report, ed. J. Tustin
(2007), pp. 4–19. Available at: http://www.ieabioenergy.com/DocSet.aspx?id=5566&
ret=lib. Accessed 3 May 2011.

12. R. C. Brown and J. Holmgren, Fast pyrolysis and bio-oil upgrading (2011).
Available at: http://www.ars.usda.gov/sp2UserFiles/Program/307/biomasstoDiesel/
RobertBrown&JenniferHolmgrenpresentationslides.pdf. Accessed 27 April 2011.

13. Dynamotive. Dynamotive biooil information booklet 2011, (2011). Available at:
http://www.dynamotive.com/assets/resources/PDF/PIB-BioOil.pdf. Accessed 2 May
2011.

14. E. Taylor, Coproducts and byproducts of woody biorefinery processing, in *Proceedings
Of The Role Of Extension In Energy*, eds. B. C. English, J. Menard and K. Jensen,
30 June–1 July (Little Rock, AR, USA, 2009), pp. 62–63. Available at: http://
www.farmfoundation.org / news / articlefiles / 1704-Final%20proceedings%209-14-09.
pdf. Accessed 2 May 2011.

15. J. Scahill, Pyrolyis — A promising new path to liquid fuels, Presented at *Can
Forests Meet Our Energy Needs? The Future of Forest Biomass*, 21 February 2008.
Colorado State University, Fort Collins, CO, USA. Available at: http://digitool.
library.colostate.edu///exlibris/dtl/d3_1/apache_media/L2V4bGlicmlzL2R0bC9kM18
xL2FwYWNoZV9tZWRpYS80OTI2.pdf. Accessed 2 May 2011.

16. USDOE/EERE, Using heat and chemistry to make fuels and power: Thermochemical conversion, U.S. Department of Energy, Energy Efficiency & Renewable
Energy, Biomass Program (2010). Available at: http://www1.eere.energy.gov/library/
asset_handler.aspx?src=http://www1.eere.energy.gov/biomass/pdfs/thermochemical_
four_pager.pdf&id=4374. Accessed 4 May 2010.

17. L. Kangmin and M. W. Ho, Biogas China. ISIS Report 02/10/06. Institute of Science
in Society (2006). Available at: http://www.i-sis.org.uk/BiogasChina.php. Accessed
28 April 2011.

18. W. F. Lazarus, Anaerobic digester technology, in *Proceedings of the Role of Extension
in Energy*, eds. B. C. English, J. Menard, and K. Jensen, 30 June–1 July (Little
Rock, AR, USA, 2009) pp. 42–45. Available at: http://www.farmfoundation.org/news/
articlefiles/1704-Final%20proceedings%209-14-09.pdf. Accessed 2 May 2011.

19. J. Crenshaw, What's a digester cost these days? Presented at the 2009 AgSTAR
National Conference, 24–25 February 2009, Baltimore, MD, USA. Available at:
http://www.epa.gov/agstar/documents/conf09/crenshaw_digester_cost.pdf. Accessed
2 May 2011.

20. A. Aden, M. Ruth, K. Ibsen, J. Jechura, K. Neeves, J. Sheehan and B. Wallace, Lignocellulosic biomass to ethanol design and economics using co-current dilute acid pretreatment and enzymatic hydrolysis for corn stover. NREL/TP-510-32438 (National Renewable Energy Laboratory, Golden, CO, USA, 2002).

21. D. Humbird and A. Aden, Biochemical production of ethanol from corn stover: 2008 State of Technology Model, NREL/TP-510-46214. (National Renewable Energy Laboratory, Golen, CO, USA, 2009). Available at: http://www.nrel.gov/biomass/pdfs/46214.pdf. Accessed 4 May 2011.

22. A. Dutta, N. Dowe, K. N. Ibsen, DJ Schell and A. Aden, An economic comparison of different fermentation configurations to convert corn stover to ethanol using *Z. mobilis* and *Saccharomyces*, *Biotechnol. Prog.* **26**(1) (2009), pp. 64–72.

23. D. Humbird, L. Tao, A. Aden and D. Dudgeon, Abstract: 2011 update of the NREL biomass-to-ethanol process design report, Presented at the 33rd Symposium on Biotechnology for Fuels and Chemicals, 2–5 May (Seattle, WA, USA, 2011). Available at: http://sim.confex.com/sim/33rd/webprogram/Paper17696.html. Accessed 10 May 2011.

24. E. D. Jarvis, Aquatic species program: Lessons learned, Presentation at the AFSOR Workshop on Algal Oil for Jet Fuel Production, 19–21 February (Washington, DC, USA, 2008). Available at: http://www.nrel.gov/biomass/pdfs/43232.pdf. Accessed 29 April 2011.

25. J. Sheehan, T. Dunahay, J. Benemann and P. Roessler, A look back at the U.S. Department of Energy's Aquatic Species Program — Biodiesel from algae, NREL/TP-580-24190 (National Renewable Energy Laboratory, Golden, CO, USA, 1998).

26. C. S. Byrt, C. P. L Grof and R. T. Furbank, C_4 plants as biofuel feedstocks: optimising biomass production and feedstock quality from a lignocellulosic perspective, *J. Integr. Plant Biol.* **53**(2) (2011), pp. 120–135.

27. D. J. Parrish and J. H. Fike, The biology and agronomy of switchgrass for biofuels, *Crit. Rev. Plant Sci.* **24**(2005), pp. 423–459.

28. R. Mitchell, K. Vogel, M. Schmer and D. Pennington, Switchgrass for biofuel production (2010). Available at: http://www.extension.org/pages/Switchgrass_for_Biofuel_Production. Accessed 9 May 2011.

29. K. P. Vogel and R. B. Mitchell, Heterosis in switchgrass: Biomass yield in swards, *Crop Sci.* **48** (2008), pp. 2159–2164.

30. A. F. Turhollow, E. G. Webb and M. E. Downing, Review of sorghum production practices: Applications for bioenergy, (ORNL/TM-2010/7, Oak Ridge National Laboratory, Oak Ridge, TN, USA, 2010).

31. V. S. Belum Reddy, S. Ramesh, P. Sanjana Reddy, B. Ramaiah, P. M. Salimath and Rajashekar Kachapur, Sweet sorghum — A potential alternative raw material for bioethanol and bio-energy, *Sorghum Millets Newslett.* **46** (2005), pp. 79–96.

32. V. S. Belum Reddy, Willaim D. Dar, P. Parthasarathy Rao, P. Srinivasa Rao, A. Ashok Kumar and P. Sanjana Reddy, Sweet sorghum as a bioethanol feedstock: Challenges and opportunities, International Conference on Sorghum for Biofuel, 19–22 August (Houston, TX, USA, 2008). Available at: http://www.ars.usda.gov/meetings/Sorghum/presentations/Reddy.pdf. Accessed 2 May 2011.

33. Thomas L. Tew, Robert M. Cobill and Edward P. Richard, Jr. Evaluation of sweet sorghum and sorghum x sudangrass hybrids as feedstocks for ethanol production, *Bioenergy Res.* **1** (2008), pp. 147–152.

34. S. R. Schill, Tall Texas sorghum, *Ethanol Producer Magazine*, November 2007. Available at: http://www.ethanolproducer.com/articles/3424/tall-texas-sorghum. Accessed 2 May 2011.

35. S. C. Aravindhakshan, F. M. Epplin and C. M. Taliafarro, Economics of switchgrass and miscanthus relative to coal as feedstock for generating electricity, *Biomass Bioenerg.* **35**(1) (2011), pp. 308–319.

36. A. K. Jain, M. Khanna, M. Erickson and H. Huang, An integrated biogeochemical and economic analysis of bioenergy crops in the Midwestern United States, *Global Change Biol. Bioenerg.* **2**(5) (2010), pp. 217–234.

37. M. Khanna, B. Dhungana and J. Clifton-Brown, Costs of producing Miscanthus and switchgrass for bioenergy in Illinois, *Biomass Bioenerg.* **32**(2008), pp. 482–493.

38. D. G. Christian, A. B. Richie and N. E. Yates, Growth, yield and mineral content of Miscanthus x giganteus grown as a biofuel for 14 successive harvests, *Ind. Crop Prod.* **28**(2008), pp. 320–327.

39. L. Ercoli, M. Mariotti, A. Masoni and E. Bonari, Effect of irrigation and nitrogen fertilization on biomass yield and efficiency of energy use in crop production of Miscanthus, *Field Crops Res.* **63**(1999), pp. 3–11.

40. B. L. Legendre and D. M. Burner, Biomass production of sugarcane cultivars and early-generation hybrids, *Biomass Bioenerg.* **8**(2) (1995), pp. 55–61.

41. D. E. Riemenschneider, H. E. Stelzer and G. S. Foster, Quantitative genetics of poplars and poplar hybrids, in *Biology of Populus and its Implications for Management and Conservation*, eds. R. F. Stettler, H. D. Bradshaw, Jr, P. E. Heilamn and T. M. Hinkley, Part I, Chapter 7 (NRC Research Press, National Research Council of Canada, Ottawa, ON, Canada, 1996), pp. 159–181.

42. R. S. Zalesny Jr., R. B. Hall, J. A. Zalesny, B. G. McMahon, W. E. Berguson and G. R. Stanosz, Biomass and genotype x environment interactions of Populus energy crops in the Midwestern United States, *Bioenerg. Res.* **2** (2009), pp. 106–122.

43. H. G. Adegbidi, T. A. Volk, E. H. White, R. D. Briggs, L. P. Abrahamson and D. H. Bickelhaupt, Biomass and nutrient export by willow clones in experimental bioenergy plantations in New York, *Biomass Bioenerg.* **20** (2001), pp. 389–398.

44. M. Labrecque and T. I. Teodorescu, High biomass yield achieved by Salix clones in SRIC following two 3-year coppice rotations on abandoned farmland in southern Quebec, Canada, *Biomass Bioenerg.* **25** (2003), pp. 135–46.

45. H. G. Adegbidi, R. D. Briggs, T. A. Volk, E. H. White and L. P. Abrahamson, Effect of organic amendments and slow-release nitrogen fertilizer on willow biomass production and soil chemical characteristics, *Biomass Bioenerg.* **25** (2003), pp. 389–398.

46. T. A. Volk, L. P. Abrahamson, C. A. Nowak, L. B. Smart, P. J. Tharakan and E. H. White, The development of short-rotation willow in the northeastern United States for bioenergy and bioproducts, agroforestry and phytoremediation, *Biomass Bioenerg.* **30** (2006), pp. 715–727.

47. T. A. Volk, B. D. Kiernan, R. Kopp and L. P. Abrahamson, First- and second-rotation yields of willow clones at two sites in New York State, Fifth Biomass Conference of the Americas, 17–21 September (Orlando, FL, USA, 2001). Available at: http://www.brdisolutions.com/pdfs/bcota/abstracts/22/z146.pdf.

48. T. Bucholz and T. A Volk, Improving the profitability of willow biomass crops — identifying opportunities with a crop budget model, *Bioenerg. Res.* **4**(2) (2011), pp. 85–95.

49. M. Langholtz, D. R. Carter, D. L. Rockwood and J. R. R. Alavalapati. The economic feasibility of reclaiming phosphate mined landswith short-rotation woody crops in Florida, *J. Forest Econ.* **12**(4) (2007), pp. 237–249.

50. M. Rahmani, A. W. Hodges, J. Stricker and C. F. Kiker. Economic analysis of biomass crop production in Florida. Making a buiness from biomass in energy, environment, chemical, fibers and materials, eds. R. P. Overend and E. Chornet, *Third Biomass*

Conference of the Americas, 24–29 August (Elsevier Science, New York, NY, USA) pp. 91–99.

51. W. Dar, Research needed to cut risk to biofuel farmers, Science and Development Network, 2007. Available at: http://www.scidev.net/en/opinions/research-needed-to-cut-risks-to-biofuel-farmers.html. Accessed 6 May 2011.

52. USDA/FAS, EU-27 oilseeds and products annual, GAIN Report Number: E60016, U.S. Department of Agriculture, Foreign Agricultural Service, 2011. Available at: http://gain.fas.usda.gov / Recent%20GAIN%20Publications / Oilseeds%20and%20Products%20Annual_Berlin_EU-27_4-4-2011.pdf. Accessed 29 April 2011.

53. USDA/FAS, Malaysia oilseeds and products annual. GAIN Report Number: MY1004. U.S. Department of Agriculture, Foreign Agricultural Service, 2011. Available at: http://gain . fas . usda . gov / Recent % 20GAIN % 20Publications / Oilseeds % 20and % 20Products%20Annual_Kuala%20Lumpur_Malaysia_4-22-2011.pdf. Accessed 29 April 2011.

54. USDOE/EERE, National algal biofuels technology roadmap, U.S. Department of Energy, Energy Efficiency & Renewable Energy, Biomass Program, 2010. Available at: http://www1.eere.energy.gov/biomass/pdfs/algal_biofuels_roadmap.pdf. Accessed 25 April 2011.

55. S. C. Davis and S. W. Diegel, *Transportation Energy Handbook*, ORNL-6974 (Oak Ridge National Laboratory, Oak Ridge, TN, USA, 2006).

56. United Nations Development Programme, United Nations Department of Economic and Social Affairs and World Energy Council, *World Energy Assessment Overview 2004 Update*, Table 1, eds. J. Goldenberg and T. B. Johansson (United Nations Development Programme, New York, NY, 2004).

57. G. Berndesa, M. Hoogwijk and R. van den Broek, The contribution of biomas in the future global energy supply: A review of 17 studies, *Biomass Bioenerg.* **25**(1) (2003), pp. 1–28.

58. M. Hoogwijk, A. Faajia, R. van den Broek, G. Berndes, D. Gielen and W. Turkenburg, Exploration of the ranges of the global potential for biomass for energy, *Biomass Bioenerg.* **25**(2) (2005), pp. 119–113.

Appendix

Fuel Heating Values

Table A1:   Fuel Heating Values.[55]

Fuel	Units	Higher heating value	Lower heating value
Wood	MJ per dry Mg	19.77	
Grass	MJ per dry Mg	17.45	
Ethanol	MJ per L	23.44	21.19
Gasoline	MJ per L	34.79	32.06
Biodiesel[55]	MJ per L	35.18	32.64
Diesel	MJ per L	38.66	35.79
Bio (pyrolysis) oil	MJ per L	16.00–19.00	
Biogas	MJ per m^3	19.00–29.00	
Natural gas	MJ per m^3	38.24	

Chapter 12

Geothermal Energy

Gordon Bloomquist
Center for Distributed Generation and Thermal Distribution (Retired)
Washington State University, Pullman, WA 98504, USA
Gordon_Bloomquist@hotmail.com

John Lund
Geo-Heat Center, Oregon Institute of Technology
Klamath Falls, OR 97601, USA
John.Lund@oit.edu

Magnus Gehringer
1901, 12th Street NW, Washington, DC 20009
magnus@imago.is

Geothermal energy is the natural heat from the earth that can be used to produce electricity or for direct heating of buildings, greenhouses, swimming pools, and aquaculture ponds, and for industrial processes. Geothermal resources have been identified in nearly 90 countries and utilized to produce electricity in 24 countries and for direct heating in 72 countries. The worldwide installed capacities for electricity generation and direct heating are approximately 11 GWe and 16 GWt, respectively. Generally, geothermal projects require a high initial capital investment but low annual operating costs. The common types of power plants are high-temperature flash steam and low-temperature binary cycle types. The greatest risk in developing a geothermal project is in locating, defining and developing the resource resulting in the drilling of a number of production and injection wells to produce the necessary hot water and/or steam. Most geothermal resources are found along tectonic plate boundaries and volcanic regions where high-temperatures resources are near the surface. Geothermal projects must consider the quality of the geothermal fluid that may cause scaling or corrosion problems. Generally environmental impacts are small and can be mitigated. Other applications of geothermal fluids included co-produced and enhanced geothermal systems, and geothermal heat pumps.

1 Introduction to Geothermal Energy and Its Utilization

Heat is constantly produced within the Earth from decay of radioactive material in the core of the planet. A tiny fraction of this heat can, or could, be used by mankind to generate electricity and for other purposes requiring thermal energy — this heat energy is called "Geothermal Energy." The heat is moved to the surface through

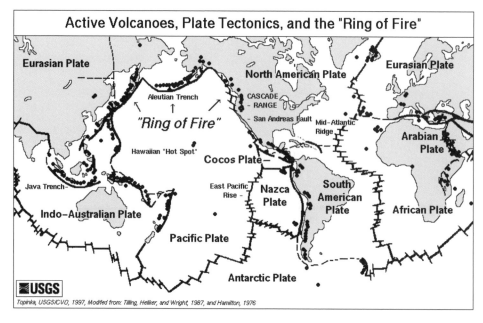

Fig. 1.   World map of plate boundaries.

conduction and convection. In the crust, the temperature gradient[a] is typically
30°C/km but can be as high as 150°C/km in some geothermal areas.

The technical potential[b] of the resource in question is inexhaustible, which
makes it a renewable energy source. However, it has proven to be difficult to harness
even the smallest share of this resource, due to reasons discussed in detail in the
following sections.

Geothermal fields are generally located around volcanically active areas often
located close to boundaries of the tectonic plates. Figure 1 provides a world map
with the main plate boundaries showing the most important geothermal areas.[1]

Geothermal energy has been used since the early 20th century for electricity
generation and even longer for direct uses.[c] The installed capacity of geothermal
power generation has increased from about 2 to around 11 GW electric[d] (GWe),
and direct heat use has increased from about 2 to around 16 GW thermal (GWt)
(2011) during the last three decades.

---

[a]A temperature gradient describes the changes in temperature at a particular location. In geophysics, it is usually measured in degrees Celsius per vertical kilometer (°C/km).
[b]Technical potential contains all projects that could be implemented globally, if all geothermal
resources could be found and utilized. The economical potential refers to those projects that would
be economically and financially viable within a given timeframe.
[c]The term "direct use" refers to applications other than power generation e.g. space heating,
industrial processes, greenhouse and pond heating cooling, etc.
[d]In generating electricity, the output power is often denoted as Megawatt electric (MWe) or
Gigawatt electric (GWe) as opposed to the input power in the form of heat which is designated as
Megawatt thermal (MWt) or Gigawatt thermal (GWt). (See also Chapter 2.)

2 World Overview of Utilization

Geothermal resources have been identified in nearly 90 countries, and there are quantified records of geothermal utilization in approximately 72 countries. As of 2010, electricity is produced by geothermal energy in 24 countries. Iceland and El Salvador have the highest share of geothermal power in their country energy mix, generating about 25% of their electrical power from geothermal resources. USA and Philippines have the greatest installed geothermal capacity, 3000 Megawatts electric (MWe) and 1900 MWe, respectively.

The 24 countries using geothermal resources for power generation are outlined in Fig. 2. In total, about 38 countries are considered to possess significant geothermal potential that could be developed to augment and, in some developing countries, increase their current power generating capacities and make them less dependent on energy imports.

2.1 *Direct Use of Geothermal Resources*

The direct use of geothermal resources has been reported in 78 countries, with a total installed capacity of 15,359 Megawatts thermal (MWt) and an annual energy use of 223,682 TJ (62,139 GWh) (excluding geothermal heat pumps). Installed capacity has increased 2.26 times or 5.57% annually, and the annual energy use has increased 2.29 times or 5.67% annually over the past 15 years.[3] In terms of energy utilization (TJ/yr), bathing and swimming accounts for 49% of the use, space heating

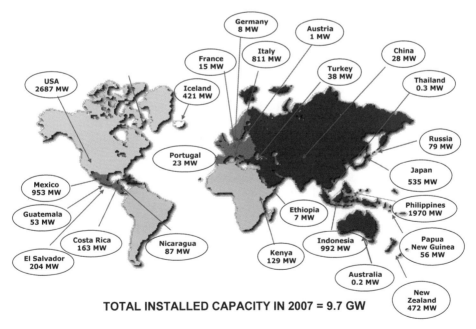

Fig. 2.   Installed capacity worldwide.[2]
Note: The units in the figure refer to electrical output which is normally designated MWe.

Table 1a: Leading Countries in Terms of Direct-Use Installed Capacity (>300 MWt).

Country	MWt	Major use(s)
China	3688	Bathing, district heating
Japan	2086	Bathing, space heating
Turkey	2046	District heating, bathing
Iceland	1822	District heating
Italy	636	Space heating, bathing
Hungary	615	Bathing, greenhouses
USA	612	Space heating, aquaculture
New Zealand	386	Industrial
Brazil	360	Bathing
Russia	307	Greenhouses, space heating

Table 1b: Leading countries in Terms of Annual Direct Geothermal Energy Use (>3000 TJ/yr).

Country	TJ/yr	Major use(s)
China	46,313	Bathing, district heating
Turkey	36,349	Bathing, district heating
Japan	25,630	Bathing, space heating
Iceland	24,341	District heating
New Zealand	9513	Industrial
Hungary	9249	Bathing, greenhouses
USA	9152	Bathing, space heating
Italy	8980	Space heating, aquaculture
Brazil	6622	Bathing
Mexico	4023	Bathing
Slovakia	3054	Bathing, space heating
Argentina	3048	Bathing

for 28% (of which district heating is 85%), greenhouse heating for 10%, industrial applications for 5%, aquaculture pond heating for 5%, cooling and snow melting for 1%, agricultural drying for 1%, and other uses for 1%.

The leading countries[3] using geothermal energy for direct-use are shown in Tables 1(a) and 1(b).

In terms of the contribution of geothermal direct-use to the national energy budget, two countries stand out: Iceland and Turkey. In Iceland, geothermal provides 89% of the country's space heating needs, which is important since heating is required almost all year and saves approximately US$100 million per year in imported oil.[4] Turkey has increased their installed capacity over the past five years, from 1495 to 2084 MWt, mostly for district heating systems.[5] Other countries having a significant contribution from geothermal energy are: Tunisia where greenhouse heating has increased from 100 hectares[e] (ha) to 194 ha over the past five year; Japan, where over 2000 hot spring resorts (*onsens*), over 5000 public bath houses,

---

[e] 1 hectare (ha) = 10,000 m^2.

and over 15,000 hotels using geothermal energy are visited by 15 million guests annually; France, where district heating supplies geothermal heat to 150,000 housing units, mainly in the Paris and Aquitaine basins; New Zealand, where geothermal energy from the Taupo Volcanic Zone is used for greenhouse heating, prawn farming, kiln drying of timber, for special tourism development by the Maoris and at Kawerau, where pulp and paper manufacturing accounts for 56% of the national geothermal direct-use; and China, where geothermal energy is utilized for fish farming, heating greenhouses, agricultural crop drying, industrial process heat, district heating, and for bathing and swimming. China is the largest user of geothermal energy in the world, counting for 20% of the annual energy use.[3]

2.2 *Geothermal Direct Utilization Technologies*

Standard off–the-shelf equipment is used in most direct-use projects, provided allowances are made for the chemical and thermal properties of the geothermal water and steam. Corrosion and scaling caused by the sometimes unique chemistry of geothermal fluids, may lead to operating problems with equipment components directly exposed to flowing water and/or steam. Care should be taken to not introduce dissolved oxygen into the system (which is absent in most geothermal waters), and to isolate hydrogen sulfide gases as they will attack solder and copper. The isolation of the detrimental effects of geothermal fluids can be accomplished by

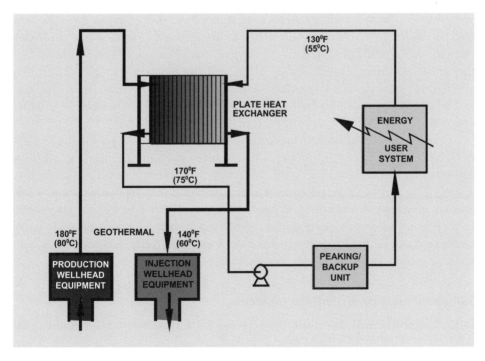

Fig. 3.   Geothermal direct-use system using a plate heat exchanger.

installing heat exchangers and circulating a clean secondary fluid through the use side of the system, as shown in Fig. 3.

The primary components of most low-temperature direct-use systems are downhole and circulation pumps, transmission and distribution pipelines, peaking or back-up plants, main heat exchangers (typically a plate heat exchanger), and various forms of heat extraction equipment[6] as shown in Fig. 3. Fluid disposal is either surface or subsurface (injection), depending on local regulations and/or need for pressure support within the reservoir. A peaking system allows for meeting maximum load without overtaxing the geothermal reservoir and with optimal efficiency of the distribution piping system.

2.2.1 *District cooling*

The use of geothermal heat to feed cooling systems is an often-overlooked and under-valued potential market for geothermal energy. Waste energy/fluids from geothermal power plants as well as geothermal fluids from low- or medium-temperature resources can be deployed for cooling applications and even refrigeration. To date, direct use of geothermal fluids for district heating has dominated direct use applications due to its high efficiency. However, the utilization of hot fluids for cooling can increase the load factor from approximately 30% to in excess of 50%, thus improving overall project economics.

In combined heat and power (CHP) applications, the thermal energy can be used for power generation as well as both district heating and cooling. In terms of economics, district cooling can have substantially higher economic benefits as it directly offsets the use of costly electricity for cooling and in areas where electricity rates contain significant peak demand charges, overall economic benefit can be further enhanced through incorporation of thermal storage (e.g. through refrigeration allowing the creation of super chilled water or brine) into the district cooling system which, in addition, also carries with it other overall system benefits to building operators.

Cooling can be provided by deployment of absorption cooling technologies such as advanced lithium bromide systems that achieve optimum efficiency at geothermal input temperatures as low as 90°C and achieve output temperatures of approximately 5°C. Ammonia-based systems are capable of providing temperatures below freezing, making thermal energy storage attractive. Another interesting alternative to absorption cooling systems is the use of binary turbine generated shaft horse power that can be used to directly drive a conventional refrigeration unit of scroll, screw or centrifugal configuration design.

2.3 *Economics of Direct-Use Systems*

Generally, geothermal direct-use projects require a relatively large initial capital investment, with small annual operating and maintenance costs thereafter. Thus, a district heating project, including producing wells, transmission and distribution pipelines, heat exchangers, and injection wells, may cost several million dollars. In contrast, the initial investment in a fossil fuel system includes only the costs of the central boiler and distribution lines. The annual operation and maintenance costs for the two systems are similar, except that the fossil fuel system continues to pay for the fuel at an ever-increasing rate, while the cost of the geothermal fuel is relatively stable. The two systems, one with a high initial capital cost and other with high annual cost, must be compared in order to determine the most economically viable alternative.

Geothermal resources can fulfill many energy needs, from power generation to heating of buildings and industrial processes. Considered individually; however, some of these uses may not promise an attractive return on investment because of the high initial capital cost. Thus, there is a need to consider using a geothermal fluid several times to maximize benefits. This multistage utilization, where decreasing water temperatures are used in successive steps, is called cascading or waste heat utilization. A simple form of cascading employs waste heat from a power plant for direct use projects referred to as a combined heat and power applications, as illustrated in Fig. 4.

Examples[3,7] of the average cost of direct-use systems in the USA for 2005 are illustrated in Table 2.

2.4 *Future Utilization Scenario for Power Generation and Direct Use*

Direct Use: Future development of direct-use projects will most likely occur under the following conditions:

- Co-located resource and use (<10 km apart)
- Sites with high heat and cooling load density (>36 MWt/sq.km)

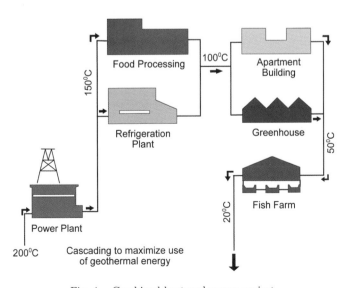

Fig. 4. Combined heat and power project.

Table 2:   Typical Cost (US$) of Direct-Use Systems in the US for 2005.

Applications	Capital US$/kW	Cost/year US$/kWyr	O&M US$/kWyr	Total US$/kWyr	Capacity factor
Residential space heating*	2500	71.1	7.1	78.2	0.29
Comm./Inst. Space heating*	1500	44.4	4.4	48.8	0.25
District heating	2000	57.7	5.8	63.5	0.37
Greenhouse heating	750	22.2	2.2	24.4	0.48
Aquaculture heating(uncovered)	75	17.8	1.8	19.6	0.56

*Assumes one production and one injection well for a single building.
Note: Based on 30-year life at 8.0% interest and O&M at 10% of capital cost. The above costs include a shallow well (<300 m) and no retrofit costs; however, cost can vary by as much as 100% depending upon local geology, hydrology, building construction and infrastructure.

- Food and grain dehydration (especially in tropical climates where spoilage is common)
- Refrigeration and freezing in warm climates
- Greenhouses in colder climates
- Aquaculture to optimize growth — even in warm climates
- Combined heat and power installations

Areas where there will be increased use of geothermal energy for direct-use over the next five to 10 years, will be in China, Europe (especially in France, Germany, Poland, Hungary and Slovakia), in Kenya and other countries of East Africa, Japan, India and New Zealand.

Power Generation: Based on information on currently planned projects and those that are actually under construction, the worldwide increase in the field of geothermal power generation until the year 2020, from hydrothermal resources only, could be expected to come from mainly Indonesia (2,500 MWe), followed by Kenya (1,500 MWe), the Philippines (500 MWe), Ethiopia, Djibouti and Rwanda (400 MWe in total), Central-America and Mexico (800 MWe in total), and the USA (800 MWe). These figures total 6,500 MWe of new installed electrical capacity worldwide until 2020. If added to the 11,000 MWe installed in 2011, this results in an estimate of 17,500 MWe for 2020.[8]

3 Geothermal Geology

Figure 5 shows the components of a typical hydrothermal (steam or water based) volcanic-related geothermal system, which are, from bottom to top:

- The magmatic intrusion, also called hot rocks, where hot magma intrudes exceptionally far into the Earth's crust, is often caused by movements of the continental plates;
- The actual geothermal reservoir where steam and/or hot water are trapped under a tight, non-permeable layer of rocks (the cap rock) and heated by the hot body below;

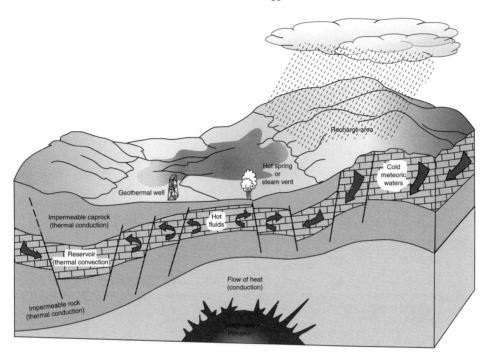

Fig. 5. Schematic view of an ideal geothermal system.[9]

- Water/precipitation coming from recharge areas like lakes, rivers or the seas provide cold waters slowly seeping down through the ground to lower layers through cracks and along faults in the rocks; and
- The geothermal wells tap into the geothermal reservoir and access the hot steam or water, then transfer it from the wells through pipelines to the power plant or direct use application, from where the spent fluids are usually returned to the reservoir via an injection well.

4 Development of Geothermal Power Generation Projects

Geothermal electrical generation projects have seven key phases of project development before actual operation and maintenance (O&M) commences. According to the schedule shown in Fig. 6, it takes approximately seven years to develop a typical full-size geothermal project from a completely undeveloped site. However, depending on the relevant country's institutional and regulatory framework, geological conditions, location and availability of financing, the project development time could either be reduced or prolonged by several years.

Each phase of the geothermal project development consists of several tasks. After each milestone, the relevant developer — either a project company or a country's institution — will have to decide whether to continue developing the project or not and whether to assume the risks of the next phase. The first three phases,

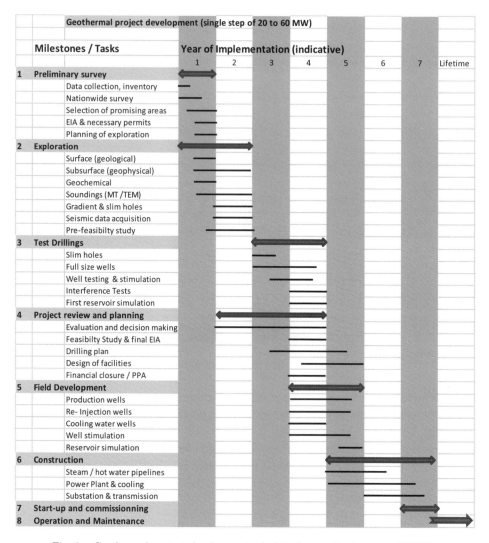

Fig. 6. Geothermal project development schedule for a unit of approx. 50 MWe.

or milestones, take the developer from early reconnaissance to actual site specific scientific research through test drillings. This first part of the project development (which could be broadly called the exploration stage) either confirms the existence of a geothermal reservoir suitable for economically viable power generation or not; therefore, it is usually seen as the riskiest part of project development and therefore also the most difficult to finance. If the results from the first three phases, including the test drillings, are positive and the geothermal potential is confirmed, phase 4 is initiated with the actual design of the power project, including the feasibility study, engineering design and financial closure. Phases 5 to 7 comprise the development of the project itself, consisting of the drilling of geothermal production and injection wells, construction of pipelines, the power plant and its connection to the transmission system. The following sections describe the development phases in more detail and provide indicative figures for a costs analysis.

4.1 *Exploration*

Phase 1, the preliminary survey, includes a first reconnaissance of a geothermal area based on a nationwide or regional study. If no geothermal master plan studies are available, developers usually conduct their own studies based upon available literature and data, or execute their own reconnaissance work to select the areas where to apply for exploration concessions. Methods for this purpose include geological ground studies as well as satellite and airborne research. Phase 1 is important to establish the rationale and assess the need for the project in question and at the same time to find a justification to enter into investments induced by the exploration and test drillings (Phases 2 and 3). Costs for this first phase can be estimated as US$0.5 up to 1.5 million, which shows that they can vary considerably according to the data available, the documentation required and the size and accessibility of the area being considered for geothermal power generation. Phase 1 usually takes from several months up to one year to complete.

Phase 2, the exploration phase,[f] consists of detailed geophysical surveys and may include the drilling of temperature gradient or core holes to further confirm the preliminary resource assessment and starts as soon as the project developer is satisfied with the results of phase 1. In total, the second phase can take up to two years, depending on the size and accessibility of the geothermal field and the data already available.

Costs for the activities under Phase 2 can be significant, from 5% to 10% (indicative figures) of the total project costs, depending on the project size. Costs for conducting transient electromagnetic and magnetic telluric resistivity measurements, seismic surveys, gravity measurements or drilling gradient or core holes depend on the accessibility of the geothermal site and the availability of tools, equipment and capacity of staff to operate the equipment and interpret the results. While minimum exploration costs for a geothermal site would in many cases be US$1 to 2 million, every single gradient well could add US$0.5 to 1 million to that figure. Since all geothermal fields and projects are different, it is difficult to generalize the required investment costs for Phases 1 and 2.

Phase 3, the test drilling phase, is the last of the exploratory phases. At the end of this phase the project developer should be able to decide, based on scientific evidence and characteristics of the reservoir, whether or not the project warrants

---

[f]In this context, the term "exploration phase" refers to the second phase in the detailed breakdown of the project cycle. This usage is distinct from "exploration" in a broad sense, which consists of the first three phases including the test drilling phase. The latter usage is more common for the oil and gas industry.

being continued, i.e., it will be economically viable to build and operate a power plant, or the project should be abandoned.

In the beginning of this phase, a drilling program is designed to target and to confirm the existence, the exact location and the potential of the reservoir. Usually a set of three to five full-size geothermal wells[g] are drilled. In some locations, due to accessibility issues, and availability of resources such as water and electricity at the geothermal field, it may be necessary to start with the drilling of slim holes (i.e. holes with a smaller diameter) that can be drilled with lighter equipment (drillings rigs) than full-size wells and at lower cost. Slim holes drilled to penetrate the reservoir will provide much needed information relative to depth, temperature and may allow for some level of well testing to determine the potential for productivity. Cost for a three-well reservoir confirmation drilling program (full sized production wells) could run or even exceed US\$20 million, depending upon location, accessibility, depth and completion diameter. Upon completion of the drilling program, a long-term flow test as well as injection tests should be conducted to determine the potential for interference between production and injection wells. The completion of a reservoir engineering report at this point will be critical to determining the conversion technology to be employed, the optimum size of the power plant and as a prerequisite for the obtaining of mezzanine financing that will allow for further field development, completion of feasibility studies and the start of engineering design.

4.2 Drilling and Well Testing

Despite the fact that a discovery well or wells have been successfully completed and tested during Phase 3, risk associated with drilling out the well field (production and injection wells) to support the planned power generation facility are still considerable and should not be taken lightly. In fact, even in well-developed and producing fields, 20% or more of all wells drilled are unsuccessful. The outcome of drilling of a well can be defined over the entire spectrum from success to failure. Success is usually defined as a well capable of an energy output suitable to meet the intended utilization. A successful well would have the following attributes:

- High flow rate
- High enthalpy of the geothermal fluid
- Low non-condensable gas content
- Low potential for scaling
- Low potential for corrosion

---

[g]For example, a full size well could be 1.5 to 3.5 km deep and have a bottomhole diameter of 7 to 8 inches. The top (surface) diameter can be over 20 inches.

Unfortunately, even if only one of the attributes is unfavorable, the well may be a failure for the purpose of meeting the intended use. An important consideration is the fact that the first wells drilled in a new field tend to be less productive than those drilled at later stages of well field development due to increased understanding of the subsurface and the drilling techniques that are best suited for the particular geological environment.

Well output can vary considerably from field to field and even from well to well. The worldwide average output per well is just over 4.0 MWe per well. However some fields have been successfully developed when output is as low as 2.0+ MWe per well and in highly unusual circumstances average well output can exceed 7–8 MWe with some wells producing in excess of 40 MWe.

It is thus critically important that as much information be gained from the drilling of each well as possible in order to continually improve drilling success as well as output per well. Every effort should be made to incorporate such information as quickly as possible into the drilling plan as it has a major impact upon overall project economic viability.

Critical components of the drilling plan include rig specifications, casing set points, the cementing program, maximum angle of deviation (especially critical in the upper most sections of wells drilled for binary plants in order to ensure ability to set pumps), angles of deviation to intersect faults and fractures at optimum angles, program for over pressure and/or under pressure drilling, the mud program, drill bit selection, completion type, logging program etc. Other critical consideration include blow out prevention and contingencies for bring a well back into control if a problem should occur.

4.2.1 *Well testing*

Well testing is the process used to obtain the data needed to fully evaluate the performance of a geothermal reservoir, including long-term productivity potential and the potential for well field interference between production and injection wells and resulting in premature temperature declines.

Much of the needed information critical to any well test is obtainable during drilling operations and thus diligently gathering and recording such data is critical to future interpretation of test data. Such data include indications of the depth and nature of production intervals, lost circulation zones, temperature anomalies, rapid changes in drilling rates, evidence of factures and/or alteration minerals in drill cuttings.

The first test to be conducted upon completion of drilling operations is a rig test. It is of limited duration (several hours to one to two days) conducted while the drilling rig is still on the hole and is primarily to clean out the hole and to provide initial estimates of well productivity and allow for the sampling of reservoir fluids for chemical analyses. The test data is of critical importance for making decisions regarding further drilling and completion of the well and provides the basis for planning future tests of the well.

Following the completion of the well, a single-well controlled flow production test of several days to a week or more is run. The single-well test provides information relative to the wells productivity and allows for temperature and pressure surveys to be run under stabilized and controlled conditions. Pressure and/or water lever draw down and build-up data allows for the making of initial estimates of reservoir parameters.

Longer-term production and interference tests of approximately 30 days duration are generally run to determine reservoir parameters, detect boundaries or recharge sources, evaluate scaling and corrosion potential, assess reservoir capacity and identify any factors that could result in a need to modify the design of the utilization system. In interference tests, pressure or water level changes in observation well are monitored while producing from the production well. Injection tests serve the same purpose as production tests i.e. to determine well performance, locate injection zones and determine various reservoir parameters. It must be stressed that reservoir properties obtained from injection tests are not necessarily applicable to the same reservoir undergoing production testing. This is due primarily due to the thermal dependence of such parameters as fluid density, formation porosity and fracture aperture.

It may be required to inject during a production test due to the need to dispose of the fluids produced and if this is the case, careful planning should allow the production and injection tests to be run concurrently.

Of primary concern in long-term reservoir productivity and optimization planning is the potential for premature thermal break through resulting in temperature degradation of the production fluids and resulting in decreased power output. It is thus important to include tracer testing in any comprehensive well testing program. Results of tracer tests can help in the sitting of injection wells and in the management of production and injection.

5 Geothermal Power Generation Technologies

The standard power plant classification defines five different types: Binary, single flash, double flash, back pressure, and dry steam. The relative share in power generation (2010) of each of these technologies is reflected in Fig. 7.

Large-scale electricity generation mainly takes place through the use of conventional steam or binary turbine plants, depending on the thermal, pressure and chemical characteristics of the geothermal resource.

5.1 *Flash Plants, Condensing Units*

Commonly built in sizes from 25 to 60 MWe, the "condensing unit," also called conventional steam cycle, is the standard application to use for resource temperatures above 200°C (see Fig. 8).

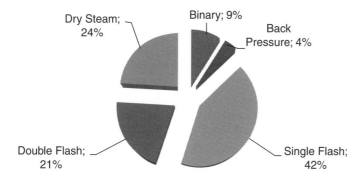

Fig. 7. Relative share of various technologies in world-wide geothermal power generation.

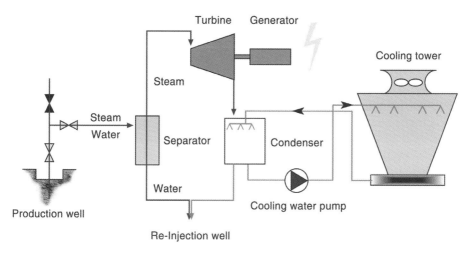

Fig. 8. Concept of condensing geothermal power plant.

The most common version of the condensing unit is the "Flash" steam plant and it usually is the most economical choice for liquid dominated resources.[7] The hot water or liquid vapor mixture produced from the wellhead is directed into a separator, where the steam is separated from the liquid by reducing the pressure. The steam is expanded through a turbine and then after condensing, the portion not required for cooling tower make up is usually re-injected, together with the separated brine, back into the reservoir. The brine could, however, be used in a bottoming cycle[h] and/or in another application, such as heating, cooling or other multiple use.

---

[h] "Bottoming cycle" is discussed in more detail in the next section on Binary Cycles.

A double flashed steam cycle differs from a single flash cycle in that the hot brine is passed through successive separators, each at a lower pressure. The steam is directed to a dual-entry turbine, with each steam flow flowing to a different part of the turbine. The advantage is increased overall cycle efficiency and better utilization of the geothermal resource but at an overall increase in cost. The decision as to whether or not a double flash plant is worth the extra cost and complexity can only be made after a thorough economic evaluation based on the cost of developing and maintaining the geothermal resource, or cost of purchasing geothermal steam from a resource company, plant costs, and the value of the electricity to be sold.

5.2 *Binary Cycles*

Generating electricity from low-to-medium temperature geothermal fluids ($<200°C$) and from the waste hot fluids coming from the separators in high-temperature liquid-dominated geothermal fields has made considerable progress due to improvements made in binary technology. Binary units are now available in size ranges from a mere 250 kWe to approximately 20 MWe. *Binary plants* utilize a secondary working fluid, usually an organic fluid (typically *n*-pentane), that has a low boiling point and high vapor pressure at low temperatures when compared to steam. The secondary fluid is used in a conventional Rankine cycle: the geothermal fluid yields heat to the secondary fluid through heat exchangers, in which this fluid is heated and vaporizes; the vapor produced drives a turbine, is then cooled and condensed, and the cycle begins again (see Fig. 9).

By selecting suitable secondary fluids, binary systems can be designed to utilize geothermal fluids of temperatures well below 100°C, although such low temperatures

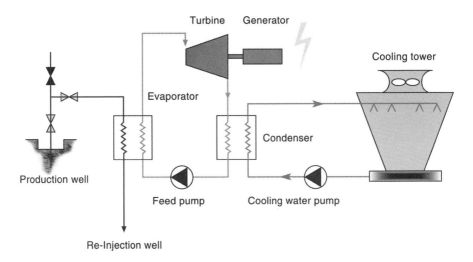

Fig. 9.   Concept of typical binary power plant, ORC or Kalina.

would severely impact the financial viability of projects, depending on their location, direct use options and the power tariff offered. Binary plant technology is a cost-effective and reliable means of converting into electricity the energy available from liquid-dominated geothermal fields with temperatures up to 200°C.

Competing with the above-mentioned Organic Rankine Cycle (ORC) plants, another binary system, the Kalina cycle, utilizes a water–ammonia mixture as the working fluid. This technology was developed in the 1990s.

Binary power plants are also commonly used as "bottoming units" — in these applications, the binary plant uses the waste fluids coming from the separators as well as the residual heat from the main power plant. For example, steam of 250°C utilized by the main power plant (usually a conventional steam (flash) plant) can, depending on fluid chemistry, have a temperature of 120°C to 170°C after expansion when leaving the turbine. Instead of condensing this steam by means of air-cooling or cooling towers, it can be efficiently used to generate more power by the bottoming unit and thereby increase the overall efficiency and economics of the entire power plant unit. However, bottoming units add significantly to the complexity of and the total project cost. These costs will be reflected in the plant's actual power generation costs (per kWh) and might reduce the margin between generation cost and the power tariff as paid by the off-taker/utility company. The resulting reduction of operational profit is the reason why project developers may choose not to deploy bottoming units in all cases. From the country perspective on the other hand, producing 10% to 20% more power out of the same resource could be very attractive, because after having installed a few of these bottoming cycles, their output can easily add up to building an entire new power plant. Second, due to the fact that each geothermal reservoir has a limited potential for steam production, it could be worth considering using the resource as efficiently as possible although the potential for premature cooling of the reservoir must also be fully considered.

Binary units can be produced in very small sizes (250 kWe to 2.5 MWe), even as container module units. Small mobile plants, conventional or not, can not only reduce the risk inherent to drilling new wells but also, more importantly, help in meeting the energy requirements of isolated non–grid-connected areas. The standard of living of many communities could be considerably improved were they able to draw on local sources of energy.[9]

5.3 *Additional Technologies*

Dry steam: In a few areas of the World, namely in California (USA), Italy, Indonesia and to a lesser extent in Japan and New Zealand, geothermal reservoirs can be found that produce pure hot steam due to the low reservoir permeability. These resources can be utilized by a condensing power plant, as introduced above, but in this case there is no need for a separator to separate fluids and steam. Generally Dry Steam units are large (>50 MWe) and operate with high efficiency.

Back pressure units: Back pressure units are steam turbines that exhaust the incoming steam, whether dry or wet, directly into the atmosphere. This makes them compact, simple to install and run and the cheapest choice available. However, they are usually only used for a limited amount of time e.g. as test units or wellhead generators, until a more appropriate solution can be implemented. Back pressure units have a relatively low efficiency as compared to the other technologies mentioned above — generally the electrical output is about 50% of the other technologies, which means they generate significantly less power out of the same amount of steam, and they can, depending on the chemical composition of the fluids/steam, be hazardous to the environment as they lack equipment to deal with non-condensable gasses and in addition generate a great deal of noise pollution.

5.4 *Power Plant Condensers*

Two major categories of condensers are used with steam cycles: the surface condenser and the direct contact condenser. In a surface condenser, the cooling water is circulated through the inside of the heat exchanger tubes with the steam condensing on the outside of the tubes. In a direct contact condenser, the cooling water is sprayed into the condenser, where it directly contacts the steam from the turbine discharge. Non-condensable gas must be removed from the condenser to reduce back pressure and optimize steam use efficiency. If the non-condensable gas has a significant contribution of hydrogen sulfide, the use of an abatement system is required resulting in a significant increase in the capital cost of the plant and increasing operation and maintenance costs as well.

In binary power plants, the use of air-cooled condensers is often required due to unavailability of cooling water as in arid regions. Where water is available, the condenser usually is of a tube and shell configuration.

5.5 *Power Plant Cooling*

Several cooling tower designs are available for supplying cooling water to the plant condenser, including cross-flow, cross-flow with high efficiency fills, and counter flow. The availability of cooling water is another important consideration in plant design. In a condensing direct steam or flash power plant, the condensate is available and used for cooling tower water make up. The plant can thus take advantage of the low wet bulb temperature that may be present even though the ambient dry bulb may be quite high. Where cooling water is unavailable or available in only limited quantities (especially true for binary cycle plants if an outside source of water is not available) dry cooling employing an air-cooled condenser will be required. In such systems, the condensing working fluid is directed through the heat transfer tubes of the condenser and air forced across the tubes by mans of fans to remove the heat. Where some water is available, the air-cooled condenser performance can be optimized through the use of enhanced evaporative cooling, where water is either injected as a mist directly into the air flow of the air cooled condenser or used to wet a fibrous (fiber glass) material used to enclose the walls of the cooling tower and through which air is then drawn thus lowering the wet bulb temperature. The use of enhanced evaporative cooling with a binary cycle can improve summer efficiency of a power plant by more than 25%.

5.6 *Constructing Power Plants According to the Stepwise Approach*

In cases where the exploration program for a given geothermal field shows very positive results, it is tempting for the developer to assume that it should be possible to build a large (>50 MWe) power plant as a first step. Several projects have done exactly this and in some of these cases the outcome was a partial or total loss of the investment. The reason is clear: Even after successful test drillings and reservoir testing, any developer's knowledge about the long-term reservoir productivity, in terms of steam production and power generation capacity, is limited. Therefore, geothermal experts have learned, sometimes the hard way, to deploy a stepwise approach for the development of geothermal power projects. To minimize the risks of depleting the geothermal reservoir and the resulting pressure drops, geothermal projects are usually developed in steps of for example 25 to 50 MWe at a time. This means that geothermal power projects cannot be a quick fix for any country's power shortages, it is rather more appropriate as a long-term strategy. However, several power projects could be developed in parallel, thus increasing the supply more quickly and probably with less geological risks. Conventional steam (flash) power plants as well binary plants tend to have a minimum installed capacity of 20 MWe due to economies of scale considerations; however, small modular power units and favorably located, smaller power plants can also make an economically justifiable project dependent upon reservoir characteristics, drilling depths and price for power. After having operated the first unit for some years and confirming the potential of the reservoir, subsequent stages can be added to the first power unit. The stepwise approach and the parallel development approach are important methods for risk mitigation in geothermal power projects.[8]

5.7 *Determination of Power Plant Size by Demand Analysis*

Two factors strongly determine the highest possible installed capacity, and thereby power generation, of a geothermal power plant: (i) The demand for electricity in the country or within the system and (ii) The potential of the geothermal reservoir.

The electric load within a country depends on power generation on one hand and on demand for power on the other. The system only functions optimally if generation and demand are nearly the same at all times.

Figure 10 presents a typical country load curve, in this case with two peaks over the day. Depending on the country, load curves have different shapes, according to the system demand they reflect and demand may also vary considerably from season to season as well as over the course of a day.

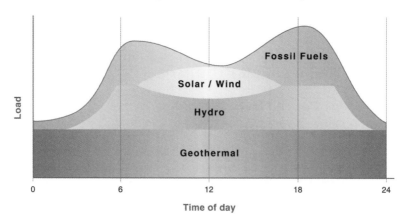

Fig. 10.   Simplified load curve with typical fuel sources.[8]

It is a common and justifiable view that geothermal power plants cannot be adapted to following system demand, because the power plant depends on hot steam coming from the ground, which is a steady source and somewhat difficult to control, except in the case of binary systems, where the wells are generally pumped. Therefore, geothermal power plants are usually deployed to provide base load to the system, as shown in Fig. 10. Other power sources can adapt more quickly to the demand; this applies e.g. to diesel generators and hydropower plants. Along with gas power plants, these sources can be used to track the load within the system. Dispatching various power sources depends on whether they can be used for base load, for peaking operations and how fast they can adapt to changes in the system demand. Another argument against operating a geothermal plant in a load following or dispatchable mode is the high up-front capital cost and low fuel cost as well as O & M costs, all which favor base load operation. It is common practice to grant priority dispatch to geothermal power as well as to most other forms of renewable energy, in order to decrease the use of fossil fuels and make the water for hydro power in its reservoirs available for use over a longer period of the year. For this reason, the size of a geothermal power plant should not exceed the minimum system demand.

6 Economics of Power Generation

Table 3 presents an indicative cost analysis for geothermal project development of a typical 50 MWe project in a geothermal field with no initial development and with wells of around 2 km in depth.[6] Power plants of 25 to 50 MWe can often be a suitable first step unit, which can be expanded or multiplied at a later stage, or as final unit. Wells generally range in depth from between 1.5 and 3 km, with an international average of around 2 km, which will be used in the following calculations. The cost figures include all exploration and drilling costs, as well as estimated financing costs for the development of a hydrothermal reservoir for power generation.

Table 3: Indicative Costs for Geothermal Development (50 MWe Generator Capacity) in Million US$.

Phase/activity		Low estimate	Medium estimate	High estimate
1	Preliminary survey, permits, market analysis[i]	1	2	5
2	Exploration[j]	2	3	4
3	Test drillings, well testing, reservoir evaluation[k]	11	18	30
4	Feasibility study, project planning, funding, contracts, insurances, etc.[l]	5	7	10
5	Drillings (prod and inject) (20 boreholes)[m]	45	70	100
6	Construction (power plant, cooling, infrastructure, etc.)[n]	65	75	95
	Steam gathering system and substation, connection to grid (transmission)[o]	10	16	22
7	Start-up and commissioning[p]	3	5	8
	TOTAL	142	196	274
	In million US$ per MWe installed	2.8	3.9	5.5

[i]Costs for survey depend heavily on size and accessibility of area. Costs for EIA depend on country regulations.
[j]Depending on methods used and accessibility and size of area.
[k]For three to five drillings with variable depths and diameter, from slim hole to full-size production wells.
[l]Studies and contracts provided by external suppliers or own company. Conditions and regulations of relevant country.
[m]Depending on depth, diameter and fluid chemistry, casings and wellhead requirements in terms of pressure and steel material/coating. Also influenced by underground and fractures (drilling difficulty and time).
[n]Power plant prices vary by system used and supplier, but most impact comes from infrastructure (roads, etc.) and cooling options (water or air).
[o]Depending on distance from plant to transmission grid access point, and on distance between boreholes and power plant.
[p]Standard industrial process. Power plant may need fine tuning for some time and minor adaptations. For high estimate, major changes, repairs and improvements are needed to supply power according to PPA.
Source: Gehringer and Loksha.[8]

Given this variability of geothermal investment costs, a useful question to ask is: how high can the investment cost of geothermal become before it ceases to be economically competitive? This can be accomplished by comparing geothermal with other base-load technologies, such as steam turbines on heavy fuel oil (HFO) or coal, medium-speed diesels (MSD) on HFO, and eventually large hydropower plants.[10]

Based on an oil price of around US$75 per barrel, the economic break-even investment costs for geothermal would be:

- US$8900 per kWe installed, as compared to steam turbines on HFO
- US$7000 per kWe as compared to MSD
- US$5200 per kWe as compared to steam turbines on coal
- US$4400 per kWe as compared to large hydropower facilities with capacity factor 60%

Geothermal energy is an unusual case in that the main source of uncertainty lies in the investment cost. The only comparable case is hydro, where investment costs may vary according to how geological characteristics develop during construction; however, *a priori* determination of expected costs can be gauged with some accuracy, whereas in the case of geothermal, the actual exploration cost is a major factor in the economics of a potential project. Computer models that accurately take into account this source of uncertainty to quantify the tradeoffs with competing resources have yet to be developed.

6.1 *Risk/Cost Profiles and Financing Options*

Figure 11 shows the correlation between the project cycle, risks and cumulative cost for a typical medium-sized geothermal power project (50 MWe). While

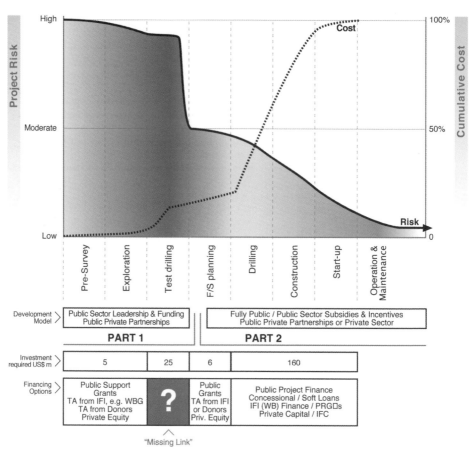

Fig. 11.   Financing a typical medium size geothermal power project.[11]

the regulatory and legal framework of developed countries usually offers various means to support and even finance geothermal project, the situation is different for developing countries. Here it is beneficial to split the project cycle in two parts; Part 2 may use project finance as most other power projects (coal, hydro, etc.), but Part 1 usually requires some government or other public support, because of the high investment and geological risks. For developing countries, the test drillings, shown in Fig. 11 as the "missing link" is the most difficult hurdle to surpass, since risk capital is often unavailable or private investors would charge a high risk premium for funding this part of the project. In general, it seems beneficial and will often be imperative for governments to provide substantial support to the first three project phases in order to attract the private sector to further develop and operate the project.

6.2 *Incremental Costs*

Several countries with geothermal resources have other energy resources which can be used for power generation at a lower cost. Examples are Canada for hydropower and Indonesia for coal. The incremental costs of geothermal power generation can present a barrier to the deployment of geothermal power plants in these countries, especially in those cases where the power price paid by the end consumer is low. In the example of Indonesia, coal power plants situated within the coal mine area usually generate electricity at a lower cost than geothermal plants. In addition, sometimes public utilities do not want or are not allowed by the government to raise consumer tariffs.

Incremental costs can also result from the location of a geothermal site and the necessary investment for new transmission lines. Typically, areas promising a viable geothermal resource will not coincide with electric load centers. Cities with large populations are not generally built on active geological faults or on the flanks of active volcanoes that are likely to be able to support large-scale geothermal power plants (i.e. flash plants). This introduces the additional risk of finding geothermal reservoirs of sufficient size to build power plants large enough to justify the cost of transmission lines to the load center. Small geothermal binary plants, which take advantage of lower geothermal temperatures found at greater distances from primary geothermal areas, may have applications when they are situated closer to load centers.

To summarize, even in countries richly endowed with geothermal resources suitable for electricity production, geothermal power generation is clearly not always the cheapest option. In spite of its usually attractive power generation costs of US 5 to 10 cents per kWh on average and a capacity factor of at least 80%, the economic attractiveness of geothermal power projects need to be carefully assessed compared to other competing sources of electricity (e.g. hydropower, gas, coal, etc.) available in countries producing exportable fossil fuels.

7 Other Geothermal Resource Types and Applications

In addition to those mentioned earlier, there are a number of other geothermal resource types, technologies for utilization and benefits that can be realized from development. A full coverage is beyond the scope of this chapter, but still worth mentioning in order to provide the reader with an understanding of the breath of the resource as well as the multitude of economic benefits that can be derived from geothermal exploitation. These include geopressured resources, Co-produced resources, Enhanced Geothermal Systems (EGS), mineral extraction and geothermal heat pumps.

7.1 *Geopressured Resources*

Geopressured resources occur in deep sedimentary basin environments where deeply buried fluids are heated by the normal or enhanced geothermal gradient. The fluids contained in these permeable sedimentary rocks are tightly confined by surrounding impermeable rocks and have pressures much above hydrostatic levels. The sources of energy include heat, mechanical energy and methane. The heat, ranging from 90°C to more than 200°C can be converted to electricity using a binary cycle (see Section 5) or used directly for heating or cooling. The hydraulic energy can be converted to electricity through use of a hydraulic turbine. The methane gas can be separated and used to generate electricity or burned to provide heat energy. Geopressurred resources are found in several locations in the USA as well as in other parts of the world. A recent estimate of USA's potential by the National Renewable Energy Laboratory (NREL) determined that 50,000 to 60,000 MWe could be developed[12] by the year 2025. In addition, geopressure development could lead to significant development of presently uneconomic gas fields thus greatly increasing natural gas reserves.

7.2 *Co-Produced Resources*

Co-produced resources are thermally heated water associated with oil and gas wells. Historically this abundant source of hot water (some 25 billion barrels per year in the USA) has been little more than an inconvenience requiring expensive disposal. However, the hot water is now increasingly being looked upon as a source for power generation through the use of binary systems or simply for various heating and/or cooling applications requiring moderate temperatures. NREL estimates placed the US developable resource potential[12] as high as 10,000 to 20,000 MWe by 2025. Co-production will have the added benefit of greatly improving the economic viability of what now must to considered marginal oil and gas well; thus adding significantly to the productive life of known and producing oil and gas fields.

7.3 *Enhanced Geothermal Systems*

Enhanced geothermal systems (EGS), also referred to as engineered geothermal systems, and often in the literature as hot dry rock, could well make geothermal development for power generation nearly universal. EGS are engineered reservoirs created to produce energy from geothermal resources that are or would not otherwise be economical to develop due to a lack of fluids and/or permeability. Through EGS technology, heat is extracted by means of engineered pathways created for fluid-flow through hot rocks. The pathways are created by stimulating fractures with cold water injected into a well at high pressures and utilizing a number of techniques to maintain those pathways. The EGS power cycle requires continuous injection of water down a well and through the pathways created where it is heated and then brought back to the surface using a second well where the heat is extracted through a binary cycle to generate electricity. The water is then re-injected to form a continuous loop. EGS techniques for stimulation of fracture pathways may also significantly improve the productivity and heat extraction capabilities of hydrothermal systems that lack adequate permeability.

Considerable work on development of EGS is ongoing in Australia, USA, Germany, Japan and the UK.[13]

A 2007 study led by the Massachusetts Institute of Technology estimated that, with adequate investments to fully develop the technology, 10% of USA's demand for electricity could be met by EGS within 50 years.[14] Another important aspect of EGS is that, in the future, Engineered Geothermal Systems may be locatable near end users, thus increasing its benefits to major population centers worldwide.

As part of a recent trend, shale gas reservoirs are being developed by oil and gas companies in various parts of the world. The methodology behind these developments is similar to EGS as it includes intensive fracking with highly pressurized water, and these shale gas reservoirs could possibly be used for EGS projects, after most of the gas has been extracted. In these cases, the drillings, the grid connection and the reservoir would be ready to use for the EGS developer. However, it remains to be seen how economical these projects will turn out to be.

7.4 *Mineral Extraction*

The extraction of minerals from geothermal brines could also be referred to as natural solution mining.[15] Geothermal fluids as the percolate through hot rocks and through interaction with the rocks become increasingly saturated with various minerals. As a rule, the higher the temperature at which the interaction takes place the greater the chemical content of the brine. In some geothermal fields, the total dissolved solid (TDS) content can exceed 300,000 parts per million (ppm). Such high levels of TDS create major engineering challenges to deal with severe corrosion and/or scaling problems in pipelines and within the power plant itself.

However, many of the chemical constituents in geothermal brine are a potential source of valuable minerals and metals. Minerals of primary interest include silica, zinc, lithium, manganese and a number of rare earth, for example, cesium and rubidium. There may also be significantly high concentrations of precious metals such as silver, gold, palladium and platinum.

Silica is one of the most common and ubiquitous components of most geothermal brines and a major source of scale, limiting the amount of heat that can be removed prior to the onset of precipitation and thus power output per unit of geothermal flow. Silica is also potentially one of the most valuable minerals contained in geothermal brine. Silica unfortunately interferes with the extraction of other minerals such as zinc, manganese, lithium, etc. Thus the successful removal of silica can not only provide an economically attractive product and allow for the extraction of other minerals but also allow for greatly increasing power output per unit of brine removed from the reservoir.

Although considerable research has gone into mineral extraction, including the construction of pilot plants, no sustained commercial operation has yet been achieved. However, the future is promising and in the summer of 2011 a contract for lithium extraction was awarded[16,17] to start-up company in the USA.

7.5 *Geothermal Heat Pumps*

Geothermal heat pumps allow for the almost universal application of geothermal energy since wells drilled to depths of only a few tens of meters to a few hundred meters anywhere on earth will provide sustainable temperatures of 10 to 15°C, temperatures ideal for heat pump applications.

Geothermal Heat Pumps (GHP) are refrigeration devices that can be run in reverse using the constant temperature of the earth as a heat source (in heating) or sink (in cooling). These devices produce the hot and cold water that is used to provide heating/cooling to a building, and they can also be used to produce some of the hot water needed for occupant use. They use ground or ground-water temperatures between 5 and 30°C with a Coefficient of Performance (COP) around 4.0 (the ratio of the heating or cooling energy output divided by the electrical energy input into the compressor). There are essentially four types of GHPs: closed-loop vertical, closed-loop horizontal, closed-loop lake, and open loop using ground water or lake water. The most common type is the closed-loop system, in which a closed-loop of high density polyethylene pipe is filled with a water-antifreeze solution that is pumped through the pipes, absorbing or rejecting heat through the pipe walls into the earth, and is then transferred to the heat pump. These loops are either installed and grouted in holes 50 to 100 m deep, or in trenches approximately two meters deep. Open-loop systems use ground-water or lake water directly to provide heating or cooling through the heat pump. Approximately three million units have been installed in over 40 countries[3] with a total installed capacity of around 34 GW and annual energy production of 200×10^{15} Joule.

8 Environmental Impacts, Mitigation Measures, and Benefits

The utilization of geothermal power instead of fossil fuels such as oil, gas, and coal, can have a tremendous impact on reducing CO_2 emissions. Data from 85 geothermal plants (operating capacity 6.648 MWe) in 11 countries, representing 85% of

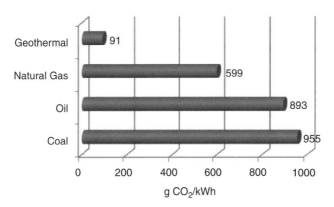

Fig. 12. CO_2 emissions by energy source in the US.

global geothermal capacity in 2001, indicate a weighted average of CO_2 emissions of 122 g/kWh.[18] In USA, which is the largest producer of geothermal energy in the world, CO_2 emissions were reported at 91 g/kWh.[19] Figure 12 compares CO_2 emissions various sources of energy in the USA.[20]

There are, however, some environmental impacts and risks to manage. As with all human intervention, development and exploitation of geothermal energy has an impact on the environment, though experts usually agree that it is one of the most environment-friendly and responsible forms of energy. Some important environmental safeguard issues and risks to be considered by a geothermal project developer include the following:

- The first perceptible effect on the environment is that of *drilling*, whether the boreholes are shallow ones for measuring the geothermal gradient in the study phase, or exploratory/producing wells. Installation of a drilling rig and all the accessory equipment entails the construction of access roads and a drilling pad. Drilling may also produce considerable noise pollution, which may impact nearby residences. Drilling itself has the potential to contaminate ground water resources if wells are not properly cased and cemented to prevent flow along the bore hole. In a worst case scenario a well blow-out could have a significant environmental impact and properly applied blow-out prevention equipment is a requirement. However, the impact on the environment caused by drilling mostly ends once drilling is completed.
- The next stage, installation of the pipelines that will transport the geothermal fluids, and construction of the power plant and associated transmission facilities, will also affect animal and plant life and the surface morphology. There can also be impacts on archeological or historically significant areas. However, the impacts are local and usually limited or avoidable.
- Environmental problems also arise during plant operation. Geothermal fluids (steam or hot water) usually contain *gases* such as carbon dioxide (CO_2), hydrogen sulphide (H_2S), ammonia (NH_3), methane (CH_4), and trace amounts of other

272 *G. Bloomquist, J. Lund and M. Gehringer*

gases, as well as *dissolved chemicals* whose concentrations usually increase with temperature. Installation of hydrogen sulfide abatement equipment is required wherever significant amounts of H_2S are encountered.

- The waste water from cooling towers has a higher temperature than the environment and therefore constitutes a potential thermal pollutant. Cooling tower drift can also be a problem and impact vegetation if for example significant amounts of boron are present.
- *Discharge of waste fluids* is also a potential source of chemical pollution. After having passed the through the plant, geothermal fluids with high concentrations of chemicals such as sodium chloride, mercury, boron, fluoride or arsenic should either be treated or re-injected into the reservoir. Fluids coming from low-to-medium temperature geothermal fields, as used in most direct-use applications, generally contain low levels of chemicals.
- The withdrawal and/or reinjection of geothermal fluids may trigger or increase the frequency of *seismic events* in certain areas. However these are primarily micro-seismic events that can only be detected by means of sensitive instrumentation. Exploitation of hydrothermal resources is unlikely to trigger major seismic events, and so far has never been known to do so. Withdrawal of geothermal fluid can also cause local subsidence that could impact some land uses.
- The noise associated with operating geothermal plants could be a problem to nearby populated areas, where the plant in question generates electricity. During the production phase, there is the higher pitched noise of steam traveling through pipelines and the occasional vent discharge as well as noise from cooling towers.[7]
- Other areas of environmental concern include social impacts, resettlement issues, compatibility with other land uses, impacts upon protected areas such as national parks and wildlife reserves, and protection of archeological, historical, and religious areas.

References

1. United States Geological Survey. Available at: www.cnsm.csulb.edu. Accessed 22 December 2011.
2. R. Bertani, World Geothermal Generation in 2010, *Proc. WGC 2010*, Bali. Available at: http://www.wgc2010.org/pdf/WGC2010_Daily_News_1stEdition.pdf. Accessed 17 October 2011.
3. J. W. Lund, D. H. Freeston and T. L. Boyd, Direct utilization of geothermal energy 2010 worldwide review, *Geothermics* **40** (2011), pp. 159–180.
4. A. Ragnarsson, Geothermal development in Iceland 2005–2009, *Proc. 2010 World Geothermal Congress*, Paper No. 0124. (International Geothermal Association, Bali, Indonesia, 2010), p. 12.
5. O. Mertoglu, S. M. Simsek, H. Dagistan, N. Bakir and N. Dogdu, Geothermal country update report of Turkey (2005–2010), *Proc. 2010 World Geothermal Congress*, Paper No. 0119. (International Geothermal Association, Bali, Indonesia, 2010), p. 9.
6. J. W. Lund, Geothermal direct-use equipment overview, *Quarterly Bulletin* **19**(1) (1998), pp. 1–6. Geo-Heat Center.

7. O. Edenhofer, R. P. Madruga and Y. Sokona (eds.), Geothermal energy, in *Renewable Energy Sources and Climate Change Mitigation, Special Report of the Intergovernmental Panel on Climate Change*, Chapter 4 (Cambridge University Press, Cambridge, 2012), pp. 401–436.

8. M. Gehringer and V. Loksha, Geothermal *Handbook: Planning and Financing Power Generation* (ESMAP Publication, World Bank, Washington DC, 2012).

9. M. H. Dickson and M. Fanelli, What is Geothermal Energy? Istituto di Geoscienze e Georisorse (CNR, Pisa, 2004).

10. R. G. Bloomquist and G. Knapp, Economics and financing, in *Geothermal Energy Resources for Developing Countries*, (eds.), D. Chandrasekharam and J. Bundschuf, Chapter 18A.A. Balkema, Publishers, Tokyo 2002).

11. M. Gehringer, Cost comparison of geothermal and other technologies, *GRC Proc.* 25 October (San Diego, 2011).

12. NREL, Geothermal Energy Production with Co-produced and Geopressured Resources, US Department of Energy, Fact Sheet, July 2010 Energy Efficiency and Renewable Energy.

13. Wikipedia, Enhanced Geothermal System www//en/Wikipedia.org/wki/Enhanced_geothermal_Systems. Accessed 2 July 2012.

14. J. W. Tester *et al.*, *The Future of Geothermal Energy: Impact of Enhanced Geothermal Systems (EGS) on the United States in the 21st Century* (Massachusetts Institute of Technology, Cambridge, MA, 2006).

15. R. G. Bloomquist, Economics of mineral extraction from geothermal brines, *International Conference on Economic Benefits of Mineral Extraction from Geothermal Brines*, Tucson, Arizona. Prepared by Washington State University Energy Program, Olympia, WA, 2006.

16. M. L. Wald, Start-up in California plans to capture lithium and market share, *The New York Times*, 29 September 2011.

17. P. Patel (2011). Start-up to capture lithium from geothermal plants, *Technology Review* 16 November 2011.

18. R. Bertani and I. Thani, Geothermal power generating plant CO_2 emission survey, *IGA News, Italy* 49 (2002), pp. 1–3. Available at: http://iga.igg.cnr.it. Accessed 16 October 2011.

19. K. K. Bloomfield, J. N. Moore and R. N. Neilson, Geothermal energy reduces greenhouse gases, *Geothermal Resources Council Bulletin* **32** (2003), pp. 77–79.

20. I. B. Friðleifsson, The possible role and contribution of geothermal energy to the mitigation of climate change, Report for IPCC, Reykjavik, Iceland, February 2008.

Chapter 13

Hydropower and Pumped Storage

Torbjørn K. Nielsen

Department of Energy and Process Engineering
Norwegian University of Science and Technology
N-7491 Trondheim, Norway
torbjorn.nielsen@ntnu.no

Hydropower has a very long tradition in many countries and was first used to provide working power for grinding corn, sawing timber, and other previously manual tasks. The development of the modern turbine took a big step forward in the 17th century when Leonard Euler presented his turbine theory. Later, hydropower became one of the main resources for electrical energy. Hydropower is the most effective method of energy transformation. The efficiency of a modern Francis turbine is above 95%. There is still a huge potential for hydropower to be utilized throughout the world. High-head hydropower plants are unique in respect to their governing ability. Therefore, hydropower plants play an important role in balancing power demands in countries dominated by thermal power. For renewable energy sources, hydropower is also essential for balancing power demands.

1 Introduction

Hydropower was known both in ancient China and ancient Egypt more than 4,000 years ago. The main purpose of these machines was to utilize mechanical power for milling grain and pumping water. Old sketches of these machines show a remarkable similarity to some of the turbines used today. The modern era of hydropower really began when Leonard Euler, in 1750–1554, developed his turbine theory. He presented the equations that still have significant use in hydroturbine design today.

However, hydropower is not only about turbines. The whole system, consisting of dams, conduits, pressure shafts, surge arrangements, turbines, electric generators, and transformers must work together in an optimized manner to produce electric power for the grid (Fig. 1). The grid is essential for distributing the energy.

Hydraulic power is obtained by multiplying the flow and the head available from nature. The head is the difference in elevation between the lower and the upper reservoir. The design flow is selected based on a hydrologic survey. Also, the total amount of energy available is defined by the hydrology of the area. The reservoir

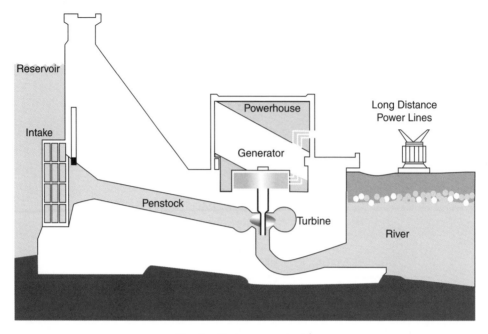

Fig. 1. Hydropower plant.[1]

serves as a collector of rainwater, and the utilization of the energy is balanced to meet the demand for electricity at all times.

2 Global Hydropower Resources

Global hydropower resources are huge and might make a difference in meeting the world's energy needs if fully utilized. The industrial countries have already generally developed the resources that are politically acceptable. In spite of the fact that hydropower is a rather clean source of energy, there are some negative environmental impacts. Huge reservoirs are especially controversial. Also run-of-the-river plants could have negative effects on fish and plants. However, there are ways to build even large reservoir power plants with minimum environmental impact.

To determine the remaining potential hydropower is rather difficult, and figures from various sources differ. One question is the overall potential, but in addition, the economic feasibility has to be considered, as shown in Fig. 2. According to a World Bank document, which refers to the *International Journal of Hydropower and Dams* and *World Atlas 2006*, the remaining hydropower resources that are economically feasible but are still undeveloped in the different regions are:

- 93% in Africa
- 82% in East Asia and the Pacific
- 79% in the Middle East and North Africa
- 78% in Europe and Central Asia

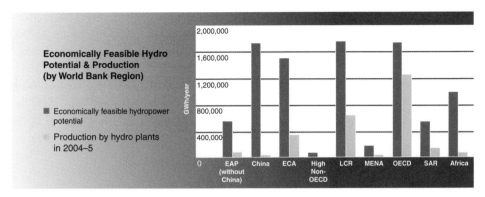

Fig. 2.   Unexploited and exploited hydropower potential.[2]

Table 1:   Hydroelectric Production from the Most Intensive Countries in TWh/Year from 2003–2009.[3]

Country	2003	2004	2005	2006	2007	2008	2009
Canada	334.8	337.0	361.2	352.9	367.8	408.8	396.9
US	277.2	269.7	271.9	290.8	249.0	256.1	273.7
Argentina	33.9	30.4	34.8	43.1	37.4	37.0	40.5
Brazil	304.5	319.4	336.2	347.2	372.2	367.8	389.4
Columbia	35.6	39.6	39.6	42.7	41.8	45.8	40.9
Venezuela	60.3	70.0	77.4	81.4	82.7	86.2	85.8
Austria	35.6	33.9	33.9	31.7	33.9	34.8	36.5
Italy	36.5	42.2	38.3	37.0	32.6	41.4	46.2
France	59.4	59.4	51.9	55.9	60.3	60.3	57.6
Norway	105.6	108.7	136.0	119.2	134.6	139.9	126.7
Sweden	53.2	55.9	72.6	61.6	66.0	68.6	65.6
Switzerland	34.8	33.4	31.2	30.8	35.2	36.1	35.6
Turkey	35.2	45.8	39.6	44.0	35.6	33.0	35.6
Russia	161.9	179.5	174.2	174.2	177.8	165.9	175.1
China	282.5	352.0	395.1	433.8	483.1	582.6	612.9
India	69.1	83.6	96.8	111.8	121.9	114.4	105.6
Japan	92.8	92.8	76.1	86.2	74.4	73.9	73.5
World Total	**2,624.6**	**2,785.6**	**2,897.4**	**3,010.9**	**3,063.7**	**3,218.2**	**3,257.3**

- 75% in South Asia
- 62% in Latin America and the Caribbean

3   Worldwide Use of Hydropower

Hydropower is widely used in the industrialized countries, depending on the topology and local hydrology. Table 1 shows the utilization of hydropower in some of the most hydropower-intensive countries and shows the electricity production from 2003 to 2009. The previously industrialized countries have nearly no new development, but countries like China and India are heavily exploiting the potential of hydropower. The total electricity production in the world is around 18,000 TWh/year, but only about 3,000 TWh/year comes from hydropower. The technical-economical potential of hydropower is said to be approximate 9,000 TWh/year.[3]

In Norway, nearly 100% of the electricity demand comes from hydropower, along with 30% of the electricity demand in Sweden. In the industrial countries, the development of new hydropower sources is politically difficult. The public believes that hydropower is not compatible with the preservation of nature. In Norway, more than 30% of the economically feasible hydropower resources have not been developed for this reason.

In China, building hydropower plants is a major priority. China has a huge undeveloped potential for hydropower, coupled with a need for energy. The same is true in India and Nepal. Chile, Brazil, and Venezuela have a long tradition in utilizing hydropower. There is also tremendous potential in these countries, coupled with a growing wish to preserve the environment.

4 Hydropower and Turbines

4.1 *Basic Equations*

The use of the appropriate fluid mechanics equations is necessary in designing hydropower plants. The most important of these is the continuity equation and the conservation of energy. In hydropower, the common pressure unit is meter of water column, dimension (m). The connection between the pressure dimension (N/m^2) and (m) is easily seen by a simple force balance, as shown in Fig. 3:

$$p = \rho g h. \tag{1}$$

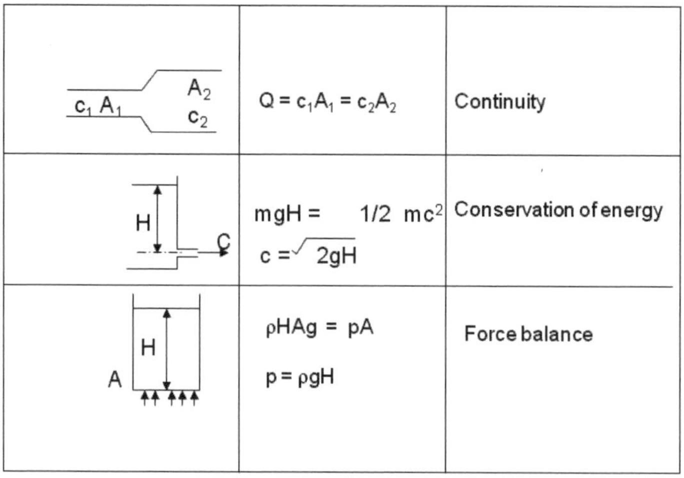

$\overline{c_1\ A_1}$ $\begin{array}{c}A_2\\ c_2\end{array}$	$Q = c_1 A_1 = c_2 A_2$	Continuity
H $\quad \c$	$mgH = \tfrac{1}{2}\,mc^2$ $c = \sqrt{2gH}$	Conservation of energy
A $\quad$ H	$\rho H A g = p A$ $p = \rho g H$	Force balance

Fig. 3.   Basic equations.

Another basic equation is the Bernoulli equation, stating that in steady state flow, energy is conserved when a fluid flows in a closed conduit from one position to another. However, friction losses will remove useful energy, so in practical use, the total energy at one position equals the energy at the next position plus the losses, hence:

$$z_1 + h_1 + \frac{c_1}{2g} = z_2 + h_2 + \frac{c_2}{2g} + \sum \text{loss}. \tag{2}$$

The equation has the dimension of meter, m. The first term z is the static head or the elevation measured from a given datum. The second term is the hydraulic pressure and the third term is the velocity head. According to Bernoulli, the hydraulic head increases when the velocity head (a function of the square of the velocity) decreases, and vice versa. Indices 1 and 2 denotes two positions in the conduit as illustrated in Fig. 4.

The head H is the sum of the hydraulic pressure h and the elevation z, i.e:

$$H = h + z. \tag{3}$$

Hydraulic pressure is defined relative to the atmospheric pressure i.e. the atmospheric pressure is set equal to zero, and hence hydraulic pressure is defined as the gauge pressure. At the reservoir, the hydraulic pressure is zero, which means that $H = z$ at the reservoir.

The tunnel and the penstock transport the water to the turbine. The energy available for the turbine is defined by the effective head, or the net head, H_n.

Formally, the net head is defined as:

$$H_n = z_1 - z_2 + h_1 - h_2 + \frac{c_1^2}{2g} - \frac{c_2^2}{2g}. \tag{4}$$

Index 1 denotes the inlet of the spiral chasing and Index 2 denotes the outlet from the draft tube, see Fig. 5.

Examining the formula, using the illustration in Fig. 5, shows that H_n is the head difference over the turbine, which is exactly the same as the gross head, H_G

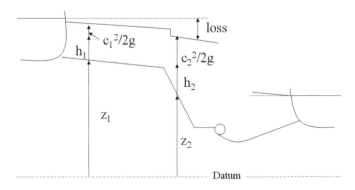

Fig. 4.   Bernoulli's equation in a hydropower plant.

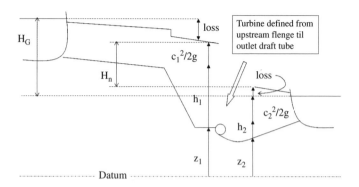

Fig. 5.   Definition of the net head, H_n.

minus all the hydraulic losses in the conduits, both before and after the turbine, hence:

$$H_n = H_G - \sum \text{losses} \tag{5}$$

where H_n is the design head of the turbine. The losses in the conduit system should not be blamed on the turbine.

4.2 *Hydraulic Loss*

Losses in fluid flow are of two types: Friction loss and singular loss or minor loss.

The friction losses in fluids are highly dependent on the turbulence. The turbulence level is defined by the Reynolds number:

$$R_e = \frac{cD}{v}, \tag{6}$$

where c is the velocity, D is a diameter, and v is the kinematic viscosity; see the definition of symbols in Table A1.

If the Reynold's number is high, which is the case for most flow in hydropower plants, the flow is said to be turbulent. In that case, the friction loss is proportional to the square of the velocity. The loss is expressed in the Darcy–Weisbach formula:

$$h_f = f\frac{L}{D}\frac{c^2}{2g}. \tag{7}$$

The friction factor f is dependent on the Reynolds number and the roughness of the conduit or pipe. The friction factor can be found in the Moody diagram (see Fig. 6).

For channel flows and for flows in tunnels, it is common to use the Manning number as the friction factor and the Manning formula:

$$h_f = \frac{Lc^2}{MR_h^{4/3}}, \tag{8}$$

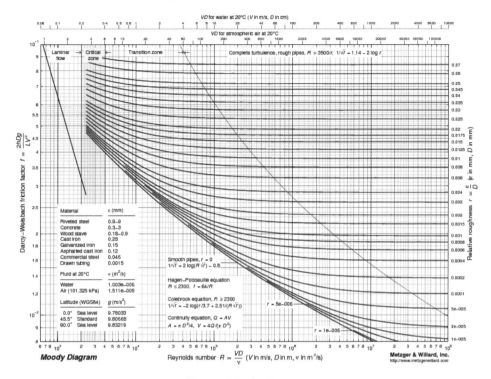

Fig. 6.   Moody diagram.

where R_h is the hydraulic radius defined as the area divided by the wetted perimeter. The formula is not non-dimensional and the Manning number varies from 30 for small tunnels to above 40 for big tunnels and also with the roughness. This makes it difficult to use.

A better way is to use the Darcy–Weisbach formula for tunnels and channels with non-circular area, replacing the diameter D with the hydraulic diameter D_h, defined as four times the area divided by the wetted perimeter.

The singular losses, or minor losses, are losses connected to change in velocity or change in direction at a certain position of the conduit. Examples are loss in valves, through gates, inlet and outlet losses from reservoirs, bends, etc. Also, these losses are proportional to the square of the velocity according to the equation below:

$$h_s = k_c \frac{c^2}{2g}. \qquad (9)$$

The term k_c is a geometry-dependent loss factor that can be found in various handbooks.[4] In Fig. 7, some loss coefficients are shown. The loss coefficients are found by experiments. Various handbooks define the loss coefficients differently, so it is important to make sure that the equation for the loss corresponds to the definition of the loss coefficient.

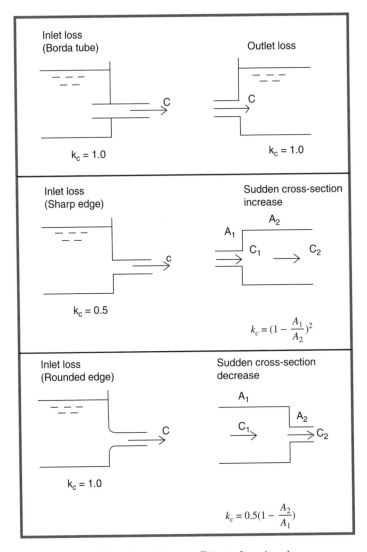

Fig. 7. Examples of loss coefficients for minor losses.

4.3 *Turbine Types*

There are, in principle, two kinds of turbines, the reaction turbine and the impulse turbine. Reaction turbines are characterized by a pressure difference across the turbine runner. The transformation of hydraulic power to rotating mechanical power is based on the reaction forces that are obtained both from the pressure difference and by the change of velocity through the runner.

Impulse turbines have no pressure difference across the runner. Normally, the runner rotates in atmospheric pressure. The hydraulic pressure is transformed into velocity, and the change in direction of velocity through the runner gives a reaction

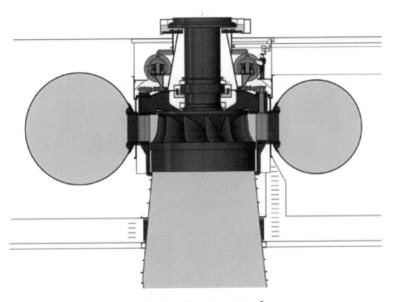

Fig. 8. Francis turbine.[5]

force and thereby a rotating torque on the turbine shaft. The most commonly used turbines are the Kaplan, Francis, and Pelton turbines, named after their inventors. The first two are reaction turbines, whereas the Pelton turbine is an impulse turbine. The Kaplan turbine is a low-head, high-flow turbine, and on the other hand, the Pelton turbine is a high-head, low-flow turbine.

Francis (see Fig. 8) and Kaplan turbines have a set of guide vanes upstream from the runner. A mechanism makes sure that all the guide vanes can be adjusted simultaneously and with the same angle. The whole mechanism is referred to as the wicket gate. By altering the wicket gate, the flow will be adjusted, and hence the power output.

A Kaplan turbine runner has the shape of a propeller, as shown in Fig. 9. Because of this beneficial geometry, it is possible to adjust the runner angles. This will improve the efficiency when the guide vane angles vary. In a Kaplan turbine, the flow through the turbine is axial i.e. the water flows parallel to the shaft. For a different head/flow ratio, the Euler equations results in a more semi-axial form and the turbine will be a Francis.

For both these turbines, the flow from the pressure shaft enters into a spiral-formed chasing. The spiral form is the optimum form for distributing the flow equally around the runner.

In a reaction turbine, a spin or swirl is established by the spiral chase and the wicket gate. The turbine transforms this spin to rotating mechanical power, so that the spin or swirl at the outlet of the turbine is zero. This is only obtainable at the design point, the best efficiency point (BEP).

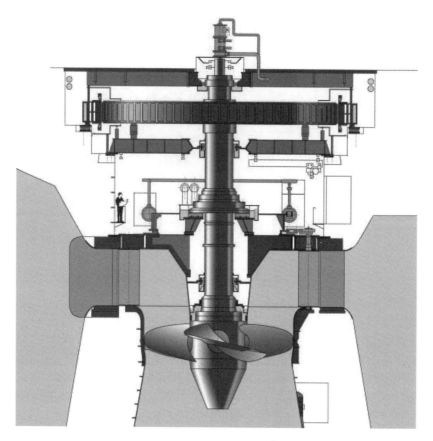

Fig. 9.   Kaplan turbine.[6]

The turbine outlet is connected to the draft tube. The goal is to have an unbroken hydraulic connection all the way from the head reservoir to the tail water reservoir in order to utilize the whole head. The draft tube has an expanding form so that the velocity is reduced and hence the velocity energy is transformed to pressure energy. Without the draft tube, the velocity head would be lost. The velocity head at the outlet of the draft tube corresponds to lost energy.

The Pelton turbine is an impulse turbine, which means that the head is transformed into velocity through a needle valve, which forms a water jet hitting the runner blade. The reaction force which gives torque and rotation to the runner, is only because of the change of direction of the velocity vector. A Pelton turbine may have up to six jets all connected to the pressure shaft by means of the distributer, as shown in Fig. 10. Each of the jets can be separately adjusted controlling the flow.

The Pelton runner must run freely in atmospheric pressure above the tail water level. The water ends up in the sump beneath the runner.

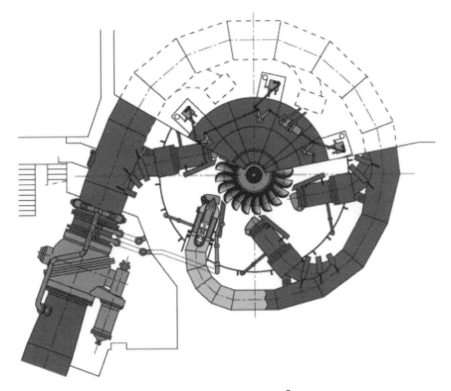

Fig. 10.   Pelton turbine.[7]

4.4 *Turbine Theory*

Hydropower is about transforming power from one energy regime to another. The hydraulic power is pressure multiplied by flow. Pressure means the head difference over the turbine H_n. The hydraulic power available for the turbine is:

$$P = \rho g Q H_n,$$ (10)

where, as earlier mentioned:

$$H_n = H_G - \sum \text{losses}.$$ (11)

The turbine transforms the hydraulic power to rotational mechanical power, $\omega \mathrm{T}$, to the turbine shaft. The generator transforms the mechanical rotational power to electric power, UI (see Fig. 11).

The aim is to make the power transformations as effective as possible. The conduit system must have low losses, which is dependent on the velocity, wall roughness, and hydraulic design. The turbine and generator design are optimized by minimizing the losses which are highly dependent of the design. The losses involved are friction losses, losses due to secondary flows, and swirls in the turbine runner channels, leakage losses, etc.

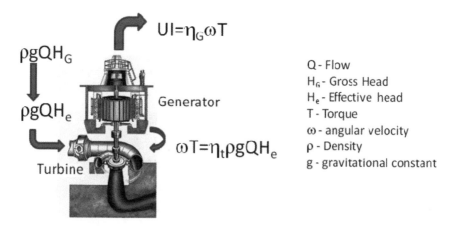

Fig. 11.   Power transformations in hydropower.

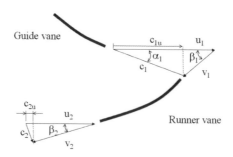

Fig. 12.   Velocity vectors for a turbine runner of Francis-type.

According to Leonard Euler, the reaction power extracted is dependent on the difference in peripheral speed multiplied by the absolute speed's component in the peripheral speed direction, hence:

$$P = \rho Q(u_1 c_{1u} - u_2 c_{2u}) = \omega T. \tag{12}$$

The design challenge is to optimize the geometry of the turbine according to this equation.

The velocities u_1 and u_2, are the peripheral velocities at the inlet and at the outlet of the runner, respectively. The velocities c_{1u} and c_{2u} are the absolute velocity's component in the peripheral direction at the inlet and outlet of the runner. Figure 12 shows the velocity vectors at the inlet and outlet of a simplified turbine blade. In reality, a turbine blade is highly three-dimensional and difficult to illustrate. Still, the power output of a turbine is a matter of how these velocity vectors changes in direction and magnitude through the runner. One of the design criteria for a Francis-type of turbine is that c_{2u} should be zero. This follows partly from the Euler equation, but it is also important that the water enters the draft tube with no swirl in order to retrieve pressure energy when the water flows through the draft tube. The velocity head at the outlet of the draft tube will be lost.

For an impulse turbine like Pelton, the head is transformed to velocity head through the jet. According to Toricelli's theorem, the velocity is only dependent on the head, hence

$$c = \sqrt{2gH_n}. \tag{13}$$

The flow through the nozzle will at full opening of the needle valve be:

$$Q_j = \frac{\pi}{4} d_j \sqrt{2gH_n}, \tag{14}$$

where d_j is the diameter of the jet.

The optimum speed of rotation is given by the fact that the maximum power transformation will occur when the peripheral speed is half of the absolute water velocity from the jet. By adjusting the needle valve, the outlet area of the valve will be adjusted giving the appropriate flow.

The design of an impulse turbine also requires the Euler equation. In order to produce as much reaction force as possible, the Pelton shovel (or bucket) turns the water through almost 180°, a little less so that the water leaving the bucket does not hit the back side of the following bucket.

The velocity vectors are shown in Fig. 13. When the water leaves the nozzle, the whole head has been transformed into velocity, c_1. The optimal speed of rotation is when the peripheral speed of the runner is half of the velocity from the nozzle. Also for a Pelton turbine, one of the design criteria is that c_{2u} should be zero.

4.5 *Efficiency*

Efficiency is defined as the power out divided by the power in. The hydraulic efficiency for a turbine indicates how efficiently the hydraulic power is transformed to rotational mechanical power according to the Euler equation for that particular turbine:

$$\eta_h = \frac{\rho Q(u_1 c_{1u} - u_2 c_{2u})}{\rho g Q H_n} = \frac{\omega T}{\rho g Q H_n}. \tag{15}$$

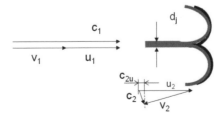

Fig. 13.   Velocity vectors for a Pelton-type turbine.

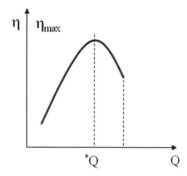

Fig. 14.   Turbine efficiency as a function of flow.

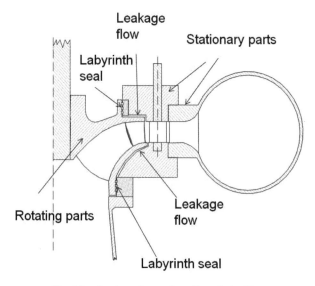

Fig. 15.   Leakage losses in a Francis turbine.

For a turbine with fixed runner blades, the optimum design point is reached for only one particular set of the parameters H_n, Q, and n.

In order to control the power output, the turbines must be equipped with a mechanism for adjusting the flow. When adjusting the flow, the turbine geometry is not longer optimal in accordance with the Euler equation, and the efficiency will be lower. Figure 14 shows a typical efficiency curve for a Francis-type turbine.

The Kaplan turbine has the benefit of adjustable runner blade angles so that the velocity vectors become more optimal even for another wicket gate position.

However, there are also friction losses and leakage losses to account for. In a Francis turbine, the runner rotates freely in a stationary mount. There must therefore be a certain clearance between the rotating and the stationary part (see Fig. 15). Some of the water leaks through this clearance and gives a leakage loss.

This leakage loss is expressed by the volumetric efficiency:

$$\eta_\nu = 1 - \frac{Q_{\text{leakage}}}{Q}.$$

Francis turbines are equipped with labyrinth seals in order to minimize the leakage.

There will also be friction losses in the bearings and disk friction loss due to rotation of the runner surrounded by water in the clearance space. Friction losses will give less torque and is expressed in the mechanical efficiency:

$$\eta_\nu = 1 - \frac{\Delta T}{T}.$$

In a Kaplan turbine, there has to be some clearance between the propeller-shaped runner and the surrounding tubing. Some of the water will leak in between, and there will also be a swirl at the tip of the runner stealing energy.

For a Pelton turbine, in addition to being dependent on an optimized bucket design, the efficiency is equally dependent on the design of the distributor, the nozzles to make uniformed jets and the housing around the runner. Ideally, the water leaving the buckets should drop freely to the sump. If the housing is too small, there will be water particles all around the runner, which will give high disk friction losses. Some of the available head is lost because the runner has to be mounted above the tail water level in order to run freely.

4.6 *Classification of Turbines*

The correct type of turbine should be used in a given hydropower plant. For example, it is not optimal to use a Francis turbine where a Pelton should be used. The specific speed of the turbine, size of the head, flow, and speed of rotation specifies the form of the turbine. The specific speed is a classification number that comes forth by identifying certain dimensionless numbers. Two turbines are similar if the velocity vectors through the turbines are similar. In that case, the Euler turbine equation will result in more or less the same optimized geometry, the main shape will be the same although the real dimensions are different. Two turbines with the same specific speed will be geometrically similar.

There are several ways of defining a classification number. The most common is the so-called specific speed:

$$n_q = n\frac{\sqrt{Q}}{H_e}, \tag{16}$$

where n is speed of rotation (RPM); Q is the flow (m^3/s); and H_n is the effective head (m).

The head and the flow are given by nature, while the speed of rotation is selected to be synchronous i.e in order to get the correct frequency on the grid, the angular speed of rotation must be the grid frequency divided by the pole pairs of

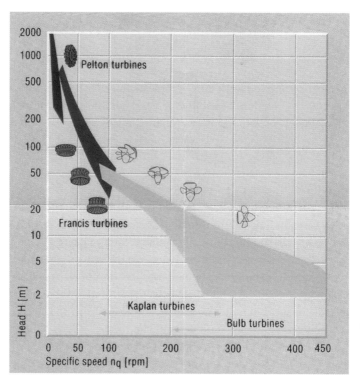

Fig. 16.   Turbine types versus specific speed.

the generator:

$$n = \frac{60f}{P} \quad \text{where } f = 50 \text{ or } 60\,\text{Hz.} \qquad (17)$$

The turbine types versus specific speed are shown in Fig. 16.

4.7 *Cavitation*

Cavitation is a phenomenon that can occur in all types of fluid flow. When the absolute pressure becomes lower than the vapor pressure, the liquid changes to vapor. If the vapor bubbles enters a region with higher pressure, the vapor bubbles will collapse and cause erosion of the material.

Reaction turbines like Kaplan and Pelton are equipped with a draft tube. At the outlet of the turbine, just before entering the draft tube, the velocity is relatively high and the pressure is low. If the absolute pressure becomes lower than the vapor pressure, cavitation will occur. Cavitation and the collapse of the vapor bubbles cause cavitation erosion, especially at the outlet edge of the runner blades. However, cavitation may also occur elsewhere in the turbine. By lowering the turbine centre relative to the tail water level, the absolute pressure in the turbine can be increased and thereby cavitation can be avoided.

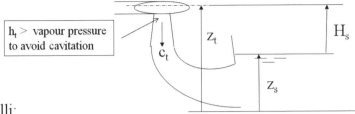

Bernoulli:

$$h_t + H_{atm} + z_t + \frac{c_t^2}{2g} = z_s + H_{atm} + loss$$

$$h_{t(abs)} = \underbrace{z_s - z_t}_{} + H_{atm} - \frac{c_t^2}{2g} + loss > h_{vapour}$$

$$-H_s$$

$$H_s < \underbrace{H_{atm} - h_{vapour}}_{\approx 10} - \underbrace{\frac{c_t^2}{2g} + loss}_{\sigma H_n}$$

$H_s < 10 - \sigma H_e$

σ – Thoma cavitation number

Another notation NPSH:

$$NPSH = \sigma H_e$$

Fig. 17.   Establishing the turbine center in order to avoid cavitation.

By using the Bernoulli equation and demanding the pressure at the turbine outlet to be above vapor pressure (see Fig. 17), the following equation will give the position of the turbine centre:

$$H_s < 10 - \sigma H_n, \tag{18}$$

where σ is the Thoma cavitation number, which is dependent on the specific speed of the turbine as shown in Fig. 18. The curve is based on experimental data, and there are several versions of this curve, depending on both turbine and draft tube design.

An alternative expression is the Nett Pressure Suction Head (NPSH). The conduit system has a $NPSH_{available}$. In order to avoid cavitation, $NPSH_{required} > NPSH_{available}$.

$$NPSH_{available} = H_{atm} - h_{vapor} - H_s, \tag{19}$$

$$NPSH_{required} = \frac{c_t^2}{2g} + loss, \tag{20}$$

$$NPSH_{required} > NPSH_{available}. \tag{21}$$

Hence

$$H_s < H_{atm} - h_{vapor} - \left(\frac{c_t^2}{2g} + Loss \right) = 10 - NPSH_{required}. \tag{22}$$

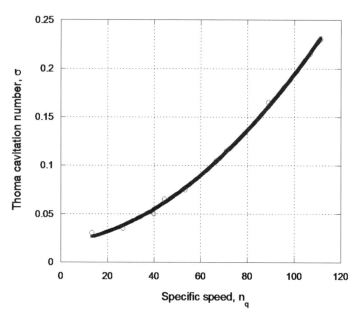

Fig. 18.   Thoma cavitation number as a function of specific speed.

The connection to the Thoma cavitation number is:

$$NPSH_{\text{required}} = \sigma H_n. \tag{23}$$

Again, NPSH$_{\text{required}}$ is dependent on the turbine's specific speed.

4.8 Technical Advances in Turbine Design

The turbine geometry has to comply with the fluid dynamics originally described by Leonard Euler. The turbine geometry is of course three dimensional, which makes it a challenge to optimize the flow through the turbines. The design methods of today include advanced use of Computational Fluid Dynamics (CFD) programs for solving the Navier–Stokes equations. However, in the initial design, the turbine geometry must be defined and described more or less analytically and entered into the computer program. With that given geometry, all kind of simulations can be performed. The designer will, based on the simulations, analyze the results and make the design changes appropriate for optimizing the efficiency. The use of CFD programs in turbine design has increased their efficiency by 2% to 4%, from late 1970s up to the present. The understanding of turbine behavior outside the BEP has also benefited from the use of CFD tools.

The recent design challenges are due in many cases to the variation in performance. In practical operation, turbines with high efficiency at part and full load operation are needed. The turbine manufacturers compete not only on the maximum efficiency but also on how flat the efficiency curve is. Some turbines have severe pressure pulsation phenomena especially at low load. This is partly due to the flow condition in the runner and also due to the flow condition in the draft tube. Advanced use of CFD tools helps the manufacturers to make optimal design of the whole turbine element.

The turbine price is strongly related to the price of materials. Better structural analyses by means of element-method programs have resulted in more lightweight construction. However, in recent years, many manufacturers have experienced breakage caused by fatigue. Better material quality is needed.

5 Hydropower Plant Performance

Hydropower requires more than just turbines. The conduit design is essential both for optimal performance and safety. The design of the hydropower system, as well as the turbine is individually constructed using practical considerations regarding effective operation, maintenance, and safety. However, there are some main elements that normally are present in all hydropower systems.

Starting from the upper reservoir (see Fig. 19), a trash rack is mounted for preventing debris from entering the system. After the trash rack, there is normally an intake gate which is used for emptying the system and making it possible to inspect and maintain the tunnel. A sand trap in the tunnel collects sand and gravel, and an additional trash rack is often mounted at the entrance to the penstock to prevent gravel from entering the turbine. The penstock is either a steel tube or a steel-lined tunnel. The pressure becomes higher as the water flows toward the turbine. In front of the turbine, a valve is mounted. In high-head power plants, it is a spherical valve, and in lower head power plants, a butterfly valve. A butterfly valve is cheaper but is not able to withstand the highest pressure. The main valve and actuator are designed to close at full flow. After the turbine, the draft tube and sometimes a tail water tunnel, connects to the lower reservoir. Just after the draft tube, a gate is mounted, again for closing off the water for inspection and

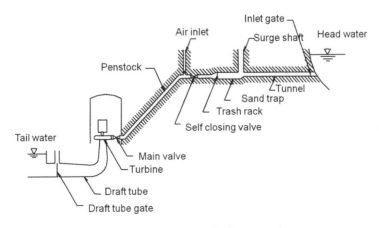

Fig. 19. Main elements in a hydropower plant.

maintenance. In many power plants, the tail water level is above the turbine level in order to avoid cavitation. Surge shafts, both upstream and downstream of the turbine, are sometimes necessary for handling stability and pressure transients, as described in the following chapters. For filling up the penstock, an arrangement for air inlet is necessary.

The procedure for starting a turbine is to open the main valve and then slowly open the turbine's wicket gate. The turbine begins to turn, and when it reach the synchronous speed of rotation, the face angle between current and voltage must be adjusted so that it corresponds with the grid. Then the generator is connected to the grid. The wicket gate opening is increased for delivering the needed power output.

5.1 *Steady State Performance*

In steady state performance, the main issue is to produce electrical energy as efficiently as possible, taking into consideration variation in reservoir elevation, conduit system, and turbine performance. In many power plants, several turbines are installed. The power must be optimally distributed among the running turbines. Maintenance of the power plant is essential. Trash racks must be cleared of debris of different kinds, tunnels must be kept in order, and steel-lined pressure shafts must be regularly renewed to control corrosion. Nowadays, most power plants are remotely controlled. If anything goes wrong, it may not be discovered in time to avoid a catastrophe in the plant.

In most of the industrialized countries, there is a free energy market i.e. the utility companies compete on price. Running a power plant is no longer a question of only optimizing the energy transformation but also a question of competing in the market. In systems dominated by thermal energy sources, hydropower plants are used for balancing the power according to the demand. Both the market and the demand for balancing the power load have presented new challenges to the hydropower industry. Turbines are exposed to more starts and stops and the turbines are often run outside the BEPs exposing the material to more pressure pulsations and vibrations. This has resulted in a significant increase in dynamic loads resulting in more wear, fatigue problems, etc. Variation caused by more frequent and bigger load variation produces bigger flow variation in the conduit system, resulting in more sand transportation and erosion.

5.2 *Governing Power*

Each turbine connected to the grid participates in the governing of the power. The turbine governor's main task is to keep the speed of rotation constant in order to maintain the required grid frequency at all times. In most countries, the frequency is 50 Hz, although some use 60 Hz.

When the electricity demand increases, the turbine speed of rotation will decrease, and using signals from the governor, the turbine will increase the wicket gate opening in order to deliver more power. This is a continuous ongoing process that makes sure that the hydraulic power is utilized in balance with the demand for electricity.

The quality requirements for keeping the frequency correct at all times is high, and the challenge is to achieve the required stability. This is a question of designing the hydraulic system with appropriate dynamic properties. The hydraulic inertia must be kept low i.e. not too long a penstock and tunnel, compared with the rotating masses represented by the generator. If the penstock and head race tunnel is too long, a surge chamber is needed. By this means, the hydraulic inertia immediate affecting the turbine is reduced. This again gives rise to so-called U-tube oscillations between the head reservoir and the surge shaft. Briefly, the whole system must be simulated and analyzed with respect to dynamic behavior. Taking the dynamic properties into consideration in the early stage of the design process is essential. The turbine must work together with the attached conduit system in an optimized way, both dynamically and in the steady state.

5.3 *Transient Behavior*

Power plants are designed to be able to close down in a matter of seconds. If there is a grid failure, the generator will lose load and veer dangerously toward increased speeds. The turbine must close the wicket gate as soon as possible to reduce the hydraulic power. This is a situation that might happen at any time.

Normal start and stop will be more controlled. However, there is a desire to carry out these operations as quickly as possible. Closing the wicket gate will stop the water in the pressure shaft and the retardation forces will produce a pressure rise. If the tunnel and pressure shaft are long, the hydraulic inertia is high, and in accordance with Newton's second law, the pressure rise will be high. The way to reduce the retardation force is to add surge shafts, hence the water in the tunnel will move up the surge shaft and reduce the transient pressure at the turbine.

Another rather dangerous phenomenon is referred to as the water hammer. If the water flow is stopped too fast, elastic pressure waves will occur and might lead to severe damage, even ruptured penstocks. Several accidents have happened due to water hammer phenomenon. In August 2009, Russia's largest hydropower plant, the Sayano–Shushenskaya plant in Siberia,[8] experienced a massive explosion and flood, initiated by water hammer, which killed at least 74 people. Another water hammer incident occurred in the Bieudron hydropower plant in Switzerland in December 2000, rupturing a penstock, with fatal results.[9]

6 Small-Scale Hydropower

In recent years, interest in small hydropower plants has been increasing mainly because small hydropower plants have less impact on the environment than large plants. Small hydropower plants are often run-of-the-river types, where the turbine utilizes the hydraulic energy that is available continuously. Run-of-the-river plants have very little reservoir capacity. It is, however, necessary to have a minimum size reservoir in order to define the pressure for the turbine. The small reservoir capacity also implies that the governing ability is poor. In that respect, small hydropower plants operate by adding essentially non-governed power to the grid in the same way as wind, tidal, and wave power plants.

Small-scale hydroplants are often classified as follows:

Small hydro $1-10\,\text{MW}$
Mini hydro $100-1,000\,\text{kW}$
Micro hydro $<100\,\text{kW}$

The turbine types used are often simplified versions of Kaplan, Francis, and Pelton turbines. However, there are some turbine types that have different geometry as the ones described in the following subsections.

6.1 *Cross-Flow Turbines*

The geometry of a cross-flow turbine as shown in Fig. 20 is very simple. The water passes a single flapper valve, which functions as the flow regulator. The jet hits the runner and crosses through the runner hub and is supposed to give extra torque flowing through the runner vanes diametrical opposite. The runner has the shape of a cylinder with runner vanes attached.

The cross-flow turbine is widely used in developing countries. The efficiency is poor but the maintenance is easy. The turbine is easy to make and is quite reliable. The cross-flow turbine is also called the Osborn turbine or Michel turbine.

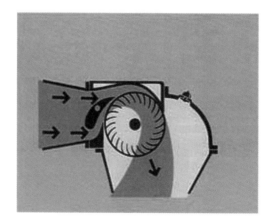

Fig. 20.   Cross-flow turbine.

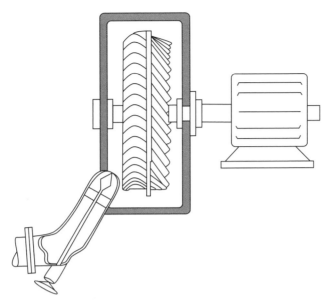

Fig. 21.   Turgo turbine.

6.2 *Turgo Turbine*

The Turgo turbine (see Fig. 21) is an impulse turbine similar to the Pelton turbine, but the buckets have a different design. The jet is mounted so that it hits the outer rim of the bucket, and the water makes a nearly 180° turn before the outlet. The efficiency is much lower than for a Pelton turbine.

6.3 *Centrifugal Pumps Run as Turbines*

Centrifugal pumps have a form very similar to a Francis turbine. When used as a turbine, the flow direction is opposite, rotating an electric generator. These pumps, when used as turbines, are able to produce power with reasonable high efficiency. A centrifugal pump is not equipped with a wicket gate. Therefore, there will be no flow regulation. However, for a run-of-the-river plant, using centrifugal pumps as turbines is a cheap and reliable solution.

7 Issues with Hydropower as a Future Component of Renewable Energy

Hydropower is a well known and fully developed technology with a very competent industry supporting it. Hydropower equipment is reliable and based on proven technologies. Compared to other renewable energy sources, hydropower is very competitive. The variety of manufacturers ensures a competitive market. The interest in small hydropower is increasing and has resulted in various, although not always experienced, companies entering the market. Knowledge about building hydropower plants exists in nearly all relevant countries.

In addition to being very reliable and competitive in price, the most important property of hydropower is the ability to produce the required electric power instantaneously. In combination with other renewable sources of electric power, hydropower can provide the governing, allowing other energy sources to deliver energy when it is available. (A wind turbine is only able to produce energy when the wind blows.) However, the governing ability of a hydropower plant is highly dependent on the reservoir capacity. The main issue is to build reservoirs in a way that is environmentally acceptable.

Less than 10-MW hydropower plants have recently been regarded as producing "green" energy. However, building ten 10-MW plants is not necessary better as regards environmental impact than building one 100-MW plant. Another issue is what is regarded as a small turbine and therefore described as "green" energy. A high-head turbine of 10 MW is small. A low-head 10-MW turbine is huge. The term "small hydro" should really be applied to the size of the plant rather than to the power output.

8 Pump Storage Plants

8.1 *Introduction*

In pumped hydroelectric storage plants, the hydraulic system is connected to aggregates that can work in two directions: (1) by utilizing the water from the hydraulic system to generate electric power and (2) by utilizing electric power from the grid to pump the water back into the upper reservoir. To achieve this result, one may either install two separate aggregates, one pump and one turbine, or use a machine that runs both ways, a reversible pump turbine, (RPT) Alternative arrangements for reversible pump storage plants are shown in Fig. 22.

Basically, a pumped storage plant is attached to an upper reservoir and a lower reservoir by a conduit system consisting of headrace tunnel and pressure shaft, draft tube and tail race tunnel, in the same way as in conventional hydropower.

Traditionally, these kinds of machines have been used for storing energy i.e. pumping water to the high reservoir to make sure that there is enough water available when the energy demand is high. In a free market for electricity, the common

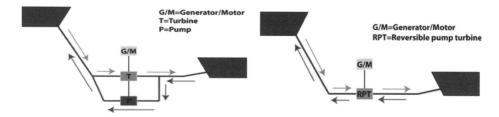

Fig. 22.   Alternative pumped storage plants.

strategy is to utilize energy from the reservoir when the price is high during the day and to pump water back to the reservoir when the price is low during the night.

However, the main benefit of pumped storage plants is their ability to maintain the reliability of the grid. The quality of the grid is defined by the ability to keep both constant frequency and constant voltage. The frequency changes because the demand for electric power varies. If there is a surplus of generated power, the frequency will increase, and conversely if the generated power decreases, the frequency will decrease. Using turbine and pumping modes of operation alternatively, the power balance to the grid is maintained.

The voltage also varies by the ratio of active to reactive power. The generator must be able to produce exactly the same ratio. Voltage regulation is achieved by adjusting the magnetic field of the generator. In order to regulate the voltage, one has to have available generators connected to the grid, even if they are only idling. This is the condensing mode of operation.

Plants that are devoted to grid support are equipped with special starting devices in order to be able to change rapidly between turbine and pump modes of operation. It is also possible to run these turbines in idle speed.

As mentioned earlier, a pumped storage plant has essentially the same design as a conventional hydropower plant, but plants used for grid maintenance are usually equipped with very small reservoirs. Depending on the timeframe, they are designed to be run in either mode of operation. These plants have no ability to store energy, only to balance the power on the grid.

The design challenges of pumped storage plants are mainly connected to the increased number of starts and stops and alternating electricity production. Especially in the turbine mode of operation, these machines must be able to operate both at low load and high load. Their dynamic behavior provides requirements both to the conduit system and to the performance characteristics of the machines. In steady state operation, stability in the pumping and the turbine modes of operation are essential at best operational points as well as in off-design operations. Important issues here are noise, vibrations, and pressure pulsation.

During starts and stops, the pressure transients must be acceptable. Instability and pressure transients can cause failure in operation, but also fatigue breakage with catastrophic consequences.

8.2 *Separate Pump and Turbine*

For a system with separate pump and turbine, it is a question of selecting the correct machines with correct specific speed, a function of design head, flow and speed of rotation. Moreover, the design must agree with the standard pump and turbine theory.

It should be noted that, put into a system with the same reservoir levels, the pump and the turbine will experience different heads. The turbine is taking energy out of the hydraulic system and the hydraulic losses will lower the head available

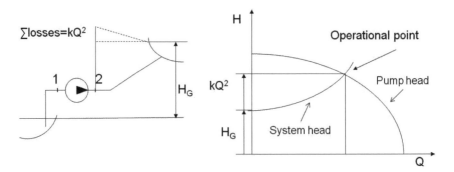

Fig. 23.   Definition of the pump head and the operational point.

for the turbine. As previously described, the turbine net head will be:

$$H_n = z_1 - z_2 + h_1 - h_2 + \frac{c_1^2}{2g} - \frac{c_1^2}{2g} = H_G - \sum \text{losses.} \qquad (24)$$

The pump takes energy from the electric grid and increases the energy in the hydraulic system. In addition to the reservoir head, the hydraulic losses in the system must be overcome by the pump head. The required pump head will be:

$$H_P = z_2 - z_1 + h_2 - h_1 + \frac{c_2^2}{2g} - \frac{c_1^2}{2g} = H_G + \sum \text{losses,} \qquad (25)$$

where 1 denotes pump inlet and 2 denotes pump outlet (see Fig. 23).

For a centrifugal pump, which is the most common pump type in pumped storage plants, the head delivered by the pump is a function of the flow Q and the speed of rotation. In most plants, the speed of rotation is constant. The system head is the gross head H_G plus the sum of the losses in the conduit system. The losses are proportional with the square of the velocity, which means they are also proportional to the square of the flow. When connected in a system, the pump head equals the system head, hence:
The system head:

$$H = H_G + kQ^2. \qquad (26)$$

The pump head:

$$H = f(Q). \qquad (27)$$

The point of operation is found by setting these heads equal:

$$H_G + \sum \text{losses} = f(Q). \qquad (28)$$

This is all according to common pump theory. The pump's Q-H characteristic, shown in Fig. 23, is typical for a centrifugal pump. The form will vary with, among other design parameters, the pump's outlet blade angle. By adjusting this angle, the characteristic will be steeper.

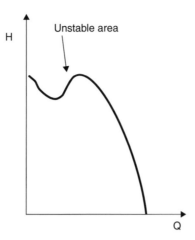

Fig. 24. Unstable pump characteristic.

In order to achieve stable pump operation, the pump characteristics must have a negative gradient in the whole operation area. Steeper characteristics are beneficial for stable operation. Many pumps have an unstable area at low flow as shown in Fig. 24, where the pump must not be allowed to operate. However, especially in pumped storage plants, the reservoir levels might vary so much that it is difficult to avoid low flow rates. Filling up the upper reservoir after it has been emptied for some reason, requires special attention.

With a frequency-governed electric motor, the speed of rotation and, thereby, the flow can be adjusted. The pump characteristic is proportional to the speed of rotation. With such equipment, stable operation is easier to achieve and, of course, it might also be beneficial to be able to govern the pumping flow for other purposes. But this is a rather expensive solution.

Cavitation is also an issue for pumps. At the suction side of the pump, the absolute pressure can be lower than the vapor pressure and cavitation will occur. With reference to Fig. 25, the criterion for avoiding cavitation is the same as for turbines:

$$NPSH_{\text{available}} > NPSH_{\text{required}}, \tag{29}$$

where

$$NPSH_{\text{available}} = H_{\text{atm}} - h_{\text{vapor}} - H_s - \Delta h_s,$$

$$NPSH_{\text{required}} = \frac{D_0}{2} + \frac{c_0^2}{2g} + \Delta h, \tag{30}$$

where Δh_s is the loss in the draft tube, and Δh is the interna losses in the pump. For a pump with horizontal shaft, the $D_o/2$ arises for geometrical reasons.

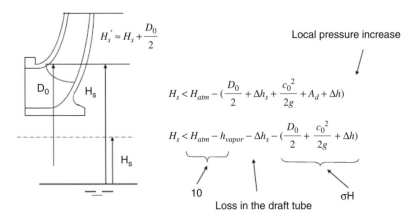

Fig. 25.   The cavitation criterion for a centrifugal pump.

8.3 Reversible Pump Turbine

A reversible pump turbine (RPT) is a compromise between an optimal pump and an optimal turbine. The design challenge is to achieve a stable pump as well as a stable turbine. A stable pump implies that the pump's F-H characteristics must decrease over the whole operational region, as mentioned earlier.

However, the pump design will influence the turbine performance, producing too steep Speed–Flow Characteristics. Some RPTs in turbine mode of operation tend to be unstable, especially at idling speed. Instability during idling produces problems when interfacing with the grid. Steep speed–flow characteristics may also produce enhanced water hammer amplitudes at start-stop.

Compared with a turbine with the same n_q, the RPT will necessary have a bigger outer diameter in order pump the same flow, but against a higher head, (see Eqs. (25) and (26)). Figure 26 illustrates the difference between a turbine and an RPT with the same specific speed.

The turbine characteristics show Flow versus Speed of rotation for given wicket gate positions. For a high-head Francis turbine, the characteristics are typically decreasing. The flow decreases as the speed of rotation increases. This has consequences for the dynamic behavior of the turbine. For frequency governing, it normally has a positive effect, a self-governing effect. Frequency governing means keeping the frequency constant. If the RPM increases, the flow has to be decreased by changing the wicket gate position. The effect of the decreasing turbine characteristics will therefore help to govern frequency. However, if the characteristics are too steep, the amplification between wicket gate position and change in flow will be too high, and lead to instability. Due to the increased diameter, the centrifugal forces in an RPT is even greater than for a normal Francis turbine, and due to hysteresis effects, the characteristics might even have an S-form. In Fig. 27, the turbine characteristics of a high head Francis and a RPT are compared.

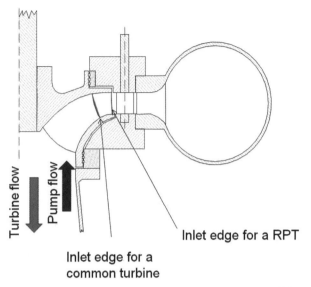

Fig. 26. Comparison between a pump and a RPT. The outer diameter will have to be increased because of pump performance.

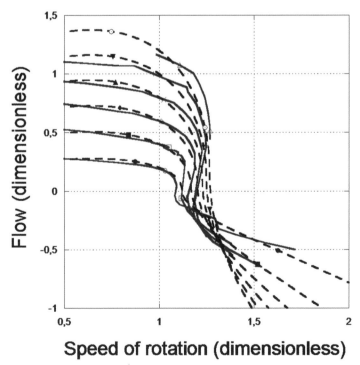

Fig. 27. Comparison of turbine characteristics of RPT and normal high-head Francis turbine (dotted lines).

For achieving stable pump characteristics, the pump outlet blade angles are a design parameter, as previously described. The blade angle will, however, have an influence on the turbine characteristics. The designer has to compromise, altering the blade angle to achieve pump stability might give rise to turbine instability.

Another design challenge is that, in order to achieve the desired pump discharge with the given head, the turbine will be too big i.e. the BEP is at too high a flow rate, which means that the turbine will have to be run at low load and therefore with less efficiency.

8.4 *Control and Electrical Interface*

The electric machine (either run as motor or as generator) is normally a synchronous machine. Starting up as a turbine is rather conventional. By using the wicket gate for regulating power, the machine can be attached to the grid, when stable operation at synchronous speed of rotation is achieved. Starting up as a pump is more complicated. A synchronous machine of, say, 300 MW may not be immediate connected to the grid as the starting torque will be tremendous.

There are several methods used for pump startup. The traditional method is a back-to-back start i.e. using a turbine (devoted to this purpose) electrically or mechanically connected to the pump shaft. By starting the turbine, the pump starts to rotate, and when the synchronous speed of rotation is achieved, the pump motor is connected to the grid and the devoted turbine disconnected. In order to reduce the starting torque, it is common to evacuate the water from the pump runner using a compressor to blow out the water. When the runner has achieved the correct RPM, the valves are opened, and the runner starts to pump.

In recent years, frequency transformers of sufficient capacity have been developed, so modern RPTs are more often equipped with frequency transformers, allowing the RPT to start at low RPM and slightly increase speed until synchronous speed is achieved. However, the price of these transformers is strongly dependent on the power output, hence, evacuating the water before starting is economically beneficial. This will reduce the needed capacity of the frequency transformer to about 20% of the pumping power.

Acknowledgments

Most of the background material for this chapter is based on lecture notes and information prepared in collaboration with colleagues at the Waterpower Laboratory at the Norwegian University of Science and Technology in Trondheim.

References

1. USGS. Available at: http://ga.water.usgs.gov/edu/hyhowworks.html. Accessed 15 December 2011.
2. The World Bank Group (2009). Directions in Hydropower. Available at http://site resources.worldbank.org/INTWAT/Resources/Directions_in_Hydropower_FINAL.pdf. Accessed 15 September 2011.

3. BP (2010), *BP Statistical Review of World Energy 2010.*

4. I. E. Idelchik, *Handbook of Hydraulic Resistance* (CRC Press, Florida, 1994).

5. Wikipedia, Francis Turbine: Available at: http://en.wikipedia.org/wiki/File:M_vs_francis_schnitt_1_zoom.jpg. Accessed 15 August 2012.

6. Wikipedia, Kaplan Turbine. Available at: http://en.wikipedia.org/wiki/File:S_vs_kaplan_schnitt_1_zoom.jpg. Accessed 15 August 2012.

7. Wikipedia, Pelton Wheel. Available at: http://commons.wikimedia.org/wiki/File:S_vs_pelton_schnitt_1_zoom.png. Accessed 15 August 2012.

8. The Big Picture. Available at: http://www.boston.com/bigpicture/2009/09/the_sayanoshushenskaya_dam_acc.html. Accessed 15 December 2011.

9. Wikipedia. Available at: http://en.wikipedia.org/wiki/Bieudron_Hydroelectric_Power_Station. Accessed 15 December 2011.

Appendix

Table A1: List of Symbols.

Symbols	Terms	Dimension	Typical value
Q	Flow	m^3/s	
H	Head	m	
H_n	Net head	m	
z	Elevation	m	
h	Hydraulic pressure	m	
c	Absolute velocity	m/s	
u	Peripheral velocity	m/s	
v	Relative velocity	m/s	
η	Efficiency		
η_h	Hydraulic efficiency		0.80–0.94
η_v	Volumetric efficiency		0.92–0.98
η_m	Mechanical efficiency		0.94–0.98
	Gravitational constant	m/s^2	9.81
ρ	Density	kg/m^3	1000
υ	Kinematic viscosity	m^2/s	10^6
σ	Thoma cavitation number		
NPSH	Net Pressure Suction Head		
k_c	Loss factor, minor losses		
f	Friction factor		

Table A2: Handy Equations.

What?	Equation	Dimensions comments
Hydraulic power	$P = \rho g Q H$	W
Velocity head	$\dfrac{c^2}{2g}$	m
Head	$H = z + h + \dfrac{c^2}{2g}$	m

(*Continued*)

Table A2:   (*Continued*)

What?	Equation	Dimensions comments
Effective head	$H_n = z_1 - z_2 + h_1 - h_2 + \dfrac{c_1^2}{2g} - \dfrac{c_2^2}{2g}$ $H_n = H_G - \sum \text{losses}$	m
Mechanical rotating power	$P = \omega T$	W
Continuity	$Q = c_1 A_1 = c_2 A_2$	m^3/s
Velocity trough a nozzle	$c = \sqrt{2gh}$	m/s
Bernoulli's equation	$z_1 + h_1 + \dfrac{c_1^2}{2g} = z_2 + h_2 + \dfrac{c_2^2}{2g} + \sum \text{loss}$	m
Reynold number	$R_e = \dfrac{cD}{v}$	Turbulent flow For $R_e > 2{,}300$
Friction loss	$h_f = f \dfrac{L}{D_h} \dfrac{c^2}{2g}$	f from the Moody diagram
Singular loss	$h_f = K_c \dfrac{c^2}{2g}$	K_c from hydraulic handbooks
Euler equation	$P = \rho Q (u_1 c_{1u} - u_2 c_{2u})$	W
Hydraulic efficiency	$\eta_h = \dfrac{\omega T}{\rho Q (u_1 c_{1u} - u_2 c_{2u})}$	
Angular speed of rotation	$\omega = \dfrac{2\pi n}{60}$	rad/s
Synchronous speed	$n = \dfrac{60f}{P}$ f = 50 Hz or 60 Hz	RPM
Peripheral speed	$u = \omega \dfrac{D}{2}$	m/s
NPSH$_{available}$	$NPSH_{\text{available}} = H_{\text{atm}} - h_{\text{vapor}} - H_s$	Available NPSH from system
NPSH$_{required}$	$NPSH_{\text{required}} = \dfrac{c_t^2}{2g} + \text{loss}$	Required by the machine
Cavitation avoides if	$NPSH_{\text{available}} > NPSH_{\text{available}}$	
Turbine centre relative to the tail water level	$H_s < 10 - \sigma H_n$	σ is Thoma cavitation number

Chapter 14

Wind Energy

Jos Beurskens[*,‡] and Arno Brand[†,§]

SET Analysis, Kievitlaan 26, 1742 AD Schagen, the Netherlands
†Energy Research Centre of the Netherlands ECN, Unit Wind Energy
P.O. Box 1, 1755 ZG Petten, the Netherlands
‡hjmbeurskens@gmail.com
§brand@ecn.nl

Over the years, wind energy has become a major source of renewable energy worldwide. The present chapter addresses the wind resource, which is available for exploitation for large-scale electricity production, and its specific physical properties. Furthermore, the technical options available to convert the energy of the air flow into mechanical energy and electricity are described. Specific problems of large-scale integration of wind energy into the grid as well as the present and future market developments are described in this chapter. Finally, environmental aspects are discussed briefly.

1 Introduction

Despite the fact that only a small portion of the solar radiation that reaches the earth is converted to wind energy, the potential of the wind to contribute to the world's energy needs is significant. About 0.7% of the captured solar radiation is converted into the earth's pressure system. The pressure differences cause the air to flow. About 0.8% of the energy contained in the pressure system is converted into kinetic energy in the lower boundary layer of the earth. If 10% of this energy were converted to electricity (this would be an extremely high degree of utilization of the earth's winds), it could meet the world's current total electricity demand. However, the distribution of wind energy is highly uneven across the world.

Wind has been used for about 15 centuries to generate useful energy. The first applications (Persia, 8th century BC) were very basic. A drag-driven vertical axis rotor was used to drive a scoop wheel for water pumping and to drive milling stones for grinding cereals. In 1759, John Smeaton of the UK performed the first systemic scientific experiments on a model of a (classical) windmill rotor, mounted at the tip of a rotating beam, to determine the maximum "effect" as a function of the rotor blade configuration. His results are still valid to this day.[1]

The history of wind energy can be divided in two main periods. The first period, which spans from the ancient times until the 1800s, is characterized by classical windmills. During the classical period, wind "devices" (windmills) converted the kinetic energy of moving air into mechanical energy. Only when electric generators (dynamos and alternators) were invented and introduced for public electricity supply, were windmills used for generating electricity. This development effectively started in the late 1800s and became a great commercial success after the energy crisis of 1973. For instance, at the end of 2011, wind energy generated 6.3% of Europe's electricity consumption.[2] In Denmark, wind energy already meets over 25% of the electricity demand. Wind energy has become a booming multibillion dollar worldwide business, with annual growth rates varying from over 30% about a decade ago down to 10% at present.

2 Wind Resource

2.1 *The Origin of the Wind and its Variations*

The sun emits 63.2 MW of power per unit area of its surface,[3] a small fraction (1367 W per sqm of surface on average) of which reaches the earth. Solar radiation heats the surface of the earth, causing the warmer, and thus lighter, air in the atmosphere to rise. Inhomogeneities in this process subsequently produce pressure differences, which result in the displacement of air parallel to the earth's surface. The speed of the horizontal air flow is much larger than that of the vertical flow. The characteristics, speed, and direction of wind vary with time and place. The motion of the air is influenced by the rotation of the earth (Coriolis forces) which, depending on the latitude and the velocity of the air, exerts an extra force on the moving air.

Second, the actual value of the power per unit area received from the sun depends on the day of the year and the latitude of the location. In addition, it depends on the hour of the day and the composition of the atmosphere. Third, the surface roughness of the earth varies between locations, from low for seas and oceans to high for urban and industrial regions. Fourth, the surface heating and the heat flux generally depend on the heat capacity of the surface layer, ranging from small (deserts) to large (oceans). And finally, the elevation of the terrain affects the motion of the air. Together, these variations of the drivers of the wind are responsible for the complex patterns and variations in the wind that is observed over the earth's surface.

2.2 *Power of the Wind*

One of the physical properties of moving air is its kinetic energy. The kinetic energy of the wind is a function of the speed and the density of the air.

The power in the wind per unit of area perpendicular to the direction of the wind is:

$$p_{\mathrm{w}} = \frac{1}{2} \cdot \rho \cdot v^3, \tag{1}$$

where p_w is the power density, ρ is the air density, and v is the momentary value of the wind speed. Air density depends on the air temperature ($1.225 \, \text{kg/m}^3$ at sea level and 15°C, $1.29 \, \text{kg/m}^3$ at 0°C, and $1.16 \, \text{kg/m}^3$ at 30°C) and pressure ($\rho = 0.73 \, \text{kg/m}^3$ at 4000 m height and 20°C). The nonlinear relationship between power and wind speed [see Eq. (1)] has many important consequences for the establishment of the wind power potential at a certain location, for the design of a wind turbine, and for the control of the turbine.

In order to determine the wind power available at a certain height and location, the long-term frequency distribution of the values of the wind speed needs to be known. The most common method to describe the frequency distribution is the Weibull function.

$$f(v) = k/a \cdot (v/a)^{k-1} \cdot \exp[-(v/a)^k]. \qquad (2)$$

Here $f(v)$ is the probability density of the wind speed v. The parameter k is the shape factor, which determines the shape of the frequency distribution (see Fig. 1). For coastal areas, such as in North Western Europe, the value of k is approximately 2. The Weibull function with $k = 2$ is also called the Rayleigh distribution. And a is the scale factor, which is proportional to the average wind speed (see Fig. 2.)

By means of the Weibull function, the average wind power per m^2 area, perpendicular to the wind speed direction p_w, can be represented as follows:

$$\bar{p}_w = \int_0^\infty \frac{1}{2} \cdot \rho \cdot v^3 \cdot f(v) dv = k_E \cdot \frac{1}{2} \cdot \rho \cdot \bar{v}^3 \quad [W/m^2]. \qquad (3)$$

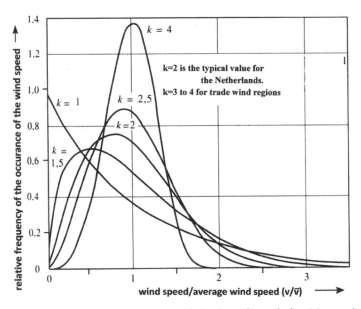

Fig. 1. Weibull distributions for various values of the shape factor k: $k=4$ is a typical value for the trading winds, $k=2$ is typical for moderate coastal areas e.g. in North Western Europe, and $k=1$ for polar areas.[4]

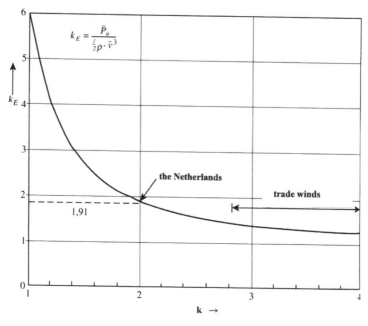

Fig. 2. The energy pattern factor k_E as a function of the Weibull shape parameter k.[4]

The last part of this equation resembles the expression for the momentary wind power [Eq. (1)]. The real content of the wind power, thus taking into account the larger contribution of the high wind speed, can be established by replacing the momentary wind speed by the average value and by including the factor k_E. This factor has a unique relationship with the shape factor k (see Fig. 2). The sensitivity of k_E for variations of k decreases with increasing values of k larger than, say, 2.

This approach enables a quick estimation of the wind energy potential at a site when the climatological regime and the (annual) average wind speed are known.

2.3 Variability of the Wind

As described in Sec. 2.1, the wind varies at several time and spatial scales.

Following Orlanski,[5] the wind characteristics can be represented at different time scales: inter-annual, intra-annual, diurnal, inter-minute, and intra-minute scale, but also with variations at scales ranging from a few meters to several thousand kilometers. These processes are sketched in Fig. 3.

Processes at the largest or terrestrial scale last the longest and span the earth. They are relevant for wind energy exploitation because they are the main drivers for the secondary circulations and allow for an assessment of the energy production in the long term, typically 30 years.

The terrestrial wind systems are not constant with respect to time or place. They may have changed position in the past and may change position in the future under the influence of changes in the global temperature. This change will have an

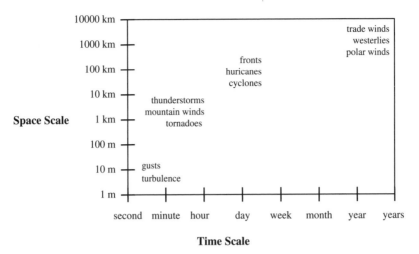

Fig. 3.   Processes in the planetary boundary layer according to the classification by Orlanski.[5]

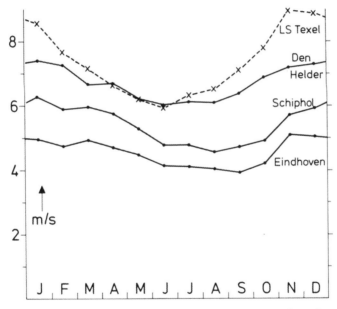

Fig. 4.   Average monthly change of the wind speed at four metrological stations in the Netherlands.[7]

impact on weather in general and wind in particular.[6] Overall, the changes in the wind speed are expected to be regional and small. On the other hand, the impact on the distribution of the wind, in particular, the likelihood of the occurrence of very small and very large wind speeds, is significant.

The secondary circulations are an order of magnitude faster and smaller than the terrestrial wind systems. Variations at the scale of the secondary circulations are relevant for the design and operation of wind farms. Atmospheric systems, such as fronts, occur quite often, typically once per day, and affect the diurnal energy production and the ultimate loading of the wind turbines. Hurricanes and cyclones occur less often, and as a result, have little effect on the energy production. They might however cause maximum mechanical loading of a wind turbine.

Also, average annual wind speeds vary considerable compared to the long-term (typically 10 years) average value. Maximum annual deviations can be less or more than 10% of the long-term average value.[7]

Tertiary circulations are an order of magnitude faster and smaller. Variations at the scale of the tertiary circulations are relevant for the design and operation of single wind turbines. Relatively persistent but not very frequent processes, such as thunderstorms, mountain winds, and tornadoes, determine the extreme events that might cause maximum mechanical loading of wind turbine components. And the always present short-lived and small-sized events, such as gusts and turbulence, are responsible for the fatigue loading of the wind turbine structure and the decay of the turbine wake.[8]

As far as a wind turbine is concerned, extreme events are related to extreme wind speeds, extreme wind speed changes, and extreme wind direction changes. The extreme wind speed is expressed in terms of the reference wind speed, which is the extreme 10-min average wind speed at turbine hub height with a recurrence period of 50 years. The reference wind speed is expected to increase under the influence of climate change. For more information on extreme events the reader is referred to the Chapters 3 and 7 of *European Wind Turbine Standards II*.[9]

Since the wind is slowed down by the roughness of the earth's surface, generally the wind speed increases with the distance above the surface.

The vertical wind gradient (wind shear) depends on the terrain roughness and can, for example, be calculated by means of the following experimental expression from the potential wind speed at 10 m height. The potential wind speed is a fictitious

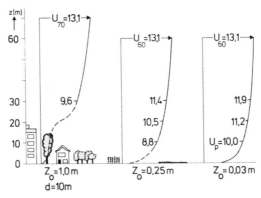

Fig. 5.   Vertical wind shear for various values of the terrain roughness z_0 for the same value of the meso wind speed $U_m = 13.1\,\mathrm{m/s}$.[7]

Table 1:   Relation Between Terrain Type and Roughness Length.[17]

Class	Brief description of terrain	Roughness length z_0 (m)
1	Open sea, fetch at least 5 km	0.0002
2	Mud flats, snow, no vegetation, no obstacles	0.005
3	Open flat terrain, grass, few isolated obstacles	0.03
4	Low crops, occasional large obstacles	0.1
5	High crops, scattered obstacles	0.25
6	Parkland bushes, numerous obstacles	0.5
7	Regular large obstacle coverage (suburbs, forest)	1.0 (R&D intensified)
8	City center with high- and low-rise buildings	>2 (R&D ongoing)

wind speed, which would have been observed if the surrounding terrain were flat and open with a terrain roughness class of 3 (see Table 1). In order to achieve the real wind speed, corrections for actual roughness and obstacles have to be made.

$$v/v_p = 1.3084 \ln(z/z_0)/\ln(60/z_0), \quad (4)$$

where v is the average wind speed at the height z (mostly taken as the hub height of the wind turbine) and v_p is the potential wind speed.

Terrain roughness has been classified according to various types of terrain, which are listed in the Table 1.[7]

This approach is an example that can be applied in relatively uniform and flat terrain. If the terrain is complex, advanced modeling, based on Computational Fluid Dynamics (CFD), must be applied and has to be supplemented with actual long term (>1 year) measurements in the field.

In addition to causing wind shear, friction due to surface roughness is the main source of turbulence. Another source, or sink, of turbulence is provided by the vertical heat flux. This heat flux can, depending on the direction of the heat flow, effectively increase or decrease the vertical gradient of the wind speed as well as the variations in the wind speed. The roughness of the surface and the rotation of the earth together also affect the way the wind direction depends on the distance above the surface.

2.4 World and Regional Wind Potential

The patterns in the wind can be observed either on a global or regional scale or in the long or short term. A wind pattern can be revealed by averaging the wind over climatologically relevant periods and geographically homogeneous regions. Such a pattern is called a wind map.

In general, there are two types of wind maps. One type gives a rough guide to the wind resource in a location as expressed by the long-term mean of the wind speed at a height of 10 m. Another type does the same for the prevailing wind direction. Usually this information is presented visually. Long term here indicates the period in which the climate, more specifically, the mean air temperature at a height of 2 m, is supposed to be constant. It is a meteorological standard to consider

J. Beurskens and A. Brand

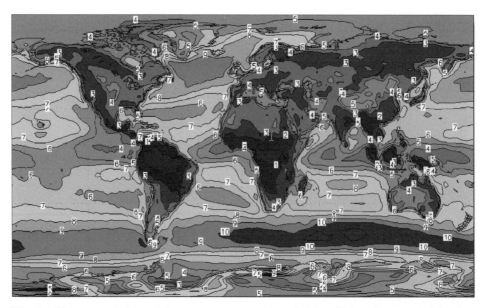

Fig. 6.   The mean wind speed in meters per second at a height of 10 m above ground level or mean sea level for the period 1976–1995, according to the NCEP/NCAR reanalysis data set.[10,11]

periods of 30 years and to update the data every 10 years. The height of 10 m is another standard from the meteorological sector. Figure 6 shows a global wind map.

From Eq. (3) and Fig. 1, it appears that knowing the average wind speed at a location is not sufficient to determine the wind energy resource of that location. For accurate wind resource assessment, the frequency distribution of the wind speed and the wind direction, averaged over periods of 1 h (minimum requirement) or 10 min (preferred) are needed. In the past, wind maps were made by analyzing long-term records from meteorological stations. This resulted in the wind maps for the maritime sector in the 19th century, revealing the trade winds over the oceans.

Later, in the 1980s, wind maps dedicated to wind energy applications appeared. Among the first were overviews of the global wind climate[12] and the *Danish Wind Resource Atlas*.[13] These maps were followed by wind atlases of the US[14] and Europe.[15] The information in these first maps, valid for onshore locations only, is given for a relatively coarse grid and were given for the hub heights of wind turbines mostly used at that time. Subsequently, maps for offshore regions and maps with higher resolutions and/or for larger heights were made. An overview of all maps and atlases is given in *Wind Atlases of the World*.[10]

Currently, the source of information on which wind maps are based is shifting from data from meteorological stations toward data from atmospheric models and new instruments, although the measured data is expected to play a role for some time to come. The atmospheric model data includes the reanalysis data sets from NCEP/NCAR[11] and the ECMWF.[16] An example of a new instrument is the NASA

Modern-era Retrospective Analysis for Research and Application,[17] which combines information from various sources.

In the near future, wind maps will, besides providing information on wind speed and direction, also include information on turbulence intensity and atmospheric stability. This will open the door to the use of wind maps for site-specific design of wind turbines.

3 Wind Turbines

3.1 *Drag Machines and Lift Machines*

The type of aerodynamic forces that the Wind Energy Conversion System (WECS) utilizes to generate energy is the basis for the classification of wind turbines. Two main categories can be distinguished: Drag machines and Lift machines.

Drag, the driving force of a drag machine, moves the active part of the rotor in the direction of the wind speed, whereas the driving force of a lift machine is the (component) of the aerodynamic force which is perpendicular to the incident flow. The basic difference in these two types of wind turbines is explained in Fig. 7.

With the axial momentum theory (see Fig. 8), the maximum power coefficient C_p for both types of wind turbines can be derived. C_p is defined as the quotient of the power produced by the machine and the undisturbed wind energy flow through the area which is swept by the wind turbine rotor.

$$C_p = P / \left(\frac{1}{2} \rho v^3 A \right), \tag{5}$$

where A is the rotor swept area.

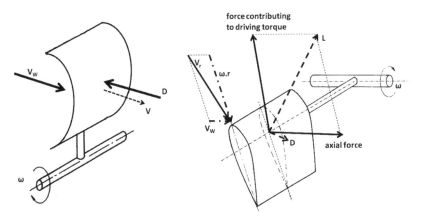

Fig. 7. The working principle of a drag-driven wind turbine (left) and a lift-driven wind turbine (right). The essential difference between the two types of machines is the movement of the blade compared to the driving force ($V_R // D$) (left) and Lift ($V_R^{\perp} L$) (right). (D = drag, L = lift force, ω = rotational speed, r.ω = translational speed of blade segment, v_w = local wind speed, v_r = relative incident flow speed). [Figure 7 by Jos Beurskens based on information from Erich Hau, *Wind Turbines: Fundamentals, Technologies, Application, Economics.*[18]]

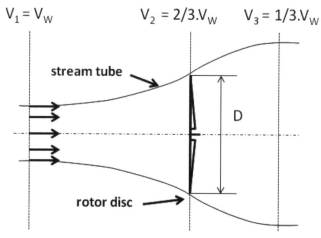

Fig. 8. The flow tube of a wind turbine rotor, explaining the momentum flow theory. The volume behind the rotor is called the wake.

For drag-type wind turbines, the maximum value of C_p is 4/27 C_D and for lift-type wind turbines, this value is 16/27 (if there are no radial forces acting on the flow, e.g. by a diffuser). Assuming that the value of C_D is 1.4 (for a concave cup), than the maximum value of C_p is 0.25. In practice however, this value seldom exceeds 0.15 because of energy losses caused by the reverse movement of the active area against the wind.

Apart from the low values of C_p of the drag-driven machine, the amount of material to construct the rotor of a drag-driven machine is much larger than for a lift machine. The rotor swept area of a drag machine has to be fully covered with material, whereas the rotor area of a lift machine only has to be covered partly. The solidity σ (the projected area occupied by the blades compared to the rotor swept area) for a drag machine is 1, whereas σ for a lift machine can be as low as 0.03.

For these reasons only lift machines are used for generating electricity on a competitive basis. Most lift machines have a horizontal axis, with the exception of (variations of) Darrieus rotors (egg-beater type of rotors) (see also Fig. 12).

3.2 *Rotor Characteristics*

Flow conditions are optimal (maximum lift and minimum drag) only for a specific angle of the rotor profile compared to the local velocity (Fig. 7, right). As the angle of attack of the local wind speed varies along the radius of a blade, the angle of the blade section compared to the plane of rotation has to vary as well. At the section of the blade which is closest to the hub, the root, this angle is much larger than that at the blade tip. In the case where the flow conditions deviate from the optimum configuration, the power output will be lower than the maximum achievable value. Figure 9 represents an example of the power output of a wind turbine rotor as a function of the rotor speed, at various constant values of the wind speed. These

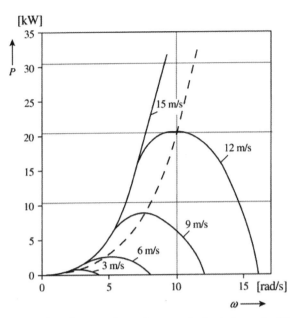

Fig. 9. Power characteristics of a rotor for various constant wind speeds. ($P =$ power output at the rotor shaft, $\omega =$ rotational speed of the rotor shaft. The dotted curve ($P \sim \omega^3$) indicates the point of maximum efficiency for various wind speeds.[4]

curves result, for example, from wind tunnel measurements. Flow phenomena are usually characterized by dimensionless parameters.

For wind turbines, the independent dimensionless parameters that characterize the rotor are:

- the power coefficient C_p [see Eq. (5)]
- the tip speed ratio $\lambda = v_{\text{tip}}/v$, where v_{tip} is the absolute value of the speed of the rotor tip and v is the undisturbed wind speed

As the behavior of a wind turbine system is almost completely independent of viscous effects, the Reynolds number is not relevant to characterize the behavior of a wind turbine. In order to characterize the power performance of a rotor, the $C_p - \lambda$ curve is constructed from the power curves according to Fig. 9. From such a curve, of which Fig. 11 is an example, the maximum rotor efficiency and the tip speed ratio at which maximum efficiency is realized, can be determined. It is obvious that the ratio between wind speed and rotational speed for a given wind turbine has to be kept constant in order to let the wind turbine rotor operate at maximum efficiency in a wide range of wind speeds.

The value of the tip speed ratio at which C_p has its maximum value is used to divide rotors into different types, such as slow runners, fast runners, etc. (see Fig. 13). Slow-running rotors are used for driving loads which need a high torque, combined with low rotational speed (like a piston pump), while fast-running rotors

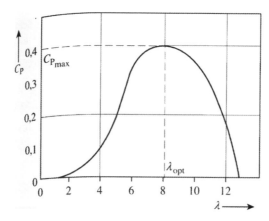

Fig. 10.   The $C_p - \lambda$ curve of a rotor with an optimum tip-speed ratio of 8.[4]

λ = 1   λ = 3   λ = 8   λ = 10   λ > 15

Fig. 11.   Various types of rotors. On the left a slow-running wind pump. The other wind turbines are electricity-producing machines with increasing tip speed ratio. Three-bladed machines are the most applied ones because of the tradeoff between dynamic mechanical loading and rotational speed of the main shaft. Photographs 1,3,4 and (from left to right): Jos Beurskens.

are used to drive electric generators which require high speeds at a relatively low torque.

3.3 *Energy Conversion and Control*

More than 99% of the present wind turbines are used to generate electricity. The first wind turbines to generate electricity date from the late 1800s [Blyth (UK), Brush (US), La Cour (DK)].[1] The very first wind turbines generated electricity for battery charging and the electrolysis of water (La Cour). As the appearance of the first modern wind turbines coincided with the introduction of the commercial dynamos (DC) and alternators (AC) and the first electricity plants (New York, 1880) feeding small grids, wind turbines were also used to feed electricity into the grid. The generators which were driven by the wind rotors were induction machines which were running at a frequency that was slightly higher than the grid frequency, thus at almost constant rotational speed. The rotor blades were fixed to the hub and had no pitch control. As the wind rotor is able to produce much more energy than the maximum rating of the generator, a control system is needed to limit the maximum rotor power to the rated power of the generator. The first wind turbines used stalling of the flow around the blade profiles to achieve this effect. As the wind speed increased and the rotational speed of the rotor was kept constant, the angle of attack of the blade profile at higher wind speed became so large that the flow around the profile detached and the lift force stabilized or dropped but the drag forces increased significantly. As a result, the aerodynamic efficiency of the rotor became so low that the rotor power did not increase anymore with increasing wind speed. This passive control system had the advantage of having no moving parts in the hub structure.

The disadvantages, however, included load fluctuations, as the occurrence of stall occurred suddenly and was difficult to control. Furthermore, the tip speed ratio in the operational wind speed range of the wind turbine could not be kept constant for maximum efficiency. So with the appearance of power electronic convertors and the introduction of the insulated gate bi-polar transistors (IGBTs) as an alternative electronic switching device for thyristors, the power quality of electronic power invertors was revolutionized. The extremely high efficiency of IGBTs and the high switching frequency enabled power conversion from a variable frequency and voltage generator output via a rectifier stage into fixed frequency AC output with very low harmonic distortion and with adjustable reactive power production. As a result, all modern wind turbines are equipped with power electronic invertors, fed by generators, operating at variable rotational speed (either synchronous or doubly fed induction generators), and allowing the wind turbine to operate independent from the grid. The inverter actually forms an interface between the grid and the wind turbine generator. The rotor blades have adjustable pitch. This system thus allows operation with constant tip speed rotation in the operational wind speed range and limiting rotor power to the rated generator power.

With the increasing number of parameters which could be adjusted, control actions also increased (see Table 2).

Table 2: Controlling Wind Turbines (First and Second Columns are not Correlated).

Control actions	What is controlled?
Blade pitch (individually or collectively	Power efficiency
Yaw Angle	Aligning wind turbine nacelle with the wind direction
Magnetic Field of Generator	Limiting power output to rated generator power
Firing Angle of electronic switching devices	Avoiding peaks resulting from wind gusts
	Active structural vibration control
	limiting fatigue and ultimate loads
	Supervision for safety
	Stops, starts, and grid connection

The architecture of the electric conversion systems varies considerably. The major difference between the concepts is the path the generated electricity is following from generator to grid. In the oldest concept, i.e. the direct grid-connected induction generator, the electricity is directly fed into the grid. Applying a synchronous generator running at variable rotational speed generates AC with a variable frequency and voltage. After rectifying the electricity, it is converted in a grid-compatible AC. All generated electricity has to pass through the AC–DC–AC convertor. The rotational speed interval is relatively large. A somewhat cheaper solution is using a doubly fed induction generator in combination with an invertor. The generator is directly connected to the grid, where most of the electricity flows. In order to minimize losses caused by the rotor of the generator, this rotor has windings which are connected to an AC–AC convertor. Now only a minor part of the generated electricity (about 30%) has to pass through the convertor, which saves investment cost. The rotational speed range of this concept is smaller than the previous full convertor design. The first concept requires fixed blades, whereas the second and third concepts need pitchable blade angles.

The generators first used in wind turbines were derivatives of commercial high-speed electric motors and generators. As the rotational speed of the wind rotor was usually much lower than the required generator speed, gearboxes were needed in between the rotor and the generator. In practice, it appeared that the gearbox was expensive and sensitive to damage (not fully understood dynamic loading). As a natural strategy, direct-driven (DD) multi-pole generators were introduced. The fact that DD wind turbines were lacking one major component promised higher levels of reliability, especially in areas with extreme external conditions, like offshore. An important disadvantage of the first-generation DD generator was its size and weight. DD generators come in various concepts. The magnetic field can be generated by electric coils (the oldest types), and at present more and more DD generators are equipped with permanent magnets. As the prices in world markets for copper and the (rare earth) feed stock material for the strongest permanent magnets (neodymium) are highly volatile, an additional concept is being developed. The research community and the industry are now exploring the use of high-temperature superconductors (HTSs) to generate extreme strong magnetic fields for wind turbine generators. By applying either permanent magnets or HTSs, DD generators can be built very compact and will decrease considerably the tower head mass of wind turbines.

Figure 12 illustrates two types of modern wind turbine nacelles with rotor and drive train.

3.4 *Power Curves and Energy Output*

Combining the power curve, as indicated in Fig. 13, with the load curve of the generator or pump results in the wind turbine power curve. Each type of wind turbine comes with such a curve.

Fig. 12.   A classical drive train (left) (*Source*: Neg-Micon) and a DD drive train (right) (*Source*: GE).

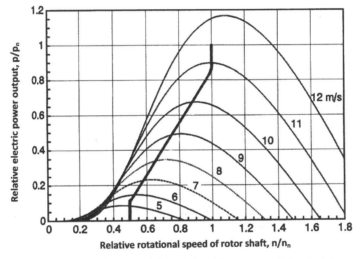

Fig. 13.   Power–rotational speed curves of the wind turbine rotor and the electric generator load. From the points where both curves cross, the power–wind speed curve of a wind turbine can be constructed (see Fig. 15).[19]

Multiplying the power curve with the wind speed frequency distribution curve results in the annual energy output of a wind turbine:

$$\bar{P}_{\mathrm{T}} = \int_{0}^{\infty} P_{\mathrm{T}}(v) \cdot f(v) \cdot dv, \qquad (6)$$

where $f(v)$ is the frequency distribution curve according to (2) and $P_T(v)$ the power–wind-speed curve according to Fig. 14.

Similar to reducing the wind speed frequency curves in a region to one curve with the Weibull expression (2), it is possible to generalize the power–wind-speed curves of wind turbines by taking constant values of the cut-in speed and rated

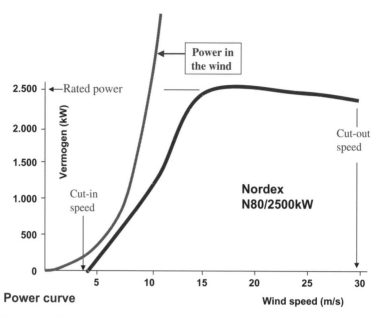

Fig. 14. Power–wind-speed curve as provided by the wind turbine manufacturer. These curves are being established according to standardized methods by independent testing stations.

wind speeds divided by the local average wind speed at hub height. By doing so, a rule of thumb can be derived for the annual average long-term energy production E of a wind turbine at 100% technical availability.

$$E = b\bar{v}^3 A [\text{kWh}]^4, \tag{7}$$

where $\bar{v}$ is the annual average wind speed at hub height. In this expression, b is the output-quality factor.

In the pioneering phase of modern wind energy technology (late 1970s and early 1980s), the value of b was about 2. By applying advanced control conversion and control systems in combination with improved aerodynamic efficiency of rotors, the value has increased to 3.5 at present. Note that this value only applies for wind regimes with a Weibull shape factor of 2 (although the output is not very sensitive to values slightly deviating from 2). It must also be mentioned that the actual output of a wind turbine is the potential energy production E multiplied by the technical availability. Modern land based wind turbines have long-term availability values exceeding 0.95.

The requirement that the ratio between rated wind speed and average wind speed has to be constant, also implies that there is a specific value for the rated generator power per m^2 rotor swept area for various wind regimes. The lower the local average wind speed at hub height is, the lower the specific power has to be

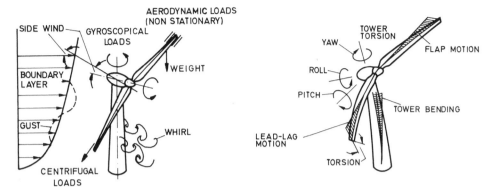

Fig. 15.   Mechanical excitations and responses of a wind turbine structure.

and vice versa. Typical values for specific power are:

$300\,\text{W/m}^2$ for low wind speeds (e.g. wind class IV)

$450\,\text{W/m}^2$ for medium wind speeds (e.g. wind class III), and

$600\,\text{W/m}^2$ for very high wind speeds (e.g. class I)

For the exact definition of wind classes, see Ref. 20.

Installing a wind turbine with a high value of the specific power in a low-wind regime leads to loss of energy, but worse, to a low level of equivalent full load hours. The amount of wind power which can be connected to the grid is primarily determined by the rated power of the wind turbine and not by the number of kWh absorbed by the grid. To make wind energy as grid friendly as possible, one should actually apply wind turbines with system parameters designed for low-wind regimes, but mechanically adopted to high winds.[4,21] Varying the specific power for a given location within a limited wind speed interval has a big impact on the equivalent full load hours (or capacity factor) and only a minor effect on the long-term energy output.

3.5 *Concepts and Structural Aspects*[18,19,22−24]

Apart from being a flow device and an electricity plant, a wind turbine is a highly compliant structure for which the integrity and safe operation has to be guaranteed. Figure 15 illustrates the sources of mechanical excitations and the various responses of a wind turbine structure.[5]

Actually a wind turbine is a highly dynamically loaded structure taking into account the number of cyclic loads it experiences during its lifetime and the amplitude and irregularity of the loads (see Fig. 16).

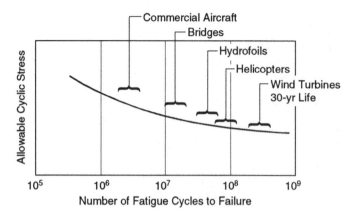

Fig. 16.   Fatigue loading of a wind turbine compared with other structures and machines.[25]

With increasing size, the intrinsic flexibility of a wind turbine increases and structural dynamic design becomes more and more critical. Because of the introduction of offshore wind energy systems, both the size of wind farms and the wind turbines themselves have increased spectacularly (see Sec. 6).

Upscaling does not automatically lead to cheaper electricity from a particular wind turbine as appears from the so-called cubic upscaling law. Expressed in economic terms, it postulates that the mass, but also the initial investment cost of the entire structure, is proportional to the cube of the rotor diameter D. The electricity-generating capacity is proportional to the rotor swept area, thus to the rotor diameter squared. The cost of energy (COE) is approximately proportional to the quotient of investment cost and energy output and thus proportional to $D^3/D^2 = D$. If the wind turbine concept is not changed when upscaling, the cost would increase proportionally to D. Nevertheless, upscaling takes place throughout the entire wind energy sector. The main driver is cost reduction at other places in a wind energy project. For example, in offshore technology, the cost of support structures is dominant and relatively independent of the load-carrying capacity. So offshore designers want wind turbines to be as large as possible. Furthermore, there are cost reductions of operation and maintenance (O&M) if fewer units per MW of installed power have to be maintained. Besides, the COE per turbine can be reduced or kept constant by using materials with a better strength to mass ratio. Optimized control strategies, built-in structural compliance, and smart rotors with distributed aerodynamic blade control have the potential to further reduce costs.

A basic difference between offshore and onshore concepts is the foundation. As the tower for offshore turbines is higher because of the water body between the foundation and the tower structure, the total moment of the wind turbine structure with the interface between the foundation & the sea floor, is significantly larger than that for land-based wind turbines. The moment of force is caused by the axial force on the turbine rotor. At optimum working conditions, the tower top "feels" a force which is equal to the axial force of the wind encountering a solid disk with the size of 8/9 of the swept rotor area! The moment depends on the water depth. That is why the type of foundations for offshore wind turbines primarily — but not solely — depends on the water depth. If water depths exceed about 50 m, floating structures have to be used. Reference 27 gives, among others, an overview of foundations which have been used up till now.

At present, most modern wind turbines have three blades. The main reasons are:

- The blades of a three-bladed rotor have lower tip speeds than the blades of a two-bladed rotor and thus have lower acoustic noise emission levels. (Acoustic noise levels are proportional to, among other things, the absolute value of the tip speed.)
- Dynamic loading on the hub and shaft of a three-bladed rotor is less than that of a two-bladed one. In order to reduce cyclic loading of two-bladed turbines, these often have complex hubs (teetering hub, individually hinged blades, etc.).
- Visual impact of three-bladed wind turbines is less severe than that for two-bladed wind turbines.

4 Wakes and Clusters

4.1 *Clusters of Wind Turbines: Wind Farms*

A wind farm is a cluster of wind turbines that is operated as a single electricity-generating unit, connected to the electricity grid. In the past, wind farms were only placed onshore as single units or in small wind farms, but over the years, more offshore wind farms have been commissioned (see Fig. 17). The size of the wind turbines and the wind farms has increased considerably. The size of the area matters because, together with the size of the wind turbines and the number of wind turbines, it determines the array efficiency. The array efficiency is the collective effect of the wind turbines on their power production and mechanical loading.

Fig. 17.   160 MW Horns Rev I offshore wind farm, consisting of 80 wind turbines of 2 MW each, near Esbjerg (Denmark). (Photo: Elsam, Vestas).

The distance between the wind turbines in terms of rotor diameters is the most important parameter that determines the array efficiency. If the separation distance is sufficiently large, say more than 20 rotor diameters, the wind turbines can be considered to be aerodynamically independent of each other. In that case, a wind turbine in the cluster has no effect on the power production and the mechanical loading of another wind turbine. If, on the other hand, the turbine separation distance is smaller, a wind turbine might find itself operating in the wake (see also Sec. 3.1) of one or more other wind turbines. This may lead to lower power production and a higher mechanical loading.

The aerodynamic aspects of the wakes are discussed in the following order:

- Wake of a single wind turbine
- Internal wakes inside a wind farm
- Wake of a wind farm
- Effect of wind farm wakes on wind farm clusters

4.2 Single Wind Turbine Wakes

The separation distance at which significant wake interaction between wind turbines is observed depends on the upwind wind conditions: the wind velocity and the turbulence intensity. The velocity in the wake of a wind turbine depends on the amount of power the rotor is extracting from the wind compared to the maximum possible power extraction (the value of C_p from Sec. 3). In other words, it depends on the aerodynamic state of the wind turbine. The aerodynamic state also determines the static mechanical loading and the initial values of the velocity deficit in the wake. The turbulence intensity in the wake is also increased compared to the turbulence intensity of the incident flow. The turbulence intensity of the flow in the rotor plane determines the fatigue loading of the wind turbine.

As a result of these processes, Milborrow and Elliott, who are among the pioneers in wind turbine cluster research, distinguish two types of wind turbine wakes: the velocity wake and the turbulence wake.[26,28] In the velocity wake, the mean velocity is equal to the sum of the upwind mean velocity and the velocity deficit. In the turbulence wake, the velocity variance is equal to the sum of the variances of the upwind and the added turbulences. The velocity deficit, as well as the added turbulence, is reduced with increasing distance from the wind turbine. This is caused by mixing of the wake flow and the undisturbed flow surrounding the rotors under the influence of the turbulence. The upwind turbulence originates from two independent sources: a mechanical source driven by the surface roughness and a thermal source, or sink, driven by the vertical temperature gradients (see Sec. 2).

More information about modeling the turbulence wake can be found in Brand et al.[29]

Production

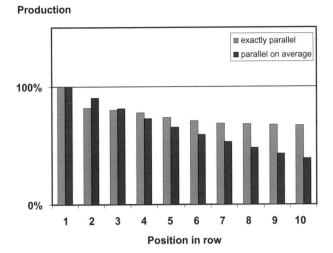

Fig. 18. Production along a row of wind turbines if the wind is blowing (a) exactly parallel to the row and (b) parallel to the row on average only.

4.3 *Internal Wakes Inside a Wind Farm*

Depending on the wind direction, a wind turbine can be completely or partially positioned in the wake of one or more other wind turbines. In such a multiple wake situation, the wind field a wind turbine is exposed to is composed of the velocity deficits and the added turbulence of the other wakes. The consequences for power production and the fatigue loading are severe.

To illustrate the effect of wakes on power production (see Fig. 18), the normalized power of a row of wind turbines is shown for the wind blowing exactly parallel to a wind turbine row. The second turbine in the row only produces a fraction of the power of the most upwind turbine, whereas the other turbines more or less produce the same. If, on the other hand, the wind direction changes so that only on average the wind is parallel to the row, a more gradual decrease of the production along the row results, as shown in Fig. 18.

4.4 *Wind Farm Wakes*

A wind farm is a flow disturbing object which exerts a force on and extracts energy from the flow of the lower part of the atmospheric boundary layer (Fig. 19). As a result, a wind farm decreases the velocity and increases the turbulence at the downwind side of the wind farm.[30−34] The key parameter determining the initial value of the velocity deficit and the added turbulence due to a wind farm is the thrust force per unit horizontal area, covered by the wind farm. It loosely translates into the power density being the aerodynamic power of the turbines per unit horizontal area. If the thrust per unit area is large, the initial values of the velocity deficit and the added turbulence are large, and vice versa. The initial velocity deficit and the

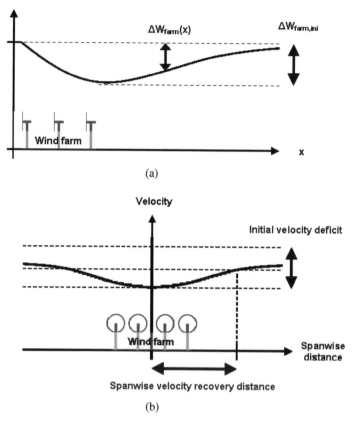

Fig. 19.   Recovery of the velocity wake behind (a) and besides (b) a wind farm.[32]

initial added turbulence decay under the influence of the turbulence in the wake,
which is the sum of the upwind and the added turbulence. The key parameters
determining these decays are the downstream distance behind the wind farm, to be
distinguished in the streamwise distance and the spanwise distance, and the decay
parameter.

With the appearance of large offshore wind farms, the possibility has been
created for measurements to verify and validate analytical models and to further
develop them.

4.5 *Wind Farm Clusters*

Wind farms form a cluster if they are located close together and are operating in
the same microclimate.

The placing of wind farms close together started with the huge onshore clusters
in the windy valleys in California in the 1980s and is evolving into a trend to concentrate as many wind farms as possible in designated offshore areas. Examples of this
trend can be found in the North Sea with the Horns Rev I and II (separation 23 km)

wind farms in the Danish Exclusive Economic Zone (EEZ) of the North Sea, the OWEZ and Q7 wind farms in the Dutch EEZ (separated by 15 km), and the Alpha Ventus and MEG Offshore I wind farms in the German EEZ (separated by 4 km).

As explained in Sec. 4.4, each wind farm produces a wake of its own, dispensing the velocity deficit and the added turbulence over the downwind wind field. As a result, a cluster of wind farms has a complex system of internal velocity and turbulence wakes, similar in appearance but of a larger scale than the internal wake system of a single wind farm. Because of the strong impact of a wind farm cluster on the wind, such a cluster is expected to change the local wind climate. This will result not only in modifications in the sectorwise Weibull distribution of the mean wind speed in a given location but also in modifications in the distribution of the turbulence intensity.

To date, reliable measuring data from the few clustered offshore wind farms are lacking, but this is expected to change with the commissioning of the new wind farms in the North Sea.

Apart from influencing the wind, some researchers argue that large concentrations of wind turbines may affect the local weather, in particular, temperature or precipitation.[34] The rationale is that the added turbulence dissipates into extra heat, which causes an increase in temperature and humidity and eventually more precipitation. The magnitude of these changes, however, is expected to be small, as is the expected effect on the local climate.

While information on observed wind farm wake effects is scarce, predicted effects are available from a number of models. Wind farm wake modeling requires simulation of mesoscale atmospheric flow together with energy extraction/redistribution due to wind turbines. A critical review of the different approaches is given in Ref. 32.

5 Grid Integration

5.1 *Introduction*

When feeding electricity from wind energy systems into the electricity supply system on a large scale, the challenge of balancing total electricity supply and demand at all times becomes greater as wind energy is a variable source and wind farms output is only controllable to a limited degree. Various means are available for balancing: back-up by, for instance, fossil-fuel and hydro plants, load matching, storage, and curtailing wind farms output. At present, these actions are controlled by price setting in the electricity market at various time scales. Grid integration of wind energy and balancing has a hardware side as well as a "software" side. A lack of appropriate hardware, such as an offshore electrical infrastructure, may be considered as the most serious showstopper for large-scale introduction of wind energy. That is why grid integration of wind energy has become a prominent topic in project development and research agendas.[35,36] To solve the urgent problem of creating an electricity transport network in the North Sea, for instance, at least 15 concepts circulate. The dilemma is that investors will make financial resources for the infrastructure available only if the offshore wind power sites are legally identified and the financial market structure is known far into the future. On the other hand, projects will only be built if the infrastructure is available. Otherwise, individual project designers will create their own transport structure to connect to the grid. This is widely considered as a sub-optimal solution. Another challenge for the grid operators is the sheer capacity of wind farms which need to be connected in the medium term. The peak power of these variable output plants is of the same order of magnitude as the presently installed capacity of conventional power. Transmission Systems Operators (TSOs) have never been faced with the task of integrating this type of generating capacity into their systems. For a deeper understanding of the hardware side, see Chapter 20 of this book and Refs. 36–38.

In the next section, the "software" side of grid integration is addressed as far as it is related to meteorology.

5.2 *Grid Requirements*

5.2.1 *System balance*

A wind turbine or a wind farm is just another electricity generator which has to meet grid code requirements. These codes are aimed at maintaining balance in the electricity system. If balance is not maintained, a blackout may occur.

There are specific rules for frequency control, voltage control, reactive power control, and energy flows. Each rule is valid for a specific time scale, ranging from sub-second (frequency and voltage control) to 15 min and longer (energy control). Primary services are aimed at delivering active power to the grid, whereas ancillary services provide reactive power.

5.2.2 *Program imbalance*

Requirements on wind energy generators differ between countries, but in general electricity from the wind has to contribute to the imbalance control at a time scale of 15 min. This is referred to as program imbalance control. Program balance control may either be the responsibility of market parties or the grid operator. Imbalance control at the sub-15-min scale, on the other hand, usually is the responsibility of the grid operator but may be subcontracted to market parties. The same goes for the ancillary services.

In program imbalance control, each so-called program-responsible party has to issue schedules for the production and the consumption of electricity in its portfolio. Such a schedule is called the energy program and is valid for each 15-min period of the day before actual production/consumption. At each 15-min period, the scheduled production must be equal to the scheduled consumption of energy. The energy program must be submitted to the grid operator usually — gate closure times differ between countries — by noon on the day before production/consumption. On the day of production/consumption, the program-responsible party may not be able to balance production and consumption in its portfolio on its own. If the production is larger than the consumption, the party feeds surplus energy into the grid. On the other hand, if the production is smaller than the consumption, it contributes to a shortage of energy supply in the grid. In both cases, the difference, called the imbalance energy, is supplied by the grid operator. Usually the grid operator charges the party for the settlement of imbalance energy against the price for imbalance energy. Actually there are two prices, one for feeding energy into the grid (positive program imbalance) and the other for extracting energy from the grid (negative program imbalance). Both prices may be negative and are highly volatile as they differ from one 15-min period to the other.

For a program-responsible party with wind energy in its portfolio, program imbalance control boils down to making a schedule for the production of wind energy every 15 min of the day of delivery. This is the day-ahead forecast, which usually is to be issued before noon of the day before delivery. On the day after delivery, the program imbalance, being the difference between the observed and the forecasted energy, is assessed.

A program-responsible party can select from several instruments to minimize program imbalance. These include short-term forecasts, short-term forecast updates, aggregation, storage, fast start-up units, and wind farm shut-down strategies.[39] Although falling beyond the scope of the system of program imbalance, the instrument of short-term forecasting of wind energy supply has opened the door to trading wind energy at the day-ahead market.

5.3 *The Natural Variability and the Limited Predictability of Wind Energy*

5.3.1 *Variability*

Natural variations in the wind[5,40−42] are either mitigated or augmented by a wind turbine.[43] Since the output of a wind turbine is zero if the wind speed is too low (cut-in speed) or too large (cut-out speed) and because the output is approximately proportional to the third power of the wind speed (see Sec. 3.2), the output of a single wind turbine can change significantly both during a 15-min period and during an hour or a day.

The output variations are smoothed if the output of more than one wind turbine is fed into the same electricity grid as one electricity plant. Generally, the very short-term fluctuations in the minute and faster time-frame are small and have little coherence. Longer-term variations in the hourly or daily time frame can be of the order of the turbine capacity. These fluctuations have medium coherence and can be mitigated by aggregating from an area proportional to the secondary weather

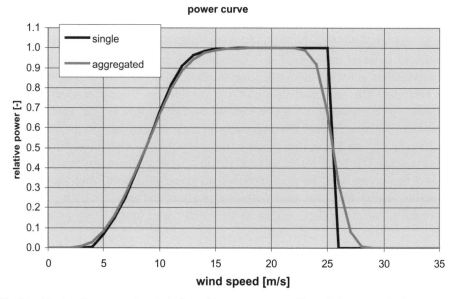

Fig. 20.   Regionally averaged and single turbine power curve. The relative power is the power as normalized to the rated power.

systems. (The secondary weather systems have a time scale of several hours and a spatial scale of several hundreds of kilometers, see Sec. 2.) Aggregating on a still larger scale smoothes modulations at seasonal and annual time scales.

The level of smoothing increases with the size of the region where the turbines are placed. This is because the correlation between the wind speed in two points decreases with the distance between those points. The number of turbines in the region also has an impact on the level of smoothing.

The output from the wind turbines operating in a specific region can be represented in a regionally averaged power curve (Fig. 20). Regional averaging provides the average power output of a cluster of wind turbines in an area where the wind climate is known, assuming the turbines do not interact with each other. The regionally averaged power curve is created by applying a Gaussian filter to a single-turbine power curve[44–47] within the same wind climate zone. One of the important parameters in this analysis is D_{decay}, which is a measure of the distance over which the wind is correlated. Typical calculated values for the characteristic distance D_{decay} are: 500 km, based on Danish and Scandinavian data, respectively,[39] 610 km, based on data in the Netherlands[48] and 723 km, based on 60 locations spread throughout the European Union.[47]

The variability of wind energy is relevant for the grid because all variations must be compensated for by either the other generators or demand side management. The level of the extra capacity required for balancing is a matter of dispute but in general depends on the layout of the electricity system. For example, in a thermal system with high ramping rate capability, the requirements for extra capacity are small, in particular if the penetration of wind energy is relatively low. An electricity system with a high penetration of low ramping-rate plants, may require a significant amount of extra capacity. Note that in both cases, the capability to exchange energy with connected grids must be taken into account.

5.3.2 *Predictability*

Apart from predictions for the long term, to estimate annual energy production, predictions are needed at the inter-day (day before delivery) and the intra-day (day of delivery) levels. The requirements originate from program imbalance control. It essentially consists of a mandatory forecast of the per 15-min period, to be issued before noon of the day before delivery, in combinations with the possibility to change the forecast no later than 1 h before delivery.

Wind power forecast consists of the expected value of the wind power output in a given 15-min period, together with the confidence interval. Although other methods exist, for example, methods based on persistence of the actual wind power output, in general a wind power forecast is made by transforming the relevant output of a numerical weather prediction model.[39,49,50] To do so, there are two different methods: the physical and the statistical forecasting methods.

A physical method, as its name suggests, uses physical models of the local atmospheric processes in combination with specific information from the site. The relevant output of the numerical weather prediction model consists of the expected values of the wind speed, the wind direction, the air temperature, and the air pressure, up to two days ahead in time. The physical models describe the processes relevant to a wind turbine or a wind farm, such as the wind shear due to the surface roughness near the site or the vertical temperature differences. The site-specific information consists, for example, of descriptions of the surface roughness near the site, the power curve of the wind turbine or wind farm and a compensation mechanism for systematic forecasting errors. Eventually, this gives the expected value of the production up to two days ahead in time.

A statistical method uses data from a numerical weather prediction model too but relies on statistical models to make the translation to wind power. Again, the data from the numerical weather prediction model consist of the expected values of the relevant physical quantities, but now estimated relations are employed in order to obtain the expected values of the wind power.

Physical as well as statistical methods in general not only provide the expected value of the wind power but also specify a measure of the uncertainty of the prediction. In addition, both methods need some measured data in order to allow the correlation between predictions and measurements to be established (physical methods) or to train the estimated relations (statistical methods).

Over the years, quite a large number of wind power forecasting methods and systems have been developed. The market does not seem to be saturated as new methods and tools are still being introduced.[51] A typical wind power forecast is shown in Fig. 21.

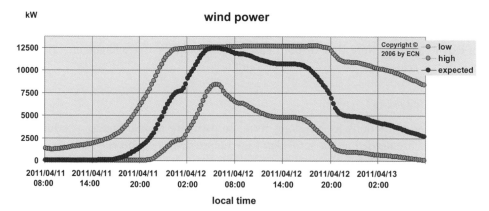

Fig. 21.   Typical wind power forecast, consisting of the expected value and the confidence interval of the wind power. This forecast is valid for a wind farm with a nominal power of 12.5 MW and a turbine hub height of 70 m.

The wind power forecasting error is the difference between the observed and the forecasted power and is of two types: the systematic and the stochastic forecasting errors. The systematic error originates from processes that are not represented in the forecasting method but remain the same over time. This error can be compensated for by correlating the observed and forecasted productions. The stochastic forecasting error, on the other hand, originates from non-modeled processes that vary from one moment to the other. This error implies that the probability of a correct forecast is less that 100%. For this reason, a confidence interval is used in order to specify the range in which the observation will fall with a given probability.

The main source of wind power forecasting error is the numerical weather prediction model.[51] These models are not so much inaccurate in predicting the weather patterns but generally fail to predict the moment an event occurs. And predicting the exact time is essential in the energy sector as accurate point estimates at the 15-min level are required.

The wind power forecasting error is directly related to the program imbalance. A positive value of the forecasting error indicates surplus energy and a negative forecasting error is a shortage of energy supply, as explained in Sec. 5.2.2.

6 Market Developments[52]

The wind energy market has developed very quickly since the beginning of the 21st century. The market is very dynamic in terms of technology supply, regional development, and industrial development.

At the end of 2012, the total installed wind power in the world amounted to 241 GW and was estimated to provide 2.26% of the world's electricity demand. Short-term forecasts predict that this percentage will be 8% by 2021. In some countries, wind energy is already (at the end of 2011) a significant electricity provider. Some examples[36] at the top of the list include Denmark (25.9%), Spain

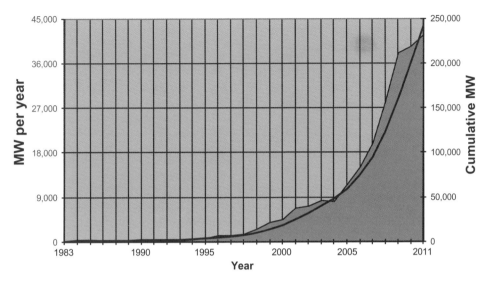

Fig. 22. Historic development of installed wind power in the world.[52]
Source: BTM Consult — A Part of Navigant — March 2012.

(15.9%), Portugal (15.6%), Ireland (12.0%), and Germany (10.6%). The average annual growth from 2007 until 2011 was 22.7%. The annual growth rate, however, is decreasing, and for the years 2012 to 2016 it is estimated at 10%, with a slight increase during the period thereafter (see Fig. 22).

Most of the variations are due to regional developments. In some regions (Netherlands, Denmark, Northern Germany), wind energy on land is reaching its limit because of public resistance or for technical reasons like reaching maximum grid penetration levels. The alternative to land-based projects, offshore wind energy, began 10 years ago, but is not yet at full implementation speed to maintain the earlier growth rates. Other regions in the world, notably the new economies of countries like China, India, and Brazil, are installing wind energy systems on a very large scale as part of their strategy to compensate for a shortage of electricity supply as soon as possible. The global turnover of the wind energy sector in 2011 was €52.2 billion and is expected to increase to €86.3 billion.[52]

Table 3 illustrates the distribution of installed wind power in the world in 2010 and 2011. At the end of 2011, Europe still had the largest amount of installed wind power, but the highest growth rate took place in South and East Asia.

We can see a similar shift of suppliers of wind turbine systems from Europe and the US toward Asia. In the recent past, more than 80% of wind turbines were manufactured in the western world. As can be seen from Table 4, in 2011, the balance is beginning to tip toward Asia.

In the middle of 2012, with a total amount of installed offshore power of 4.1 GW (3.9 GW in Europe and 0.2 GW in China), offshore wind power is only at the beginning (1.7% of the total installed wind power) of its expected growth curve

Table 3:   Regional Distribution of Wind Power Around the World in 2010 and 2011.[52]

	Installed MW 2010	Cumulative installed MW 2010	Installed MW 2011	Cumulative installed MW 2011	% of installed MW 2011
Total Americas	6,639	46,990	9,573	56,563	22.9%
Total Europe	10,980	87,565	10,226	97,563	24.5%
Total South and East Asia	21,130	58,277	21,005	79,282	50.4%
Total OECD Pacific	478	5,368	694	6,062	1.7%
Total Africa	98	1,112	133	1,245	0.3%
Total other continents and areas	79	208	81.6	290	0.2%
Annual MW installed capacity	39,404		41,712		
Cumulative MW installed in the world		199,520		241,029	

Source: BTM Consult — A part of Navigant — March 2012.

Table 4:   Top 10 Global Suppliers of Wind Turbines in 2011 as a Percentage of the Total Installed Power of 40.4 GW in 2011.[52]

Suppliers (region)	Percentage of total installed power
Vestas (Germany)	12.9%
Goldwind (China)	9.4%
GE Wind (US)	8.8%
Gamesa (Spain)	8.2%
Enercon (Germany)	7.9%
Suzlon Group (India)	7.7%
Sinovel (China)	7.3%
United Power (China)	7.1%
Siemens (Germany)	6.3%
Mingyang (China)	2.9%
Others	21.5%

(see also Table 5). On average, offshore wind energy is still more expensive than wind energy onshore by a factor of 1.5 to 2. Experience with the 50 existing offshore projects and technical innovations promise to bring the cost down by 30% to 50% within a period of about 10 years. Offshore applications actually form the backbone of almost all national R&D programs in the world. As already mentioned in Sec. 3, offshore is the driver for upscaling the size of wind turbines. This is well illustrated by Fig. 23, where the product cycle of wind turbines in Germany is depicted. The development in Germany may be considered to be representative for the world. In Fig. 24, this trend is illustrated in more detail. It is most likely that upscaling of wind turbines for land-based application will end at levels of 130-m diameter due to logistical limitations. Offshore up scaling will continue to dimensions of well over 130-m diameter. The largest wind turbines which are now being designed and built have diameters of over 150-m. Technically, it is feasible to construct wind turbines of 200-m diameter and 20 MW installed power.[53] Economic feasibility, of course, still has to be proven.

Table 5: Distribution of Offshore Wind Power in the World in 2010 and 2011.[52]

Country	Installed MW 2010	Accu. MW 2010	Installed MW 2011	Accu. MW 2011
Belgium	165	195	0	195
P.R. China	39	102	107.9	209.9
Denmark	207	832.9	0	832.9
Germany	108	168	30	198
Ireland (Rep.)	0	25	0	25
Netherlands	0	246.8	0	246.8
Norway	0	2.3	0	2.3
Portugal	0	0	2	2
Sweden	0	163.3	0	163.3
UK	925	1775	330	2105
Total capacity — World	1,444	3,510	470	3,980

Source: BTM Consult — A Part of Navigant — March 2012.

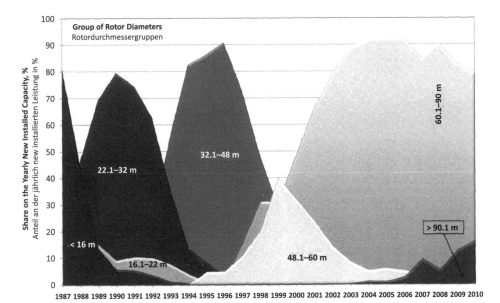

Fig. 23. Product cycles of commercial wind turbines based on rotor diameters.[55]

The developments sketched earlier could be summarized in a forecast for the growth of the market for the next (starting point 2012) 10 years, divided between onshore and offshore applications (see Fig. 25). The regional breakdown is presented in Fig. 26.

Apart from the technical and economic issues, other institutional and environmental issues play a crucial role in further expanding wind energy capacity. Although wind energy in general is an environmentally sound technology — illustrated, for instance, by a energy payback time (time to produce an equivalent amount of energy to that needed to produce the wind turbine) of three to nine

Global Average Annual WTG in kW

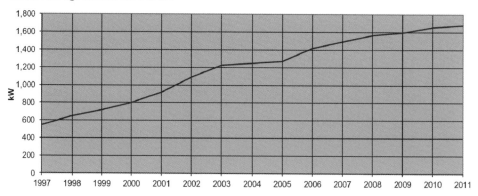

Fig. 24.   Global average installed power per wind turbine within one year.[52]

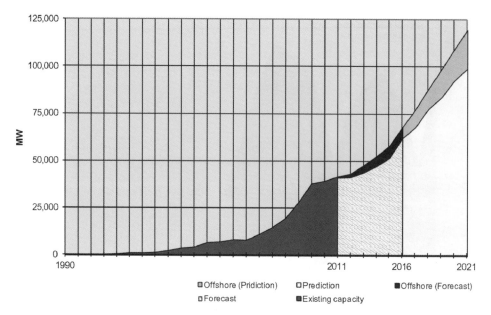

Fig. 25.   Forecast of annual wind power installations.[52]
Source: BTM Consult — A Part of Navigant — March 2012.

months — the following, most important, aspects need to be taken into account in
the planning process:

- Visual impact (landscape)
- Visual flickering effect by shadow of blades
- Acoustic noise emission
- Impact on birds (collisions, habitat, migration routes, and forage areas)
- Collision with bats

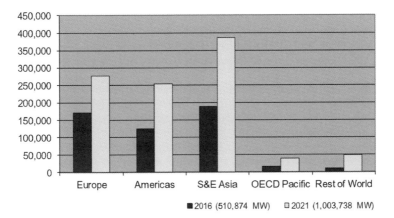

Fig. 26. Regional distribution of accumulated installed wind power until 2021.[52]
Source: BTM Consult — A Part of Navigant — March 2012.

- Effect of construction noise on hearing capability of sea mammals
- Effect on Fish (many positive effects have been observed as well)
- Offshore shipping safety

In many cases, but not always, the problems can be mitigated by careful planning or temporary mitigating measures, such as high-frequency pingers to induce porpoises to leave the construction areas offshore during construction work and stopping wind turbines for short periods during the passage of migrating birds.

Finally it has to be kept in mind that lack of electrical infrastructure, legal provisions, and public and political support, based on perceived negative effects of wind turbines, in practice can be showstoppers for further expansion of wind power. For offshore effects, see Ref. 27 for further reading.

References

1. J. Beurskens, L. Dannenberg, T. Faber, F. Fuchs, K. Rave, S. Siegfriedesen, A. Schaf-farczyk, H. van Radecke, R. Schütt, and S.Wanser, *Einführung in die Windenergi-etechnik*, (in German) (Hanser Verlag, Munich, forthcoming 2012).
2. EWEA, Wind Energy — the Facts: A guide to the Technology, Economics and Future of Wind Power (2009).
3. C.-J. Winter, R. L. Sizman, and L. L. van 't Hull, *Solar Power Plants* (Springer Verlag, Berlin, 1990).
4. J. Beurskens and G. van Bussel (2008): *Energie zakboek*. Vols. I and IV (in Dutch), (ed.) P. H. H. Leijendeckers (Reed Business, Doetinchem (NL), 2008), pp. II/140–II/150; pp. IV/136–IV/165.
5. I. Orlanski, A rational subdivision of scales for atmospheric processes, *Bull. Am. Met. Soc.* **56**(5) (1975), pp. 527–530.
6. S. C. Pryor and R. J. Barthelmie, Climate change impacts on wind energy: A review, *Renew. Sust. Energ. Rev.* **14** (2010), pp. 430–437.
7. J. Wieringa and P. Rijkoort, Windklimaat van Nederland. *Staatsuitgeverij* (1983).

8. A. J. Brand, J. Peinke, and J. Mann, Turbulence and wind turbines. *13th European Turbulence Conference*, Warsaw, Poland, 2011.

9. J. W. M. Dekker and J. T. G. Pierik (eds.), *European Wind Turbine Standards II.*, Book ECN-C–99-073 (Energy Research Centre of the Netherlands, the Netherlands, 1999).

10. *Wind Atlases of the World* (2011). Available at: http://www.windatlas.dk. Accessed 16 July 2011.

11. R. Kistler *et al.* The NCEP-NCAR 50-year reanalysis. *Bull. Amer. Meteor. Soc.* **82** (2001), pp. 247–268.

12. D. L. Elliot, N. J. Cherry, and C. I. Aspliden, World-wide wind energy assessment. Energy and Special Applications Programme, Report No. 2. World Meteorological Organisation, 1981.

13. E. L. Petersen, I. Troen, S. Frandsen, and K. Hedegaard, Danish Wind Atlas: A rationale method of wind energy siting. Risø National Laboratory. Report Risø-R-42, 1981.

14. D. L. Elliot, C. G. Holladay, W. R. Barchet, H. P. Foote, and W. F. Sandusky, *Wind Energy Resource Atlas of the United States.* Dept of Energy. Report DOE/CH 10093-4, 1986.

15. I. Troen and E. L. Petersen, *European Wind Atlas* (Risø National Laboratory, Roskilde, 1989).

16. S. M. Uppala *et al.* The ERA-40 reanalysis. *Q. J. R. Meteorol. Soc.* **131** (2005), pp. 2961–3012.

17. M. M. Rienecker *et al.* MERRA — NASA Modern-era Retrospective Analysis for Research and Application, *J. Climate* **23** (2011), pp. 5041–5064.

18. E. Hau, *Wind Turbines: Fundamentals, Technologies, Application, Economics* (Springer, Berlin, Heidelberg, 2005).

19. S. Heier, *Grid Integration of Wind Energy Conversion Systems* (Wiley, Berlin, Heidelberg, 2006).

20. IEC 61400-1, Ed. 2.(1998): Wind Turbine Generator Systems — Part 1: Safety Requirements.

21. J. P. Molly, Rated power of wind turbines: What is the best? *DEWI Magazine* February, pp. 49–57, 2011.

22. P. Jamieson, *Innovation in Wind Turbine Design* (Wiley, Chichester, UK, 2011).

23. R. Harrison, E. Hau, and H. Snel, *Large Wind Turbines* (Wiley, Chicester, UK, 2000).

24. J. F. Manwell, J. G. McGrowan and A. L. Rogers, *Wind Energy Explained: Theory, Design and Application*, 2nd edn. (Wiley, Chicester, UK, 2011).

25. D. A. Spera (ed.), *Wind Turbine Technology* (ASME Press, New York, 1995).

26. D. L. Elliott, Status of wake and array loss research, Windpower, 91, Palm Springs, California, 1991.

27. J. Beurskens *et al. Converting Offshore Wind into Electricity* (Eburon Academic Publishers, The Netherlands).

28. D. J. Milborrow, The performance of arrays of wind turbines, *J. Ind. Aerod.* **5** (1980), pp. 403–430.

29. A. J. Brand and J. W. Wagenaar, Turbulent wind turbine wakes in a wind farm, *Progress in Turbulence and Wind Energy IV, Springer Proceedings in Physics*, **141**(8) (2012), pp. 231–234.

30. M. B. Christiansen and C. B. Hasager, Wake studies around a large offshore wind farm using satellite and airborne SAR. *31st Int Symp Remote Sensing of Environment*, St Petersburg, Russian Federation, 2005.

31. R. Barthelmie *et al.* Efficient Development of Offshore Windfarms (ENDOW). Risø National Laboratory Report Risø-R-1407(EN).

32. A. J. Brand, Wind Power Plant North Sea — Wind farm interaction. Energy Research Centre of the Netherlands. Report ECN-E-09-041, 2009, p. 53.

33. S. Frandsen *et al.* The necessary distance between large wind farms offshore — Study. Risø National Laboratory Report Risø-R-1518(EN), 2004.

34. P. Rooijmans, Impact of a large-scale offshore wind farm on meteorology — Numerical simulations with a mesoscale circulation model, Masters thesis, Universiteit Utrecht, 2004.

35. F. van Hulle, *Large Scale Integration of Wind Energy in the European Power Supply: Analysis, Issues and Recommendations* (EWEA, Brussels, 2005).

36. EWEA, *Powering Europe: Wind Energy and the Electricity Grid* (EWEA, Brussels, 2010).

37. Th. Ackermann, *Wind Power in Power Systems*, 2nd edn. (Wiley, Chichester, UK, 2012).

38. 3E (Coordinator), Offshore Electricity Grid Infrastructure in Europe. Final Report. Brussels (2011).

39. L. Landberg, M. A. Hansen, K. Vesterager, and W. Bergstrom, Implementing wind forecasting at a utility, Risø National Laboratory Report Risø-R-929(EN), 1997.

40. M. Courtney and I. Troen, Wind speed spectrum from one year of continuous 8 Hz measurements. *Ninth Symp. Turbulence and Diffusion* (1990), pp. 301–304.

41. J. Gjerstad, S. E. Aasen, H. I. Andersson, I. Brevik, and J. Løvseth, An analysis of low-frequency maritime atmospheric turbulence, *J. Atmos. Sci.* **52** (1994), pp. 2663–2669.

42. I. van der Hoven, Power spectrum of horizontal wind speed in the frequency range from 0.0007 to 900 cycles per hour. *J. Meteorol.* **14** (1957), pp. 160–164.

43. IEA, Variability of wind power and other renewables (International Energy Agency, 2005).

44. P. Norgard and H. Holttinen, A multi-turbine power curve approach, *Nordic Wind Power Conf.* Göteborg, (2004), p. 5.

45. M. Gibescu, A. J. Brand, and W. L. Kling, Estimation of variability and predictability of large-scale wind energy in the Netherlands, *Wind Energy* **12**(3) (2009), pp. 241–260.

46. H. Holttinen, Hourly wind power variations in the Nordic countries, *Wind Energy* **8** (2005), pp. 173–195.

47. G. Giebel, On the benefits of distributed generation of wind energy in Europe, PhD Dissertation, Carl von Ossietzky University, Oldenburg, 2000.

48. A. J. Brand, M. Gibescu, and W. W. de Boer, Variability and predictability of large-scale wind energy in the Netherlands, in *Wind Power* (ed.) S. M. Muyeen (INTECH, Croatia, 2010), pp. 259–288.

49. M. Lange and U. Focken, *Physical Approach to Short-Term Wind Power Production* (Springer Verlag, Berlin, 2005).

50. H. Madsen, P. Pinson, G. Kariniotakis, H. A. Nielsen, and T. S. Nielsen, Standardizing the performance evaluation of short-term wind power prediction models, *Wind Eng.* **29**(6) (2005), pp. 475–489.

51. L. Landberg, J. Collins, and J. Parkes, Taking the guesswork out of wind power forecasting, *Proc. EWEA* Copenhagen, 2012.

52. BTM Consult/Navigant, World market update 2011. Forecast 2012–2016, 2012.

53. EC-FP-6, UpWind: Design limits and solutions for very large wind turbines (EWEA, 2011).

54. EWEA (2012): Wind in Power; 2011 European statistics. Available at: www.ewea.org.

55. C. Ender, Wind Energy Use in Germany — Status 31.12.2010. *DEWI Magazine* February, pp. 36–48, 2011.

Chapter 15

Ocean Energy

Ian Bryden
School of Engineering, University of Edinburgh
The Kings Buildings, Mayfield Road
Edinburgh, EH9 3JL, UK
Ian.Bryden@ed.ac.uk

Marine (or ocean) sources are now being seriously considered for the production of renewable electricity. This chapter covers waves, tidal entrainment, and tidal currents and gives a brief overview of the history of efforts to exploit these potential resources. As well as discussing the technical issues associated with development of the necessary technology, the nature of the resources are also briefly considered. Discussions on the exploitation options include introduction to the fundamental device concepts, which are representative of the vast majority of devices under development.

In the later sections, the author considers the present status of development of the marine resources, including ongoing projects, drawing largely from his experience in the UK sector and how future developments might proceed.

1 Introduction

The earth's surface is mostly covered by water and is exposed to a complex mix of external forces: winds drive ocean currents and cause waves on the surface of the seas and oceans; tidal effects result in the cyclic rise and fall of the ocean surface and associated currents, some of which can be extreme. Both waves and tides offer the prospect of sustainable energy supplies.

The rise and fall of sea level, driven by the tides has been utilized to grind cereal to make flour since the times of the Roman Empire[1] and much more recently, barrages have been constructed to entrain tidal waters and allow electricity to be generated by water flowing through low-head turbines.[2,3] Technology is also now available which can convert energy in waves and tidal current into electricity.

Just how large the marine resources are is still a subject of research. However, there is now sufficient understanding to allow assessment of the available energy at selected sites, provided there is sufficient survey data available. However, securing sufficiently robust data remains a challenge. Published assessments of resources are often contradictory, frequently as a result of failure to specify what aspect of the marine energy resource is being reported. The "Theoretical" resource is a description of the total energy which could be extracted subject to the laws of physics. More challenging, however, is the determination of the "Technical" resource, which is that proportion of the theoretical resource exploitable using achievable technology. More restrictive still is the "Practical" resource. This represents the proportion of the technical resource deemed exploitable after consideration of external constraints such as grid accessibility, competing use (shipping lanes, etc.) and environmental sensitivity.

2 Wave Energy

2.1 *The Technology*

Serious consideration of ocean waves as an energy resource is generally believed to have begun in the 1970s,[4] although there are patents for wave energy devices dating back to the 19th century.[5]

All wave energy systems can be considered as possessing some fundamental properties:

The machines will each include a ***prime interface***, which is an element of the system which responds to forces resulting from the waves. The relative motion between the prime interface and some other ***reference element*** will allow the driving of the ***Power Takeoff System*** (PTO). This will essentially convert mechanical energy into a form suitable for export, which would generally be in the form of electricity, but it is conceivable that a wave energy system might export some high-energy product such as compressed air or desalinated water as an alternative. The wave energy technology must also be fixed in place, otherwise it will move as a result of the wave forces. If the system is floating, then the fixing will involve some form of mooring system, otherwise the device may be rigidly attached to the sea bed or to a cliff.

A variety of so-called fundamental device concepts have evolved for the artificial extraction of energy from the waves. These include:

Floating responders, in which the prime interface is in the form of a floating surface buoy that can respond readily to the incoming waves. The reference element may be the sea bed, in which case the power take off system might be included within the mooring system as shown in Fig. 1, or the buoy might incorporate an internal

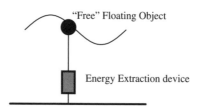

Fig. 1.   Floating buoy with the PTO included in the mooring system.

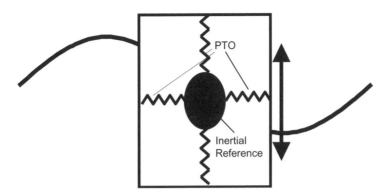

Fig. 2.   Floating buoy with inertial reference.

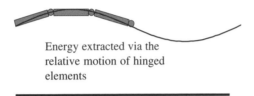

Fig. 3.   The articulated relative motion concept.

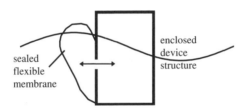

Fig. 4.   The flexible membrane concept.

mass so that the power takeoff is driven by the relative motion between the buoy exterior and the internal inertial element, as shown in Fig. 2.

The use of an inertial reference decouples the PTO from the mooring, which might offer advantages under some conditions.

Relative Motion Devices are closely related conceptually to inertial reference machines. The best known of such devices utilize articulated structures incorporating multiple prime interfaces as shown in Fig. 3. Such a device would almost certainly be used as a floater, although it is conceivable that one end could be fixed to a harbor wall or some other fixed structure such as the tower of an offshore wind turbine.

Flexible Membrane Devices use the pressure under a wave to compress air, which is then driven through low pressure turbines, as shown in Fig. 4. The membrane represents the prime interface.

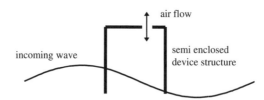

Fig. 5.   The enclosed water column concept.

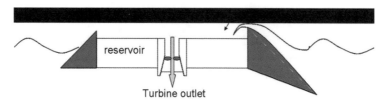

Fig. 6.   The wave dragon "overtopper" concept.

Enclosed Water Column Devices, as shown in Fig. 5, are philosophically related to the flexible membrane types, in that they use water pressure to drive air through a turbine but they rely on the air water interface itself to act as the pressure mechanism.

Overtopping devices involve devices that are shaped to funnel waves so that their energy directly lifts water into an elevated storage pond before flowing back down to sea level through low-head turbines. The best known of such devices is the Wave Dragon, shown in Fig. 6. Interestingly, the water itself is the prime interface in this concept.

It is common to consider the effectiveness of a device in absorbing wave energy in terms of a parameter called the absorption width. Devices that have a physical width less than the length of the waves can have an absorption width greater than their physical width. This is referred to as the point absorber effect.

2.2 *Resource*

The energy in waves is generated by the wind passing over the ocean surface, so that energy is transferred through frictional mechanisms into propagating instabilities in the free surface. It is these traveling instabilities that we refer to as waves. The resource is greatest where the predominant wind directions pass over substantial ocean areas prior to heading toward coastal regions.

The wave resource is probably the best understood of all the marine energy resources. When considering the wave energy incident toward a coastline, it is conventional to express the resource in terms of the energy flux across fixed-depth contours. This might be in terms of the annually averaged power crossing, for example, 1 m of the 50 m water depth contour. It should be appreciated, however, that waves in the real world are complex multidirectional entities and overly simple approaches can be misleading. Most energy analyses, for example, are based upon the principles of linear wave theory, which is known to be too simple to describe real wave behavior, especially under energetic conditions. Nor do real waves conveniently cross-contour lines perpendicularly!

Researchers have adopted various empirical methods[6] to determine wave energy flux densities, such as Eq. (1). The use of such methods has allowed approximate assessments of the incoming energy flux from limited wave data. More detailed analysis, which takes cognizance of the performance characteristics of devices, requires appreciation of the relationships between wave period, directionality and wave height in a particular sea state. Generally, this uses the principle of spectral density, which is also an essentially linear concept.

$$\text{Power (kilowatt)} = 0.55\,H_{\text{Sig}}^5 T_z, \tag{1}$$

where, H_{Sig} is the significant wave height (metre) and T_z is the mean zero crossing period (second)

The global theoretical wave energy flux is believed to fall between 1000 and 10,000 GW. The exact figure is subject to scientific and philosophical debate relating to the meaning of the concept itself! For example, should the energy flux present in the mid-oceans be included? In such locations, there might be a dynamic equilibrium between wave generation and dissipation. If energy were to be extracted, the equilibrium would be upset and the balance could pass once again to wave formation. Should the theoretical resource estimates take into account such disruption?

In many parts of the world, such as the eastern boundary of the North Atlantic, the theoretical flux density can exceed an annual average of 60 kW/m in deep water but is, of course, seasonal and intermittent. In Northern Europe, for example, the resource is greatest in the winter months, which are the times of greatest electrical demand.

2.3 *The Status*

Prototype wave energy systems are now being tested at locations such as the European Marine Energy Centre (EMEC) in Orkney, UK. The Pelamis concept, shown in Fig. 7, is an example of a floating jointed relative motion technology in which the device is aligned with its long axis parallel to the incoming principal direction of the waves.

The Pelamis P1 devices was tested extensively at EMEC between 2003 and 2007 and was used in the unsuccessful Aguçadoura Wave Farm in Portugal, which was a victim of the economic crisis affecting the world rather than technical failure.

The Oyster device, which is a bottom-mounted flap device, has also been tested extensively at EMEC, as shown in Fig. 8.

At the time of writing (August 2011), the device developers, Aquamarine Ltd are in the process of replacing the original Oyster 1 with their next-generation Oyster 2 machine. Similarly, it is anticipated that there will soon be testing of

I. Bryden

Fig. 7. Pelamis P1 under test in Orkney.

Fig. 8. The Oyster 1 device under test in Orkney.

commercial scale arrays of devices, which may be closely based on those devices that are presently under test. These may be conducted at test centers such as EMEC, but if the technology is to be truly commercial, then it will be necessary for developers to develop sites that are sufficiently large to allow arrays of many tens of MW capacity. At present, the cost of generation using wave energy technology is too great to be considered economical and requires a reduction of the order of five before it becomes directly competitive with onshore wind power generation. Costs at present are dominated by capital and installation. Both must be reduced if the resource is to become competitive. Industry optimists suggest that 2016–2017 could, if present trends continue, be when wave (and tidal current) technology becomes competitive with wind.

3 Tidal Current Energy

3.1 *The Technology*

The physics of the conversion of energy from tidal currents is superficially very similar to the conversion of kinetic energy in the wind. Therefore, many devices bear resemblance to wind turbines. There is, however, no total agreement on the form and geometry of the technology itself. Wind systems are generally horizontal axis rotating turbines, as shown in Fig. 9, with the axis of rotation parallel to the direction of flow. Many tidal developers also favor this approach, as can be seen in the images of prototype systems.

However, vertical axis systems, in which the axis of rotation is perpendicular to the direction of flow, as shown in Fig. 10, have not been rejected.

In addition, devices that utilize the venturi effect to drive a secondary flow circuit, as shown in Fig. 11, have been proposed, as have devices which mimic the behavior of a whale's tail (Fig. 12).

Variations in each of these prototypes have also been proposed, including the horizontal cross-flow system, which combines features of the horizontal and vertical axis devices; the open center horizontal axis devices, in which the turbine is

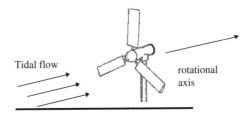

Fig. 9.   Horizontal axis concept.

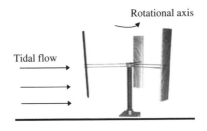

Fig. 10.   Vertical axis concept.

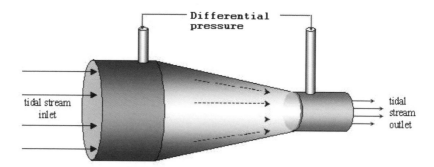

Fig. 11. Venturi secondary flow concept.

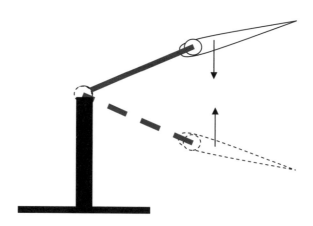

Fig. 12. Linear lift concept.

supported by its rim and water is free to flow through the centre of the device unimpeded, as shown in Fig. 13, and the ducted horizontal axis systems, in which some of the features of the horizontal axis and venture systems are combined.

3.2 The Resource

High speed sites tend to be geographically compact, although the energy flux densities can be considerable. Figure 14 shows the distribution of energetic sites around Scotland.

It is very tempting to draw analogies between wind power and tidal current power. In both, it is possible to calculate the kinetic energy carried by a moving fluid and to relate this to an energy flux density. This would represent, however, only part of the energy available in a tidal current. Tidal devices certainly respond to the kinetic flux, defined in terms of Joules/second in Eq. (2).

$$\text{Kinetic Flux (Joules/second)} = {}^{1}\!/_{2}\, \rho \int_{A} U^{3} \mathrm{d}A, \tag{2}$$

Fig. 13.   The open hydro horizontal axis, rim-mounted system.

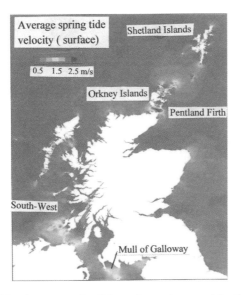

Fig. 14.   Average spring tide peak speeds around Scotland.

where ρ = water density $(\mathrm{kgm^{-3}})$; A = channel cross-section $(\mathrm{m^2})$; U = local flow speed $(\mathrm{ms^{-1}})$.

Even this simple equation hides the spatial and temporal variability of the flow. The true resource itself is a significantly more complex mixture of kinetic flux, potential flux and frictional dissipation, all of which must be considered in any determination of the potential for a site to deliver energy.[7] High speed of flow at any location is a necessary condition for exploitation and must be associated with a water depth and lateral extent that is sufficiently large to allow the installation of technology. There are many sites that experience extreme tidal currents but are too shallow to represent any significant potential. Also, there are many sites which, due to their size, are host to prodigious energy overall fluxes but have kinetic flux densities that would make exploitation unrealistic.

When energy is extracted from a site, the hydraulic nature of the local environment changes and the overall flow speed and volume flux will diminish. This is analogous to, but more complex than, the introduction of resistors into an electrical circuit.

In a three-dimensional flow environment, there may be complex alterations to the flow velocity distribution. It is relatively straightforward to determine the ultimate energy extraction potential, or consequences of more modest extraction, for a simple tidal channel in which all of the boundaries are well understood. In many cases, it can be readily shown that the ultimate extractable energy can greatly exceed the total kinetic energy flux. More complex, multiple connected environments require substantial survey information associated with complex numerical modeling if the consequences of local energy extraction are to be understood. Even the determination of the "theoretical resource" for a well-defined site is, therefore, a substantial task. It would, in most cases, prove to be of limited use, especially if no attempts were made to discriminate against sites that might be insensitive to energy extraction, resulting in substantial apparent theoretical resources but with such low kinetic flux densities that conversion would be unrealistic. Unfortunately, many theorists fail to realize that flow speed is vital for conversion and the resource potential relates to other hydraulic consideration. In most real locations, it makes more sense to determine the consequences of specific extraction scenarios for their acceptability or to determine the "practical resource" or the "technical resource", subject to the usual uncertainties in the factors governing them. This would involve, however, the use of highly complex numerical methods to assess the influence of energy extraction at an array and at a regional scale.

3.3 The Status

Like wave energy, tidal current devices are in the process of being tested at full scale. Figure 15 shows the SeaGen device being tested in Northern Ireland at the time of writing.

There are other devices under test at EMEC and at test sites in, for example, Canada. The next stage will be the testing of arrays of devices at a commercial scale. In the UK, the Crown Estate has issued licenses for the development of the energetic waters in and around the Pentland Firth, which lies between Orkney and The Scottish mainland. These licenses are for both wave and tidal developments, but it is for tidal current development that the Pentland Firth has received most consideration. Figure 16 shows the areas for which development licenses have been granted.

Fig. 15.   The SeaGen device in Strangford Lough, Northern Ireland.

At present, as for wave energy, the economics of tidal current systems still looks unattractive in comparison with onshore wind, with the generation costs being, perhaps, three to five times more expensive. However, developers are confident that costs will come down as the systems move toward multiple deployments and take advantage of the economies associated with large-scale manufacture.

4 Tidal Entrainment

4.1 *The Technology*

Tidal entrainment systems operate by using the rise and fall of the sea level under tidal influences. Essentially, the approach is always the same. An estuary or bay with a large natural tidal range is identified and then artificially enclosed with a barrier. This would typically also provide a road or rail crossing of the gap in order to maximize the economic benefit. The electrical energy is produced by allowing water to flow for one side of the barrage, through low-head turbines, to generate electricity. There are a variety of suggested modes of operation, such as ebb generation, flood generation and multiple basin operation.

A tidal barrage power plant has been operating at La Rance in Brittany since 1966.[6] This plant is capable of generating 240 MW and incorporates a road crossing of the estuary. It has recently undergone a major 10-year refurbishment program.

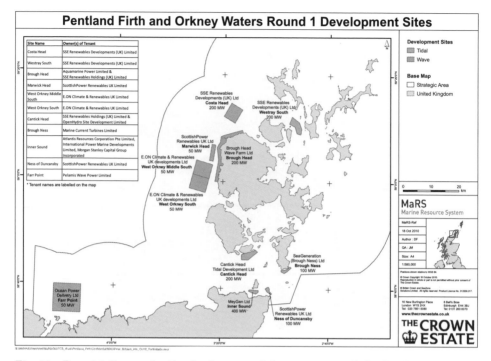

Fig. 16.   Round 1 licences for the development of the waters around the Pentland Firth and orkney.

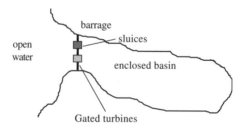

Fig. 17.   Hypothetical tidal barrage configuration.

Barrages have also been constructed at Annapolis Royal in Nova Scotia (18 MW), The Bay of Kislaya, near Murmansk (400 kW), and at Jangxia Creek in the East China Sea (500 kW). Schemes have been proposed for the Bay of Fundy and for the Severn Estuary but have never been built. However, a new large plant at Shiwa Lake in Korea has been recently commissioned.

4.1.1 *Single basin tidal barrage schemes*

These schemes require a single barrage across an estuary, as shown in Fig. 17. There are three different methods of generating electricity with a single basin. All involve a combination of sluices which, when open, allow water to flow relatively freely through the barrage and/or gated turbines, which generate electricity when required.

Ebb Generation: Electricity is generated only during the ebb tide. During the flood tide, water is allowed to flow freely through sluices into the barrage. At high tide, the sluices are closed and water retained behind the barrage. When the water outside the barrage has fallen sufficiently to establish a significant head between the basin and the open water, the basin water is allowed to flow out though low-head turbines and to generate electricity. The system behavior can be considered in phases. These phases are represented in Fig. 18 to show the periods of generation within the tidal cycle.

Typically the water will only be allowed to flow through the turbines once the head is approximately half the tidal range. This method will generate electricity for, at most, 40% of the tidal range.

Flood Generation: The sluices and turbine gates are initially kept closed during the flood tide to allow the water level to build up outside of the barrage, as shown in Fig. 19. As with ebb generation, once a sufficient head has been established, the turbine gates will be opened and water can flow into the basin generating electricity.

This approach is generally viewed less favorably than the ebb method, as keeping a tidal basin at low tide for extended periods might have detrimental effects on

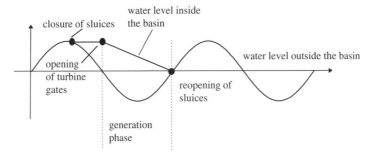

Fig. 18. Water levels in an ebb generation scheme.

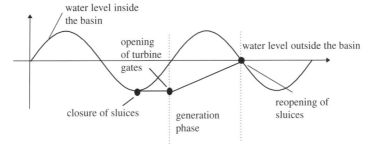

Fig. 19. Water levels in a flood generation scheme.

the environment and shipping. However, the existing land use and environmental considerations, as well as power generation in the Korean Sihwa project, have made the flood mode the preferred option even if it is not the most efficient.[8]

Two-Way Generation: It is possible, in principle, to generate electricity in both ebb and flood. Unfortunately, computer models do not generally suggest a major increase in the energy production. In addition, there would be further costs associated with requirements for either two-way turbines or a double set to handle the two-way flow. However, advantages include reduced periods with no generation and lower peak power, allowing reduction in generator costs.

4.1.2 Double basin systems

Single basin systems can deliver energy only during parts of the tidal cycle. Double basin systems, as shown schematically in Fig. 20, might allow energy storage and enhanced control over output power levels.

The main basin might behave like conventional ebb or flood generation single-basin system. Some of the electricity generated could be used to pump water to and from the second basin to ensure a more robust generation capability.

It is not likely that multiple basin systems will become popular, as low-head turbines are unlikely to be sufficiently efficient to allow economic storage of energy. The overall efficiency of such low-head storage is unlikely to exceed 30%. Conventional pump storage systems, with overall efficiencies exceeding 70%, are more attractive, especially as they are proven technology (see also Chapter 13).

4.2 The Resource

It is probably meaningless to attempt to assess the national and international theoretical resource for tidal entrainment. Data on tidal ranges are available for much of the world's coastlines, although for most of the coastline, the use of entrainments through barrages or lagoons would be prohibitively expensive or unacceptable for

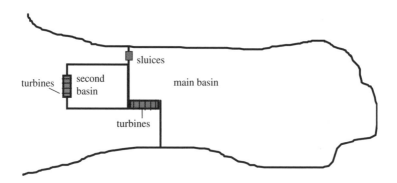

Fig. 20.   Hypothetical two basin system.

environmental or social reasons. However, it is possible to gain a superficial estimate of the energy potential for a particular site.

For example, the theoretical maximum energy delivery over one tidal cycle can be calculated for a vertical sited basin using Eq. (3). However, more detailed analysis is necessary for serious resource assessment.

$$E = \frac{1}{2}\, \rho g A R^2 \,(\text{Joule}), \tag{3}$$

where:

R = range (height) of tide (in m)
A = area of tidal pool (in km^2)
$g = 9.81\,\text{m/s}^2$ = acceleration of gravity
$\rho = 1{,}025\,\text{kg/m}^3$ = density of seawater

Worldwide, there are a considerable number of sites technically suitable for development, although whether the resource can be developed economically is yet to be conclusively determined. As the accessibility of the resource is highly dependent upon local geography, it is not possible to give general assessments of the regional potential for exploitation. However, some areas are well known as potential sites for exploitation, as given in Table 1.

4.3 *The Status*

A list of all currently operating and proposed tidal power stations with capacity greater than 10 MW is given in Table 2. Only three power stations are currently in operation. Some of the proposed systems are very large, with generating capacity in excess of 5,000 MW.

Although the UK government was showing increased interest in developing the Severn Estuary resources toward the end of last decade, this has now waned and it appears unlikely that there will be any significant development of large barrage sites in the UK for many years. However, the South Korean government has been continuing the development of the Lake Sihwa resource and this has now replaced La Rance as the world's largest tidal energy development, with a capacity of 254 MW. There is an even larger system under construction at Incheon in South Korea, with a capacity of 824 MW.[10]

Table 1: Potential Sites for Tidal Entrainment Systems.

Site	Mean tidal range (metre)	Barrage length (metre)	Estimated annual energy production (GWh)
Severn Estuary (UK)	7.0	17,000	12,900
Solway Firth (UK)	5.5	30,000	10,050
Bay of Fundy (Canada)	11.7	8,000	11,700
Gulf of Cambay (India)	6.1	25,000	16,400

Table 2:   Operating and Proposed Tidal Power Plants.[9]

Power plant	Capacity (MW)	Country	Completed date	Operational(O), under construction (C), proposed (P)
Rance River Tidal Power Plant	240	France	1966	O
Annapolis Royal Generating Station	18	Canada	1984	O
Sihwa Lake Tidal Power Plant	254	South Korea	2011	O
Incheon Tidal Power Plant	820	South Korea	2015	C
Severn Barrage Power Plant	8,640	Great Britain		P
Tugurskaya Tidal Power Plant	3,640	Russia		P
Mezenskaya Tidal Power Station	10,000	Russia		P
Penzhinskaya Tidal Power Station	87,100	Russia		P
Dalupiri Blue Energy Project	2,200	Philippines		P
Gulf of Kutch Project	50	India		P

It is unclear what future developments are likely worldwide, as tidal barrage developments are large, with long lead times from investment to cost recovery, which discourages private investment. It is likely that future developments will require economic conditions that encourage long-term investments.

References

1. Available at: http://s277147633.websitehome.co.uk/Elingmill.html. Accessed 1 September 2011.
2. S. H. Salter, Wave power, *Nature* **249** (1974), pp. 720–724.
3. J. Falnes, A review of wave-energy extraction, *Marine Struct.* **20**(4) (2007), pp. 185–201.
4. A. Stahl, The utilization of the power of ocean waves, *Trans. Am. Soc. Mech. Eng.* **13** (1892), pp. 438–506.
5. J. Twidell, T. Weir, *Renewable Energy Resources* (E&FN Spon, London, 2000), pp. 323–326.
6. J. P. Frau, La Rance, a successful industrial-scale experiment, *IEEE Transac. Energy Conv.* **8**(3) (1993), pp. 552–558.
7. I. G. Bryden, T. Grinsted and G. T. Melville, Assessing the potential of a simple tidal channel to deliver useful energy, *Appl. Ocean Res.* **26**(5) (2005), pp. 200–206.
8. International Water Power and Dam Construction Feb 2009. Available at: http://www.waterpowermagazine.com/story.asp?sc=2052179.
9. Wikipedia, Tidal Power Stations. Available at: http://en.wikipedia.org/wiki/List_of_tidal_power_stations.
10. Tidal Today, Asia's Cinderella renewable energy. Available at: http://social.tidaltoday.com/industry-insight/tidal-asia%E2%80%99s-cinderella-renewable-energy.

Chapter 16

Ocean Thermal Energy Conversion

Gérard C. Nihous
Department of Ocean and Resources Engineering
University of Hawaii, Honolulu, HI, USA
nihous@hawaii.edu

Stable temperature differences of the order of 20°C between the surface of most tropical oceans and water depths of about 1 km can be used to drive a heat engine. Formulated in the late 19th century, the concept is known as Ocean Thermal Energy Conversion (OTEC). While deep water marine technologies are challenging *per se*, the small temperature difference available to OTEC systems results in the need for large components and high capital costs. OTEC was first tested at sea by Georges Claude in the 1920s. Further research and development took place in the late 20th century, mostly in the US, but this work fell short of establishing OTEC on a firm commercial footing. While renewed efforts are under way to deploy and operate floating OTEC pilots of 5 to 10 MW, this vast renewable resource with exceptional baseload capabilities remains untapped.

1 Basic Concept of Ocean Thermal Energy Conversion

Wherever seawater temperature varies sharply with depth and the stratification of the water column is stable, it is theoretically possible to extract mechanical power in a heat engine. This basic concept of Ocean Thermal Energy Conversion (OTEC) was first articulated in 1881 by d'Arsonval,[1] although several decades would pass before significant field tests of OTEC would take place.

Ocean temperature differences of the order of 20°C can be found within the upper kilometer of most tropical oceans, between latitudes 30°N and 30°S. This is illustrated in Fig. 1. Strong cold currents and upwellings along the western coasts of Africa and America locally reduce the zonal extent of the OTEC thermal resource. Other specific regional factors such as the Red Sea outflow also affect the availability of vertical temperature differences exceeding 20°C. Yet, the area of interest for OTEC extends over 100 to 120 million km².

The existence of steep stable thermoclines in tropical oceans may be taken for granted but is far from obvious. That intense solar radiation would warm up the surface layer is clear. In the absence of another fundamental physical process, however, the downward diffusion of heat would tend to homogenize the water column, as happens in the great lakes of Africa. Instead, a vast network of planetary

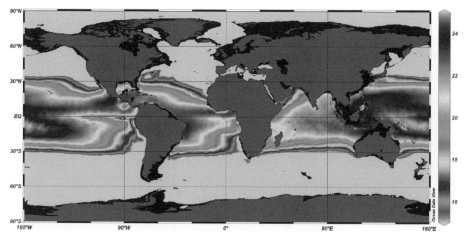

Fig. 1. Yearly average seawater temperature difference between 20 m and 1,000 m water depths (°C — grayscale palette between 15 and 25).[2]

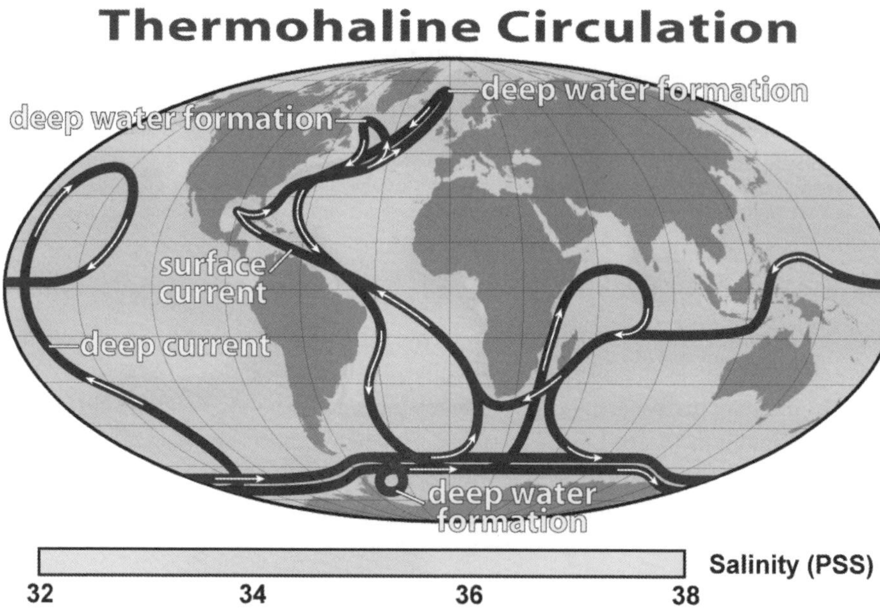

Thermohaline Circulation

Fig. 2. Sketch of the thermohaline circulation.[3]

currents is responsible for supplying deep cold seawater of polar origin throughout the world's oceans. Technically known as the thermohaline circulation, this Great Ocean Conveyor Belt is sketched in Fig. 2.

The heat engine envisioned by d'Arsonval is similar to any machine that cyclically receives heat from a hot reservoir at temperature T_1 and rejects heat to a cold reservoir at temperature T_2. In other words, the fundamental thermodynamic

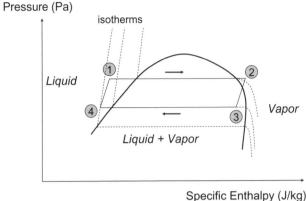

Fig. 3. Schematic diagram of a closed-cycle Rankine process.

description of an OTEC plant is similar to that of other thermal power plants. In the simplest version of OTEC, an auxiliary working fluid like ammonia follows a Rankine cycle in a closed loop schematically depicted in Fig. 3. Between Points 1 and 2, the working fluid warms up and boils in an evaporator fed by warm surface seawater; it then expands through a turbine between Points 2 and 3, where mechanical work is produced; the vapor leaving the turbine enters a condenser fed by deep cold seawater where it condenses to a liquid and cools to the state represented by Point 4. The cycle is completed by pumping the liquid from Point 4 back to Point 1.

A simplified heat-and-mass balance for a typical OTEC plant that would produce about 2 MW of net electrical power is shown in Fig. 4. The working fluid closed loop is drawn in grey. It is worth noting that the mechanical power consumed by the working fluid pump is negligible when compared to the mechanical power produced by the turbine. Using the Clausius–Clapeyron equation, it can be shown that the ratio of these terms is approximately equal to the working fluid vapor density divided by the working fluid liquid density.

Although OTEC systems are not conceptually different from other thermal power plants, in-plant power consumption is much larger, with values of the order of 30% of the turbo-generator output. These parasitic pumping power losses P_P occur *outside* of the working fluid loop as a result of the significant seawater flow rates needed in the heat exchangers. Because surface seawater is more accessible than cold seawater drawn from depths of about 1,000 m, plant optimization generally leads to a higher surface seawater flow rate.

If one considers the seawater temperatures available to an OTEC plant, they are the same as those involved in room-temperature refrigeration. In this case, however, mechanical power is consumed in a compressor (driven by an electric motor) to artificially maintain a temperature difference that would not exist otherwise. Hence, the working fluid loop in Fig. 4 essentially describes a refrigerator that would be run in reverse. Not surprisingly, the same substances that are used in moderate

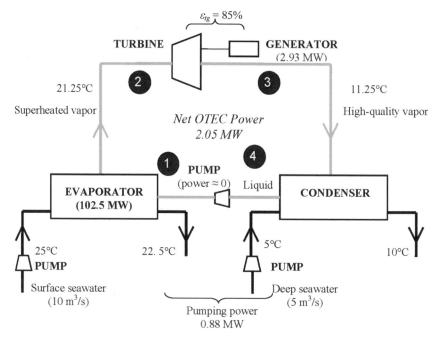

Fig. 4. Simplified heat-and-mass balance of a typical OTEC plant.

refrigeration are adequate for OTEC. From an engineering point of view, these fluids have high enthalpies of condensation per unit volume of vapor in the temperature range of interest, which is equivalent to having steep saturation curves. With excellent heat transfer characteristics to boot, ammonia is an excellent choice.

Heat engines are often evaluated by their thermodynamic efficiency η i.e. the ratio of power produced by the working fluid through the cycle divided by the heat flow rate from the hot reservoir. By considering an engine operating reversibly between two reservoirs at absolute temperatures T_1 (higher) and T_2 (lower), Carnot established the upper limit $\eta_C = 1 - T_2/T_1$. Such an ideal machine, however, would not produce any power because reversible heat exchange would (have to) be exceedingly slow. An analysis of the maximum output of semi-ideal (or endoreversible) engines that exchange heat irreversibly at constant temperature, but operate reversibly otherwise, yields the surprisingly elegant result[4] $\eta_E = 1 - \sqrt{T_2}/\sqrt{T_1}$. This is applicable to a Rankine cycle because heat exchange at constant pressure *also* takes place, for the most part, at constant temperature as the working fluid changes phase (e.g. between Points 1 and 2, or Points 3 and 4 in Figs. 3 and 4). In the particular case of OTEC, $T_1 \approx T_2$ (in Kelvin) so that we have $\eta_E \approx 1/2 - T_2/(2T_1)$. In other words, only half of the available seawater temperature difference $(T_1 - T_2)$ should be used across the turbine while the rest would allow the optimal transfer of heat. This is graphically illustrated by the temperature ladder shown in Fig. 5. The analysis of semi-ideal machines also allows an estimation of the terms x and y,

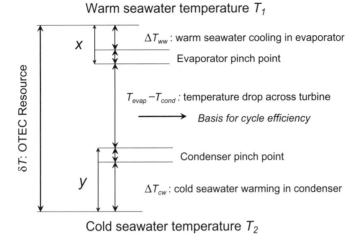

Fig. 5.   OTEC temperature ladder.

as well as of the cycle (endoreversible) power P_E. The latter is proportional to $(\sqrt{T_1} - \sqrt{T_2})^2$, which for OTEC conditions is about $(T_1 - T_2)^2/(4T_1)$.

To accurately describe OTEC systems, additional irreversibilities have to be taken into account. The turbine and working fluid pump are not ideal isentropic machines, and the conversion of mechanical power to electrical power is imperfect; this can be quantified by specific efficiencies. Moreover, working fluid pressure drops in the heat exchangers should be minimized since they will diminish the actual pressure difference across the turbine.

With seawater temperature differences $(T_1 - T_2)$ of the order of 20°C to 24°C, values of η_E (3%–4%) are about 10 times smaller than the thermodynamic efficiencies of thermal power plants that produce electricity from fossil or nuclear fuels. While much has been made of this fact *per se*, its meaning should be interpreted cautiously because OTEC resources are abundant and renewable. The limited thermal potential difference available to OTEC power plants does affect their design in many ways, however. This will be discussed in Sec. 3.

An immediate consequence of the dependence of P_E on the square of $(T_1 - T_2)$ is the sensitivity of power output on available seawater temperatures. Namely, a change of 1°C in $(T_1 - T_2)$ will correspond to a variation of about 10% for P_E or P_G, the gross power produced by the turbogenerator (which is essentially P_E multiplied by turbine and generator efficiencies). This 10% rule of thumb is illustrated in Fig. 6, where time histories of P_G and $(T_1 - T_2)$ are displayed for an experimental land-based OTEC plant operated through the early 1990s in Hawaii; in this particular instance, a large warm eddy moving along the coast rapidly affected T_1 and as a result, P_G.

Matters are actually more acute for the *net* power $P_{\text{net}} = P_G - P_P$ produced by OTEC plants, because the parasitic seawater pumping power P_P remains essentially

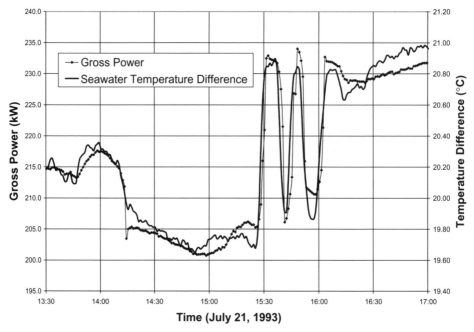

Fig. 6.   Sample time histories of gross power and available seawater temperature difference from an experimental OTEC plant.[5]

the same when $(T_1 - T_2)$ varies (there is only a small margin for operational optimization). With P_P about 30% of P_G, a change of 1°C in $(T_1 - T_2)$ will correspond to a variation of about 15% for P_{net}.

Finally, a unique OTEC cycle that does not resemble a refrigeration system in reverse is briefly discussed here. The Rankine cycle is closed, with the working fluid moving in a sealed loop. In the 1920s, the brilliant engineer and successful entrepreneur Georges Claude conceived instead a process known as Open-Cycle (OC) OTEC, where the working fluid consists of water vapor continuously produced in the evaporator and consumed in the condenser after its passage through a turbine.[6] In Figs. 3 and 4, this would correspond to a removal of the step between Points 4 and 1; otherwise, the thermodynamic analysis of OC-OTEC is not fundamentally different. The boiling of about 0.5% of the surface seawater in a flash evaporator relies on very low pressures of less than 3 kPa throughout the system. With an OC plant, large and costly metal heat exchangers are replaced by simple vacuum chambers, unless fresh water production in a standard condenser is desirable. This also reduces or eliminates the need for seawater treatment to prevent biofouling, the buildup of bacterial mats on heat conducting metal surfaces.[7] Such benefits are offset by the need for large low-pressure turbines as well as for vacuum compressors. The latter must continuously remove the non-condensable gases released from seawater exposed to very low pressures. Owing to the perseverance and wealth of Claude, OC-OTEC was first tested in the field in the late 1920s. It is

Fig. 7.   Aerial view of an experimental OC-OTEC plant in Kailua-Kona, Hawaii; the heat exchangers and turbine were enclosed inside the concrete cylindrical vacuum structure in the lower left corner.

also the OC-OTEC system shown in Fig. 7 that was last operated in the 1990s after setting a number of world records[8] (power production, length of operation, grid connection and desalination potential); the data in Fig. 6 were generated by this experimental plant.

2 Available OTEC Resources

Proponents of OTEC technologies have routinely used the enormous amount of solar power absorbed by the upper layer of tropical oceans as a starting point to evaluate OTEC resources. Not surprisingly, this leads to the conclusion that when sized against mankind's current total energy use of 1.4×10^{14} kWh per year (16 TW), OTEC resources are virtually unlimited in spite of low thermodynamic efficiencies.

A more sober approach consists in viewing OTEC as a true energy conversion technology. The thermal potentials provided by the temperature stratification of the water column in tropical regions are very much like elevation differences in a hydropower plant. What is not obvious for OTEC is an analog for the river flow rate that also is a fundamental factor in determining hydropower resources. The driving mechanism for the existence of a stable pool of deep cold seawater everywhere is the planetary thermohaline circulation depicted in Fig. 2. It has been suggested that the rate of deep water formation in polar areas is a natural flow rate scale for sustainable OTEC power extraction on a massive scale. Based on a cold seawater flow rate intensity of $3\,\mathrm{m}^3/\mathrm{s}$ per net OTEC megawatt and a thermohaline circulation of the order of $30\,\mathrm{Sv}$ ($1\,\mathrm{Sv} = 1$ million m^3/s), Cousteau and Jacquier[9] argued that OTEC resources were as large as 10 TW.

Because the seawater needs of OTEC systems are great, however, simple one-dimensional analyses of the water column with OTEC showed that it was theoretically possible to reach a point where the available vertical temperature difference itself could be eroded. In such scenarios, overall OTEC power actually reaches a maximum of about 3 to 5 TW when the combined OTEC cold seawater demand is roughly equal to the rate of deep water formation.[10] Such one-dimensional models are likely too conservative because the great geographic extent of the area favorable for OTEC only represents about one-third of the overall oceanic surface. Therefore, horizontal diffusive and advective transport mechanisms that are inherently much stronger in the ocean cannot be properly accounted for. Better models should include three-dimensional ocean dynamics and a detailed heat budget at the ocean–atmosphere interface.

In the absence of widespread OTEC development, the question of an accurate estimation of worldwide OTEC resources may seem rather academic, although there is also a regional aspect to it (local maximum OTEC power production density). Background uncertainties related to global warming and the possibility of changes in the thermohaline circulation would add significant complexity to the problem.[11] Until much better modeling approaches are implemented, it is safe to say that OTEC is a terawatt-size global resource.

3 Advantages and Disadvantages of OTEC

As suggested by Fig. 1 and following the discussion in Sec. 2, OTEC represents a vast potential contributor of renewable power. However, the fact that OTEC development has not proceeded beyond the pre-commercial stage warrants some examination.

Offhand, it is clear that marine technologies are more difficult to implement that their land-based counterparts or competitors. This was clearly demonstrated by the history of offshore oil and gas exploration and production, which only took off in the 1970s. Such a development was spurred by OPEC's decision to sharply increase the price of crude oil. More recently, the commercial success enjoyed by the wind power industry also shows that a move from land-based systems to shallow offshore farms has been both difficult and costly. In fact, OTEC is not only a marine renewable technology but also requires deep waters where even wind farms have yet to be deployed. The remoteness of offshore OTEC plants offers, of course, a few conceptual benefits. There would be no interference with land and coastal communities and ecosystems.

Another unique attribute of OTEC is its geopolitical distribution in tropical regions. Although this point is seldom mentioned, it is believed to represent a strong disadvantage. Essentially located across higher latitudes, the wealthiest and most technologically advanced nations may be reluctant to commit their taxpayers' contributions toward developing scarce domestic OTEC resources.

From an engineering point of view, the OTEC zone includes the area where powerful tropical cyclones occur since OTEC and these storms share a need for warm surface seawater. Hence, the design of offshore OTEC systems would have to follow very strict standards. OTEC resources that are safe from hurricanes and typhoons lie along the Equator, where the Coriolis force becomes negligible, and in the tropical South Atlantic, where high-altitude wind shear prevents the formation of strong cyclones.

A remarkable quality of OTEC among renewable energy technologies is a potential for very high capacity factors. Within the reach of submarine power cables, OTEC would be capable of supplying baseload power. This would be an exceptional advantage for isolated electrical grids. It was mentioned in Sec. 1 that OTEC power output is very sensitive to available seawater temperatures. From the point of view of an electrical utility, the definition of baseload power might therefore hinge on some minimum expected OTEC power output. In this respect, a map of yearly average seawater temperature differences, such as Fig. 1, is not sufficient because it does not show seasonal changes of the thermal resource. Fortunately, a very large fraction of the region deemed favorable for OTEC in an average sense also exhibits low seasonal variability (e.g. within 3°C to 4°C). In Fig. 8, historical monthly seawater

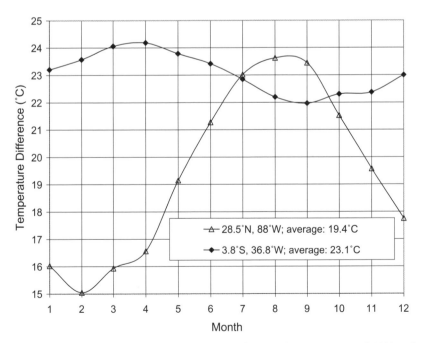

Fig. 8. Examples of monthly seawater temperature differences between 20 m and 1000 m depths. *Source*: 2005 World Ocean Atlas.[12]

temperature differences at either an excellent site, off of Fortaleza (Brazil), or at an unfavorable site, off of Mobile (US Gulf Coast), show the need to assess variability; designing practical or cost-effective OTEC systems in the northern Gulf of Mexico would be very difficult.

Beyond the practical reach of large submarine power cables, the development of OTEC resources would have to rely on the *in situ* manufacture of liquid fuels. Typically, hydrogen could be produced from electrolysis and stored and transported as such or used to manufacture ammonia. Taking into account power losses and process efficiencies, a commercial OTEC plant ship producing 100 MW of electrical power could manufacture 30 metric tons of hydrogen per day (delivered to shore). Given the energy density of hydrogen, this would be equivalent to about 750 barrels of oil per day. Though substantial, this figure may be put in perspective by comparison with the 1999 peak North Sea oil production of 6 million barrels per day.[13]

The most definite consequence of the small temperature differences that characterize OTEC processes is the need for large systems. Heat exchanger surfaces of the order 10,000 m^2 per net MW typically are required either for the evaporator or for the condenser. High seawater flow rate intensities are necessary as well, with cold seawater needs of 2.5 to 3 m^3/s per MW and somewhat larger warm seawater flow rates (as a result of better accessibility). Since the cooling fluid is drawn from depths of about 1,000 m, the size of the cold water pipe (CWP) presents a substantial technological challenge. Resorting to high pumping velocities is not a viable option because parasitic losses would rapidly reach unacceptable levels; for a given flow rate, friction losses, for example, depend inversely on the fifth power of the pipe diameter. Therefore, one has to consider CWP diameters as large as 12 m for a 100 MW OTEC plant; in truth, this is what limits the probable size of a single OTEC system.

The size of OTEC hardware and its offshore location result in high capital costs. From available design studies and best estimates, a steep economy of scale is also expected since several critical components are not modular and there are some substantial fixed costs involved as well. This point is clearly illustrated in Fig. 9 where capital costs per unit kW drop sharply; here, OTEC power production is assumed to take place under some standard conditions (e.g. temperature difference of 22°C). Figure 10 shows the cost breakdown for a typical commercial 100 MW unit; the overall investment burden would be $750 million. The most expensive items consist of the heat exchangers and the floating vessel, although the CWP and its installation present the highest risk and would define a technological frontier even today. Submarine power cable costs depend almost linearly on distance while the influence of power-carrying capacity is weaker. For a commercial OTEC plant or a pilot, installed cable costs of $1700/m and $1300/m, respectively, are typical. Hence, the relative burden imposed by power transmission to shore for a plant located 100 km at sea would rise to nearly 20% of a significantly higher investment.

OTEC by-products have received much attention as well, in attempts to boost projected revenues when estimating cost effectiveness. The best known is desalinated

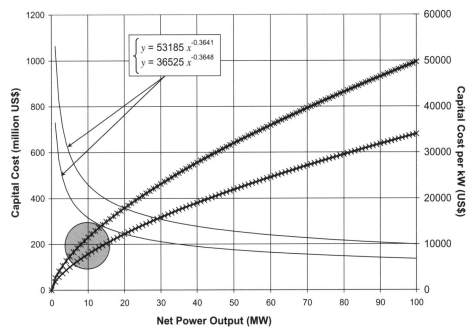

Fig. 9. Capital cost estimates for OTEC plants[14]; the smooth lines define a range for cost per kW; the circle suggests desirable pre-commercial pilot projects.

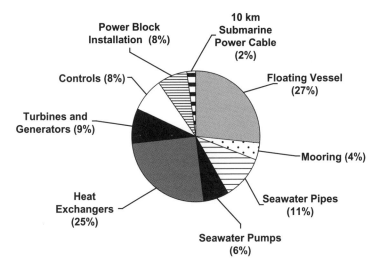

Fig. 10. Estimated capital cost breakdown for a 100 MW commercial OTEC system (overall cost of US$750 million).[14]

water for OC-OTEC. Roughly speaking, a fresh water output of 23 kg/s per net MW could be generated in a typical OC-OTEC system; using the residual temperature difference of the plant's seawater effluents in an additional (second-stage) desalination module could double that figure (note that the residual thermal resource would be too small for additional electricity generation). Therefore, a single-stage 100 MW commercial OC-OTEC plant could produce about 200,000 m^3/day. This is a significant amount, but it must ultimately be evaluated by taking into account the market value of fresh water and the need for storage and transport from offshore OTEC plants.

Other touted by-products hinge on the specific properties of deep seawater alone. Cold, clean, and nutrient-rich deep seawater can be used in air-conditioning applications and aquaculture operations and in the production of high-value items like nutraceuticals. There have been a few successful implementations of coastal seawater air-conditioning (SWAC) systems, and many technology incubators patterned after the Natural Energy Laboratory of Hawaii Authority (NELHA) at Keahole Point, Hawaii, have been built around the world, notably in Japan. Generally speaking, however, there is a disconnect between the coastal nature of most by-product technologies and the offshore character of commercial-scale OTEC power production, as well as a mismatch between the great seawater flow rate intensity of OTEC systems and the seawater needs of by-product technologies.

The only idea that fully relies on massive offshore seawater flow rates is artificial upwelling (AU): deep-seawater nutrients (nitrates, phosphates etc.) brought to the surface for OTEC and released in the plant's effluents could spur phytoplankton growth and boost the marine food web if the discharge plume settled within the photic zone, where sufficient solar radiation is available. AU is conceptually possible but practically untested. For lack of better knowledge, existing permitting frameworks for OTEC projects are likely to adopt a strict cautionary principle that would aim to prevent potential environmental effects from OTEC effluents on the marine environment. From a wider perspective, the overall legal context that would apply to the large-scale development of offshore OTEC resources remains uncertain.

4 Status of OTEC Development

Globally, the historical development of OTEC proceeded in two distinct phases (details can be found elsewhere).[15] From 1926 to 1933, Claude relied on his creativity, tenacity, and wealth to single-handedly test OTEC systems at sea. He experienced many difficulties and setbacks but also achieved OTEC power production while solving daunting technical problems.[6] His efforts, however, eventually bankrupted him in the midst of the Great Depression. Following the so-called oil shocks of the 1970s, and with the benefit of four decades of technological advances in the fields of ocean engineering and material science, OTEC saw a renaissance when governments in the US and Japan launched research and development initiatives that led to the execution of several small but high-profile projects in Hawaii and Nauru. As the price of oil stabilized and political priorities in the US changed after 1980, this effort came to a premature end by the mid 1990s. The responsibility of further OTEC development implicitly was left to the private sector.

The ambitious programs that aimed to establish the technical viability of pre-commercial OTEC systems have not yet been satisfactorily completed. Floating OTEC pilots of 5 to 10 MW that would be operated for a few years still need to be constructed. Otherwise, it is unlikely that a private company would assume the financial risk of investing in commercial OTEC plants while the technology has an insufficient track record. Figure 9 shows, however, that from an economy-of-scale point of view, pilots remain unfavorably situated. A lack of commercial maturity and large investment needs (at about $200 million for a significant pilot) clearly suggest that only a public effort is likely to move OTEC technologies forward. This will require a strong political will. However the geographic location of OTEC resources may represent a serious handicap from the perspective of most wealthy technologically developed nations. Following a sharp rise in the prices of fossil fuels through the first decade of the 21st century, several industrial and governmental stakeholders have recently been trying to realize the objective of deploying OTEC pilot plants, but their efforts have yet to proceed beyond design stages.

References

1. J.-A. d'Arsonval, Utilisation des forces naturelles — Avenir de l'électricité. *Revue Scientifique*, **17** (1881), pp. 370–372 (in French).
2. G. C. Nihous, Mapping available Ocean Thermal Energy Conversion resources around the main Hawaiian Islands with state-of-the-art tools, *J. Renew. Sustain. Energ.* **2**(043104) (2010), p. 1–9. Available at: http://jrse.aip.org/resource/1/jrsebh/v2/i4/p043104_s1?view=fulltext. Accessed 23 December 2011.
3. Available at http://www.globalwarmingart.com/images/b/b0/Thermohaline_circulation.png. Accessed 23 December 2011.
4. A. De Vos, Efficiency of some heat engines at maximum-power conditions, *Am. J. Phys.* **53**(6) (1984), pp. 570–573.
5. G. C. Nihous and L. A. Vega, Performance Test Report: Analysis of representative time history records obtained at the USDOE 210 kW OC-OTEC Experimental Facility in the power production mode, U.S. Department of Energy Report, DE-AC36-92CH10539, December 1996, p. 81.
6. G. Claude, Power from the tropical seas, *Mech. Eng.* **52**(12) (1930), pp. 1039–1044.
7. L. R. Berger and J. A. Berger, Countermeasures to microbiofouling in simulated ocean thermal energy conversion heat exchangers with surface and deep ocean waters in Hawaii, *Appl. Environ. Microbiol.* **51**(6) (1986), pp. 1186–1198.
8. L. A. Vega, The 210 kW Open-Cycle OTEC experimental apparatus: Status report, *Proc. Oceans '95 Conference*, San Diego, 1995, p. 6.
9. J. Y. Cousteau and H. Jacquier, *Énergie des mers: plan-plan les watts*. Chapter 9 in *Français, on a volé ta mer* (R. Laffont, Paris, 1981) (in French).
10. G. C. Nihous, A preliminary assessment of Ocean Thermal Energy Conversion (OTEC) resources, *J. Energ. Resour. Technol.* **129**(1) (2007), pp. 10–17.
11. G. C. Nihous. An estimate of Atlantic Ocean Thermal Energy Conversion (OTEC) resources. *J. Ocean Eng.* **34** (2007), p. 2210–2221.
12. R. A. Locarnini, A. V. Mishonov, J. I. Antonov, T. P. Boyer and H. E. Garcia, *World Ocean Atlas 2005*, Vol. 1: Temperature. NOAA Atlas NESDIS 61, ed. S. Levitus (U.S. Government Printing Office, Washington, 2006), p. 182.

13. Wikipedia. Available at: http://en.wikipedia.org/wiki/North_Sea_oil. Accessed 23 December 2011.
14. L. A. Vega, The economics of Ocean Thermal Energy Conversion, *4th Energy Ocean Conference*, Turtle Bay Resort, Oahu, Hawaii, 2007.
15. H. A. Avery and C. Wu, *Renewable Energy from the Ocean — A Guide to OTEC* (Oxford University Press, New York, 1994), 446 pp.

Chapter 17

Capacitive Electric Storage

Lu Wei*,† and Gleb Yushin*,‡

*School of Materials Science and Engineering
Georgia Institute of Technology
Room 288, 771 Ferst Drive NW, Atlanta
GA 30332-0245, USA

†School of Materials Science and Engineering
Northwestern Polytechnical University
Xi'an, Shaanxi 710072, PR China
‡yushin@gatech.edu

Capacitors are devices for storing electric charge. Several types of capacitors have been developed. Conventional dielectric and electrolytic capacitors, storing charge on low-surface-area plates, deliver limited capacitance but can be operated at high voltages. As an emerging technology, electrochemical capacitors (also called supercapacitors), storing charge in an electric double layer or at surface reduction–oxidation (Faradaic) sites, currently fill the gap between batteries and conventional capacitors. They store hundreds or thousands of times more charge than conventional capacitors because of a much higher surface area and nanoscale charge separation. However, they have a lower energy density than batteries and commonly lower power than traditional capacitors. Impressive enhancements in their performance have been demonstrated in the past decade due to the discovery of new electrode materials and improved understanding of ion behavior in small pores, as well as the design of new hybrid systems combing Faradaic and capacitive electrodes. This chapter presents the classification, construction, modeling, advantages, and limitations of capacitors as electrical energy storage devices. The materials for various types of capacitors and their current and future applications are also discussed.

1 Introduction

Capacitors are passive devices for storing electric charge and therefore energy. In their most basic form, capacitors consist of two conductors separated by a dielectric. An individual capacitor (previously called "a condenser") can be characterized by a single quantity, the capacitance, with the unit Farad, which is the ratio of the charge on one of the conductors divided by the voltage across the dielectric.

Capacitors can be generally divided into dielectric, electrolytic, and electrochemical capacitors (Table 1). The electrochemical capacitors are, in turn, subdivided into electrochemical double-layer capacitors (EDLCs), pseudocapacitors, and hybrid capacitors. As conventional capacitors, the dielectric capacitors are

373

Table 1:   Classifications of Capacitors.

Classification	Basis of charge or energy storage
(i) Dielectric capacitors	Electrostatic
(ii) Electrolytic capacitors	Electrostatic
(iii) Electrochemical capacitors	
(a) Electrochemical double-layer capacitors	Electrostatic
(b) Pseudocapacitors	Faradaic charge transfer
(c) Hybrid capacitors	Faradaic charge transfer

fundamental electrical circuit elements that store electrical energy (on the order of micro-Farads, μF) and assist in filtering. Electrolytic capacitors are the next-generation capacitors which can be commercialized. They are similar to batteries in cell construction but the anode and cathode materials remain the same. The third-generation evolution is the EDLC, where the electrical charge is stored at a metal/electrolyte interface, and the main component in the electrode construction is activated carbon. Although this concept was initialized and industrialized some 40 years ago, there was stagnancy in research until recently. The need for the revival of interest arises from the increasing demands for electrical energy storage in certain current applications, such as digital electronic devices, implantable medical devices, industrial lifts and cranes, stop/start operation in vehicle traction, as well as the electrical grid, all of which need long cycle life and very short high-power pulses that could be fulfilled by EDLCs. Although EDLCs can deliver high power density and have very long cycle life, they suffer from low energy density. To rectify this problem, recently researchers are trying to develop novel carbon electrodes with enhanced energy storage capacity and incorporate transition metal oxides and conductive polymers along with carbon into the electrode materials. When the electrode materials consist of transition metal oxides and many conductive polymers, the reduction–oxidation (redox) processes may enhance the value of specific capacitance by up to 100 times, depending on the nature of the electrode and electrolyte system. In such a situation, the electrochemical capacitor is called a pseudocapacitor, which is the fourth-generation capacitor.[1−3] The latest important alternative approach to produce high energy density is to develop hybrid (asymmetric) capacitors. This approach can overcome the energy density limitation of the EDLCs because it employs a hybrid system of a battery-like (Faradaic) electrode and a capacitor-like (commonly non-Faradaic) electrode, producing higher working voltage and capacitance.[4]

The Ragone plot in Fig. 1 shows the comparison of the different storage technologies. There are a number of desirable qualities that make capacitors a valuable option as energy storage devices. For example, they have reversible storing and releasing charge capabilities that allow them to withstand a large number of charge/discharge cycles and also charge/discharge more quickly than batteries. The advantages and limitations of capacitors[5] as electrical energy storage devices are listed in Table 2.

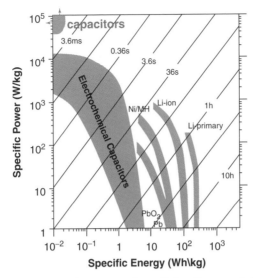

Fig. 1. Specific power versus specific energy, also called a Ragone plot, for various electrical energy storage devices.[6] If an electrochemical capacitor is used in an electric vehicle, the specific power shows how fast one can go, and the specific energy shows how far one can go on a single charge.

Table 2: Advantages and Limitations of Capacitors as Electrical Energy Storage Devices.

Advantages

Long cycle life, $> 100,000$ cycles, some systems up to 10^6;

Quick charge/discharge rate;

Selected devices can operate at ultra-low ($< -50°$C) and ultra-high ($> 100°$C) temperatures;

Good power density (under certain conditions, limited by IR, or equivalent series resistance (ESR) complexity of equivalent circuit);

Simple principle and mode of construction;

High working voltage (for electrolytic capacitors);

Cheap materials (for aqueous embodiments);

Combines state-of-charge indication, $Q = f(V)$;

Can be combined with rechargeable battery for hybrid applications (electric vehicles).

Limitations

Limited energy density;

Poor volume energy density;

Non-aqueous embodiments require pure, H_2O-free materials, more expensive;

Requires stacking for high potential operation (electric vehicles);

Hence, good matching of cell units is necessary.

2 Dielectric Capacitors

Dielectric capacitors are typically made of two conductors (parallel plates) separated by an insulating material (dielectric), as shown in Fig. 2. When there is a potential difference (voltage) across the conductors, a static electric field develops across the dielectric, causing positive charge to collect on one plate and negative charge on the

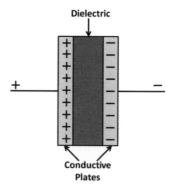

Fig. 2.   Schematic of a dielectric capacitor.

other plate. Energy is stored in the electrostatic field. The basic equation for the capacitance of such a device is:

$$C = \varepsilon S/d, \tag{1}$$

where C is the capacitance, ε is the dielectric constant of the dielectric, S is the surface area of the electrode, and d is the thickness of the dielectric.[7]

Dielectric capacitors yield capacitance in the range of 0.1 to 1 μF, with a voltage range of 50 to 400 V. Various materials such as paper, paraffin, polyethylene, insulated mineral oil, polystyrene, ebonite, polyethylene tetraphtharate, sulfur, mica, mylar, steatite porcelain, Al porcelain, plastics (polymers), and glass are used as dielectrics.[1] Dielectric capacitors are mainly used in:

(a) electronic circuits for blocking direct current while allowing alternating current to pass;
(b) filter networks for smoothing the output of power supplies; and
(c) resonant circuits that tune radios to particular frequencies.

3 Electrolytic Capacitors

The next generation of capacitors is an electrolytic capacitor, which uses an electrolyte as an ion conductor between the dielectrics and a metal electrode. There are three types of electrolytic capacitors: aluminum (Al), tantalum (Ta), and ceramic. The capacitances of these electrolytic capacitors are in the range of 0.1 to 10 μF, with a voltage profile of 25 to 50 V.[8]

A typical aluminum electrolytic capacitor is constructed from two conducting aluminum foils, one of which is coated with an insulating oxide layer, and a paper spacer soaked in electrolyte. The foil insulated by the oxide layer is the anode while the liquid electrolyte and the second foil acts as the cathode (Fig. 3). The layer of insulating aluminum oxide on the surface of the aluminum foil acts as the dielectric, and it is the very low thickness of this layer that allows a relatively high capacitance to be achieved in a small volume. Aluminum oxide has a dielectric constant

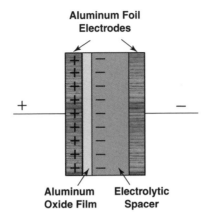

Fig. 3.   Schematic of a typical aluminum electrolytic capacitor.

of 10, which is several times higher than most common polymer insulators. It can withstand an electric field strength of the order of $25\,\text{MV}\,\text{m}^{-1}$, which is an acceptable fraction of that of common polymers. This combination of high capacitance and reasonably high voltage result in high energy density. Electrolytic aluminum capacitors are mainly used as power supplies for automobiles, aircraft, space vehicles, computers, monitors, motherboards of personal computers, and other electronics. However, aluminum electrolytic capacitors have several drawbacks which limit their use, including relatively high leakage current and inductance, poor tolerances and temperature range, and short lifetimes.[1,9,10]

Tantalum electrolytic capacitors have a lower energy density and are produced to tighter tolerances than aluminum electrolytic capacitors. The anode electrode is formed of sintered tantalum grains, with the dielectric electrochemically formed as a thin layer of oxide. The thin layer of oxide and high surface area of the porous sintered material gives this capacitor type a very high capacitance per unit volume. The cathode electrode is formed either from a liquid electrolyte connecting the outer can or from a chemically deposited semi-conductive layer of manganese dioxide, which is then connected to an external wire lead. A further development of this type of capacitors replaces the manganese dioxide with a conductive plastic polymer (polypyrrole), which reduces the internal resistance and eliminates a self-ignition failure. Compared to aluminum electrolytic capacitors, tantalum electrolytic capacitors have very stable capacitance, little DC leakage, very low impedance at high frequencies, and smaller size for a given capacitance. However, unlike aluminum electrolytic capacitors, they are intolerant of positive or negative voltage spikes and can be destroyed (often exploding violently) if connected in the circuit backward or exposed to spikes above their voltage rating. Moreover, tantalum capacitors are more expensive than the aluminum ones and generally are only available in low-voltage versions. In order to improve the characteristics of electrolytic capacitors, oxides of niobium and titanium with high dielectric constants and a lower cost than tantalum oxide have been investigated. The application of oxides of niobium and titanium can produce a considerable decrease in the size of capacitors, an increase in the range of working temperatures, an improvement in electrical characteristics, and an increase in the reliability and lifetime of capacitors during storage and operation.[2,11]

A ceramic capacitor is a capacitor constructed of alternating layers of metal and ceramic, with the ceramic material acting as the dielectric. Multilayer ceramic capacitors (MLCs) typically consist of ~100 alternate layers of electrode and dielectric ceramics sandwiched between two ceramic cover layers. They are fabricated by screen-printing electrode layers on dielectric layers and co-sintering the laminate. The dielectric ceramic materials that have been identified and used include $BaTiO_3$, $CaZrO_3$, $MgTiO_3$, $SrTiO_3$, etc., and Ag, Pd, Mn, Ca, etc., are the internal electrodes used. Ceramic capacitors are in widespread use in electronic equipment, providing high capacity and small size at low price compared to the other types.[12–14]

4 Electrochemical Capacitors

Electrochemical capacitors (ECs), also called supercapacitors or ultracapacitors, are power devices that can be fully charged or discharged in seconds. While their energy density (about $5\,Wh\;kg^{-1}$) is lower than that of batteries, they offer a much higher power delivery or uptake ($10\,kW\;kg^{-1}$) and can be charged or discharged in a short time (commonly 1–30 s).[6] They play an important role in complementing or replacing batteries in the energy storage field, such as for uninterruptible power supplies and load-leveling applications.

Several types of ECs can be distinguished depending on the charge storage mechanism (Table 1) as well as the active materials used (Table 3). EDLCs, the most common devices at present, use carbon-based active materials with high surface area.[15–24]

A second group of ECs, known as pseudocapacitors or redox supercapacitors, uses fast and reversible surface or near-surface reactions for charge storage. Transition metal oxides as well as electrically conducting polymers are examples of pseudocapacitive active materials.[25,26]

Hybrid capacitors, combining a capacitive or pseudo-capacitive electrode with battery electrode, are the latest kind of EC, which benefit from both the capacitor and the battery properties.[27,28]

4.1 *EDLCs*

EDLCs store the charge electrostatically using reversible adsorption of ions of the electrolyte onto active materials that are electrochemically stable and have high accessible specific surface area (SSA).[7] Figure 4(a) presents the schematic of an EDLC. Charge separation occurs through polarization at the electrode–electrolyte interface. A typical cyclic voltammogram of a two-electrode EDLC laboratory cell

Table 3:   Electrode Materials for ECs.

Types of ECs	Electrode material
EDLCs	Activated, templated, and carbide-derived carbons; Carbon fabrics, fibers, nanotubes, aerogels, onions, and nanohorns; Graphene and graphene-based materials.
Pseudocapacitors	Surface compounds, such as hydrogen adsorption on platinum or lead adsorption on gold; Metal oxides, such as RuO_2, Fe_3O_4, MnO_2, NiO_x, $Ni(OH)_2$, Co_3O_4, SnO_2, In_2O_3, Bi_2O_3, TiO_2, V_2O_5, $BiFeO_3$, $NiFe_2O_4$, and their mixtures; Conductive polymers, such as polyacetylene, polyparaphenylene, polyaniline, polypyrrole, polythiophene, and their derivatives; Nanostructured redox-active materials (metal oxides and electrochemically active polymers) deposited on high-surface-area porous carbons; Functionalized graphene, carbon nanotubes, and porous carbons.
Hybrid capacitors	Pseudo-capacitive metal oxides (such as NiO_x, $Ni(OH)_2$, MnO_2, VO_2, V_2O_5, RuO_2, and PbO_2) or conductive polymers as one electrode combined with a capacitive porous carbon as another electrode; Lithium-insertion materials (such as $Li_4Ti_5O_{12}$, $Li_4Mn_5O_{12}$, Li_2FeSiO_4, and pre-lithiated carbon) as one electrode combined with a capacitive porous carbon as another electrode.

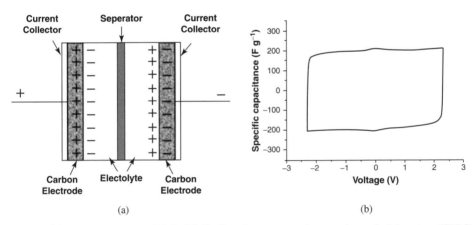

Fig. 4.   (a) Schematic of an EDLC; (b) Cyclic voltammetry of a two-electrode laboratory EDLC cell in 1M tetraethylammonium tetrafluoroborate ($TEABF_4$) solution in acetonitrile (AN), containing activated carbon electrodes on aluminum current collectors. Cyclic voltammetry was recorded at room temperature and potential scan rate of $10\,mV\,s^{-1}$.

is presented in Fig. 4(b). Its rectangular shape is characteristic of a pure double layer capacitance mechanism for charge storage according to:

$$I = C\,dV/dt, \qquad (2)$$

where I is the current, (dV/dt) is the potential scan rate, and C is the double layer capacitance.

The double-layer capacitance is between 5 and $20\,\mu F\ cm^{-2}$, depending on the electrolyte used. Specific capacitance achieved with aqueous alkaline or acid solutions is generally higher than for organic electrolytes or ionic liquids. But organic electrolytes are more widely used as they can sustain a higher operating voltage (up to 2.7 V).[6] Because the energy of an EDLC is proportional to the square of the highest operational voltage according to

$$E = \frac{1}{2} \, CV^2 \qquad\qquad (3)$$

replacing the aqueous electrolytes with organic electrolytes allows an increase in the operating voltage (V) by up to a factor of three to four, which results in an order of magnitude increase in the energy stored at the same capacitance. However, organic solvents do not meet the highly desired requirements of environmental compatibility and safety because they exhibit high vapor pressure and suffer from inherent flammability and potential explosion risks.

Recently, ionic liquids have been explored as electrolytes in certain advanced EDLCs with improved energy and power densities, operational safety, and lifetime. This is because ionic liquids have a large electrochemical window, wide liquid phase range, non-volatility, non-flammability, non-toxicity, and environmental compatibility with respect to conventional aqueous and organic electrolytes.[29,30]

The key to reaching high capacitance by charging the double-layer capacitors is in using high SSA and electronically conducting electrode materials with an optimized porosity. Porous carbons satisfy all the requirements for this application, including high conductivity, electrochemical stability, and open porosity. Activated carbons (Table 4) are the most widely used materials in commercial EDLCs today, because of their high SSA and moderate cost. Activated carbons are derived from carbon-rich organic precursors by carbonization (heat treatment) in an inert atmosphere with subsequent selective oxidation in CO_2, water vapor, or KOH to increase the SSA and pore volume. Natural materials, such as coal, tar, pitch, nut shells, wood, corn grain, sucrose, banana fibers, and other agricultural and forest residues, or synthetic materials, such as polymers, can be used as precursors. A porous network in the bulk of the carbon particles is produced after activation, and micropores (pores <2 nm in size), mesopores (2–50 nm), and macropores (>50 nm) can be created in carbon grains. Accordingly, the porous structure of carbon is characterized by a broad distribution of pore size. Longer activation time or higher temperature leads to larger mean pore size. The double layer capacitance of activated carbon reaches 70–250 F g^{-1} in organic electrolytes, and 80–300 F g^{-1} in ionic liquids (at high temperatures). This value can even exceed 150–300 F g^{-1} in aqueous electrolytes, but at a lower cell voltage because the electrolyte voltage window is limited by water decomposition.

As previously mentioned (Table 3), many carbons have been tested for EDLC applications, and some of these carbons are shown in Figs. 5(a)–(f). The SSA and capacitances of some of the carbons are shown in Table 5. Templated and carbide-derived carbons offer good pore size control and exhibit attractive properties, but their production is very limited. Activated carbon fabrics can reach the same capacitance as activated carbon powders, as they have similar SSA, but the high

Table 4: Activated Carbon Materials for Electrodes of EDLCs.

Carbon source	Activation method	S_{BET} (m² g⁻¹)	Capacitance (F g⁻¹)	Electrolyte			Ref.
				Aqueous	Organic	Ionic liquid	
Banana fibers	ZnCl$_2$	1097	74	1 M NaSO$_4$			36
Sucrose	CO$_2$	2102	163	1 M H$_2$SO$_4$			37
Sucrose	CO$_2$	1941	148			EMImBF$_4$	29
Starch	KOH	1510	194	30 wt.% KOH			38
Wood	KOH	2967	236		1 M TEABF$_4$/AN		39
Corn grain	KOH	3199	257	6 M KOH			40
Pitch	KOH	2660	299	1 M H$_2$SO$_4$			41
Pitch	KOH	2171	140		1 M Et$_4$NBF$_4$/PC		42
Sugar cane bagasse	ZnCl$_2$	1788	300	1 M H$_2$SO$_4$			43
Coal	KOH	3150	317	1 M H$_2$SO$_4$			41
Coal	KOH	>2000	220		1 M LiClO$_4$/PC		44
Apricot shell	NaOH	2335	339	6 M KOH			45
Coffee grounds	ZnCl$_2$	1019	368	1 M H$_2$SO$_4$			46
Wheat straw	KOH	2316	251		1.2 M MeEt$_3$NBF$_4$/AN		47
Polyfurfuryl alcohol	KOH	1140	160			PYR14TFSI	48
Polypyrrole	KOH	3432	300			EMImBF$_4$	49
Polyaniline	KOH	1976	455	6 M KOH			50
Polyaniline	K$_2$CO$_3$	917	210	6 M KOH			51

AN: acetonitrile; PC: propylene carbonate; TEABF$_4$/AN: tetraethylammonium tetrafluoroborate salt solution in acetonitrile; Et$_4$NBF$_4$/PC: tetraethylammonium tetrafluoroborate in propylene carbonate; PYR14TFSI: *N*-methyl-*N*-butyl-pyrrolidinium bis(trifluoromethanesulfonyl)imide; EMImBF$_4$: 1-ethyl-3-methylimidazolium tetrafluoroborate.

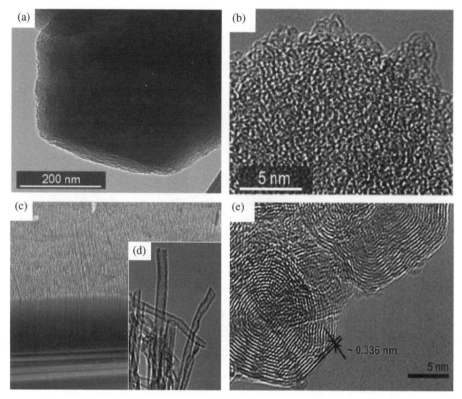

Fig. 5.   The structures of carbons used as EDLC electrode materials. (a) Transmission electron microscopy (TEM) image of zeolite Y-templated carbon particles, from Ref. 52; (b) TEM image of SiC-derived carbon particles, from Ref. 62; (c) Scanning electron microscopy image of a plasma-etched aligned carbon nanotube electrode, from Ref. 78; (d) TEM image of carbon nanotubes, after plasma etching, from Ref. 78; (e) TEM image of carbon onions, from Ref. 17.

Table 5:   Comparison of the EDLCs Based on Different Carbon Materials.

Carbon electrode material	S_{BET} ($m^2 g^{-1}$)	Capacitance ($F g^{-1}$)	Ref.
Activated carbon	1,000–3,500	70–450	36–51
Templated carbon	1,000–2,000	130–300	52–57
Carbide-derived carbon	600–3,000	100–200	16, 58–67
Carbon fabric	1,340	<100	68–70
Carbon fiber	1,000–3,000	100–200	71–73
Carbon nanotube	400–1,200	150–400	74–79
Carbon aerogel	600–1,600	90–190	80–82
Carbon anion	400–550	40–100	17, 83, 84
Graphene-based material	300–3,000	100–300	31–35, 85–89

price limits their use to specialty applications. Carbon nanotubes and macroporous carbon aerogels with large external surface area commonly offer high rate capability but low capacitance per unit mass and, more importantly, volume. Vertically aligned carbon nanotubes can deliver high specific capacitance and power density after plasma-etching treatment and thus induced functional groups on their surface, but the preparation requires a complicated synthesis procedure and high stability of such materials has not been reached so far. Graphene, a two-dimensional nanosheet of graphite, has recently received rapidly growing attention in supercapacitors as it possesses superior electrical conductivity, a high theoretical surface area of over $2600 \, m^2 g^{-1}$, and chemical tolerance. However, due to the unavoidable aggregation of graphene nanosheets, the surface area of graphene is usually much lower than the theoretical one and its capacitance is generally in the range of $100–200 \, F \, g^{-1}$. More importantly, due to very high pore volume and thus low density, graphene electrodes commonly offer volumetric capacitance $<20 \, F \, cm^{-3}$.

Although the SSAs of graphene-based materials (GBMs) are usually much lower than that of a single-layer graphene because of the π-stacking of graphene sheets, many strategies have been applied to increase the porosities or SSAs of GBMs, and the capacitance of EDLCs with GBM electrodes can reach $300 \, F \, g^{-1}$.[31-35]

4.2 *Pseudocapacitors*

Pseudocapacitors store electric charge at the expense of both Faradaic pseudo-capacitance of sufficiently reversible and fast redox surface reactions and the EDL capacitance. The principal difference between a pseudo-capacitor and a double-layer capacitor is that in the former there is a net ion exchange between the electrode and the electrolyte during the charge and discharge processes. It should be noted that the EDL capacitance in such systems always co-exists along with a pseudo-capacitance. Nevertheless, the numbers of ions that take part in the EDL generation process are usually smaller than the number of ions (such as protons) being exchanged between the two electrodes. Hence, the pseudo-capacitance in the same system is much greater than the EDL capacitance. Several types of Faradaic processes occur in the electrodes of pseudocapacitors, for example, adsorption of hydrogen or lead on the surface of platinum or gold, redox reactions of transition metal oxides, and reversible processes of electrochemical doping–dedoping in electrodes based on conductive polymers. The specific pseudo-capacitance exceeds that of carbon materials using double-layer charge storage, justifying interest in these systems. But because of the redox reactions involved in the charge storage, pseudocapacitors, like batteries, often suffer from a lack of very long-term stability during cycling and rarely withstand $>100,000$ deep charge-discharge cycles.

4.2.1 *Pseudocapacitors with surface compounds*

In pseudocapacitors, reversible Faradaic surface redox reactions can occur with chemical compounds that undergo electrosorption or desorption with a charge transfer. Electrochemical processes of hydrogen adsorption on platinum and lead adsorption on gold are well studied.[90-92] In the case of the lead adsorption on gold, the following reaction takes place[2]:

$$Au + xPb^{2+} + 2xe \leftrightarrow AuxPb, \tag{4}$$

where, x represents the stoichiometry of chemisorbed lead atoms, which corresponds to a partial coverage of the surface by lead. The capacitance for reaction (4) is defined as follows:

$$C_F = q_{Pb} d\theta_{Pb}/dE, \tag{5}$$

here $q_{Pb} = 400\,\mu C\ cm^{-2}$ is the charge required for a complete monolayer coverage of the gold surface by lead. The capacitance C_F as defined by (5) corresponds to a Faradaic reaction and is called a pseudo-capacitance. Hence, similarly to the EDL capacitance, C_F is not an electrostatic capacitance.

The existence of functional groups on the surface of high SSA carbon materials also often leads to pseudocapacitance,[93–98] where the chemical moieties are involved in the redox processes with the electrolyte. This pseudocapacitance effect is most notably seen in aqueous electrolytes but is believed to occur in organic and IL electrolytes as well.

4.2.2 *Pseudocapacitors with metal oxides*

A literature survey of supercapacitors shows that a variety of metal oxides have been employed. These include of RuO_2, MnO_2, NiO, In_2O_3, Co_3O_4, V_2O_5, Fe_3O_4, Bi_2O_3, IrO_2, $NiFe_2O_4$, and $BiFeO_3$, to name a few. Some of the metal-oxide–based supercapacitors and their main electrochemical properties are shown in Table 6.

Ruthenium oxide (RuO_2) has been widely studied as one of the transition metal oxides for electrodes of pseudo-capacitors because it is electrically conductive and has three distinct oxidation states accessible within 1.2 V.[6] The pseudo-capacitive behavior of RuO_2 in acidic solutions can be described as a fast, reversible electron transfer together with an electro-adsorption of protons on the surface of RuO_2 particles, according to Eq. (6):

$$RuO_2 + xH^+ + xe^- \leftrightarrow RuO_{2-x}(OH)_x, \tag{6}$$

where $0 \leq x \leq 2$. The continuous change of x during proton insertion or de-insertion occurs over a window of about 1.2 V and lead to a capacitive behavior with ion adsorption following a Frumkin-type isotherm. Specific capacitance of more than $600\,F\,g^{-1}$ has been reported.[99,100] However, Ru-based electrochemical capacitors are expensive, which restricts their broad practical use. Less expensive oxides of iron, vanadium, nickel, and cobalt have been tested in aqueous electrolytes, but none has been investigated as much as manganese oxide.[128] The charge storage mechanism is based on surface adsorption of electrolyte cations C^+ (K^+, Na^+...) as well as proton incorporation according to the reaction:

$$MnO_2 + xC^+ + yH^+ + (x+y)e^- \leftrightarrow MnOOC_xH_y. \tag{7}$$

MnO_2 films show a specific capacitance of about $700\,F\,g^{-1}$ in neutral aqueous electrolytes within a voltage window of $<1\,V$.[106]

Table 6: Metal Oxide–Based Supercapacitors and their Main Electrochemical Properties.

Electrode material	Electrolyte	Voltage (V)	C_S (F g^{-1})	Cycles	Ref.
RuO$_2$film	0.5 M H$_2$SO$_4$	0 to 0.6	788	—	99
RuO$_2$ thin film	0.5 M H$_2$SO$_4$	−0.1 to 0.6	1,190	1,000	100
RuO$_2$ on CNTs	0.5 M H$_2$SO$_4$	0 to 1	1,170	—	101
Ru-Ir oxides	0.5 M H$_2$SO$_4$	0 to 1	367.6	—	102
Ru-Sn oxides	1 M H$_2$SO$_4$	0 to 1	930	1,000	103
RuO$_2 \cdot x$H$_2$O	H$_2$SO$_4$	0 to 1	768	60,000	104
MnO$_2$ thin film	0.1 M Na$_2$SO$_4$	0 to 0.9	1370	—	105
MnO$_2$ film	0.1 M Na$_2$SO$_4$	0 to 0.9	698	1,500	106
Tetrapropylammonium/ Mn oxide	0.1 M LiClO$_4$	0 to 0.9	720	1,500	107
Mn-Ni-Cu oxides	6 M KOH	−0.4 to 0.4	490	500	108
Mesoporous Co$_3$O$_4$	2 M KOH	−0.2 to 0.5	202.5	1,000	109
Mesoporous α-MoO$_3$ film	1 M LiClO$_4$/PC	1.5 to 3.5	605	—	110
NiO film	1 M KOH	−0.1 to 0.5	200–278	—	111–114
Ni-Co oxides on CNTs	1 M KOH	−0.1 to 0.5	840	1,000	115
SnO$_2$ film	0.1 M Na$_2$SO$_4$	0 to 1	285	1,000	116
SnO$_x$ on graphite	0.5 M KCl	0 to 1	298	1,000	117
In$_2$O$_3$ nanorods	1 M Na$_2$SO$_3$	−0.6 to 0.2	190	1,000	118
Bi$_2$O$_3$ thin film	1 M NaOH	−1 to 0	98	1,000	119
a-V$_2$O$_5 \cdot n$H$_2$O	2 M KCl	0 to 0.8	350	100	120
V$_2$O$_5 \cdot x$H$_2$O/CNT	1 M LiClO$_4$/PC	0.6 to 1.2	910	—	121
Fe$_3$O$_4$ thin film	1 M Na$_2$SO$_3$	−0.8 to 0.1	170	—	122
BiFeO$_3$	1 M NaOH	−1 to 0.2	81	1,000	123
NiFe$_2$O$_4$	Na$_2$SO$_3$	—	354	1,000	26
Ti-V-W-O/Ti	0.5 M H$_2$SO$_4$	0 to 1.4	125	—	124
Ti/(RhO$_x$+ Co$_3$O$_4$)	0.5 M H$_2$SO$_4$	0.4 to 1.4	500–800	—	125
Co$_x$Ni$_{1-x}$ layered double hydroxides	1 M KOH	−1.0 to 0.6	2,104	—	126
MnO$_2$/Co$_3$O$_4$ hybrid nanowire	1 M LiOH	−0.2 to 0.6	480	5,000	127

In order to improve the capacitance, nanostructured transition metal oxides and nitrides have also been studied as well.[127,129−132] Because pseudocapacitors primarily store charge in the first few nanometers from the surface, decreasing the particle size increases the active material usage. Thin MnO$_2$ deposits of tens to hundreds of nanometers were produced on various substrates such as metal collectors, carbon nanotubes or activated carbons. Specific capacitances as high as 1,300 F g^{-1} have been reported, as reaction kinetics were no longer limited by the electrical conductivity of MnO$_2$.[105] The cycling stability and the specific capacitance of RuO$_2$ nanoparticles were increased by depositing a thin conducting polymer coating that enhances proton exchange at the surface. Moreover, the charging mechanism of nanocrystalline vanadium nitride (VN) includes a combination of an electric double layer and a Faradaic reaction (II/IV) at the surface of the nanoparticles, leading to specific capacitance up to 1,200 F g^{-1} at a scan rate of 2 mV s^{-1}.[133]

Apart from the scientific quest for high capacitance of metal oxides, the long-term cycling stability of these electrodes is a technological issue that must be addressed to evaluate the commercial development of metal-oxide–based aqueous supercapacitors. Other issues, such as self-discharge, corrosion of the current collector, and low temperature performance, should also be examined. Considering that metal-oxide–based supercapacitor technology is still in its infancy, future research and development should ultimately yield high performance, low cost, and safe energy-storage devices.[26]

4.2.3 *Pseudocapacitors with conducting polymers*

One of the major electrochemical achievements of the last few years was the development of conducting polymers. They are rendered conductive through a conjugated bond system along the polymer backbone and are typically formed either through chemical oxidation of the monomer or electrochemical oxidation of the monomer. Two oxidation reactions occur simultaneously — the oxidation of the monomer and the oxidation of the polymer with the coincident insertion of a dopant/counter ion.[25] Use of some conducting polymers as electrodes for pseudocapacitors is based on the sufficient reversibility of redox reactions of their electrochemical doping–dedoping.[2,134] The process of reversible electrochemical doping–dedoping may be represented by the following reactions:

$$P_m - xe + xA^- \leftrightarrow P_m^{x+}A_x^- k, \tag{8}$$

$$P_m + ye + xM^+ \leftrightarrow P_m^{y-}M_y^+, \tag{9}$$

where, P_m is a polymer with a network of conjugated double bonds, m is the polymerization degree, A^- denotes anions, and M^+ represents cations. The doping reactions proceed from the left to the right, and the dedoping reactions, in the reverse direction. Reaction (8) is a reaction of oxidative *p*-doping, and reaction (9) is a reaction of reductive *n*-doping. While most conducting polymers can only be *p*-doped, some of them (polyacetylene, polythiophene, and their derivatives) may be reversibly *p*-doped and *n*-doped at the same time. Many kinds of conducting polymers have been tested in EC applications as pseudo-capacitive materials, the most commonly studied are polypyrrole, polyaniline, and derivatives of polythiophene, their main electrochemical properties are shown in Table 7.

Conducting polymers offer many advantages as supercapacitor electrodes. They are flexible, highly conductive, and easily processable and can be made into thin and thick porous films. Many conducting polymers exhibit high specific capacities and total capacitance, while being able to deliver energy at a relatively rapid rate. The major disadvantage of the conducting polymers used as supercapacitor electrode is poor cycle life. In general, symmetric supercapacitors based on conducting polymers will have a lower cycle life than those based on carbon. This is unavoidable, because as anions or cations are doped or dedoped into the conducting polymer, there is a corresponding volume change of the electrode when compared with carbon supercapacitors, which involve only simple ion

Table 7: Conducting Polymer Based Supercapacitors and their Main Electrochemical Properties.

Electrode material	Electrolyte	Voltage (V)	C_S ($F\,g^{-1}$)	Cycles	Ref.
Polyaniline	1 M Et_4NBF_4/AN	−2–3	107	9000	135
Polyaniline	1 M Et_4NBF_4/AN	−0.5–0.5	150	1000	136
Polypyrrole	0.5 M Na_2SO_4	−0.5–0.4	254	—	137
Polypyrrole	3 M KCl	−0.4–0.4	400	100,000	138
Polypyrrole	1 M KCl	−0.8–0.8	480	1000	139
Polypyrrole/Nafion	1 M Na_2SO_4	−0.8–0.5	380	5000	140
Polythiophene derivatives	1 M Et_4NBF_4/AN	−1.3–0.6	160	2000	141
Polythiophene-tartaric acid	PVdF-HFP in 1 M $LiPF_6$	−1–1	156	1000	142
PEDOT	0.1 M $LiClO_4$	−0.2–0.8	124	—	143
PEDOT/Polypyrrole	1 M KCl	−0.4–0.6	290	1000	144
Irradiated HCl doped polyaniline	Gel polymer electrolyte	−1.2–1.2	243	10,000	145
Non-irradiated HCl doped polyaniline	Gel polymer electrolyte	−1.2–1.2	259	10,000	145

Note: PVdF-HFP in 1 M $LiPF_6$: PVdF-co-HFP in 1 M $LiPF_6$ containing EC&PC (1:1 v/v)-based microporous polymer electrolyte; PEDOT: poly (3,4-ethylenedioxythiophene).

sorption and desorption.[25,146,147] In order to improve cycle life, research efforts with conducting polymers for supercapacitor applications are currently directed toward polymer–ceramic[148–151] and polymer–carbon[152–163] composite materials as well as hybrid systems. Selected composite materials demonstrated specific capacitance in excess of $600\,F\cdot g^{-1}$ in symmetric two-electrode configuration and no reduction in the capacitance after 10,000 charge-discharge cycles.[153]

4.3 *Hybrid Capacitors*

As the next generation of electrochemical capacitors, hybrid capacitor systems are a promising approach to meet the goal of effectively increasing the energy density. This approach combines a battery-like electrode (Faradaic) with a capacitor-like electrode (commonly non-Faradaic) in the same cell, producing high working voltage and capacitance. Compared with conventional EDLCs, hybrid capacitors can overcome the energy density limitation. However, for balance of positive and negative electrode capacities, the faradaic electrode led to an increase in the energy density at the cost of cyclability in most cases. This is certainly the main drawback of hybrid devices. Therefore, it is important to avoid transforming a good supercapacitor into a mediocre battery.

Currently, two main approaches to hybrid systems have emerged: (i) pseudo-capacitive metal oxides with a capacitive carbon electrode, such as MnO_2/AC, V_2O_5/AC, Fe_3O_4/AC, PbO_2/AC, RuO_2/AC, $Ni(OH)_2$/AC, TiO_2/AC, etc, and (ii) lithium-insertion electrodes with a capacitive carbon electrode, such as $Li_4Ti_5O_{12}$/AC, $LiNi_{0.5}Mn_{1.5}O_4$/AC, $LiCoPO_4$ nanoparticles/carbon nanofoam, $LiMn_2O_4$/AC, lithiated graphite/AC, etc. (Table 3).[6] The main electrochemical properties of some of these hybrid capacitor systems are listed in Table 8. The advent

Table 8:　Hybrid Capacitor Systems and their Main Electrochemical Properties.

Cathode	Anode	Electrolyte	Voltage (V)	C_S^a (F g^{-1})	E_d^b (Wh kg^{-1})	P_d^c (kW kg^{-1})	Cycles	Ref.
Amorphous MnO$_2$	AC	1 M KCl	0 to 2	52	20.8–28.8	0.5–8	100	164
Amorphous MnO$_2$	AC	0.1 M K$_2$SO$_4$	0 to 2	21	10	16	195,000	165
MnO$_2$ nanoparticles	AC	0.1 M K$_2$SO$_4$	0 to 2.2	31	17.3	19	5,000	166
δ-MnO$_2$ nanorods	AC	0.5 M Li$_2$SO$_4$	0 to 1.8	31	17	2	23,000	167
Amorphous V$_2$O$_5$	AC	2 M NaNO$_3$	−0.3 to 0.7	32.5	—	—	600	168
V$_2$O$_5$ nanoribbons	AC	0.5 M K$_2$SO$_4$	0 to 1.8	64.4	20.3–29	0.07–2	100	169
Fe$_3$O$_4$ nanoparticles	AC	6 M KOH	0 to 1.2	37.9	7.6	0.07	500	170
PbO$_2$ film	AC	1.28 g cm^{-3} H$_2$SO$_4$	0.8 to 1.8	34.7	11.7–7.8	0.02–0.26	200	171
Ru oxide	Modified carbon fabric	1 M H$_2$SO$_4$	0 to 1.3	159	26.7	17.3	—	172
NiO nanoflakes	Porous carbon	6 M KOH	0 to 1.3	38	10	0.01–10	1,000	173
Ni(OH)$_2$/AC	AC	6 M KOH	0 to 0.5	530	—	—	—	174
Ni(OH)$_2$/MWNTs	AC	6 M KOH	0 to 1.5	96	32	1.5	2,000	175
MnO$_2$/porous carbon	V$_2$O$_5$ nanowires/MWNTs	1 M Na$_2$SO$_4$	0 to 1.6	45	5.5–16	0.075–3.75	100	176
MnO$_2$/MWNTs	SnO$_2$/MWNTs	2 M KCl	0 to 1.7	38	20.3	143.7	1,000	177
MnO$_2$ nanowire/SWNTs	In$_2$O$_3$ nanowire/SWNTs	1 M Na$_2$SO$_4$	0 to 2	184	25.5	50.3	—	178

(Continued)

Table 8: (*Continued*)

Cathode	Anode	Electrolyte	Voltage (V)	C_S^a (F g^{-1})	E_d^b (Wh kg^{-1})	P_d^c (kW kg^{-1})	Cycles	Ref.
AC	TiO$_2$	1 M LiPF$_6$/EC/DMC	1.2 to 3.5	44	30–80	0.35	600	179
MWNTs	TiO$_2$ nanowires	1 M LiPF$_6$/EC/DEC/DMC	0 to 2.8	11.5	8–12.5	0.3–1.2	600	180
MWNTs	Fe$_2$O$_3$/MWNTs	1 M LiClO$_4$/EC/DMC	0 to 2.8	80	50	1	500	181
Graphene	MnO$_2$-coated graphene	1 M KCl	0 to 0.9	328	11.4	25.8	1,300	182
Graphene/MnO$_2$	AC nanofiber	1 M Na$_2$SO$_4$	0 to 1.8	113.5	51.1	198	1,000	183
AC	Li$_4$Ti$_5$O$_{12}$	1 M LiBO$_4$/EC/DMC	1.5 to 3	—	85	—	5,000	184
LiNi$_{0.5}$Mn$_{1.5}$O$_4$	AC	1 M LiPF$_6$/EC/DMC	0 to 2.8	32	55	—	1,000	185
LiMn$_2$O$_4$	MnO$_2$/MWNTs	1 M LiClO$_4$/PC	0 to 2.5	60	26–56	0.3–2.4	—	186
LiCoPO$_4$ nanoparticles	Carbon nanofoam	1 M LiClO$_4$/EC/PC	0 to 2	21.9	11	0.2	1,000	187
LiMn$_2$O$_4$/AC	Li$_4$Ti$_5$O$_{12}$	1 M LiPF$_6$/EC/DEC/DMC	1.2 to 2.8	59	16	2.5	1,500	188

^{a}C_S: Specific capacitance.
^{b}E_d: Energy density.
^{c}P_d: Power density.
Note: AC: Activated Carbon; EC: Ethylene Carbonate; DMC: Dimethyl Carbonate; DEC: Diethyl Carbonate; PC: Propylene Carbonate.

of nanomaterials as well as rapid advances in the area of Li-ion batteries should lead to the design of high-performance ECs. Combining newly developed high-rate conversion reaction anodes or Li-alloying anodes with a positive supercapacitor electrode could fill the gap between Li-ion batteries and EDLCs. These systems could be of particular interest in applications where high power and medium cycle life are needed.[189]

5 Promising Applications of Electrochemical Capacitors

Table 9 gives an updated view of the state-of-the-art of current ECs commercially available. The recent market introduction of low-cost ECs with large capacitance and the capability of using them at high voltage allows a large number of new applications in the following areas[8,189]:

(1) Industrial applications

- Uninterruptible power supply systems (UPS)
- Elevators and cranes

(2) Electric utility applications

- Smart grids

(3) Automotive applications

- Starting engines
- Electric and hybrid vehicles
- Heavy duty and large transport systems

(4) Electronic devices needing pulse-power (light flash, radio signal, and rapid heating)
(5) Specialized and military applications

Among them, electric vehicles (EVs), hybrid electric vehicles (HEVs), and fuel cell vehicles have received most attention. In all three cases, the EC will serve as a short-time energy storage device with high power capability and allow storing the energy obtained from regenerative braking. The energy will be reused in the next acceleration phase and boost the acceleration. It allows one to reduce the size of the primary power source [batteries (EV), internal combustion engine (HEV), fuel cell] and keep them running at an optimized operation point. Buses, delivery vans, and cars in urban areas where many go–stop intervals exist benefit the most. Many more applications can be imagined, but we emphasize that an EC is not the universal energy-storage device. Its strength is in the area shown on the Ragone plot of Fig. 1, corresponding to a ratio of energy to power of about 1. For much larger and much smaller ratios, conventional capacitors and batteries, respectively, are more adequate, unless other criteria like cycling performance or environmental requirements become important.[8,190−193]

Table 9: Technical Performance of Some Present ECs.

Company	Product	Voltage (V)	C (F)	ESR (mΩ)	Specific power (W kg^{-1})	Specific energy (Wh kg^{-1})	Cycle life	Mass (kg)
Maxwell	PC10 series	2.2–2.5	10	180	510–660	1.1–1.4	500,000	0.0063
Maxwell	HC series	2.7	1–15	14–700	1,100–3,400	0.9–4.7	500,000	0.0011–0.032
Maxwell	D cell series	2.7	310–350	2.2–3.2	4,600–6,600	5.2–5.9	500,000	0.06
Maxwell	K2 series	2.7	650–3,000	0.29–0.8	5,900–6,900	4.1–6.0	1,000,000	0.16–0.51
Maxwell	16 V small modules	16	58	22	2,200	3.3	500,000	0.63
Maxwell	16 V large modules	16	110–500	2.1–5.6	1,800–2,700	1.5–3.2	1,000,000	2.66–5.51
Maxwell	48 V modules	48	83–165	6.3–10	2,700–3,300	2.6–3.9	1,000,000	10.3–13.5
Maxwell	56 V UPS modules	56	130	8.1	2,600	3.1	1,000,000	18
Maxwell	75 V power modules	75	94	13	2,100	2.9	1,000,000	25
Maxwell	125 V transportation modules	125	63	18	1,700	2.3	1,000,000	60.5
Nesscap	Small EDLC cells	2.3–2.7	5–6	14–123	—	1.67–4.51	500,000	0.0022–0.0135
Nesscap	Medium EDLC cells	2.3–2.7	90–360	3.2–16	—	3.83–5.45	500,000	0.021–0.067
Nesscap	Large prismatic EDLC cells	2.7	600–5,000	0.25–0.64	—	2.9–5.44	500,000	0.21–0.93
Nesscap	Large cylindrical EDLC cells	2.7	650–3,000	0.22–0.5	—	3.13–5.73	1,000,000	0.205–0.535
Nesscap	Small pseudocapacitor cells	2.3	50–120	18–24	—	4.87–5.87	100,000	0.0076–0.015

(*Continued*)

Table 9: (*Continued*)

Company	Product	Voltage (V)	C (F)	ESR (mΩ)	Specific power (W kg^{-1})	Specific energy (Wh kg^{-1})	Cycle life	Mass (kg)
Nesscap	Medium pseudocapacitor cells	2.3	220–300	12–14	—	7.03–8.73	100,000	0.023–0.0252
Nesscap	Multi-cell modules	5–125	1.5–500	1.6–110	—	1.05–3.83	1,000,000	0.0034–57
ApowerCap	Ultracapacitor	2.7	55	4	5,695	5.5	—	0.009
ApowerCap	Ultracapacitor	2.7	450	1.4	2,574	5.89	—	0.057
Panasonic	EDLCs	2.5	1,200	1.0	514	2.3	—	0.34
BatScap	Individual 2600F cell	2.7	2,680	0.20	2,050	4.2	—	0.50
Asahi Glass	Supercapacitor	2.7	1,375	2.5	390	4.9	—	0.21
LS Cable	Supercapacitor	2.8	3,200	0.25	1,400	3.7	—	0.63
Power Sys.	Supercapacitor	2.7	1,350	1.5	650	4.9	—	0.21
Power Sys.	Supercapacitor	3.3	1,800	3.0	486	8.0	—	0.21
Fuji Heavy Industry	Supercapacitor	3.8	1,800	1.5	1,025	9.2	—	0.232
JSR Micro	Supercapacitor	3.8	2,000	1.9	1,038	12.1	—	0.206

6 Conclusions and Outlook

A brief overview of capacitor technology has been discussed, including its development, classification, construction, materials, and applications. Previous development efforts have been considered to indicate the current state of the technology. The significant progress in material science and in electrochemistry has made possible the realization of electrochemical capacitors with promising technical and economical properties suitable for a continuously increasing number of applications. Electrochemical capacitors are now available in multiple designs that can cover almost all the performance range existing between conventional capacitors and electrochemical batteries.

As well as the large number of applications in consumer electronics, the new application areas described refer mainly to large capacitance, power, and energy EC systems and cover various energy-intensive sectors: industrial, transport, and electricity grids. The outlook for larger diffusion of EC devices in these areas is also confirmed by market studies, forecasting an average yearly growth rate of about 27% up to 2014, with a largely increasing sharp (up to about 40%) of EC devices in large system applications.[190]

The overall prospects and the associated advantages of using EC devices in new applications will surely power future efforts in research, development, and even in production processes, to further improve technical performance in a larger spectrum of applications and significantly reduce costs.

Acknowledgments

This work was partially supported by US Air Force (AFOSR), Semiconductor Research Corporation (SRC) and Petroleum Research Fund (PRF).

References

1. M. Jayalakshmi and K. Balasubramanian, Simple capacitors to supercapacitors — An overview, *Int. J. Electrochem. Sci.* **3** (2008), p. 1196.
2. Y. M. Vol'fkovich and T. M. Serdyuk, Electrochemical capacitors, *Russ. J. Electrochem.* **38** (2002), p. 935.
3. J. Ho, T. R. Jow and S. Boggs, Historical introduction to capacitor technology, *IEEE Electr. Insul. M.* **26** (2010), p. 20.
4. K. Naoi, 'Nanohybrid Capacitor': The next generation electrochemical capacitors, *Fuel Cells* **10** (2010), p. 825.
5. B. E. Conway, *Electrochemical Supercapacitors: Scientific Fundamentals and Technological Applications* (Kluwer Academic/Plenum Publishers, New York, USA, 1999).
6. P. Simon and Y. Gogotsi, Materials for electrochemical capacitors, *Nature Mater.* **7** (2008), p. 845.
7. P. Sharma and T. S. Bhatti, A review on electrochemical double-layer capacitors, *Energ. Convers. Manage.* **51** (2010), p. 2901.
8. A. Nishino, Capacitors: Operating principles, current market and technical trends, *J. Power Sources* **60** (1996), p. 137.

9. F. Trombetta, M. O. de Souza, R. F. de Souza and E. M. A. Martini, Electrochemical behavior of aluminum in 1-n-butyl-3-methylimidazolium tetrafluoroborate ionic liquid electrolytes for capacitor applications, *J. Appl. Electrochem.* **39** (2009), p. 2315.

10. S. Niwa and Y. Taketani, Development of new series of aluminium solid capacitors with organic semiconductive electrolyte (OS-CON), *J. Power Sources* **60** (1996), p. 165.

11. A. Dehbi, W. Wondrak, Y. Ousten and Y. Danto, High temperature reliability testing of aluminum and tantalum electrolytic capacitors, *Microelectron. Reliab.* **42** (2002), p. 835.

12. Y. Sakabe, Multilayer ceramic capacitors, *Curr. Opin. Solid State Mater. Sci.* **2** (1997), p. 584.

13. J. C. Niepce, Multilayer ceramic capacitors, *Actual. Chim.* **74** (2002).

14. H. Kishi, Y. Mizuno and H. Chazono, Base-metal electrode-multilayer ceramic capacitors: Past, present and future perspectives, *Jpn. J. Appl. Phys. 1.* **42** (2003), p. 1.

15. B. Xu, H. Zhang, G. P. Cao, W. F. Zhang and Y. S. Yang, Carbon materials for supercapacitors, *Progr. Chem.* **23** (2011), p. 605.

16. B. Daffos, P. L. Taberna, Y. Gogotsi and P. Simon, Recent advances in understanding the capacitive storage in microporous carbons, *Fuel Cells* **10** (2010), p. 819.

17. C. Portet, G. Yushin, and Y. Gogotsi, Electrochemical performance of carbon onions, nanodiamonds, carbon black and multiwalled nanotubes in electrical double layer capacitors, *Carbon* **45** (2007), p. 2511.

18. L. L. Zhang and X. S. Zhao, Carbon-based materials as supercapacitor electrodes, *Chem. Soc. Rev.* **38** (2009), p. 2520.

19. M. Inagaki, H. Konno and O. Tanaike, Carbon materials for electrochemical capacitors, *J. Power Sources* **195** (2010), p. 7880.

20. A. G. Pandolfo and A. F. Hollenkamp, Carbon properties and their role in supercapacitors, *J. Power Sources* **157** (2006), p. 11.

21. P. Simon and Y. Gogotsi, Charge storage mechanism in nanoporous carbons and its consequence for electrical double layer capacitors, *Phil. Trans. R. Soc. A.* **368** (2010), p. 3457.

22. E. Frackowiak, F. Beguin, Carbon materials for the electrochemical storage of energy in capacitors, *Carbon* **39** (2001), p. 937.

23. E. Frackowiak, F. Beguin, Electrochemical storage of energy in carbon nanotubes and nanostructured carbons, *Carbon* **40** (2002), p. 1775.

24. E. Frackowiak, Carbon materials for supercapacitor application, *Phys. Chem. Chem. Phys.* **9** (2007), p. 1774.

25. G. A. Snook, P. Kao and A. S. Best, Conducting-polymer-based supercapacitor devices and electrodes, *J. Power Sources* **196** (2011), p. 1.

26. C. D. Lokhande, D. P. Dubal and O. S. Joo, Metal oxide thin film based supercapacitors, *Curr. Appl. Phys.* **11** (2011), p. 255.

27. J. H. Chae, K. C. Ng and G. Z. Chen, Nanostructured materials for the construction of asymmetrical supercapacitors, *Proc. Inst. Mech. Eng., A J. Power Energy* **224** (2010), p. 479.

28. Y. Zhang, H. Feng, X. Wu, L. Wang, A. Zhang, T. Xia, H. Dong, X. Li and L. Zhang, Progress of electrochemical capacitor electrode materials: A review, *Int. J. Hydrogen Energ.* **34** (2009) p. 4889.

29. L. Wei and G. Yushin, Electrical double layer capacitors with sucrose derived carbon electrodes in ionic liquid electrolytes, *J. Power Sources* **196** (2011), p. 4072.

30. M. Galinski, A. Lewandowski and I. Stepniak, Ionic liquids as electrolytes, *Electrochim. Acta* **51** (2006), p. 5567.

31. B. Xu, S. Yue, Z. Sui, X. Zhang, S. Hou, G. Cao and Y. Yang, What is the choice for supercapacitors: Graphene or graphene oxide? *Energy Environ. Sci.* **4**, (2011), p. 2826.
32. Y. Sun, Q. Wu and G. Shi, Graphene based new energy materials, *Energy Environ. Sci.* **4** (2011), p. 1113.
33. M. Pumera, Graphene-based nanomaterials for energy storage, *Energy Environ. Sci.* **4** (2011), p. 668.
34. L. L. Zhang, R. Zhou and X. S. Zhao, Graphene-based materials as supercapacitor electrodes, *J. Mater. Chem.* **20** (2010), p. 5983.
35. M. Liang, B. Luo and L. Zhi, Application of graphene and graphene-based materials in clean energy-related devices, *Int. J. Energ. Res.* **33** (2009), p. 1161.
36. V. Subramanian, C. Luo, A. M. Stephan, K. S. Nahm, S. Thomas, and B. Wei, Supercapacitors from activated carbon derived from banana fibers, *J. Phys. Chem. C* **111** (2007), p. 7527.
37. L. Wei and G. Yushin, Electrical double layer capacitors with activated sucrose-derived carbon electrodes, *Carbon* **49** (2011), p. 4830.
38. Q. Y. Li, H. Q. Wang, Q. F. Dai, J. H. Yang and Y. L. Zhong, Novel activated carbons as electrode materials for electrochemical capacitors from a series of starch, *Solid State Ionics* **179** (2008), p. 269.
39. L. Wei, M. Sevilla, A. B. Fuertes, R. Mokaya and G. Yushin, Hydrothermal carbonization of abundant renewable natural organic chemicals for high-performance supercapacitor electrodes, *Adv. Energ. Mater.* **1** (2011), p. 356.
40. M. S. Balathanigaimani, W. G. Shim, M. J. Lee, C. Kim, J. W. Lee and H. Moon, Highly porous electrodes from novel corn grains-based activated carbons for electrical double layer capacitors, *Electrochem. Commun.* **10** (2008), p. 868.
41. K. Kierzek, E. Frackowiak, G. Lota, G. Gryglewicz and J. Machnikowski, Electrochemical capacitors based on highly porous carbons prepared by KOH activation, *Electrochim. Acta* **49** (2004), p. 1169.
42. D. Zhai, B. Li, H. Du, G. Wang and F. Kang, The effect of pre-carbonization of mesophase pitch-based activated carbons on Their electrochemical performance for electric double-layer capacitors, *J. Solid State Electrochem.* **15** (2011), p. 787.
43. T. E. Rufford, D. Hulicova-Jurcakova, K. Khosla, Z. H. Zhu and G. Q. Lu, Microstructure and electrochemical double-layer capacitance of carbon electrodes prepared by zinc chloride activation of sugar cane bagasse, *J. Power Sources* **195** (2010), p. 912.
44. D. Lozano-Castelló, D. Cazorla-Amorós, A. Linares-Solano, S. Shiraishi, H. Kurihara and A. Oya, Influence of Pore structure and surface chemistry on electric double layer capacitance in non-aqueous electrolyte, *Carbon* **41** (2003), p. 1765.
45. B. Xu, Y. F. Chen, G. Wei, H. Zhang, G. P. Cao, Y. Yang, Activated carbon with high capacitance prepared by NaOH activation for supercapacitors, *Mater. Chem. Phys.* **124** (2010), p. 504.
46. T. E. Rufford, D. Hulicova-Jurcakova, Z. Zhu and G. Q. Lu, Nanoporous carbon electrode from waste coffee beans for high performance supercapacitors, *Electrochem. Commun.* **10** (2008), p. 1594.
47. X. Li, C. Han, X. Chen and C. Shi, Preparation and performance of straw based activated carbon for supercapacitor in non-aqueous electrolytes, *Micropor. Mesopor. Mater.* **131** (2010), p. 303.
48. M. Lazzari, M. Mastragostino, A. G. Pandolfo, V. Ruiz and F. Soavi, Role of carbon porosity and ion size in the development of ionic liquid based supercapacitors, *J. Electrochem. Soc.* **158** (2011), p. A22.

49. L. Wei, M. Sevilla, A. B. Fuertes, R. Mokaya and G. Yushin, Polypyrrole-derived activated carbons for high-performance electrical double-layer capacitors with ionic liquid electrolyte, *Adv. Funct. Mater.* 10.1002/adfm.201101866 (2011).

50. J. Yan, T. Wei, W. Qiao, Z. Fan, L. Zhang, T. Li and Q. Zhao, A high-performance carbon derived from polyaniline for supercapacitors, *Electrochem. Commun.* **12** (2010), p. 1279.

51. X. Xiang, E. Liu, L. Li, Y. Yang, H. Shen, Z. Huang and Y. Tian, Activated carbon prepared from polyaniline base by K(2)CO(3) activation for application in supercapacitor electrodes, *J. Solid State Electrochem.* **15** (2011), p. 579.

52. A. Kajdos, A. Kvit, F. Jones, J. Jagiello and G. Yushin, Tailoring the pore alignment for rapid ion transport in microporous carbons, *J. Am. Chem. Soc.* **132** (2010), p. 3252.

53. C. Portet, Z. Yang, Y. Korenblit, Y. Gogotsi, R. Mokaya, and G. Yushin, Electrical double-layer capacitance of zeolite-templated carbon in organic electrolyte, *J. Electrochem. Soc.* **156** (2009), p. A1.

54. F. Lufrano and P. Staiti, Mesoporous carbon materials as electrodes for electrochemical supercapacitors, *Int. J. Electrochem. Sci.* **5** (2010), p. 903.

55. M. Lazzari, F. Soavi and M. Mastragostino, Mesoporous carbon design for ionic liquid-based, double-layer supercapacitors, *Fuel Cells* **10** (2010), p. 840.

56. T. Morishita, T. Tsumura, M. Toyoda, J. Przepiórski, A.W. Morawski, H. Konno and M. Inagaki, A Review of the control of pore structure in MgO-templated nanoporous carbons, *Carbon* **48** (2010), p. 2690.

57. D. W. Wang, F. Li, M. Liu, G. Q. Lu and H. M. Cheng, 3D aperiodic hierarchical porous graphitic carbon material for high-rate electrochemical capacitive energy storage, *Angew. Chem. Int. Ed.* **47** (2008), p. 373.

58. J. Eskusson, A. Janes, A. Kikas, L. Matisen and E. Lust, Physical and electrochemical characteristics of supercapacitors based on carbide derived carbon electrodes in aqueous electrolytes, *J. Power Sources* **196** (2011), p. 4109.

59. M. Rose, Y. Korenblit, E. Kockrick, L. Borchardt, M. Oschatz, S. Kaskel and G. Yushin, Hierarchical micro- and mesoporous carbide-derived carbon as a high-performance electrode material in supercapacitors, *Small* **7** (2011), p. 1108.

60. E. I. Shkol'nikov and D. E. Vitkina, Nanoporous structure characteristics of carbon materials for supercapacitors, *High Temp.* **48** (2010), p. 815.

61. M. Oschatz, E. Kockrick, M. Rose, L. Borchardt, N. Klein, I. Senkovska, T. Freudenberg, Y. Korenblit, G. Yushin, and S. Kaskel, A cubic ordered, mesoporous carbide-derived carbon for gas and energy storage applications, *Carbon* **48** (2010), p. 3987.

62. Y. Korenblit, M. Rose, E. Kockrick, L. Borchardt, A. Kvit, S. Kaskel and G. Yushin, High-rate electrochemical capacitors based on ordered mesoporous silicon carbide-derived carbon, *ACS Nano* **4** (2010), p. 1337.

63. E. N. Hoffman, G. Yushin, T. El-Raghy, Y. Gogotsi and M. W. Barsoum, Micro and mesoporosity of carbon derived from ternary and binary metal carbides, *Micropor. Mesopor. Mater.* **112** (2008), p. 526.

64. R. Dash, J. Chmiola, G. Yushin, Y. Gogotsi, G. Laudisio, J. Singer, J. Fischer and S. Kucheyev, Titanium carbide derived nanoporous carbon for energy-related applications, *Carbon* **44** (2006), p. 2489.

65. J. Chmiola, G. Yushin, R. Dash and Y. Gogotsi, Effect of pore size and surface area of carbide derived carbons on specific capacitance, *J. Power Sources* **158** (2006), p. 765.

66. H. L. Wang and Q. M. Gao, Synthesis, characterization and energy-related applications of carbide-derived carbons obtained by the chlorination of boron carbide, *Carbon* **47** (2009), p. 820.

67. G. Yushin, E. Hoffman, A. Nikitin, H. Ye, M.W. Barsoum, and Y. Gogotsi, Synthesis of nanoporous carbide-derived carbon by chlorination of titanium silicon carbide, *Carbon* **44** (2005), p. 2075.
68. A. Lewandowski, A. Olejniczak, M. Galinski and I. Stepniak, Performance of carbon-carbon supercapacitors based on organic, aqueous and ionic liquid electrolytes, *J. Power Sources* **195** (2010), p. 5814.
69. Q. Zhang, J. P. Rong, D. S. Ma and B. Q. Wei, Enhanced capacitance and rate capability of graphene/polypyrrole composite as electrode material for supercapacitors, *Energy Environ. Sci.* **4** (2011), p. 2152.
70. K. S. Hung, C. Masarapu, T. H. Ko and B. Q. Wei, Wide-temperature range operation supercapacitors from nanostructured activated carbon fabric, *J. Power Sources* **193** (2009), p. 944.
71. M. K. Seo and S. J. Park, Electrochemical characteristics of activated carbon nanofiber electrodes for supercapacitors, *Mater. Sci. Eng. B* **164** (2009), p. 106.
72. J. R. McDonough, J. W. Choi, Y. Yang, F. L. Mantia, Y. Zhang and Y. Cui, Carbon nanofiber supercapacitors with large areal capacitances, *Appl. Phys. Lett.* **95** (2009).
73. B. Xu, F. Wu, R. Chen, G. Cao, S. Chen and Y. Yang, Mesoporous activated carbon fiber as electrode material for high-performance electrochemical double layer capacitors with ionic liquid electrolyte, *J. Power Sources* **195** (2010), p. 2118.
74. G. Lota, K. Fic and E. Frackowiak, Carbon nanotubes and their composites in electrochemical applications, *Energy Environ. Sci.* **4** (2011), p. 1592.
75. B. Q. Wei, Supercapacitors from carbon nanotubes, *Abstr. Pap. Am. Chem. Soc.* **241** (2011).
76. A. Izadi-Najafabadi, T. Yamada, D. N. Futaba, H. Hatori, S. Iijima and K. Hata, Impact of cell-voltage on energy and power performance of supercapacitors with single-walled carbon nanotube electrodes, *Electrochem. Commun.* **12** (2010), p. 1678.
77. H. Zhang, G. Cao and Y. Yang, Carbon nanotube arrays and their composites for electrochemical capacitors and lithium-ion batteries, *Energy Environ. Sci.* **2** (2009), p. 932.
78. W. Lu, L. Qu, K. Henry and L. Dai, High performance electrochemical capacitors from aligned carbon nanotube electrodes and ionic liquid electrolytes, *J. Power Sources* **189** (2009), p. 1270.
79. V. V. N. Obreja, On the performance of supercapacitors with electrodes based on carbon nanotubes and carbon activated material — a review, *Physica E* **40** (2008), p. 2596.
80. X. Wang, L. Liu, X. Wang, L. Bai, H. Wu, X. Zhang, L. Yi and Q. Chen, Exfoliated graphite nanosheets/carbon nanotubes hybrid materials for superior performance supercapacitors, *J. Solid State Electrochem.* **15** (2011), p. 643.
81. B. B. Garcia, S. L. Candelaria, D. Liu, S. Sepheri, J. A. Cruz and G. Cao, High performance high-purity sol-gel derived carbon supercapacitors from renewable sources, *Renew. Energ.* **36** (2011), p. 1788.
82. A. Halama, B. Szubzda and G. Pasciak, Carbon aerogels as electrode material for electrical double layer supercapacitors-synthesis and properties, *Electrochim. Acta* **55** (2010), p. 7501.
83. D. Pech, M. Brunet, H. Durou, P. Huang, V. Mochalin, Y. Gogotsi, P. L. Taberna and P. Simon, Ultrahigh-power micrometre-sized supercapacitors based on onion-like carbon, *Nature Nanotech.* **5** (2010), p. 651.
84. E. G. Bushueva, P. S. Galkin, A. V. Okotrub, L. G. Bulusheva, N. N. Gavrilov, V. L. Kuznetsov and S. I. Moiseekov, Double layer supercapacitor properties of onion-like carbon materials, *Phys. Status Solidi B* **245** (2008), p. 2296.

85. J. J. Yoo, K. Balakrishnan, J. Huang, V. Meunier, B. G. Sumpter, A. Srivastava, M. Conway, A. L. M. Reddy, J. Yu, R. Vajtai and P. M. Ajayan, Ultrathin planar graphene supercapacitors, *Nano Lett.* **11** (2011), p. 1423.

86. X. Lu, H. Dou, S. Yang, L. Hao, L. Zhang, L. Shen, F. Zhang and X. Zhang, A flexible graphene/multiwalled carbon nanotube film as a high performance electrode material for supercapacitors, *Electrochim. Acta* **56** (2011), p. 5115.

87. J. R. Miller, R. A. Outlaw and B. C. Holloway, Graphene double-layer capacitor with ac line-filtering performance, *Science* **329** (2010), p. 1637.

88. Y. Zhu, S. Murali, M. D. Stoller, K. J. Ganesh, W. Cai, P. J. Ferreira, A. Pirkle, R. M. Wallace, K. A. Cychosz, M. Thommes, D. Su, E. A. Stach and R. S. Ruoff, Carbon-based supercapacitors produced by activation of graphene, *Science* **332** (2011), p. 1537.

89. T. Y. Kim, H.W. Lee, M. Stoller, D. R. Dreyer, C. W. Bielawski, R. S. Ruoff and K. S. Suh, High-performance supercapacitors based on poly (ionic liquid)-modified graphene electrodes, *ACS Nano* **5** (2011), p. 436.

90. G. S. Attard, P. N. Bartlett, N. R. B. Coleman, J. M. Elliott, J. R. Owen and J. H. Wang, Mesoporous platinum films from lyotropic liquid crystalline phases, *Science* **278** (1997), p. 838.

91. F. G. Will and C. A. Knorr, Investigation of formation and removal of hydrogen and oxygen coverage on platinum by a new, nonstationary method, *Zeit. Elektrochem.* **64** (1960), p. 258.

92. B. E. Conway, V. Birss and J. Wojtowicz, The role and utilization of pseudocapacitance for energy storage by supercapacitors, *J. Power Sources* **66** (1997), p. 1.

93. D. Hulicova-Jurcakova, M. Seredych, G. Q. Lu and T. J. Bandosz, Combined effect of nitrogen- and oxygen-containing functional groups of microporous activated carbon on its electrochemical performance in supercapacitors, *Adv. Funct. Mater.* **19**, (2009), p. 438.

94. D. Hulicova-Jurcakova, M. Kodama, S. Shiraishi, H. Hatori, Z. H. Zhu, G. Q. Lu, Nitrogen-enriched nonporous carbon electrodes with extraordinary supercapacitance, *Adv. Funct. Mater.* **19** (2009), p. 1800.

95. M. Seredych, D. Hulicova-Jurcakova, G. Q. Lu and T. J. Bandosz, Surface functional groups of carbons and the effects of their chemical character, density and accessibility to ions on electrochemical performance, *Carbon* **46** (2008), p. 1475.

96. D. Hulicova-Jurcakova, M. Kodama and H. Hatori, Electrochemical performance of nitrogen-enriched carbons in aqueous and non-aqueous supercapacitors, *Chem. Mater.* **18** (2006), p. 2318.

97. C. O. Ania, V. Khomenko, E. Raymundo-Piñero, J. B. Parra and F. Béguin, The large electrochemical capacitance of microporous doped carbon obtained by using a zeolite template, *Adv. Funct. Mater.* **17** (2007), p. 1828.

98. H. Benaddi, T. J. Bandosz, J. Jagiello, J. A. Schwarz, J. N. Rouzaud, D. Legras and F. Béguin, Surface functionality and porosity of activated carbons obtained from chemical activation of wood, *Carbon* **38** (2000), p. 669.

99. B. O. Park, C. D. Lokhande, H. S. Park, K. D. Jung and O. S. Joo, Performance of supercapacitor with electrodepo sited ruthenium oxide film electrodes — effect of film thickness, *J. Power Sources* **134** (2004), p. 148.

100. V. D. Patake, S. M. Pawar, V. R. Shinde, T. P. Gujar and C. D. Lokhande, The growth mechanism and supercapacitor study of anodically deposited amorphous ruthenium oxide films, *Curr. Appl. Phys.* **10** (2010), p. 99.

101. I. H. Kim, J. H. Kim, Y. H. Lee and K. B. Kim, Synthesis and characterization of electrochemically prepared ruthenium oxide on carbon nanotube film substrate for supercapacitor applications, *J. Electrochem. Soc.* **152** (2005), p. A2170.
102. C. C. Hu, Y. H. Huang and K. H. Chang, Annealing effects on the physicochemical characteristics of hydrous ruthenium and ruthenium-iridium oxides for electrochemical supercapacitors, *J. Power Sources* **108** (2002), p. 117.
103. S. L. Kuo and N. L. Wu, Composite supercapacitor containing tin oxide and electroplated ruthenium oxide, *Electrochem. Solid State Lett.* **6** (2003), p. A85.
104. J. P. Zheng and T. R. Jow, High energy and high power density electrochemical capacitors, *J. Power Sources* **62** (1996), p. 155.
105. M. Toupin, T. Brousse and D. Belanger, Charge storage mechanism of $MnO2$ electrode used in aqueous electrochemical capacitor, *Chem. Mater.* **16** (2004), p. 3184.
106. S. C. Pang, M. A. Anderson and T. W. Chapman, Novel electrode materials for thin-film ultracapacitors: comparison of electrochemical properties of sol-gel-derived and electrodeposited manganese dioxide, *J. Electrochem. Soc.* **147** (2000), p. 444.
107. S. F. Chin, S. C. Pang and M. A. Anderson, Material and electrochemical characterization of tetrapropylammonium manganese oxide thin films as novel electrode materials for electrochemical capacitors, *J. Electrochem. Soc.* **149** (2002), p. A379.
108. D. L. Fang, Z. D. Chen, B. C. Wu, Y. Yan and C. H. Zheng, Preparation and electrochemical properties of ultra-fine Mn-Ni-Cu oxides for supercapacitors, *Mater. Chem. Phys.* **128** (2011), p. 311.
109. D. W. Wang, Q. H. Wang and T. M. Wang, Morphology-controllable synthesis of cobalt oxalates and their conversion to mesoporous $Co(3)O(4)$ nanostructures for application in supercapacitors, *Inorg. Chem.* **50** (2011), p. 6482.
110. T. Brezesinski, JohnWang, S. H. Tolbert and B. Dunn, Ordered mesoporous alpha-$MoO(3)$ with iso-oriented nanocrystalline walls for thin-film pseudocapacitors, *Nature Mater.* **9** (2010), p. 146.
111. K. W. Nam, W. S. Yoon and K. B. Kim, X-ray absorption spectroscopy studies of nickel oxide thin film electrodes for supercapacitors, *Electrochim. Acta* **47** (2002), p. 3201.
112. E. E. Kalu, T. T. Nwoga, V. Srinivasan and J. W. Weidner, Cyclic voltammetric studies of the effects of time and temperature on the capacitance of electrochemically deposited nickel hydroxide, *J. Power Sources* **92** (2001), p. 163.
113. V. Srinivasan and J. W. Weidner, Studies on the capacitance of nickel oxide films: effect of heating temperature and electrolyte concentration, *J. Electrochem. Soc.* **147** (2000), p. 880.
114. K. C. Liu and M. A. Anderson, Porous nickel oxide/nickel films for electrochemical capacitors, *J. Electrochem. Soc.* **143** (1996), p. 124.
115. H. Kuan-Xin, W. Quan-Fu, Z. Xiao-Gang and W. Xin-Lei, Electrodeposition of nickel and cobalt mixed oxide/carbon nanotube thin films and their charge storage properties, *J. Electrochem. Soc.* **153** (2006), p. A1568.
116. K. R. Prasad and N. Miura, Electrochemical synthesis and characterization of nanostructured tin oxide for electrochemical redox supercapacitors, *Electrochem. Commun.* **6** (2004), p. 849.
117. M. Q. Wu, L. P. Zhang, D. M. Wang, C. Xiao and S. R. Zhang, Cathodic deposition and characterization of tin oxide coatings on graphite for electrochemical supercapacitors, *J. Power Sources* **175** (2008), p. 669.

118. K. R. Prasad, K. Koga and N. Miura, Electrochemical deposition of nanostructured indium oxide: high-performance electrode material for redox supercapacitors, *Chem. Mater.* **16** (2004), p. 1845.

119. T. P. Gujar, V. R. Shinde, C. D. Lokhande and S. H. Han, Electrosynthesis of Bi2O3 thin films and their use in electrochemical supercapacitors, *J. Power Sources* **161** (2006), p. 1479.

120. H. Y. Lee and J. B. Goodenough, Ideal supercapacitor behavior of amorphous V2O5 center dot nH(2)O in potassium chloride (KCl) aqueous solution, *J. Solid State Chem.* **148** (1999), p. 81.

121. I. H. Kim, J. H. Kim, B. W. Cho, Y. H. Lee and K. B. Kim, Synthesis and electrochemical characterization of vanadium oxide on carbon nanotube film substrate for pseudocapacitor applications, *J. Electrochem. Soc.* **153** (2006), p. A989.

122. S. Y. Wang, K. C. Ho, S. L. Kuo and N. L. Wu, Investigation on capacitance mechanisms of Fe3O4 electrochemical capacitors, *J. Electrochem. Soc.* **153** (2006), p. A75.

123. C. D. Lokhande, T. P. Gujar, V. R. Shinde, R. S. Mane and S. H. Han, Electrochemical supercapacitor application of pervoskite thin films, *Electrochem. Commun.* **9** (2007), p. 1805.

124. Y. Takasu, S. Mizutani, M. Kumagai, S. Sawaguchi and Y. Murakami, Ti-V-W-O/Ti oxide electrodes as candidates for electrochemical capacitors, *Electrochem. Solid State Lett.* **2** (1999), p. 1.

125. A. R. de Souza, E. Arashiro, H. Golveia and T. A. F. Lassali, Pseudocapacitive behavior of Ti/RhOx+Co3O4 electrodes in acidic medium: application to supercapacitor development, *Electrochim. Acta* **49** (2004), p. 2015.

126. V. Gupta, S. Gupta and N. Miura, Statically deposited nanostructured coxni1-x layered double hydroxides as electrode materials for redox-supercapacitors, *J. Power Sources* **175** (2008), p. 680.

127. J. Liu, J. Jiang, C. Cheng, H. Li, J. Zhang, H. Gong and H. J. Fan, Co(3)O(4) Nanowire@MnO(2) ultrathin nanosheet core/shell arrays: a new class of high-performance pseudocapacitive materials, *Adv. Mater.* **23** (2011), p. 2076.

128. J. W. Long, A. L. Young and D. R. Rolison, Spectroelectrochemical characterization of nanostructured, mesoporous manganese oxide in aqueous electrolytes, *J. Electrochem. Soc.* **150** (2003), p. A1161.

129. J. Yan, Z. Fan, T. Wei, J. Cheng, B. Shao, K. Wang, L. Song, M. Zhang, Carbon nanotube/MnO(2) composites synthesized by microwave-assisted method for supercapacitors with high power and energy densities, *J. Power Sources* **194** (2009), p. 1202.

130. J. Hu, A. B. Yuan, Y. Q. Wang and X. L. Wang, Improved cyclability of Nano-MnO(2)/CNT composite supercapacitor electrode derived from room-temperature solid reaction, *Acta Phys. Chim. Sin.* **25** (2009), p. 987.

131. W. Sugimoto, H. Iwata, Y. Yasunaga, Y. Murakami and Y. Takasu, Preparation of ruthenic acid nanosheets and utilization of its interlayer surface for electrochemical energy storage, *Angew. Chem. Int. Ed.* **42** (2003), p. 4092.

132. T. Brezesinski, J. Wang, S. H. Tolbert and B. Dunn, Next generation pseudocapacitor materials from sol-gel derived transition metal oxides, *J. Sol-Gel Sci. Technol.* **57** (2011), p. 330.

133. D. Choi, G. E. Blomgren and P. N. Kumta, Fast and reversible surface redox reaction in nanocrystalline vanadium nitride supercapacitors, *Adv. Mater.* **18** (2006), p. 1178.

134. B. E. Conway, Transition from supercapacitor to battery behavior in electrochemical energy-storage, *J. Electrochem. Soc.* **138** (1991), p. 1539.

135. K. S. Ryu, K. M. Kim, N. G. Park, Y. J. Park and S. H. Chang, Symmetric redox supercapacitor with conducting polyaniline electrodes, *J. Power Sources* **103** (2002), p. 305.

136. F. Fusalba, P. Gouerec, D. Villers and D. Belanger, Electrochemical characterization of polyaniline in nonaqueous electrolyte and its evaluation as electrode material for electrochemical supercapacitors, *J. Electrochem. Soc.* **148** (2001), p. A1.

137. C. Shi and I. Zhitomirsky, Electrodeposition and capacitive behavior of films for electrodes of electrochemical supercapacitors, *Nanoscale Res. Lett.* **5** (2010), p. 518.

138. J. Wang, Y. L. Xu, F. Yan, J. B. Zhu and J. P. Wang, Template-free prepared micro/nanostructured polypyrrole with ultrafast charging/discharging rate and long cycle life, *J. Power Sources* **196** (2011), p. 2373.

139. L. Z. Fan and J. Maier, High-performance polypyrrole electrode materials for redox supercapacitors, *Electrochem. Commun.* **8** (2006), p. 937.

140. B. C. Kim, C. O. Too, J. S. Kwon, J. M. Bo and G. G. Wallace, High-performance supercapacitors based on poly (ionic liquid)-modified graphene electrodes, *Synthetic Met.* **161** (2011), p. 1130.

141. F. Fusalba, H. A. Ho, L. Breau and D. Belanger, Poly (cyano-substituted diheteroareneethylene) as active electrode material for electrochemical supercapacitors, *Chem. Mater.* **12** (2000), p. 2581.

142. S. R. P. Gnanakan, N. Murugananthem and A. Subramania, Organic acid doped polythiophene nanoparticles as electrode material for redox supercapacitors, *Polym. Advan. Technol.* **22**, (2011), p. 788.

143. J. H. Huang and C. W. Chu, Achieving efficient poly(3,4-ethylenedioxythiophene)-based supercapacitors by controlling the polymerization kinetics, *Electrochim. Acta* **56** (2011), p. 7228.

144. J. Wang, Y. L. Xu, X. Chen and X. F. Du, Electrochemical supercapacitor electrode material based on poly (3,4-ethylenedioxythiophene)/polypyrrole composite, *J. Power Sources* **163** (2007), p. 1120.

145. A. M. P. Hussain, A. Kumar, F. Singh and D. K. Avasthi, Effects of 160 MeV Ni12+ ion irradiation on HCl doped polyaniline electrode, *J. Phys. D* **39** (2006), p. 750.

146. L. L. Tu and C. Y. Jia, Conducting polymers as electrode materials for supercapacitors, *Prog. Chem.* **22** (2010), p. 1610.

147. D. Aradilla, F. Estrany and C. Aleman, Symmetric supercapacitors based on multilayers of conducting polymers, *J. Phys. Chem. C* **115** (2011), p. 8430.

148. Jaidev, R. I. Jafri, A. K. Mishra and S. Ramaprabhu, Polyaniline-MnO(2) nanotube hybrid nanocomposite as supercapacitor electrode material in acidic electrolyte, *J. Mater. Chem.* **21** (2011), p. 17601.

149. L. Chen, L. J. Sun, F. Luan, Y. Liang, Y. Li and X. X. Liu, Synthesis and pseudocapacitive studies of composite films of polyaniline and manganese oxide nanoparticles, *J. Power Sources* **195** (2010), p. 3742.

150. F. J. Liu, Electrodeposition of manganese dioxide in three-dimensional poly (3,4-ethylenedioxythiophene)-poly(styrene sulfonic acid)-polyaniline for supercapacitor, *J. Power Sources* **182** (2008), p. 383.

151. X. Zhang, L. Ji, S. Zhang and W. Yang, Synthesis of a novel polyaniline-intercalated layered manganese oxide nanocomposite as electrode material for electrochemical capacitor, *J. Power Sources* **173** (2007), p. 1017.

152. V. Gupta and N. Miura, Polyaniline/single-wall carbon nanotube (PANI/SWCNT) composites for high performance supercapacitors, *Electrochim. Acta* **52** (2006), p. 1721.

153. I. Kovalenko, D. G. Bucknall and G. Yushin, Detonation nanodiamond and onion-like-carbon-embedded polyaniline for supercapacitors, *Adv. Funct. Mater.* **20** (2010), p. 3979.

154. G. Zhou, D. Wang, F. Li, L. Zhang, Z. Weng, H. Cheng, The effect of carbon particle morphology on the electrochemical properties of nanocarbon/polyaniline composites in supercapacitors, *New Carbon Mater.* **26** (2011), p. 180.

155. L. Zheng, X. Wang, H. An, X. Wang, L. Yi and L. Bai, The preparation and performance of flocculent polyaniline/carbon nanotubes composite electrode material for supercapacitors, *J. Solid State Electrochem.* **15** (2011), p. 675.

156. W. X. Liu, N. Liu, H. H. Song and X. H. Chen, Properties of polyaniline/ordered mesoporous carbon composites as electrodes for supercapacitors, *New Carbon Mater.* **26** (2011), p. 217.

157. S. Konwer, R. Boruah and S. K. Dolui, Studies on conducting polypyrrole/graphene oxide composites as supercapacitor electrode, *J. Electronic Mater.* **40** (2011), p. 2248.

158. J. Ge, G. Cheng and L. Chen, Transparent and flexible electrodes and supercapacitors using polyaniline/single-walled carbon nanotube composite thin films, *Nanoscale* **3** (2011), p. 3084.

159. C. Yuan, L. Shen, F. Zhang, X. Lu and X. Zhang, Reactive template fabrication of uniform core-shell polyaniline/multiwalled carbon nanotube nanocomposite and its electrochemical capacitance, *Chem. Lett.* **39** (2010), p. 850.

160. M. Yang, B. Cheng, H. Song and X. Chen, Preparation and electrochemical performance of polyaniline-based carbon nanotubes as electrode material for supercapacitor, *Electrochim. Acta* **55** (2010), p. 7021.

161. H. Gómez, M. K. Ram, F. Alvi, P. Villalba, E. L. Stefanakos and A. Kumar, Graphene-conducting polymer nanocomposite as novel electrode for supercapacitors, *J. Power Sources* **196** (2011), p. 4102.

162. K. Zhang, L. L. Zhang, X. S. Zhao and J. Wu, Graphene/polyaniline nanoriber composites as supercapacitor electrodes, *Chemistry of Materials* **22** (2010), p. 1392.

163. Jun Yan, Tong Wei, Bo Shao, Zhuangjun Fan, Weizhong Qian, Milin Zhanga and Fei Wei, Preparation of a graphene nanosheet/polyaniline composite with high specific capacitance, *Carbon* **48** (2010), p. 487.

164. M. S. Hong, S. H. Lee and S. W. Kim, Use of KCl aqueous electrolyte for 2 V manganese oxide/activated carbon hybrid capacitor, *Electrochem. Solid State Letters* **5** (2002), p. A227.

165. T. Brousse, P. L. Taberna, O. Crosnier, R. Dugas, P. Guillemet, Y. Scudeller, Y. Zhou, F. Favier, D. Bélanger and P. Simon, Long-term cycling behavior of asymmetric activated carbon/MnO2 aqueous electrochemical supercapacitor, *J. Power Sources* **173** (2007), p. 633.

166. T. Cottineau, M. Toupin, T. Delahaye, T. Brousse and D. Belanger, Charge storage mechanism of MnO2 electrode used in aqueous electrochemical capacitor, *Appl. Phys. A* **82** (2006), p. 599.

167. Q. Qu, P. Zhang, B. Wang, Y. Chen, S. Tian, Y. Wu and R. Holze, Electrochemical performance of MnO2 nanorods in neutral aqueous electrolytes as a cathode for asymmetric supercapacitors, *J. Phys. Chem. C* **113** (2009), p. 14020.

168. L. M. Chen, Q. Y. Lai, Y. J. Hao, Y. Zhao and X. Y. Ji, Investigations on capacitive properties of the AC/V(2)O(5) hybrid supercapacitor in various aqueous electrolytes, *J. Alloys and Compounds* **467** (2009), p. 465.

169. Q. T. Qu, Y. Shi, L. L. Li, W.L. Guo, Y. P. Wu, H. P. Zhang, S. Y. Guan and R. Holze, V2O5·0.6H2O nanoribbons as cathode material for asymmetric supercapacitor in K2SO4 solution, *Electrochem. Commun.* **11** (2009), p. 1325.

170. X. Du, C. Wang, M. Chen, Y. Jiao and J. Wang, Electrochemical performances of Nanoparticle Fe3O4/activated carbon supercapacitor using KOH electrolyte solution, *J. Phys. Chem. C* **113** (2009), p. 2643.

171. N. Yu, L. Gao, S. Zhao and Z. Wang, Electrodeposited PbO(2) thin film as positive electrode in PbO(2)/AC hybrid capacitor, *Electrochim. Acta* **54** (2009), p. 3835.

172. Z. Algharaibeh, X. Liu and P. G. Pickup, An asymmetric anthraquinone-modified carbon/ruthenium oxide supercapacitor, *J. Power Sources* **187** (2009), p. 640.

173. D. W. Wang, F. Li and H. M. Cheng, Hierarchical porous nickel oxide and carbon as electrode materials for asymmetric supercapacitor, *J. Power Sources* **185** (2008), p. 1563.

174. J. H. Park, O. O. Park, K. H. Shin, C. S. Jin and J. H. Kim, An electrochemical capacitor based on A Ni(OH)(2)/activated carbon composite electrode, *Electrochem. Solid State Lett.* **5** (2002), p. H7.

175. Y. G. Wang, L. Yu and Y. Y. Xia, Investigation on capacitance mechanisms of Fe3O4 electrochemical capacitors, *J. Electrochem. Soc.* **153** (2006), p. A743.

176. Z. Chen, Y. Qin, D. Weng, Q. Xiao, Y. Peng, X. Wang, H. Li, F. Wei, Y. Lu, Design and synthesis of hierarchical nanowire composites for electrochemical energy storage, *Adv. Funct. Mater.* **19** (2009), p. 3420.

177. K. C. Ng, S. Zhang, C. Peng and G. Z. Chen, Individual and bipolarly stacked asymmetrical aqueous supercapacitors of CNTs?/?SnO2 and CNTs?/?MnO2 nanocomposites, *J. Electrochem. Soc.* **156** (2009), p. A846.

178. P. C. Chen, G. Z. Shen, Y. Shi, H. T. Chen and C. W. Zhou, Preparationl and characterization of flexible asymmetric supercapacitors based on transition-metal-oxide nanowire/single-walled carbon nanotube hybrid thin-film electrodes, *ACS Nano* **4** (2010), p. 4403.

179. T. Brousse, R. Marchand, P. L. Taberna and P. Simon, TiO(2) (B)/activated carbon non-aqueous hybrid system for energy storage, *J. Power Sources* **158** (2006), p. 571.

180. Q. Wang, Z. H. Wen and J. H. Li, A hybrid supercapacitor fabricated with a carbon nanotube cathode and a TiO2-B nanowire anode, *Adv. Funct. Mater.* **16** (2006), p. 2141.

181. X. Zhao, C. Johnston and P. S. Grant, A novel hybrid supercapacitor with a carbon nanotube cathode and an iron oxide/carbon nanotube composite anode, *J. Mater. Chem.* **19** (2009), p. 8755.

182. Q. Cheng, J. Tang, J. Ma, H. Zhang, N. Shinya and L. C. Qin, Graphene and nanostructured MnO(2) composite electrodes for supercapacitors, *Carbon* **49** (2011), p. 2917.

183. Z. Fan, J. Yan, T. Wei, L. Zhi, G. Ning, T. Li and F. Wei, Asymmetric supercapacitors based on graphene/MnO(2) and activated carbon nanofiber electrodes with high power and energy density, *Adv. Funct. Mater.* **21** (2011), p. 2366.

184. G. G. Amatucci, F. Badway, A. Du Pasquier and T. Zheng, An asymmetric hybrid nonaqueous energy storage cell, *J. Electrochem. Soc.* **148** (2001), A930.

185. H. Q. Li, L. Cheng and Y. Y. Xia, A hybrid electrochemical supercapacitor based on A 5 V Li-ion battery cathode and active carbon, *Electrochem. Solid State Lett.* **8** (2005), p. A433.

186. S. B. Ma, K. W. Nam, W. S. Yoon, X. Q. Yang, K. Y. Ahn, K. H. Oh and K. B. Kim, A novel concept of hybrid capacitor based on manganese oxide materials, *Electrochem. Commun.* **9** (2007), p. 2807.

187. R. Vasanthi, D. Kalpana and N. G. Renganathan, Olivine-type nanoparticle for hybrid supercapacitors, *J. Solid State Electrochem.* **12** (2008), p. 961.

188. X. Hu, Z. Deng, J. Suo and Z. Pan, Improved cyclability of Nano-MnO(2)/CNT composite supercapacitor electrode derived from room-temperature solid reaction, *J. Power Sources* **187** (2009), p. 635.

189. M. Al Sakka, H. Gualous, J. Van Mierlo and H. Culcu, Thermal modeling and heat management of supercapacitor modules for vehicle applications, *J. Power Sources* **194** (2009), p. 581.

190. M. Conte, Supercapacitors technical requirements for new applications, *Fuel Cells* **10** (2010), p. 806.

191. R. Kotz and M. Carlen, Principles and applications of electrochemical capacitors, *Electrochim. Acta* **45** (2000), p. 2483.

192. A. Burke, Ultracapacitor technologies and application in hybrid and electric vehicles, *Int. J. Energ. Res.* **34** (2010), p. 133.

193. A. Hammar, P. Venet, R. Lallemand, G. Coquery and G. Rojat, Study of accelerated aging of supercapacitors for transport applications, *IEEE Transact. Ind. Electron.* **57** (2010), p. 3972.

Chapter 18

Batteries

Habiballah Rahimi-Eichi* and Mo-Yuen Chow[†]

Electrical and Computer Engineering Department
North Carolina State University, Raleigh, NC, USA
*hrahimi@ncsu.edu
†chow@ncsu.edu

Batteries are a very important energy-storage technology. This chapter describes the electrochemical structure of a typical battery, including all the components and their influences on the current–voltage characteristics of the battery. Different primary and secondary battery technologies are introduced and their advantages and disadvantages, applications and costs are described. Batteries are also compared to several other energy-storage technologies. In the last section, two of the most promising areas for advanced battery technology applications, namely electric vehicles and the smart grid, are discussed, with a focus on battery technology goals and challenges of plug-in hybrid electric vehicles and plug-in electric vehicles. Finally, battery management system (BMS), as a key field of research to enhance battery applications, is briefly introduced.

1 Electrochemical Structure of a Battery

A battery is a device in which chemical energy is converted to electrical energy. Each battery pack consists of a number of cells and each cell, as shown in Fig. 1, consists of two half-cells called the anode (the negative half) and the cathode (positive half). These half-cells are connected in series by a conductive electrolyte that contains anions (negatively charged ions) and cations (positively charged ions). The anode includes the electrolyte and the electrode to which anions migrate and the cathode includes the electrolyte and the electrode to which cations migrate.[1] In the reduction–oxidation (Redox) reaction that powers the battery, cations are reduced (electrons are added) at the cathode, while anions are oxidized (electrons are removed) at the anode. The two electrodes are not physically connected to each other, but the electrical connection is made through the electrolyte. Some cells use two half-cells with different electrolytes. A separator between half-cells allows ions to flow while preventing the electrolytes from mixing.

Each half-cell has its own electromotive force (EMF), determined by its potential to flow electric current from the inside to the outside of the cell.[1] This potential is quantified by the tendency of an element used in an electrode to acquire (i.e. gain)

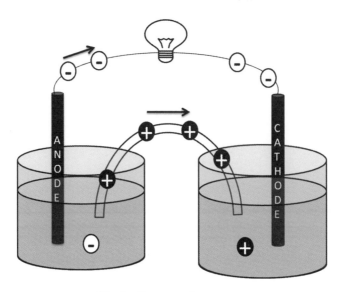

Fig. 1.   Battery cell structure.

or release (lose) electrons. The net EMF of the cell is the maximum difference between the EMFs of its half-cells. Therefore, if the electrodes have EMFs E_1 and E_2 ($E_2 > E_1$), then the net EMF is $E_2 - E_1$; in other words, the net EMF is the difference between the reduction potentials of the half-reactions. This EMF creates the open circuit voltage (OCV) of the battery, which is the voltage across the battery terminals when it is at rest (i.e. the battery is not being charged or discharged).[2] The length of time the battery needs to remain at rest before an accurate measurement of the EMF can be made depends on the chemistry and the type of the battery.

The reason why the terminal voltage of the battery is not equal to the EMF all the time is the internal dynamics of the battery. The first dynamic is the internal resistance of the battery due to the ohmic resistance of the electrolyte.[3,4] This ohmic resistance causes a voltage drop/overvoltage during the discharging/charging process. When the battery is at rest, this effect cannot be seen at the terminal voltage; instead the relaxation effect, another dynamic, causes the terminal voltage to be different than the OCV.[2,5] This effect does not allow the terminal voltage to change immediately to the final value when the charging/discharging current is switched from one value to another. The relaxation effect is caused by two major phenomena: The double-layer capacity that is formed beside both electrodes creates a short-term relaxation effect and the solid–electrolyte interface (SEI) formed around the anode is responsible for a long-term relaxation effect.[6] While the short-term relaxation is of the order of some 10 sec, the long-term relaxation effect can extend to a couple of hours. In some of the studies, it is recommended to leave the battery without charging or discharging for days.[7,8] in order to accurately measure the OCV of the battery.

2 Battery Technologies and Applications

Batteries are normally classified into two broad categories,[1] primary and secondary batteries, based on the application and the production technology.[a] Each type of battery has its advantages and disadvantages.

2.1 *Primary Batteries*

Primary batteries or disposable batteries are intended to be used only once and discarded without being charged again. Primary batteries are most commonly used in portable devices that have low current drain or in cases where other alternative power sources are only intermittently available, such as in alarm and communication circuits. Disposable primary cells cannot be reliably recharged, since the chemical reactions are not easily reversible, and active materials may not return to their original forms. Battery manufacturers strongly discourage attempts to recharge primary batteries.[1]

The common types of primary batteries include zinc–carbon batteries and alkaline batteries. In general, these batteries have higher energy densities than rechargeable batteries, but primary batteries are not good choices in applications that involve high current drain. Table 1 shows some of the most common primary battery chemistries, with their components, advantages and disadvantages, applications and costs. While the zinc–carbon batteries (Leclanché cells) are the lowest-cost primary batteries that are used in general-purpose applications, alkaline batteries are used in many applications because their longer life makes them more cost-effective. Silver–zinc batteries, although more expensive than zinc–carbon, provide a safer solution than primary lithium cells for high power density applications. Nevertheless, high energy density, low weight, small size and very long operating and shelf life all give primary lithium cells a distinct advantage compared to the other types, despite their high costs. Figures 2 and 3 provide some quantitative comparisons among these primary cells regarding EMF voltage and energy density, respectively. Figure 2 shows that the EMF of most of the primary lithium cells is around twice the EMF of other cells. In Fig. 3, both gravimetric (Wh/kg) and volumetric (Wh/liter) energy densities of different primary cells are compared. Again, lithium cells are capable of storing more energy per unit mass and per unit volume than the other cells.

2.2 *Secondary Batteries*

Secondary batteries or rechargeable batteries are intended to be recharged many times and therefore have significantly longer lifetimes than primary batteries. Secondary batteries must be charged before use. The active materials are usually

---

[a]Sometimes, flow battery is considered as a type of battery in which the active materials are stored external to the battery and are circulated through the cell stack as required to produce electric power. In this chapter, because of their similarity to fuel cells we discuss flow batteries under "other energy storage technologies" in Sec. 3.

Table 1: Properties of Primary Batteries.[9]

Cell chemistry	Structure	Pros/Cons	Applications/Costs
Alkaline cells	**Electrodes** Zinc and Manganese dioxide **Electrolyte** Potassium hydroxide (KOH).	**Advantages** Constant capacity over a wide range of current drains Suitable for high-drain-rate applications Good shelf life Better low temperature performance than zinc–carbon Less leakage than Leclanché cells Made from non-toxic chemicals **Disadvantages** Higher cost than the basic competing zinc–carbon Leclanché cells Not normally rechargeable 25% heavier than Leclanché cells	**Applications** Premium products Toys Remote controls Flashlights Clocks Consumer applications **Costs** Low cost but about 50% higher than zinc–carbon More cost-effective because of the longer life
Leclanché cells (Zinc–Carbon cells, Dry cells)	**Electrodes** Zinc and Manganese dioxide or Carbon **Electrolyte** Ammonium chloride and Zinc chloride	**Advantages** Inexpensive materials Low cost Available in a wide range of sizes Suitable for a wide range of consumer applications Interchangeable with alkaline batteries **Disadvantages** Tendency to leak Lower energy density than the competing alkaline batteries Poor low temperature performance Not rechargeable	**Applications** General purpose, low cost applications Toys Remote controls Flashlights Clocks Consumer applications **Costs** Lowest cost primary batteries
Zinc/Silver Oxide cells (Silver–Zinc cells)	**Electrodes** Zinc and Silver oxide **Electrolyte** Sodium hydroxide (NaOH) or Potassium hydroxide (KOH).	**Advantages** High capacity per unit weight Long operating life Low self-discharge and hence long shelf life Better low temperature performance than zinc–air Flat discharge characteristics Higher voltage than zinc–mercury cells **Disadvantages** Uses expensive materials Lower energy density than zinc–air Poor performance at low temperature Limited cycle life	**Applications** Miniature power sources Low power devices Submarines, missiles, underwater and aerospace applications Promoted as a safer alternative to lithium cells **Costs** Very expensive for high-power applications

(Continued)

Table 1: (*Continued*)

Cell Chemistry	Structure	Pros/Cons	Applications/Costs
		Suffers from dissolving of the zinc and the formation of zinc dendrites which pierce the separator	
Primary lithium cells	**Electrodes**	**Advantages**	**Applications**
	Lithium foil and Manganese dioxide, Sulfur dioxide, Thionyl chloride or Oxygen	High energy density, double that of premium alkaline batteries	Computer memory protection
		Low weight	Medical implants
		High cell voltage	Heart pacemakers
		Flat discharge characteristic	Defibrillators
		Low self-discharge	Utility meters
		Very long shelf life	Watches
		Very long operating life (15 to 20 years for lithium thionyl chloride)	Cameras
	Electrolyte		Calculators
			Car keys
	Lithium salts such as $LiPF_6$, $LiAsF_6$, $LiClO_4$, $LiBF_4$, or $LiCF_3SO_3$	Wide operating temperature range (-60°C to $+85^\circ$C for lithium sulfur dioxide)	Security transmitters
		Excellent durability	Smoke alarms
		Small cell size	Aerospace applications
		Disadvantage	**Costs**
		High cost	More expensive than common consumer primary Leclanché and alkaline batteries

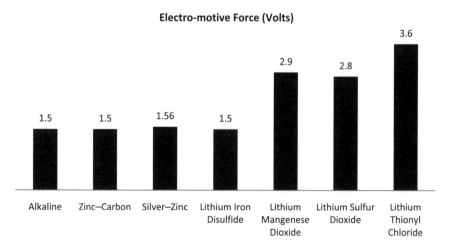

Electro-motive Force (Volts)

Alkaline	Zinc–Carbon	Silver–Zinc	Lithium Iron Disulfide	Lithium Mangenese Dioxide	Lithium Sulfur Dioxide	Lithium Thionyl Chloride
1.5	1.5	1.56	1.5	2.9	2.8	3.6

Fig. 2. Electro-motive force comparison for primary cells.[10]

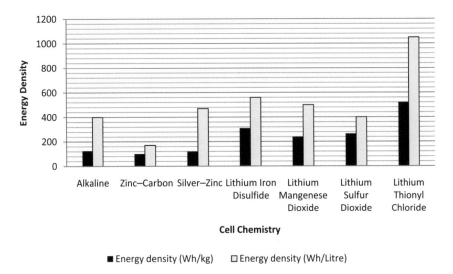

Fig. 3.   Energy density comparison of primary cells.[10]

assembled in these batteries in the discharged state. Rechargeable batteries can
be recharged by applying electric current, which reverses the chemical reactions
that occur during its discharge.[1] Chargers or rechargers are devices to supply the
appropriate current. Secondary batteries are available in the marketplace in two
forms. The first is the liquid electrolyte battery for which the lead–acid battery
is the oldest example. The valve regulated lead–acid battery (VRLA battery) is
an improved version of the lead–acid battery which is popular in the conventional
automotive industry. The second form is the battery with sealed dry cells, which
are useful in mobile appliances such as cell phones and laptops. Moreover, the
recent demand for high power and energy density batteries in different types of
electric vehicles and the smart grid has created a huge potential market for dry cell
rechargeable batteries. Batteries of this type (in order of increasing power density
and cost) include nickel–cadmium (NiCd), nickel–zinc (NiZn), nickel–metal hydride
(NiMH), and lithium-ion (Li–ion) batteries. While the Li-ion battery has the highest
share of this market currently (and it is expected to grow), NiMH has replaced
NiCd in most applications due to its higher capacity. However NiCd remains in use
in power tools, two-way radios, and medical equipment.[1]

    Table 2 describes several types of secondary batteries in terms of their chemical
structures, along with their advantages and disadvantages, applications and costs.
This table shows that lead–acid batteries, despite reliability and low cost, are not
able to provide enough energy and power density for the new applications in electric
vehicles. While NiCd and NiMH produce higher energy and power density, the presence of toxic cadmium and the capacity limitation of NiMH provide an opportunity
for lithium batteries for use in electric vehicles. This situation explains the amount
of ongoing research to improve lithium battery technology from both the energy and

Table 2: Properties of Secondary/Rechargeable Batteries.[9]

Cell chemistry	Structure	Pros/Cons	Applications/Costs
Lead–Acid	**Electrodes** Lead-dioxide and a sponge metallic lead **Electrolyte** Sulfuric acid solution	**Advantages** Low cost Reliable Tolerant to abuse and overcharging Low internal impedance Can deliver very high currents Indefinite shelf life if stored without electrolyte . **Disadvantages** Very heavy and bulky Risk of overheating during charging Not suitable for fast charging Typical cycle life 300 to 500 cycles Gassing[a] and Sulphation[b] Completely discharging the battery may cause irreversible damages Contains toxic chemicals	**Applications** Automotive and traction[c] applications Standby/Back-up/ Emergency power Submarines UPS (Uninterruptible power supplies) Lighting High current drain applications **Costs** Low cost
Nickel– Cadmium (NiCd)	**Electrodes** Nickel hydroxide $(Ni(OH)_2)$ and cadmium (Cd) **Electrolyte** Potassium hydroxide (KOH)	**Advantages** Low internal resistance High rate charge/discharge Flat discharge characteristic Wide temperature range (Up to $70°C$) Cycling life of over 500 cycles Endothermic charging process Rapid charge, typically 2 hours Can be stored in the charged or discharged state without damage Low cost electrolyte **Disadvantages** Suffers from memory effect[d] Prone overcharging damage Low cell voltage of 1.2 Volts Needs self-sealing safety vents due to overheating and pressure build up High cost and heavy Cadmium	**Applications** Motorized equipment Two way radios Electric razors Portable equipment Medical instrumentation Emergency lighting Toys **Costs** Relatively inexpensive for low power applications but between three and four times more expensive than lead–acid for the same capacity
Nickel–Metal hydride (NiMH)	**Electrodes** Nickel-hydroxide and Hydrogen absorbing alloys (like Lanthanum)	**Advantages** High energy density (Wh/kg) Low internal impedance Typical cycle life of 3,000 cycles Can be deep cycled Robust–tolerant to over-charge and over-discharge conditions Flat discharge characteristic Wide operating temperature range Rapid charge possibility	**Applications** Low cost consumer applications Portable equipment Medical instruments and equipment Automotive batteries and electric vehicles

(*Continued*)

Table 2:   (*Continued*)

Cell chemistry	Structure	Pros/Cons	Applications/Costs
	Electrolytes Potassium- hydroxide (KOH)	Environmentally friendly (No Cadmium, Mercury or Lead) Safer than Lithium based cells in case of an accident or abuse	High power static applications (Telecom, UPS and smart grid)
		Disadvantages	**Costs**
		High self-discharge rate Deteriorates after a long time storage Less tolerant of overcharging than NiCads Must incorporate safety vents to protect the cell in case of gas generation Limited supplies of rare earth element Lanthanum	Originally more expensive than NiCad cells but prices are now more in line as NiMH volumes increase and the use of toxic Cadmium based cells is deprecated
Nickel–Zinc (NiZn)	**Electrodes** Nickel Hydroxide and Zinc **Electrolytes** Potassium hydroxide (KOH)	**Advantages** High rate capability (25C) Good cycle life Fast recharge capability Can be deep cycled Uses low cost benign materials **Disadvantages** Heavy and bulky Low energy density High self-discharge rate Solubility of zinc in the alkaline electrolyte limited life cycle	**Applications** Traction applications Electric Bicycles Scooters Lawnmowers **Costs** Low
Rechargeable Lithium Cells	**Electrodes** Carbon and Lithium Cobalt dioxide, Lithium Manganese compound or Lithium Iron Phosphate **Electrolyte** Lithium salts in organic carbonates such as ethylene carbonate or diethyl carbonate	**Advantages** High cell voltage of 3.6 V No liquid electrolyte High energy and power density Low weight Can be discharged at the 40C rate[e] Fast charge possible Low self-discharge rate Very high columbic efficiency No memory effect Long cycle life 1000 to 3000 cycles **Disadvantages** Internal impedance higher than equivalent NiCds Stricter regulations on shipping Degrades at high temperatures Capacity loss or thermal runaway when overcharged Degradation when discharged below 2 V	**Applications** A wide range of consumer portable, medical and communication products Traction application especially different types of electric vehicles Standby power **Costs** The price of lithium cells continues to fall as the technology gains more acceptances. See Fig. 2 for the expected cost trend

(*Continued*)

Table 2: (*Continued*)

Cell chemistry	Structure	Pros/Cons	Applications/Costs
		Venting and possible thermal runaway when crushed Need for protective circuitry Measurement of the state of charge of the cell is more complex than for most common cell chemistries	The target commercialization price for high power cells is around US\$300/kWh. See Sec. 4.1 for more information about lithium-ion goals
Sodium Sulfur (NaS)	**Electrodes** Sodium and Sulfur **Electrolyte** Ceramic tube of beta-alumina	**Advantages** High reduction potential of Sodium: $(-2.71\,V)$ Low weight Non-toxic nature Its relative abundance and ready availability Low cost High power and energy density **Disadvantages** The sodium must be used in liquid form. Must operate at high temperatures, typically in excess of $270°C$ (problems of thermal management and safety)	**Application** High power and high energy capacity for the grid **Costs** Relatively low

[a]Production and release of hydrogen and oxygen bubbles due to the electrolysis of water in the electrolyte during the charging process that lead to the loss of electrolyte.

[b]Internal resistance increase of the battery due to the formation of large lead sulfate crystals which are not readily reconverted back to lead, lead dioxide and sulfuric acid during recharging.

[c]Traction applications are basically referred to forklifts, electric Golf carts, riding floor scrubbers, electric motorcycles, electric cars, trucks, and vans, and other electric vehicles supported by batteries.

[d]Losing the maximum energy capacity due to repeatedly being recharged after being only partially discharged.

[e]C-rate is the charge or discharge rate equal to the capacity of the battery (in Ah) divided by 1 h.

power density point of view while trying to decrease cost and increase reliability and safety. Figure 4 compares the EMF of different secondary cells and shows that different types of lithium cells have much higher (three times in some cases) EMF voltages than other cell chemistries. This can be an important factor in increasing their energy density compared to the other cells as shown in Fig. 5. This figure shows that both gravimetric (Wh/kg) and volumetric energy densities (Wh/liter) of lithium polymer are much higher than those cells with other chemistries. Although Fig. 5 shows that some lithium cells such as lithium iron phosphate have a lower energy density than NiMH, current research is trying to improve the characteristics.

Electro Motive Force (Volts)

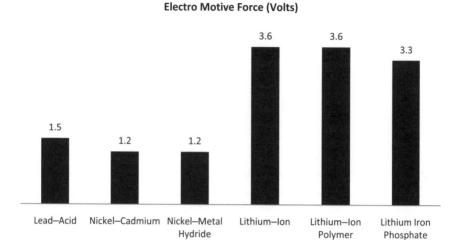

Fig. 4.   Electromotive force comparison of secondary chemistry.[10]

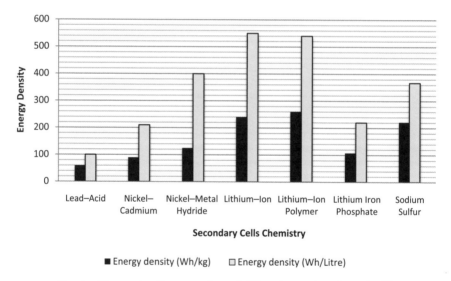

Fig. 5.   Energy density comparison of different secondary batteries.[10]

Figure 6 compares costs for different types of battery chemistry, primary and secondary and shows that lithium batteries are still more expensive than other types both for primary and secondary cells. But as indicated in Fig. 7, the price is anticipated to decrease significantly and make lithium-ion the most cost-effective battery type, at least for mobile applications.

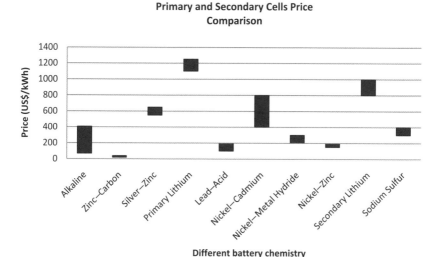

Fig. 6.   Cost comparison of different cell chemistries.[11]

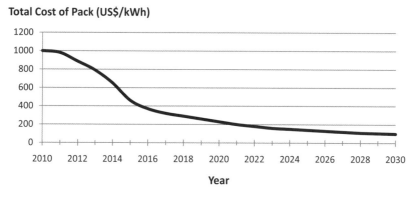

Fig. 7.   Lithium-ion batteries cost trend until 2030.

3 Batteries Compared with Other Energy-Storage Technologies

Although batteries are the most popular electrical energy-storage technology, there are many other technologies that can be used to store energy. In some cases, these technologies may be more appropriate than batteries for specific applications. Table 3 lists some of the most important technologies for energy storage and describes their advantages and disadvantages and their most suitable applications. A price range is also provided for each of the technologies, so that the costs per power unit or per energy unit can be compared to each other and to different battery chemistries. Table 3 suggests that batteries are the most appropriate energy-storage technology for mobile devices with various ranges of energy density requirements i.e. from cell phones to electric vehicles. Some of the technologies that can provide large

Table 3:　Other Energy-Storage Technologies.[13]

Energy-Storage technology	Advantages	Disadvantages	Application	Cost (US$)
Pumped Hydro-electric Energy-Storage (PHES)	Fast reaction time Environmentally friendly	Dependence on specific geological formations High capital cost	Load-leveling, Frequency Regulation, Peak Generation, Black Starts, Base-Load generating facilities during off-peak hours	$600/kW to upwards of $2000/kW
Compressed Air Energy Storage (CAES)	Fast reaction time Robust to ambient temperature	Dependence on specific geological location Large scale investment required High capital cost Young technology. (Only two CAES facilities currently exist)	Bulk energy supply and demand, frequent start-ups and shutdowns Ancillary services Wind power energy storage market	$425/kW to $450/kW Estimated maintenance between $3/kWh and $10/kWh
Flywheel Energy Storage (FES)	Extremely fast dynamic response Long life Require little maintenance Environmentally friendly	Application dependent Difficult to transfer heat out of the system Idling losses	Power quality enhancements such as Uninterruptable Power Supply (UPS), capturing waste energy (in EVs), wind turbines	$200/kWh to $300/kWh for low speed $25,000/kWh for high speed
Super Capacitor Energy Storage (SCES)	Fast charge and discharge Extremely long life No memory effect Minimal degradation due to deep discharge Do not heat up No hazardous substances	Very low energy-storage density Heavier and bulkier than conventional batteries	Hybrid cars, cellular phones, and load leveling Primarily used where pulsed power is needed in the millisecond to second time range	$12,960/kWh to $28,000/kWh
Super-conducting Magnetic Energy Storage (SMES)	High power capacity Instantaneous discharge rates	Sensitivity to temperature Rapid discharge rate limits application	Power quality market, network upgrade solution	$300/kW to $509/kW

(*Continued*)

Table 3: (*Continued*)

Energy-Storage technology	Advantages	Disadvantages	Application	Cost
Hydrogen Energy Storage (HESS)	The flexibility and potential of hydrogen to replace conventional fuel	Huge losses due to the number of energy conversions (Overall efficiency of 22% to 48%)	Integrating large quantities of intermittent wind energy	**Electrolysis** $380/kW to $1400/kW **Storage** $11/kWh to $15/kWh **Fuel Cell** $640/kW to $10,200/kW
Flow Battery Energy Storage (FBES) [Zinc Bromine (ZnBr)]	Indefinite life for the electrolyte High energy density 100% dischargeable	Limited power and energy capacity Degradation of membrane	Wind farm, Solar Panel and other renewable energies backup market	**Power capacity cost** $639/kW **Energy capacity cost** $400/kWh

energy density such as pumped hydroelectric storage (PHES) (see also Chapter 13), and compressed air electric storage requires particular locations. These technologies may be more useful than batteries in certain grid applications because of their low cost and environmentally friendly nature compared to batteries, provided the site is suitable. Some of the other technologies such as flywheel energy storage, super-capacitor energy storage (SCES), (see also Chapter 17) and superconducting magnetic energy storage, although not able to deliver a large energy density, provide a better solution for those applications that need a large amount of power in a short period of time because of their fast response power density as well as their cost effectiveness. Combining these technologies with batteries or other storage technologies with higher energy densities allows designers to optimize the size of storage devices for specific applications. Figure 8 shows the expected discharge time of different energy-storage technologies for various types of batteries versus the power that they can provide for that duration. While SCES, as mentioned earlier, can supply up to 1 MW of power in the order of seconds, PHES is able to provide 1 GW for a number of hours. In this diagram, batteries are mostly in the middle range and are suitable for providing different amount of power, up to 1 MW, for about an hour.

4 Directions and Challenges of Battery Technology

4.1 *Battery Technology Goals for PHEV/PEVs*

Plug-in Hybrid Electric Vehicle (PHEV) and Plug-in Electric Vehicle (PEV) technology is promising for automotive applications due to the enhanced fuel economy

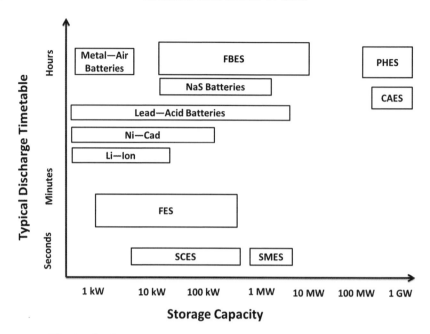

Fig. 8.   Discharge time vs. power ratings for each storage technology.

and the reduction of greenhouse gas emissions.[14] due to use of this technology.
The realization of this technology is closely tied to the recent and future improvements of battery technology as the main energy-storage mechanism. Accordingly,
several reports from different organizations[15–18] have evaluated the technical goals
and the state-of-the-art battery technology in PHEV. In this chapter, the PHEV
battery goals set by the US Advanced Battery Consortium (USABC)[18] along
with alternative analyses conducted by the Sloan Automotive Laboratory at the
Massachusetts Institute of Technology (MIT),[17] and the Electric Power Research
Institute (EPRI)[16] are examined. USABC specifies two main PHEV battery types:
A high power to energy ratio battery providing an all-electric range of 10 miles
(PHEV-10) for crossover SUVs, and a low power to energy ratio battery providing
40 miles of all-electric range (PHEV-40) for a mid-size sedan. These categories follow the California Air Resources Board's (CARB)[19] definition of PHEV-X, where X
is the number of miles the vehicle can drive in all-electric mode during a particular
drive cycle, before the gasoline engine turns on.

    The required battery characteristics for any PHEV design depends on the basic
assumptions regarding charge depleting (CD) range, CD operation mode, drive
cycle, vehicle design, and charge sustaining (CS) behavior. After specifying the basic
assumptions such as weight and volume of the battery for PHEV-10 and PHEV-40,
five major categories are indicated in Table 4 to define the battery goals for the
PHEV: power, energy capacity, life, safety and cost. There are inherent tradeoffs
between these main categories: higher power density is subject to higher voltage

Table 4: USABC Goals for Advanced PHEV Batteries.[15]

		PHEV-10	PHEV-40
Basic assumptions	Body type	Crossover SUV	Midsize Sedan
	All electric range (miles)	10	40
	Max system mass (kg)	60	120
	Max system volume (liters)	40	80
Power	Peak power (kW)	50/45	46/38
	Power density (W/kg)	830/750	380/320
Energy capacity	Available energy (kWh)	4	12
	Total energy [@70% Depth of Discharge (DOD)] (kWh)	6	17
	Total energy density (Wh/kg)	95	140
Life	Calendar life (years)	15	15
	Deep discharge cycles (cycles)	5,000	5,000
	Shallow discharge cycles (cycles)	3,00,000	3,00,000
	Temperature (°C)	−46 to +66	−46 to +66
Safety	Abuse tests	Acceptable	Acceptable
Cost	OEM[a] price (US$)	1,700	3,400
	OEM price (US$/kWh)	300	200

[a]OEM stands for original equipment manufacturers.

and higher voltage reduces the longevity and safety of the battery and increases the cost. On the other hand, increasing the energy density of the battery will lead to a decrease in the power density. Moreover, because USABC goals are highly dependent on various assumptions, MIT and EPRI alternative goals and analyses are compared to the USABC goals in Table 5 for similar categories. These five categories, peak power, energy capacity, lifetime, safety and cost are discussed in the following sections.

4.2 *Peak Power*

The USABC's peak power goals are defined based on short accelerations (pulses) between 2 and 10 sec. According to Table 5, the PHEV-10 battery is expected to provide 50 kW of power, while the PHEV-40 requires 46 kW. The slightly increased power required for PHEV-10 is due to the increased weight, rolling resistance, and frontal area (drag) of the crossover SUV compared to the PHEV-40 sedan.[15] In evaluating battery technologies, analysts typically refer to power density as the power per kilogram of the battery system (W/kg). The USABC's target weight for the PHEV-40 battery pack is 120 kg, resulting in a power density of 380 W/kg. The target weight for the PHEV-10 battery pack is 60 kg, resulting in a power density of 830 W/kg — more than double the PHEV-40 density.

Table 5:   Comparing PHEV Assumptions and Battery "Goals".[15]

	USABC		MIT	EPRI	
Vehicle assumptions					
CD range (miles)	10	40	30	20	60
CD Operation	All-electric	All-electric	Blended	All-electric	All-electric
Electricity use (kWh/mile)	0.42	0.30	0.19	0.24	0.24
Depth of discharge (percent)	70%	70%	70%	80%	80%
Body type	Cross SUV	Mid. Car	Mid. Car	Mid. Car	Mid. Car
Battery mass total (Cells only) (kg)	60 (45)	120 (90)	60 (45)	159 (121)	302 (252)
Total vehicle mass (kg)	1,950	1,600	1,350	1,664	1,780
Battery goals					
Peak power (kW)	50	46	44	54	99
Peak power density (W/kg)	830	380	730	340	330
Total energy capacity (kWh)	6	17	8	6	18
Total energy density (Wh/kg)	100	140	130	40	60
Calendar life (years)	15	15	15	10	10
CD Cycle life (number of cycles)	5,000	5,000	2,500	2,400	1,400
CS Cycle life (number of cycles)	3,00,000	3,00,000	1,75,000	<2,00,000	<2,00,000

4.3 *Energy Capacity*

Energy capacity goals, including the amount of energy stored in the battery and the battery's energy density, determine the distance that can be traveled in charge depleting mode versus the mass of the battery system. In Table 5, an important distinction is made between available and total energy. While a battery may have 10 kWh of total energy, only a portion of this capacity is available for vehicle operations.[15] This range of operation in practice is called the usable depth of discharge (DOD). The USABC values in Table 5 assume a 70% DOD, meaning that the total energy goal required for each battery is 143 (100/70)% of the required available energy. Similar to the peak power, the common metric of battery energy is energy density, measured as the total Wh/kg of the battery system. The USABC's energy density goals are 100 Wh/kg for the PHEV-10, and 140 Wh/kg for the PHEV-40. MIT's goal is within this range (130 Wh/kg) while EPRI's goals are much lower (40–60 Wh/kg). Figure 9 presents Ragone plots (power density versus energy density) of different types of battery technologies compared to the battery goals. The light gray bands present the power and energy capabilities, and tradeoffs, of lead–acid, nickel–cadmium, NiMH, ZEBRA, and Li-ion chemistries. Two major categories of battery chemistries are close to meeting the PHEV goals: Nickel–Metal Hydride (NiMH) and Lithium–Ion (Li–ion). Table 6 compares some categories of NiMH and Li–ion battery chemistries as reported by Axsen *et al.*[15] Li–ion technologies hold the promise of much higher power and energy density goals, because they use lightweight material, have the potential for high voltage, and are expected to

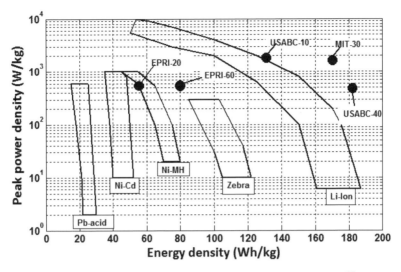

Fig. 9. Battery potential and PHEV "Goals" (Ragone Plots).[15]

Table 6: NiMH vs. Li-Ion Battery Technologies.[14]

	NiMH	Li-Ion
Power density (W/kg)	250	540
Energy density (Wh/kg)	57	94
Cycle life (cycles)	>3,000	> 3,200

have lower costs. The NiMH battery could play a temporary role in a less demanding design but it is probable that falling Li–ion battery prices may disqualify NiMH from even this role. One drawback of Li–ion batteries, however, is the need for safety as discussed in Sec. 4.5.

4.4 *Lifetime*

With use and over time, battery performance can substantially degrade, including peak power, energy capacity, and safety. Table 5 indicates four key measures of battery longevity.

1. **Calendar life** is the ability of the battery to endure degradation over time, which may be independent on how much or how hard the battery is used. The USABC goal for batteries for both vehicles is 15 years at a temperature of 35°C. MIT also targets 15 years of calendar life. EPRI uses a less ambitious target of 10 years.[15]

2. **Deep cycle life** is the number of charge/discharge cycles the battery can perform in charge depleting mode. The USABC's battery goal is 5000 deep cycles. Other studies set less ambitious targets; MIT states 2500 deep cycles for a PHEV-30,

and EPRI states 2400 and 1400 deep cycles for the PHEV-20 and PHEV-60, respectively.[15]

3. **Shallow cycles** refer to State of Charge (SOC)[b] variations of only a few percent. These smaller variations occur during charge depleting (CD) and charge sustaining (CS) modes. These frequent shallow cycles cause less degradation than deep cycles, but still influence the longevity. The USABC longevity target is 300,000 shallow cycles for both PHEV designs, again much higher than the 175,000 set by MIT, or the 200,000 set by EPRI.[15]

4. **Survival temperature range** is the range of temperatures the battery can be exposed to while not in operation, neither charging nor discharging. The USABC target range is $-46°C$ to $+66°C$, which easily covers natural conditions in most environments.[15]

4.5 *Safety*

Safety is another important issue because batteries store energy and contain chemicals that can be harmful if emitted under uncontrolled conditions. These conditions comprise short circuits, overcharging, crashes, or high temperature. In automotive applications, batteries use battery management units (BMU) that provide a higher degree of safety than typical consumer applications. The BMU, for example, monitors the cell voltage and temperature, and takes corrective action when necessary. The more advanced version of the BMU, a Battery Management System (BMS), contains many advanced features to enhance safety and also more functionality as will be explained in Sec. 4.8. Although the USABC's battery goals do not explicitly specify safety objectives, safety is implied in the goals of longevity and operating temperature. Some of the abuse tolerance tests to be performed on batteries to measure safety "comprise mechanical crashing, perforation, short circuit, overcharging, overheating, fuel fire immersion and water immersion".[15] In each test, the battery's response is recorded and assessed in regards to longevity and threats to personal safety.[15]

4.6 *Cost*

Battery cost is one of the most crucial factors affecting the commercial realization of electric vehicles. The USABC cost goals are \$1,700 and \$3,400 for the PHEV-10 and PHEV-40 battery packs, respectively, under a scenario where battery production has reached 100,000 units per year. These goals are set as costs to the original equipment manufacturers (OEMs), and do not include the markup that would be passed on to consumers.[15] To facilitate comparison, battery cost is commonly measured in dollars per total kWh (not just available kWh), which equates to \$300/kWh for

---

[b]The amount of energy left in a battery compared to the energy it has when it is full.

the PHEV-10 and \$200/kWh for the PHEV-40. In the MIT analysis, a value of \$320/kWh is assumed to be required for the commercialization of a PHEV-30.[15]

4.7 *Battery Technologies for the Smart Grid*

The role of electrochemical energy-storage systems (mainly batteries) for optimizing the use of renewable energy resources can also reduce the world's energy dependence on oil. Advanced energy storage could enable increased market penetration for many renewable energy sources such as solar and wind. The targets of this application are different than those for transportation, and alternative electrochemical energy-storage technologies need to be considered. In this application, energy density is less important than for PHEV and PEV applications. As stated by Peterson,[20]

> *The most important issues are (a) Low cost, (b) Long cycle and calendar life, (c) High system reliability, (d) Low maintenance, (e) Low self-discharge rates, and (f) High system efficiency. That is why in this area the VRLA battery is still very useful and battery chemistries with large energy capacity are more popular.*[20]

Battery energy storage can be integrated with renewable energy generation systems in either grid-connected or stand-alone applications. For stand-alone systems, batteries are essential to store electricity for use when the sun is not shining or when the wind is not blowing. For grid-connected systems, batteries support intermittent renewable resources by facilitating a better match between demand and supply.

4.8 *Battery Management System*

As mentioned in the discussion of battery safety in Sec. 4.5, the, BMS plays a crucial role in enhancing the reliability and safety and prolonging battery life. With increased demand for advanced battery technology in electric vehicles and the smart grid, an accurate and reliable BMS is vital. A typical BMS contains the following basic features[9]:

(a) *Cell Protection:* Protecting the battery from out of tolerance operating conditions. Cell thermal protection is also part of this feature.
(b) *Charge control:* Protecting the battery from inappropriate charging.
(c) *SOC Estimation:* Many applications require knowledge of the (SOC) of the battery pack or of the individual cells in the stack. Research is ongoing to give a more accurate estimation of SOC.[5,6,21−23]
(d) *SOH Estimation:* The State of Health (SOH) is a measure of a battery's capability to deliver its specified energy capacity and its ability to be charged and discharged. This is necessary for assessing the readiness of emergency power equipment and is an indicator of whether maintenance actions are needed.

Research is ongoing in this area to define and estimate the SOH[24–27] and predict the end of life (EOL)[28–30] of the battery.

(e) *Cell Balancing:* In multi-cell battery packs, small differences between cells due to production tolerances or operating conditions tend to be magnified with each charge/discharge cycle. Weaker cells become overstressed during charging, causing them to become even weaker, until they eventually fail causing premature failure of the battery. Cell balancing is a way of compensating for weaker cells by allocating the charge on all the cells in the chain and thus extending battery life.[1]

5 Summary

Several battery technologies, including their electrochemical structures, advantages and disadvantages, applications and costs, are compared. A comparison of batteries to other energy-storage technologies implies that battery technology holds promise in smart grid applications, and especially for electric vehicles. Rechargeable lithium batteries currently lead the way in approaching the PHEV battery goals. Battery production technology is moving very quickly to improve energy and power capacity and reduce costs. In addition, BMS is another rapidly evolving area designed to enhance the efficiency and reliability of using batteries in different applications by monitoring SOC, SOH and EOL and applying cell balancing.

References

1. Wikipedia (2012). Available at: http://en.wikipedia.org/wiki/Battery. Accessed 5 July 2012.
2. H. Rahimi–Eichi and M.-Y. Chow, Modeling and analysis of battery hysteresis effects, *IEEE Energy Conversion Congress and Exposition* (Raleigh, NC, 2012).
3. H. Zhang and M.-Y. Chow, Comprehensive dynamic battery modeling for PHEV applications, Paper presented at *IEEE Power and Energy Society General Meeting*, 25–29 July (Minneapolis, 2010).
4. H. Zhang and M.-Y. Chow, On-line PHEV battery hysteresis effect dynamics modeling, Paper presented at *IECON 2010 — 36th Annual Conf. IEEE Industrial Electronics*, 7–10 November (Piscataway, NJ, USA, 2010).
5. H. Rahimi-Eichi, F. Baronti and M.-Y. Chow, Modeling and online parameter identification of Li-polymer battery cells for SOC estimation, Paper presented at *ISIE 2012–2021th IEEE Int. Symp. Industrial Electronics*, 28–31 May (Hangzhou, Zhejiang, China, 2012).
6. H. Rahimi-Eichi and M.-Y. Chow, Adaptive parameter identification and State-of-Charge estimation of lithium–ion batteries, *38th Annu. Conf. IEEE Industrial Electronics Soc.* (Montreal, Canada, 2012).
7. M. Chen and G. A. Rincon-Mora, Accurate electrical battery model capable of predicting runtime and I-V performance, *IEEE Trans. Energy Conv.* **21**(2)(2006), pp. 504–511.
8. S. Abu-Sharkh and D. Doerffel, Rapid test and non-linear model characterisation of solid-state lithium–ion batteries, *J. Power Sources* **130**(2004), pp. 266–274.

9. Woodbank Communications Ltd, *Electropedia* (Battery and Energy Technologies, 2005) Available at: http://www.mpoweruk.com/. Accessed 5 July 2012.

10. ICCNEXERGY (2011). *Battery Chemistry Comparison Chart*, Available at: http://www.iccnexergy.com/battery-chemistry-comparison-chart/. Accessed 5 July 2012.

11. Battery University. Available at: http://batteryuniversity.com/learn/article/whats_the_best_battery. Accessed 5 July 2012.

12. S. Sun, Grid-scale energy storage: State of the market, Paper presented at *Bloomberg New Energy Finance Summit* (New York, 2011).

13. D. Connolly, A review of energy storage technologies for the integration of fluctuating renewable energy (University of Limerick, Limerick, Ireland, 2009).

14. W. Su, H. Rahimi-Eichi, W. Zeng and M-Y Chow, A survey on the electrification of transportation in a smart grid environment, *IEEE Trans. Ind. Inf.* **8**(1)(2012), pp. 1–10.

15. J. Axsen, Batteries for PHEVs: Comparing goals and the state of technology, in Electric and hybrid vehicles overview: Power sources, models, sustainability, infrastructure and the market, (ed.) G. Pistoia (Elsevier, Amsterdam, 2010).

16. M. Duvall, Comparing the benefits and impacts of hybrid electric vehicle options for compact sedan and sport utility vehicles (EPRI, Palo Alto, CA, 2002).

17. M. Kromer and J. Heywood, Electric powertrains: Opportunities and challenges in the U.S. light-duty vehicle fleet, LFEE 2007–2003 RP (Massachusetts Institute of Technology, Sloan Automotive Laboratory, Cambridge, MA, 2007).

18. A. Pesaran, T. Markel, H. Tataria and D. Howell, Battery requirements for plug-in hybrid electric vehicles: Analysis and rationale, *23rd Int. Electric Vehicle Symp. Exposition (EVS-23)* (Anaheim, California, 2007).

19. California Air Resources Board–CARB (2008b), Proposed Amendments to the *California Zero Emission Vehicle Program Regulations*, 8 February 2008. Available at: http://www.arb.ca.gov/regact/2008/zev2008/zevisor.pdf. Accessed 5 July 2012.

20. S. B. Peterson, J. Apt and J. F. Whitacre, Lithium–ion battery cell degradation resulting from realistic vehicle and vehicle-to-grid utilization, *J. Power Sources* **195**(8)(2010), pp. 2385–2392.

21. G. L. Plett, Extended Kalman filtering for battery management systems of LiPB-based HEV battery packs — Part 2. Modeling and identification, *J. Power Sources* **134**(2)(2004), pp. 262–276.

22. F. Zhang, G. Liu, L. Fang and H. Wang, Estimation of battery state of charge with H observer: Applied to a robot for inspecting power transmission lines, *IEEE Trans. Ind. Electron.* **59**(2)(2012), pp. 1086–1095.

23. Y. Hu and S. Yurkovich, Battery cell state-of-charge estimation using linear parameter varying system techniques, *J. Power Sources* **198**(2012), pp. 338–350.

24. E. V. Thomas, I. Bloom, J. P. Christophersen and V. S. Battaglia, Rate-based degradation modeling of lithium–ion cells, *J. Power Sources* **206**(2012), pp. 378–382.

25. J. Zhang and J. Lee, A review on prognostics and health monitoring of Li-ion battery, *J. Power Sources* **196**(15)(2011), pp. 6007–6014.

26. Y-H Sun, H-L Jou and J-C Wu, Aging estimation method for lead–acid battery, *IEEE Trans. Energy Conv.* **26**(1)(2011), pp. 264–271.

27. O. Erdinc, B. Vural and M. Uzunoglu, A dynamic lithium–ion battery model considering the effects of temperature and capacity fading, Paper presented at *Int. Conf. Clean Electrical Power, ICCEP 2009*, 9–11 June (Capri, Italy, 2009).

28. K. Goebel, B. Saha, A. Saxena, J. Celaya and J. Christophersen, Prognostics in battery health management, *IEEE Instrum. Meas.* **11**(4)(2008), pp. 33–40.

29. B. Saha, K. Goebel, S. Poll and J. Christophersen, Prognostics methods for battery health monitoring using a Bayesian framework, *IEEE Trans. Instrum. Meas.* **58**(2)(2009), pp. 291–296.
30. B. Saha, E. Koshimoto and C. C. Quach *et al*, Battery health management system for electric UAVs, Paper presented at *IEEE Aerospace Conf. 2011*, 5–12 March (2011).

Chapter 19

Fuel Cells and the Hydrogen Economy

John T.S. Irvine,* Gael P.G. Corre, and Xiaoxiang Xu
School of Chemistry
University of St Andrews, Fife, KY16 9ST, UK
*jtsi@st-andrews.ac.uk

The present chapter summarizes the current state and perspectives of fuel cells and hydrogen energy. First, the relevant fuels and fuel chemistry are discussed, focusing on hydrogen, hydrocarbons and oxygenates, followed by a discussion on the basics of fuel cell technology and applications. The different types of fuel cells are presented, with particular attention on solid oxide fuel cells, and thermodynamics and the factors influencing efficiency reviewed.

1 Introduction

The fuel cell concept, ascribed to Sir Humprey Davy, dates from the beginning of the 19th century. The first hydrogen–oxygen cell was successfully operated by Sir William Grove in 1839[1] and is generally referred to as the first fuel cell. While investigating the electrolysis of water, Grove observed that when the current was switched off, a small current flowed through the circuit in the opposite direction, as a result of a reaction between the electrolysis products, hydrogen and oxygen, catalyzed by the platinum electrodes. Grove recognized the possibility of combining several of these in series to form a gaseous voltaic battery,[2] and also made the crucially important observation that there must be a "notable surface of action" between the gas, the electrolyte and the electrode phases in a cell. Maximizing the area of contact between these three phases remains at the forefront of fuel cell research and development. Some 50 years after Grove's "gas battery," Mond and Langer introduced the term fuel cell[3] to describe their device, which had a porous platinum black electrode structure, and used a diaphragm made of a porous non-conducting substance to hold the electrolyte.

Despite the fact that the fuel cell was discovered over 160 years ago, with the high efficiencies and environmental advantages that it offered, it is only now that fuel cells are approaching commercial reality.

A fuel cell is essentially an energy conversion device that produces electricity, and may be viewed as a battery with external fuel supply. A fuel cell consists of four

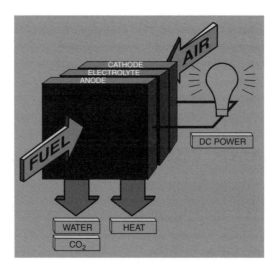

Fig. 1.   Principle of an individual fuel cell.[4]

essential components: two electrodes, the *anode* and the *cathode*, separated by an *electrolyte*, and connected by an external circuit or *interconnect*, as shown in Fig. 1 These may be further subdivided in relation to function. Fuel is fed to the anode, where it is oxidized, releasing electrons to the external circuit. Oxidant is fed to the cathode where it is reduced using the electrons delivered by the external circuit. The electrons flow through the interconnect from the anode to the cathode, producing direct-current electricity. In theory, any gas capable of electrochemical oxidation and reduction can be used as fuel or oxidant in a fuel cell. Air is the most common oxidant for fuel cells, since it is readily and economically available from the atmosphere. Hydrogen, which offers high electrochemical reactivity, is the most obvious fuel. However, fuel cells can be developed to work with alternative fuels to hydrogen.

The essential feature of a fuel cell is its high energy conversion efficiency. Because fuel cells convert the chemical energy of the fuel directly to electrical energy without the intermediate of thermal energy (unlike indirect conversion in conventional systems), their conversion efficiency is not subject to the Carnot limitation. Efficiency can be further improved to levels as high as 80%, when the produced heat is used in combined heat and power, or gas turbine applications. Besides their high energy conversion efficiency, fuel cells offer several other advantages over conventional methods of power generation. They produce much lower amount of pollutants, certainly at point of use. A fuel cell running on H_2 and air only produces water. Other significant advantages offered are modular construction and size flexibility, which makes them well suited for decentralized applications, high efficiency at part load, fuel flexibility and remote/unattended operation. Moreover, their vibration-free operation reduces noise usually associated with conventional power generation systems, although there is still some noise due to compressors and pumps. More details concerning general features of fuel cells can be found in the literature.[4,5]

2 Fuel Cell Types

There is a wide range of fuel cells in different stages of development. Although all types of fuel cells have the same basic operating principle, they have different characteristics that stem from the nature of electrolyte involved. The six main types are summarized in Table 1. The nature of the electrolyte determines the nature of mobile ions transferred and the direction of this transport, which in turn determines on which side of the electrolyte water is produced. Moreover, each electrolyte must be operated in a specific temperature range, which is a major difference in characteristics between different types of fuel cells. Molten Carbonate Fuel Cells (MCFC) and Solid Oxide Fuel Cells (SOFC) have elevated operating temperature, compared to much lower operating temperature for Alkaline (AFC), Polymer Electrolyte Membrane (PEMFC), Solid Acid (SAFC) and Phosphoric Acid (PAFC) fuel cells. The operating temperature dictates in turn the physicochemical and thermomechanical properties of materials to be used as cell component as well as the type of fuel the cell can be operated on. Moreover, this difference in operating temperatures has a number of implications for the applications for which particular fuel cell types are most suited.

Large differences exist in application, design, size, cost and operating range for the different type of fuel cells. Of the available fuel-cell technologies, PEMFCs and SOFCs are thought to have the most potential to achieve cost and efficiency targets for widespread use in power generation, and have been the most investigated types.

In general, high-temperature fuel cells exhibit higher efficiencies and are less sensitive to fuel composition. PEMFC systems require a pure H_2 fuel stream because the precious metal anode catalysts are poisoned by even low levels of CO or other compounds such as those containing sulfur. Current PEMFC systems operate below

Table 1:　Different Types of Fuel Cells and Characteristics.[4]

Fuel cells	Electrolyte	Charge carrier	Working temperatures	Anticipated scales	Anticipated electrical efficiency (hhv)
Proton exchange membrane fuel cells (PEMFCs)	Ion exchange membranes	H^+	80–120°C	1–100 kWe	40%–50%
Solid acid fuel cells (SAFCs)	Solid acids	H^+	20–250°C		
Alkaline fuel cells (AFCs)	KOH solution	OH^-	80°C	10–100 kWe	60%
Phosphoric acid fuel cells (PAFCs)	H_3PO_4	H^+	180–200°C	100–500 kWe	40%
Molten carbonate fuel cells (MCFCs)	Immobilized liquid molten carbonates	CO_3^{2-}	~650°C	300 kWe–3 MWe	45%–50%
Solid oxide fuel cells (SOFCs)	Ceramics	O^{2-} or H^+	600–1000°C	1 kWe–2 MWe	60%

$100°C$, but there is a great deal of research to find polymer electrolytes that can operate at higher temperatures since increasing the operating temperature relaxes the fuel-purity requirements relative to catalyst poisoning. In contrast, due to their high operating temperature, CO is rather a fuel than a poison for SOFC. Hence, high-temperature fuel cells can be operated on fuels other than H_2.

3 Fuels

3.1 *Hydrogen*

Although fuel cells are intimately linked with hydrogen, a major advantage is their potential for use in a wide range of fuels. This feature is particularly important for high-temperature fuel cells such as MCFCs and SOFCs, although other types of fuel cells, such as PEMFCs, are well suited to operate on fuels such as alcohol.

However, hydrogen remains the preferred fuel for many reasons. In terms of power output, hydrogen is the most attractive fuel, since it shows the highest energy density per unit of mass and its oxidation requires only the breakage of the H–H bond, allowing for fast oxidation kinetics. Fuel cells typically produce higher power output when fueled with H_2 than with any other fuels. In terms of the environmental impact, the oxidation of hydrogen produces only water, allowing for pollution-free electricity generation. Finally hydrogen does not contain any carbon. Therefore, any concern related to solid carbon deposition on the anode does not apply to hydrogen.

Because they can efficiently convert hydrogen to electricity, fuel cells are considered a keystone of a potential future economy based on hydrogen. In the hydrogen economy, hydrogen would be used as a widespread energy carrier, providing the energy source for motive power, stationary power and portable electronic devices. Due to their high efficiency and the range of applications covered, fuel cells would then play a major role in converting the hydrogen to CO_2-free electricity and possibly valuable heat in the case of high-temperature fuel cells. Nonetheless, many technological hurdles need to be overcome before the transition to any hydrogen-based economy can be considered. Important advances are required mainly in the transport, storage and production of hydrogen.

Since molecular hydrogen is not available on earth in convenient natural reservoirs (most of the hydrogen on earth is bonded to oxygen in water), it needs to be produced, and hydrogen is hence referred to as an energy vector or as energy carrier The vast majority of hydrogen (96%[6]) is currently being generated by the reforming of hydrocarbons, which has two important consequences: the feedstock used is non-renewable and the production process emits CO_2.

Hydrogen can be produced through the electrolysis of water. In this process, electricity is used to split water molecules into hydrogen and oxygen. Pollution-free electricity generated using renewable sources such as solar and wind energy can be used, meaning such a process is well suited for decentralized hydrogen production. In this respect, an interesting example is the Shetland Islands in Scotland, where wind power is used to produce hydrogen in the aim of fulfilling the community energy needs. This project is the first off-grid renewable hydrogen system in Europe and the first community-owned hydrogen production plant in the world.[7] On a larger scale, the use of nuclear energy could be considered to provide the electricity required for water electrolysis, but uranium is non-renewable and the development of this kind of energy is opposed in many countries. High Temperature Electrolysis (HTE) is another process being considered. In this process, both electricity and heat are used to convert water to hydrogen.

Another interesting alternative would the reforming of biofuels. Whilst the current reforming of fossil fuels uses non-renewable feedstock, replacing fossil fuels in this process with renewable biofuels could provide a sustainable mean of hydrogen production. The first hydrogen production plants reforming bio-ethanol have recently been built in the USA. However, if the same biofuels could be efficiently oxidized in fuel cells, both efficiency and system simplicity could be gained. Other means of hydrogen production from biomass include fermentative production and biological production.

Besides its production, many issues associated with hydrogen storage and transport need to be addressed. A transition to a hydrogen economy requires the ability to safely and efficiently transport and store molecular hydrogen. The difficulties associated with storage and transport stem from low energy density per unit of volume of hydrogen at ambient conditions. Despite showing a high energy per unit of mass, hydrogen has a low molecular weight. To be stored for practical applications, such as onboard a vehicle for motive power, hydrogen must be pressurized or liquefied to provide sufficient driving range. Achieving the high pressures required to obtain reasonable storage volumes necessitates high use of energy to power the compression. Alternatively, the use of liquid hydrogen is being considered. Liquid hydrogen possesses a higher volumetric energy density but is cryogenic and boils at $-250°C$. While being attractive in terms of weight, cryogenic storage is an energy–consuming process. Moreover, liquefied hydrogen has lower energy density by volume than gasoline by approximately a factor of 4, due to the low density of liquid hydrogen [there is actually more hydrogen in a liter of gasoline (116 g) than there is in a liter of pure liquid hydrogen (71 g)].

The inherent difficulties related to both compressed and liquefied hydrogen storage have led to the investigation of alternative storage methods. Hydrogen can be stored as a chemical hydride. In this storage method, hydrogen gas is reacted with a solid material to produce the hydride, which is much easier to transport. The hydride is then made to decompose at the point of use, yielding hydrogen gas. Current barriers to practical storage systems stem from the high pressure and temperature conditions needed for hydride formation and hydrogen release. For many potential systems, hydriding and dehydriding kinetics and heat management are also problems that need to be overcome. Another method is to absorb molecular hydrogen into a solid storage material such as carbon nanotubes. Unlike hydrides, the hydrogen does not dissociate/recombine upon charging/discharging the storage system and hence does not suffer from kinetics limitations.

The problems related to hydrogen transport basically stem from the same causes as for storage. Although the infrastructure used for the transport of natural gas could be used, hydrogen embrittlement of steel requires the pipes to be coated on the inside or new pipelines to be installed. Although expensive to install, once in place, pipelines are the cheapest way to transport hydrogen. Transport can be achieved using compressed hydrogen tanks or liquid hydrogen tanks, but involve the energy costs previously mentioned.

To summarize, the hydrogen economy could offer an interesting alternative to the current fossil fuel–based economy. The implementation of a hydrogen economy will undoubtedly accelerate the research and development of fuel cells. But before the above-mentioned technological hurdles are overcome, the use of hydrogen as an energy carrier will be limited to small scale and community projects such as the Shetlands community, the hydrogen highway in California or public transports in big cities,[6,8]

3.2 *Fuel Processing*

The mismatch between the desired fuel, hydrogen, and the available fuels, has definitely contributed to the limited commercial implementation of fuel cells so far. The important drawbacks that hinder the widespread use of hydrogen as an energy carrier have led to the development of fuel cell systems that rely on practical fuels.[9] Figure 2 illustrates the general concepts and requirements of processing gaseous,

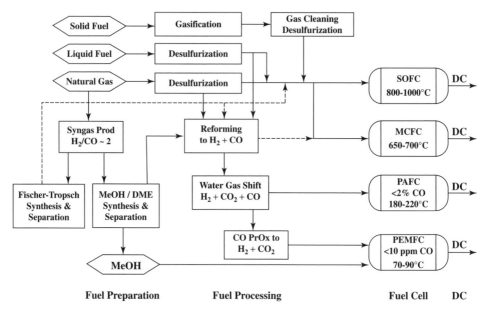

Fig. 2. The concepts and steps for fuel processing of gaseous, liquid and solid fuels for high-temperature and low-temperature fuel cells.[10]

liquid and solid fuels for fuel cell applications.[10] Different strategies must be applied according to the type of fuel cell involved. Higher temperatures release the requirements on fuel purities, and typically, less processing is required with higher temperatures.

Low-temperature fuel cells, such as PEMFC and PAFC, require many fuel processing steps. The fuel has to be converted to a fairly pure hydrogen rich gas. The conversion of hydrocarbon-related fuels to hydrogen can be done mainly by steam reforming, partial or complete oxidation, and is performed in a set of reactors external to the fuel cell stack. Catalysts are poisoned by CO at low temperatures. The CO content of the fuel entering the fuel cell stack needs to be reduced to the purity levels required by the type of cell involved. Requirements can be as strict as a few of ppm for PEMFC.

The on-site pre-processing of the fuel provides a suitable solution for supplying hydrogen to fuel cells. Transport and storage are no longer subjects of concern, since hydrogen is stored chemically in liquid fuels or gases that can be more easily transported. However, the addition of external reactors leads to significant complications in the system, by increasing system complexity and costs and decreasing efficiency.

As opposed to lowtemperature cells, CO is not a poison for SOFCs but can be oxidized and therefore acts as a fuel. SOFCs are the most flexible fuel cells with respect to their multifuel capability so that not only hydrogen and carbon monoxide but also various kinds of hydrocarbon-related species could be used in the feed. Purity requirements are less stringent for SOFCs due to their higher temperature. The desulfurization step is still required but as can be seen on Figure 2, the route from chemicals to electricity could be greatly simplified if SOFCs were to be fed directly on hydrocarbons.[9,10]

Due to the high temperature, feeding SOFCs with hydrocarbons leads to multiple and complex physicochemical processes that will considerably modify the feed composition before any electrochemical reaction take place. Taking advantage of the high temperature, the conversion of the fuel to more reactive species such as hydrogen and carbon monoxide can be performed *in situ* on the anode, which serves both as a catalyst and electrocatalyst, eliminating the need for external converters. This approach requires the fuel to be diluted with steam, carbon dioxide or oxygen to provide reactants for steam reforming and/or partial oxidation reactions and so lessens the overall efficiency by diluting the fuel.

Processes occurring within the high-temperature environment involve both homogeneous and heterogeneous chemical processes. Homogeneous processes occur regardless of the nature of the materials involved, while heterogeneous processes involve material surfaces, hence making the choice of anode materials of prime importance.

The main problem when feeding an SOFC with hydrocarbon is carbon deposition, which is a serious problem with any processes that involve hydrocarbons at high temperature. Two different types of carbon deposition can occur. The first mechanism results from reactions over a catalyst and thus belongs to the heterogeneous chemistry. This is a well-known process, which has been intensively studied over Ni, Fe and Co catalysts,[11-16] leading to the precipitation of carbon as a graphite fiber at some surface of the metal particle.[17] This deposition results in irreversible damage to the anode catalyst, but the damage can be prevented with addition of reforming agents in sufficient amounts. Thermodynamic studies reporting the amount of oxidant to be added to prevent carbon deposition have been performed for various fuels.

The second mechanism for carbon deposition occurs in the absence of a catalyst via free radical gas phase reactions.[18] Those reactions, usually initiated by the C–C bond scission, form polyaromatic compounds, perhaps best referred to as tars and quite different in form from the graphitic fibers. These polyaromatic compounds have very low vapor pressure and hence easily deposit on surfaces.[19] While the catalytic deposition irreversibly damages the active surface, the carbon deposits from gas phase reactions do not chemically interact with the anode surface. Carbon deposition by free-radical, gas-phase reaction is probably unavoidable for most hydrocarbon fuels other than methane above 700°C as shown by Gorte *et al.* in a series studies addressing the use of hydrocarbons on copper-ceria anodes.[20-23] However, the deposit layer can be controlled by appropriate choice of temperature and materials, such as ceriabased oxides, which can oxidize the deposits. Hence, a steady-state layer of deposited carbon can be achieved, which in turns can improve the performance through improve connectivity in the anode conductive phase.

Besides carbon formation, both homogeneous and heterogeneous chemistry will contribute to modify the fuel composition. The parent hydrocarbon may undergo pyrolysis. Under the effect of high temperature, hydrocarbons decompose, without any interaction either with catalytic surfaces or with additional species in the inlet stream. The importance of homogeneous pyrolysis in the operation of hydrocarbon fuelled SOFCs has been demonstrated by Dean *et al.*,[18,24] in studies of a variety of fuels under typical SOFCs conditions. As an example, at 800°C, butane undergoes full conversion for a residence time of 5 s. Oxygenates, such as ethanol, show a higher conversion than hydrocarbons, due to the presence of the OH group in that leads to a weakening of the C–H bonds. Besides, the decomposition is more efficient, as large amounts of hydrogen and CO are produced.

When the gaseous mixture reaches the surface of porous anodes, catalytic reactions will occur and play a definite role in further converting the fuel mixture before it participates in electrochemical reactions. The heterogeneous processes likely to occur are the steam reforming, CO_2 reforming, partial oxidation and autothermal reforming. These are basically the processes commonly used in the upstream processing of the fuel presented in Fig. 2. The nature of the reactions occurring along the anode chamber will be dictated by the nature of the catalysts present in the anode chamber, and the reactants. Since hydrocarbons are often diluted in stream rather than O_2 or CO_2, steam reforming will be the dominant conversion catalytic process within the anode. Even if the feed is a dry hydrocarbon, gas-phase reactions have been shown to produce steam, while electrochemical reactions at the TPB will

Table 2: Theoretical Energy Densities of Different Fuels (HHV).

Fuel	Gravimetric energy density (kWh/kg)	Volumetric energy density (kWh/L)
Hydrogen	39.406	0.003
Compressed hydrogen (700 bars)	39.406	1.250
Methane	15.445	0.011
Natural gas	14.889	0.010
CNG (250 bars)	14.889	2.500
Gasoline	12.889	9.500
Diesel	12.833	10.361
Propane	13.778	7.030
Methanol	6.361	5.037
Ethanol	8.333	6.667
Coal	9.023	20.111

also produce steam and CO_2 during operation. Hence, only a small fraction of the hydrocarbon fuel, if any at all, will be available for electrochemical reactions. H_2 and CO will likely be the oxidized species.

Specific energies of different fuels are shown in Table 2. Based on the availability of natural gas (methane with small amounts of other hydrocarbons) most stationary fuel cells have been designed for this fuel.

4 Fuel Cell Applications

The potential applications of fuel cells in society are ever increasing, driven by the various benefits that the implementation of fuel cells would bring over current technologies, such as environmental and efficiency improvements. Applications being considered range all the way from very small scale, requiring only a few watts, to larger-scale distributed power generation of hundreds of megawatts.

The small-scale power supply market is well suited for fuel cells. Indeed, fuel cells offer significantly higher power densities than batteries; also, they are smaller and lighter and have much longer lifetimes. Hence, an increasing number of applications requiring only a few watts are emerging, such as palm-top and laptop computers, mobile phones and other portable electronic devices.

Their potential high reliability and low maintenance coupled to their quiet operation and modular nature makes fuel cells well suited to localized "off grid" power generation, either for high-quality uninterrupted power supplies, or remote applications. High-temperature fuel cells (MCFC and SOFC) are suitable for continuous power production, where the cell temperature can be maintained. If the released heat is used to drive a gas turbine to produce extra energy, the system efficiency can be increased to levels as high as 80%, significantly higher than any conventional electricity generation process. Moreover, the produced heat makes SOFCs particularly suited to combined heat and power (CHP) applications ranging from less than 1 kW to several MW, which covers individual households, larger residential units and business and industrial premises, providing all the power and hot water from a single system.

The combination of their high efficiency (approaching 50% for Hydrogen PEMFC) and significantly reduced emissions of pollutants mean that fuel-cell–powered vehicles are a very attractive proposition, especially in heavily populated urban areas. The efficiency is to be compared with about 20% for a combustion engine. Low-temperature fuel cells, in particular PEMFC, are the most suited to transport applications, because of the need for short warm-up. The concept of a fuel-cell–powered vehicle running on hydrogen, the so-called "zero emission vehicle", is very attractive and currently an area of intense activity for almost all the major motor manufacturers. As an example, fuel-cell–powered buses, running on compressed hydrogen, are successfully being operated in several cities around the world.[6]

5 Proton-Conducting Electrolyte Fuel Cells

In Grove's prototype fuel cells, sulfuric acid was used as the electrolyte.[1] Although high efficiency and environmental benefits were promised by fuel cell concepts, developing the early scientific experiments into commercially industrial products proved difficult. The main problems are associated with developing appropriate materials and manufacturing techniques that enables fuel cells to become competitive with existing power generation methods in terms of cost per kWh. Three variants of proton-conducting electrolytes are considered here: polymer membrane, solid acid and phosphoric acid. All of these are currently being developed commercially.

5.1 *Proton Exchange Membrane Fuel Cells*

Proton exchange membrane fuel cells (PEMFCs) are composed of an ion-exchange membrane sandwiched between two electrode sheets. Various proton-conducting polymers could be fabricated as electrolytes in PEMFCs, such as Nafion and SPEEK. The only liquid in this kind of fuel cell is water so that corrosion problems are minimized. Water management in these systems is crucial since ion exchange membranes rely on water to be conductive but electrodes prefer a relatively "dry" condition for fuel transportations. A typical working temperature for PEMFCs is around 80°C and could be increased to 120°C at high-pressure conditions; this is mainly limited by the water balance in the membrane and electrodes. Noble metals such as platinum were used as catalysts due to their high catalytic activation at low temperatures, and high-purity hydrogen was chosen as fuel. Impurities such as H_2S and CO are detrimental to platinum catalysts and would lead to a marked degradation of fuel cell performance. This is due to the poisoning effect of CO to Pt as CO is more stably absorbed on Pt surface than H_2 at low temperatures (Fig. 3). The main problems for PEMFCs are high catalyst loading (Pt on anodes and cathodes), requisite for cooling systems and purified systems. Recent work is

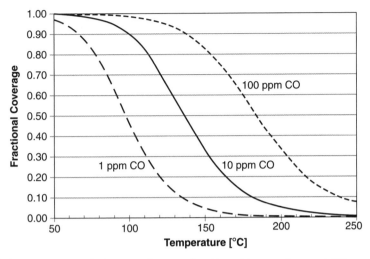

Fig. 3.   CO coverage on a platinum surface as a function of temperature and CO concentration. H_2 partial pressure is 0.5 bar.[25]

mainly focused on developing high-temperature ion exchange membranes so that fuel cells could be operated at higher temperatures. This gives the benefit of decreasing or even replacing the loading of noble metals and enables the fuel cells to be self-sustainable, avoiding the requirement of cooling and purifying systems.

5.2 *Solid Acid Fuel Cells*

Solid acids fuel cells (SAFCs) use solid acids as electrolytes. The charge carriers within the fuel cells are protons in most cases, so that fuels and products can be well separated by the electrolyte, a great advantage over oxygen ion conductors. However, some problems still need to be solved. One is the solubility of most solid acids (such as $H_3PW_{12}O_{40} \cdot \times H_2O$) in the water product that may lead to the collapse of the fuel cell infrastructures. Another is the requisite of humidified environments during fuel cell operations in many cases. Other technical problems such as thin membrane fabrication and fuel cell sealing were also encountered during fuel cell manufacturing. Recent studies are mainly focusing on synthesizing and developing stable solid acids that are independent of humidity at elevated temperatures. Figure 4 is a schematic display of a solid acid fuel cell based on CsH_2PO_4.[26] Thanks to advanced fabricating techniques, fuel cells could work with various fuels to provide promising applications in the future.

5.3 *Phosphoric Acid Fuel Cells*

Phosphoric acid fuel cells (PAFCs) are the first commercialized and widely used type of fuel cells. Hydrogen or a hydrogen-rich gas mixture is used as fuel. The

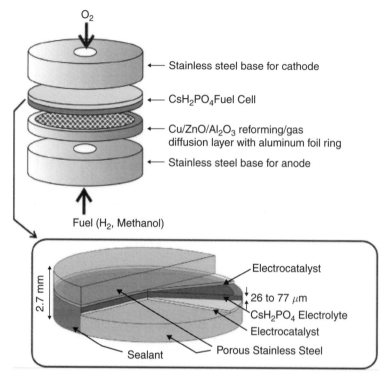

Fig. 4. Schematic display of the constitution of solid acid fuel cells.[26]

electrolyte, primarily composed of phosphoric acid (H_3PO_4), is a pure proton conductor with good thermal, chemical and electrochemical stability. The operating
temperature of the PAFCs is typically between 150–200°C and is a compromise
between the electrolyte conductivity (increases with temperature) and the cell life
(decreases with temperature).[27] The electrolyte, however, is highly corrosive so
that only noble metals such as Pt could be used. Presumably, the large capital
cost of PAFCs was expected and was only improved when Pt/carbon or graphite
electrodes were deployed in the 1960s.[28] Evolution and cell component for PAFCs
are listed in Table 3. However, one of the major problems in such fuel cells is the
accelerated carbon corrosion and Pt dissolution when cell voltages are above 0.8 V.
Therefore, long time, hot idles at open circuit have to be avoided. Another problem
usually encountered is the electrode flooding and drying and this has been recognized as one of the major causes of declining fuel cell performance. Migrations
of phosphoric acid between the matrix and the electrodes during cell load cycling
are responsible for this and an alternative matrix that is capable of maintaining
acid is still under development.[28] Recently, introducing phosphoric acid into polymers bearing basic groups such as ether, alcohol, imine, amide or imide groups has
attracted much interest. The resulting acid–base complex systems not only exhibit

Table 3:   Evolution and Cell Component Technology for Phosphoric Acid Fuel Cells.[28]

Component	ca. 1965	ca. 1975	Current status
Anode	PTFE-bonded Pt black	PTFE-bonded Pt/C Vulcan XC-72[a]	PTFE-boned Pt/C Vulcan XC-72[a]
	$9\,mg/cm^2$	$0.25\,mg\ Pt/cm^2$	$0.1\,mg\ Pt/cm^2$
Cathode	PTFE-bonded Pt black	PTFE-bonded Pt/C Vulcan XC-72	PTFE-bonded Pt/C Vulcan XC-72
	$9\,mg/cm^2$	$0.5\,mg\ Pt/cm^2$	$0.5\,mg\ Pt/cm^2$
Electrode Support	Ta mesh screen	Carbon paper	Carbon paper
Electrolyte Support	Glass fiber paper	PTFC-bonded SiC	PTFC-bonded SiC
Electrolyte	85% H_3PO_4	95% H_3PO_4	100% H_3PO_4

[a]Conductive oil furnace black, product of Cabot Corp. Typical properties: 002 d-spacing 3.6Å by X-ray diffusion, surface area of $220\,m^2/g$ by nitrogen adsorption, and average particle size of $30\,\mu m$ by electron microscopy.

a high conductivity but also possess reasonable mechanical stability at elevated temperatures ($>100°C$), such as polybenzimidazole (PBI)/H_3PO_4 systems.[29]

6 Solid Oxide Fuel Cells

6.1 *Basic Definitions*

An SOFC is defined by its solid ceramic electrolyte, which is a non-porous metal oxide. Such electrolytes are oxygen-ion (O^{2-}) conductors, impervious to gas flow and have negligible electronic conductivity. Solid oxide electrolytes require a high operating temperature to display suitable conductivities, typically in the range 700–1000°C.[30] SOFCs involve multiple complex physico-chemical processes. The principle of an SOFC, involving hydrogen as a fuel, is illustrated in Fig. 5.

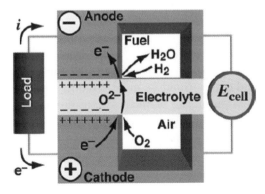

Fig. 5.   Principle of a solid oxide fuel cell involving H_2 as a fuel.[9]

Oxygen is electrochemically reduced at the cathode–electrolyte–gas interface. Electrons are delivered to the cathode through the interconnect, where they react with oxygen molecules in the gas phase to deliver oxygen ions to the electrolyte via a charge transfer reaction:

$$\frac{1}{2}O_2(g) + 2e^-(c) \Leftrightarrow O^{2-}(e). \tag{1}$$

In this equation, the three phases are denoted as (g) for the gas, (c) for the cathode and (e) for the electrolyte. Oxygen ions migrate through the electrolyte via a vacancy hopping mechanism toward the anode–electrolyte–fuel interface, where they participate in the fuel oxidation, written as follows for hydrogen:

$$H_2(g) + O^{2-}(e) \Leftrightarrow H_2O(g) + 2e^-(a), \tag{2}$$

where (a) denotes the anode. The hydrogen in the gas phase reacts with the oxygen ions provided by the electrolyte to deliver electrons to the anode. As long as a load is connected between the anode and the cathode, the electrons from the anode will flow through the load back to the cathode, and electrical current will flow through the circuit.

The overall electrochemical cell reaction in an SOFC, based on oxygen and hydrogen, is written as follows:

$$O_2 + H_2 \Leftrightarrow H_2O. \tag{3}$$

Another accurate way to represent the overall reaction occurring in the cell, regardless of the fuel involved, is to describe the oxygen transfer:

$$O_2(c) \Leftrightarrow O_2(a). \tag{4}$$

An SOFC can therefore be considered as an oxygen pump. The amount of oxygen transported to the anode will depend on the type of fuel used and the reactions occurring at the anode.

6.2 *History of SOFC*

The solid oxide fuel cell was first conceived following the discovery of solid oxide electrolytes in 1899 by Nernst.[31] Nernst discovered that the very high electrical resistance of pure solid oxides could be greatly reduced by the addition of certain other oxides. The most promising of these mixtures consisted mainly of zirconia (ZrO_2) with small amounts of added yttria (Y_2O_3). This is still the most widely used electrolyte material in the SOFC.

The first working SOFC was demonstrated by Baur and Preis in 1937, using stabilized zirconia as electrolyte and coke and magnetite, respectively, as a fuel and oxidant.[32] Unfortunately, the high operating temperature and the reducing nature of the fuel led to serious materials problems, and despite very significant efforts by Baur and other researchers, the search for suitable materials was unsuccessful.

This effectively hindered the development of solid oxide fuel cells until the 1960s, when a first period of intensive activity in SOFC development began. Intensive research programs that were driven by new energy needs for military, space and transport applications, addressed mainly the electrolyte conductivity improvement and the first steps in SOFC technology. A second period of intense activity began in the mid-1980s and still continues at present. These research efforts have led SOFC commercialization close to reality. Different companies have developed different concepts, and several demonstration units have been operated for significant amounts of times.

6.3 *Characteristics*

The advantages of the high operating temperature include the possibility of running directly on practical hydrocarbon fuels without the need for a complex and expensive external fuel reformer and purification systems. Internal reforming can be performed at high temperatures, and SOFCs are not poisoned by CO, which can be oxidized at the anode and act as a fuel. When practical fuels are used, the environmental impact is better than for combustion technologies, in the sense than less CO_2 and NO_x are produced per unit of power generated. Looking at the overall system efficiency, the high-quality exhaust heat released during operation can be used as a valuable energy source, either to drive a gas turbine when pressurized or for CHP applications.

Originally, SOFCs have been developed for operation primarily in the temperature range of 900–1000°C, which is beneficial for the fuel reforming, electrochemistry kinetics and the added value of the exhaust heat. However, some important drawbacks stem from such elevated temperatures. The materials that can be used are limited with respect to their chemical stability in oxidizing and/or reducing environment and their chemical and thermo-mechanical compatibility with adjacent components. Hence there are considerable efforts to lower the operating temperature by 200°C or more, which would allow the use of a broader set of materials, with less demands on seals and balance-of-plant components, simplifying thermal management, aiding in faster start-up and cool down, and resulting in less degradation of cell and stack compoennts.[5] Because of these advantages, activity in the development of SOFCs capable of operating in the temperature range of 600–800°C has increased dramatically in the last few years. However, at lower temperatures, electrolyte conductivity and electrode kinetics decrease significantly. In terms of applications, the length of time that is generally required to heat up and cool down the system restricts the use of SOFCs in applications that require rapid temperature fluctuations. This is a consequence of the need to use a relatively weak, brittle component as the substrate material and because of problems associated with thermal expansion mismatches. This restriction applies particularly for transport applications, where a rapid transport start-up is essential.

Reducing the cost of SOFCs is a crucial issue for their commercialization. Currently, the high cost-to-performance ratio limits SOFC's introduction to the energy market. In this respect, lower operation temperature also makes possible the use of inexpensive metallic interconnections in place of lanthanum chromite–based ceramic interconnections.

6.4 *Design*

The solid state character of all SOFC components means that, in principle, there is no restriction on the cell configuration. Instead, it is possible to shape the cell according to criteria such as overarching design or applications issues. As for other fuel cell concepts, it is necessary to stack SOFCs to increase the voltage and the power produced. A stack can, in principle, comprise any number of cells depending on the desired power, and a fuel cell plant can be designed in modules of stacks in series and parallel connections. To construct an electric generator, individual cells are connected in both electrical parallel and series to form a semi-rigid bundle that becomes the basic building block of a generator.

The most two common designs of SOFCs, the tubular and the planar, are represented in Fig. 6. In the tubular cells, the cell components are deposited in the form of thin layers on a cathode tube. In the planar design, the cell components are configured as thin, flat plates. The interconnection which is ribbed on both sides, forms gas flow channels and serves as a current conductor.

Alternative designs have been proposed, such as the planar segmented design developed at Rolls Royce,[33] or the SOFC roll developed at the University of St Andrews.[34] The latter is an innovative design that takes advantage of both planar and tubular designs.

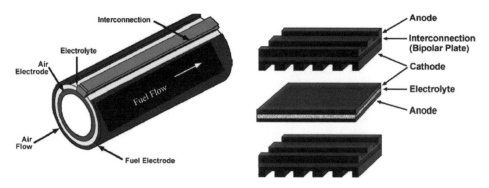

Fig. 6.   The two most common SOFC designs: Tubular (left) and planar (right).[4]

6.5 *Materials*

Electrodes, electrolyte and interconnect materials for SOFCs are limited in choice by several important requirements due to the high temperature. All materials must be non-reactive with adjacent components at the high operating temperature and must have compatible thermal expansion coefficients. Interconnects and electrolytes must be impermeable to gas, show high conductivities to minimize losses (electronic and ionic respectively) and be stable in both reducing and oxidizing atmospheres. Electrodes must show high electroctalytic activity and must be designed with an extended active surface area (Triple Phase Boundary points) since electrochemical reactions will occur only on sites possessing conductivities for those three phases detailed in Equations (1) and (2). Electrodes must fulfill some important requirements to ensure high and durable power output. To extend the TPB area, electrodes are fabricated as mixed ionic and electronic conductors (MIEC) porous ceramics or ceramic–metallic composites. An ideal microstructure would offer the highest triple-phase boundary (TPB) length for electrochemical reactions and an optimized contact between the electrolyte and the electrode, and be dimensionally stable during operation (mechanically, chemically and thermally).

The most commonly used electrolyte in SOFCs is Yttria-stabilized Zirconia (YSZ) as it possesses all required characteristics. ZrO_2, in its pure from, does not serve as a good electrolyte because of its very low ionic conductivity. The addition of certain aliovalent oxides stabilizes the cubic fluorite structure of ZrO_2 from room temperature to its melting point and, at the same time, creates a large concentration of oxygen vacancies by charge compensation. The properties of stabilized zirconia have been extensively studied, and several reviews dedicated to this material have been published.[35–37] Conventional zirconia-based SOFCs generally require an operating temperature above 850°C. This high operating temperature places severe demands on the material used as interconnects and for manfolding and sealing, and necessitate the use of expensive ceramic materials and specialist metal alloys. There is therefore considerable interest in lowering the operating temperature of SOFCs to below 750°C to enable the use of cheaper materials, such as stainless steel, and reduce fabrication costs, whilst maintaining high power outputs. Reducing the electrolyte thickness will obviously allow a reduction of the operating temperature but this approach is of course limited. The alternative route consists in developing new electrolyte materials showing higher conductivity than doped zirconia. SOFC electrolyte materials have been reviewed by Goodenough[38] and Skinner and Kilner[39] and are detailed in reviews addressing SOFCs.[8,30] Among those various materials, two promising alternative electrolytes to YSZ are gadolinia-doped ceria[40,41] and lanthanum gallate–based perovskites.[42] Both these electrolytes offer the possibility of lower temperature operation for SOFCs between 500 and 700°C. Scandia-doped zirconia has also received particular attention, since it has similar properties to YSZ but exhibits higher ionic conductivities, though it is also more expensive.

Lanthanum chromite ($LaCrO_3$) was at one stage considered as an important candidate for interconnects; however, metallic interconnects, such as chrome-based alloys, are now preferred, except in some special ceramic geometries with integrated interconnect. The main drawback associated with lanthanum chromite interconnect is their manufacturing costs due their difficult sinterability. For SOFCs operating in the intermediate temperature range of 500–750°C, it becomes feasible to use certain ferritic stainless-steel composites which fulfil the necessary criteria for the SOFC interconnect.

The standard anode material used in SOFC is the Ni/YSZ cermet, which was introduced by Spacil as a response to the failure of all-metal anodes.[43] These cermets have been extensively studied and their performance optimized. Nickel serves as an excellent reforming catalyst and electrocatalyst for electrochemical oxidation of hydrogen and the intrinsic charge transfer resistance that is associated with the electrocatalytic activity at Ni/YSZ boundary is low. Despite being used in most SOFC applications, Ni/YSZ anodes suffer from a few significant limitations. A severe limitation is their inability to operate on hydrocarbons, except when diluted with steam. The possibility to use practical fuels in SOFCs, without the need of a pre-reforming step and improved redox stability of the anode, would be a definite advantage for the development of practical systems. Therefore, much research has focused on the development of alternative anode materials that are catalytically active for the oxidation of methane and higher hydrocarbons, and inactive for cracking reactions that lead to carbon deposition. Other desirable properties include the tolerance to sulfur, for the use of practical fuels such as natural gas and biogas.

Cu-based SOFC cermet anodes have successfully synthesized and demonstrated for direct electrochemical utilization of a large variety of hydrocarbon fuels with little carbon deposition.[44] GDC anodes ($Ce_{0.6}Gd_{0.4}O_{1.8}$) have been successfully operated with steam diluted CH_4 as a fuel.[45] Perovskites have also been widely investigated as potential SOFC anode materials. Among these materials, chromites and titanates are promising SOFC anode materials.[46,47] Another important double perovskite is $Sr_2MgMoO_{6-\delta}$, which has recently been shown to offer good performance in CH_4 and good sulfur tolerance.[48]

Materials suitable for a SOFC cathode (SOFC) have to fulfill the following key requirements, high electronic conductivity, stability in oxidizing atmospheres at high temperature, thermal expansion match with other cell components, compatibility with different cell components and sufficient porosity to allow transport of the fuel gas to the electrolyte/electrode interface.

Strontium-doped lanthanum manganite ($La_{0.85}Sr_{0.15})MnO_3$, a p-type semiconductor, is most commonly used for the cathode material. Although adequate for most SOFCs, other materials may be used, particularly attractive being p-type conducting perovskite structures that exhibit mixed ionic and electronic conductivity. This is especially important for lower-temperature operation since the polarization of the cathode increases significantly as the SOFC temperature is lowered to, e.g., around 650°C. Important alternative cathodes are the perovskites, lanthanum strontium ferrite, lanthanum strontium cobalite and lanthanum strontium cobalt ferrite, which are better electrocatalysts than the state-of-the-art lanthanum strontium manganite, because they are mixed conductors.[49]

7 Molten Carbonate Fuel Cell

The second, and more commercially developed, main type of high temperature fuel cell system is the molten carbonate fuel cell (MCFC), which uses a eutectic mixture of alkali carbonates, Li_2CO_3 and K_2CO_3 immobilized in a $LiAlO_2$ ceramic matrix as the electrolyte. At high operating temperature e.g. 650–700°C, the alkali carbonates form a highly conductive molten salt, with carbonate ions, CO_3^{2-}, providing ionic conduction. Carbon dioxide (CO_2) and oxygen (O_2), must be supplied to the cathode to be converted to carbonate ions, which provide the means of ion transfer between the cathode and the anode. The anode and the cathode reactions are as follows:

$$\text{Anode:} \quad H_2 + CO_3^{2-} \rightarrow CO_2 + H_2O + 2e^- \tag{5}$$

$$\text{Cathode:} \quad \frac{1}{2}O_2 + CO_2 + 2e^- \rightarrow CO_3^{2-} \tag{6}$$

In the MCFC, CO_2 is produced at the anode and consumed at the cathode. Therefore, the MCFC systems generally feed the CO_2 from the anode to the cathode. The electrodes are nickel based, the anode usually consisting of a nickel/chromium alloy while the cathode is made of a lithiated nickel oxide. At both electrodes, the nickel phase provides catalytic activity and conductivity. At the anode, the chromium additions maintain high porosity and protect against corrosion.[49]

Because they operate at high temperatures, MCFCs have flexibility in the chosen fuel; there is no need to use precious metals as catalyst, and they have high-quality waste heat for cogeneration applications. There are also some disadvantages: a CO_2 recycling system must be implemented; the molten electrolyte is corrosive, which gives degradation issues; and the materials are relatively expensive. The main application area for this technology is distributed power generation, often utilizing biogas.

Another promising application of MCFCs that is currently being developed is the direct oxidation of carbon (direct carbon fuel cells). DCFCs are one of the most efficient electrochemical conversion systems, with practical overall achievable electrical efficiencies of 80%.[50]

8 Efficiency

8.1 *Thermodynamics of Fuel Cells*

A comprehensive study of the fuel cell thermodynamics has been performed by Kee *et al.*[9] The thermodynamic limit on fuel cell performance can be understood by considering the energy and entropy accounting associated with a generic steady flow process.

The rate at which work is done by a system, $\dot{W}$, can be obtained by combining the first and second law of thermodynamics.

$$\dot{W} = -\dot{m}(\Delta h - T_0 \Delta S) - T_0 \dot{P}_s, \tag{7}$$

where $\Delta h = h_{\text{out}} - h_{\text{in}}$ and $\Delta S = S_{\text{out}} - S_{\text{in}}$ are, respectively, the net enthalpy and entropy differences associated with the flow streams entering and leaving the system. Entropy is produced within the system due to internal irreversible processes at a rate $\dot{P}_s$. Since $\dot{P}_s$ is required to be positive by the second law of thermodynamics, the greatest power is produced by a reversible process and equals:

$$\dot{W}_{rev} = -\dot{m}(\Delta h - T_0 \Delta S). \tag{8}$$

While this general expression applies to any steady-flow process, a relevant case for fuel cells is one in which the temperature remains fixed at T_0 and the pressure is constant, but the composition changes due to internal chemical reactions. In this case, the greatest work production rate achievable is:

$$\dot{W}_{rev} = -\sum_k \Delta(\dot{N}_k \mu_k), \tag{9}$$

where μ_k and N_k are, respectively, the species chemical potentials and molar flow rate, and the sum runs over all species. If depletion effects are small enough, so that $\mu_{k,in} = \mu_{k,out}$ and if $\dot{N}_k = v_k \dot{N}$, where μ_k is the stoichiometric coefficient of species k and N is a rate of progress variable for a global oxidation reaction, Equation (9) reduces to:

$$\dot{W}_{rev} = -\dot{N} \sum_k v_k \mu_k. \tag{10}$$

If the fluid stream is an ideal gas mixture, the reversible work production rate can be written as:

$$\dot{W}_{rev} = -\dot{N}[-\Delta G^0 - RT \ln \prod_k P_k^{v_k}], \tag{11}$$

where ΔG^0 represents the free-energy change between reactants and products in the global reaction, and p_k is the partial pressure. This expression also holds in the case where one or more reactants are supplied in separate streams. All that is required is to evaluate the partial pressures of each species in the stream in which it is present.

Since the cell potential can be expressed as $E_{\text{cell}} = \dot{W}/I$, Equation (11) can be used to determine the reversible cell potential. The electric current generated in a fuel cell as a direct consequence of the reactions that result in oxidation of the fuel, is given by:

$$I = nF\dot{N}, \tag{12}$$

where F is Faraday's constant and n the number of exchanged electrons.

Hence, the potential developed by a reversible cell is:

$$E_{\text{rev}} = -\frac{\Delta G^0}{nF} - \frac{RT}{nF} \ln \prod_k p_k^{v_k}. \tag{13}$$

When using Equation (4) accounting for the oxygen transfer, the reversible potential simplifies to:

$$E_{\text{rev}} = -\frac{RT}{4F} \ln \frac{P_{O_2(a)}}{P_{O_2(c)}}. \tag{14}$$

This reversible potential is known as the Nernst potential. This ideal potential depends on the electrochemical reactions that occur with different fuels and oxygen. The Nernst equation provides a relationship between the ideal standard potential E° for the cell reaction and the ideal equilibrium potential (E) at other temperatures and partial pressures of reactants and products.

8.2 *Fuel Cell Efficiency*

Different efficiencies must be combined to produce the overall efficiency of a fuel cell. The overall efficiency is defined by the product of the electrochemical efficiency ε_E, and the heating efficiency ε_H. The electrochemical efficiency is in turn, the product of the thermodynamic efficiency ε_T, the voltage efficiency ε_V and the current or Faradic efficiency ε_J[4,9,51]

$$\varepsilon_{FC} = \varepsilon_E \varepsilon_H = \varepsilon_T \varepsilon_V \varepsilon_J \varepsilon_H. \tag{15}$$

8.2.1 *Heating efficiency*

The heating efficiency applies to cases where the fuel contains more species than the electrochemically active ones, such as gases, impurities and other combustibles. The heating value efficiency, ε_H, is defined as:

$$\varepsilon_H = \frac{\Delta H^\circ}{\Delta H_{\text{com}}}, \tag{16}$$

where ΔH° represents the amount of enthalpy in the electrochemically active species and ΔH_{com} represents the amount of enthalpy included in all combustible species in the fuel gases fed to the fuel cell. A pure fuel will obviously give a heating efficiency of 100%.

8.2.2 *Thermodynamic efficiency*

The thermodynamic efficiency of a process measures how efficiently chemical energy extracted from the fuel stream is converted to useful power, rather than heat:

$$\varepsilon_t = \frac{\dot{W}}{\dot{W} + \dot{Q}} \quad \text{or,} \quad \varepsilon_t = \frac{\dot{W}}{\dot{m}|\Delta h|}, \tag{17}$$

where $\dot{Q}$ is the heat rate production of the cell. Using the maximum work production rate achievable, which is the one of a reversible process (Equation (8)), one can write the maximum theoretical efficiency of a system as:

$$\varepsilon_t = \frac{\Delta(h - T_0 S)}{\Delta h}. \tag{18}$$

The thermodynamic efficiency is an extremely important feature when analyzing a fuel cell. This efficiency justifies the need for fuel cell development. Indeed, since the chemical energy is transferred directly to electricity, the free enthalpy change of the cell reaction may be totally converted to electrical energy. In a conventional heat engine, where only the temperature changes, ε_{rev} is limited to the familiar Carnot efficiency:

$$\varepsilon_{t,\text{carnot}} = 1 - \frac{T_0}{T}. \tag{19}$$

For a constant-temperature fuel cell, the intrinsic maximum thermodynamic efficiency is given by:

$$\varepsilon_{rev} = \frac{\Delta G}{\Delta H} = 1 - \frac{T \Delta S}{\Delta H} = \frac{\Delta G^0}{\Delta H^0} + \frac{RT}{\Delta H^0} \ln \prod_k p_k^{v_k}. \tag{20}$$

8.2.3 *Current efficiency*

The current efficiency can be commonly expressed as the fuel utilization efficiency. The efficiency of an SOFC drops if all the reactants are not converted to reaction products. For a 100% conversion of a fuel, the amount of current density, i_F, produced is given by Faraday's law:

$$i_F = zF \left(\frac{df}{dt} \right), \tag{21}$$

where (df/dt) is the molar flow rate of the fuel. For the amount of fuel actually consumed, the current density produced is given by:

$$i = zF \left(\frac{df}{dt} \right)_{\text{consumed}}. \tag{22}$$

The current efficiency, ε_J, is the ratio of the actual current produced to the current available for complete electrochemical conversion of the fuel:

$$\varepsilon_J = \frac{i}{i_F}. \tag{23}$$

This efficiency can be expressed as well in terms of fuel consumption:

$$\varepsilon_J = \frac{h_{\text{in}} - h_{\text{out}}}{h_{\text{in}} - h_{\text{ox}}}, \tag{24}$$

where h_{ox} corresponds to the enthalpy change when all the useful fuel has been consumed.

8.2.4 *Voltage efficiency*

When an electrical current is drawn from a fuel cell, part of the chemical potential available must be used to overcome the irreversible internal losses. Hence, the actual cell potential is decreased from its equilibrium potential, meaning that in an operating SOFC, the cell voltage is always less than the reversible voltage. The voltage efficiency, ε_V, is defined as the ratio of the operating cell voltage under load, E, to the reversible cell voltage, ε_R, and is given as:

$$\varepsilon_V = \frac{E}{E_r}. \tag{25}$$

A voltmeter connecting the anode and cathode can measure the cell electrical potential E, which depends on the current flow. Indeed, many irreversibilities in a fuel cell increase with current density. If no current flows, the cell voltage is the open circuit potential or open circuit voltage (OCV). In most cases, the OCV will equal the potential developed by a reversible cell. The difference between the operating cell voltage and the expected reversible voltage is termed polarization or overpotential and is represented as η. This cell overpotential comprises the total ohmic losses for the cell and the polarization losses associated with the electrodes. The useful voltage under load conditions can therefore be expressed as:

$$V = E^\circ - IR - \eta_{an} - \eta_{cath}. \tag{26}$$

In this equation, I is the current passing through the cell. The electrical resistance R encompasses the ohmic resistance of all components. The polarization resistances, η_{an} and η_{cath}, respectively, for the anode and the cathode, account for non-ohmic losses in each electrode. The polarization loss of each electrode is composed of: (i) activation overpotential due to energy barriers to charge transfer reactions, (ii) concentration overpotential associated with gas-phase species diffusion resistance through the electrodes, (iii) contact resistance which is caused by poor adherence between electrode and the electrolyte.[9,49] Although polarizations cannot be eliminated, they can be minimized by material choice and electrode designs. Figure 7 shows a typical voltage–current polarization curve for an SOFC. The voltage loss increases with the current density. Activation overpotentials contribute the most at low current, while at high currents, concentration polarizations become important. Ohmic losses dominate the losses in the intermediate currents zone.

The concave portion at low current density corresponds to activation-related losses. When the current density increases, the losses are dominated by the ohmic polarization. When the current density approaches its highest values, losses are dominated by concentration polarizations that cause the steep drop in the cell voltage. Each of the different types of losses is described hereafter.

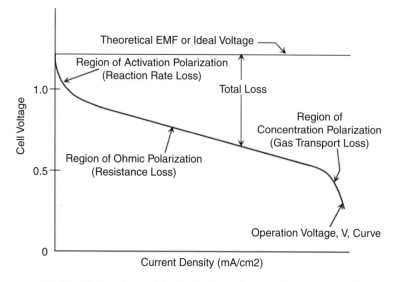

Fig. 7.   Ideal and actual fuel cell voltage/current characteristics.[4]

8.2.5 *Internal resistance*

The size of the voltage drop due to ohmic losses is simply proportional to the current:

$$V = IR. \tag{27}$$

The internal resistance R encompasses the contribution from the electrodes, electrolyte, interconnect and bipolar plates:

$$R = R_{\text{Electronic}} + R_{\text{ionic}} + R_{\text{contact}}. \tag{28}$$

In most fuel cells, the electrolyte contribution to this resistance is the most important, due to the ionic nature of its conductivity. The interconnect and bipolar plates contribution can be important as well. To minimize the ohmic losses, the preferred practice is to fabricate dense, gas-tight electrolyte membranes as thin as possible.

8.2.6 *Charge transfer or activation polarization*

The activation polarization is related to the charge transfer processes occurring during the electrochemical reactions on electrode surfaces. The losses are caused by the slowness of the reactions taking place on the surface of the electrodes. Electrochemical reactions involve an energy barrier that must be overcome by the reacting species. A proportion of the voltage generated is lost, therefore, in driving the electron transfer. This energy barrier, called the activation energy, results in activation or charge transfer polarization, η_A. Activation polarization is related to current density, i, by the Butler–Volmer equation:

$$i = i_0 \exp\left[\frac{(1-\beta)\eta_A F}{RT}\right] - i_0 \exp\left[-\frac{\beta\eta_A F}{RT}\right], \tag{29}$$

where β is the symmetry coefficient and i_0 the exchange current density. The symmetry coefficient is considered as a fraction of the change in polarization which leads to a change in the reaction rate constant. The exchange current density is related to the balanced forward and reverse electrode reaction rates at equilibrium. A high exchange current density means a high electrochemical reaction rate and, in that case, a good fuel cell performance is expected.

The exchange current density can be determined experimentally by extrapolating plots of $\log i$ versus η to $\eta = 0$. For large values of η (either negative or positive) one of the bracketed terms in Eq. (29) becomes negligible. After rearranging one obtains,

$$\eta_A = a \pm b \log i, \tag{30}$$

which is usually referred to as the Tafel equation. Parameters a and b are constants, which are related to the applied electrochemical material, type of electrode reaction and temperature.

The constant a (in the form $v = a \ln(i/i_o)$ is higher for an electrochemical reaction which is slow. The constant i_o is higher if the reaction is faster. The current density i_o can be considered as the current density at which the overvoltage begins to move from zero. The smaller the i_o value, the greater is the voltage drop.

The exchange current density is a crucial factor in reducing the activation overvoltage. The cell performance can be improved through an increase of the exchange current density. This can be done in the following ways[4]:

• raising the cell temperature
• using more effective catalysts
• increasing the roughness of the electrodes
• increasing reactant concentration e.g. pure O_2 instead of air
• increasing the pressure

In low- and medium-temperature fuel cells, activation overvoltage is the most important irreversibility and cause of voltage drop, and occurs mainly at the cathode. Activation overvoltage can be important at the anode when other fuels than hydrogen are involved.

8.2.7 *Diffusion or concentration polarization*

Concentration polarization, η_D, is related to the transport of gaseous species through the porous electrodes and, thus, its magnitude is dictated by the microstructure of the electrode, specifically, the volume percent porosity, the pore size and the tortuosity factor. It becomes significant when the electrode reaction is hindered by mass transport effects i.e. when the supply of reactant and/or the removal of reaction products by diffusion to or from the electrode is slower than that corresponding to the charging/discharging current i. When the electrode process is governed completely by diffusion, the limiting current, i_L, is reached. In such a case, the demand for reactants exceeds the capacity of the porous anode to supply them by gas diffusion mechanisms. High tortuosity (bulk diffusion resistance) is often assumed to explain this behavior.

The voltage drop due to the mass transport limitations can be expressed as:

$$\Delta V = \frac{RT}{2F} \ln \left(1 - \frac{i}{i_l} \right), \tag{31}$$

where i_L is postulated to be the limiting current density at which the fuel is used up at a rate equal to its maximum supply speed. The current density cannot rise above this value because the fuel gas cannot be supplied.

9 Summary

In general, high-temperature fuel cells exhibit higher efficiencies and are less sensitive to fuel composition. PEMFC systems require a pure H_2 fuel stream because the precious metal anode catalysts are poisoned by even low levels of CO or other compounds such as those containing sulfur. These PEMFCs are the most suited to transport applications because of the need for short warm-up. The concept of a fuel cell–powered vehicle running on hydrogen, the so-called "zero emission vehicle," is a very attractive one and is currently an area of intense activity for almost all the major motor manufacturers. The potential high reliability and low maintenance, coupled with their quiet operation and modular nature makes fuel cells well suited to localized "off grid" power generation, either for high-quality uninterrupted power supplies or remote applications. High-temperature fuel cells (MCFC and SOFC) are suitable for continuous power production.

References

1. W. R. Grove, On voltaic series and the combination of gases by platinum, *Philos. Mag.* **14** (1839) p. 127.
2. W. R. Grove, On a gaseous voltaic battery, *Philos. Mag.* **21** (1843), p. 417.
3. L. Mond and C. Langer, A new form of gas battery, *Proc. R. Soc. London* **46** (1889), p. 296.
4. EG&G Technical Services, Inc., *Fuel Cell Handbook*, 7th Edn. (US Department of Energy, 2004).
5. S. C. Singhal, Solid oxide fuel cells for stationary, mobile, and military applications, *Solid State Ionics* **152–153** (2002), p. 405.
6. IEA Technology essentials, Hydrogen production and distribution, OECD/IEA (2007).
7. Available at: www.pure.shetland.co.uk.
8. M. Ormerod, Solid oxide fuel cells, *Chem. Soc. Rev.* **32** (2003), pp. 17–28.

9. R. J. Kee, H. Zhu and D. G. Goodwin, Solid oxide fuel cells with hydrocarbon fuels, *Proc. Combustion Institute* **30** (2005) pp. 2379–2404.

10. C. Song, Fuel processing for low-temperature and high-temperature fuel cells Challenges, and opportunities for sustainable development in the 21st century, *Catal. Today* **77** (2002), pp. 17–49.

11. R. T. K. Baker, M. A. Barber, P. S. Harris, F. D. Feates and R. J. Waite, Nucleation and growth of carbon deposits from the nickel catalyzed decomposition of acetylene, *J. Catal.* **26** (1972), p. 51.

12. R. T. K. Baker, P. S. Harris, J. Henderson and R. B. Thomas, Formation of carbonaceous deposits from the reaction of methane over nickel, *Carbon* **13** (1975), p. 17.

13. R. T. K. Baker, P. S. Harris and S. Terry, Unique form of filamentous carbon, *Nature* **253** (1975), p. 37.

14. C. W. Keep, R. T. K. Baker and J. A. France, Origin of filamentous carbon formation from the reaction of propane over nickel, *J. Catal.* **47** (1977), p. 232.

15. C. H. Bartholomew, Carbon deposition in steam reforming and methanation, *Catal. Rev. Sci. Eng.* **24** (1982), p. 67.

16. R. T. K. Baker, Catalytic growth of carbon filaments, *Carbon* **27** (1989), p. 315.

17. M. L. Toebes, J. H. Bitter, A. J. van Dillen and K. P. de Jong, Impact of the structure and reactivity of nickel particles on the catalytic growth of carbon nanofibers,, *Catal Today*, **76** (2002), pp. 33–42.

18. C. Y. Sheng and A. M. Dean, Importance of gas-phase kinetics within the anode channel of a solid-oxide fuel cell, *J. Phys. Chem.* A **108** (2004), p. 3772.

19. C. H. Toh, P. R. Munroe, D. J. Young and K. Foger, High temperature carbon corrosion in solid oxide fuel cells, *Mater. High Temp.* **20** (2003), p. 129.

20. McIntosh, H. He, S.-I. Lee, O. Costa-Nunes, V. V. Krishnan, J. M. Vohs and R. J. Gorte, An examination of carbonaceous deposits in direct-utilization SOFC anodes, *J. Electrochem. Soc.* **151** (2004), pp. A604–A608.

21. T. Kim, G. Liu, M. Boaro, S.-I. Lee, J. M.Vohs, R. J. Gorte, O. H. Al-Madhi and B. O. Dabboussi, A study of carbon formation and prevention in hydrocarbon-fueled SOFC, *J. Power Sources*, **155** (2006), pp. 231–238.

22. S. McIntosh, J. M. Vohs, and R. J. Gorte, Role of hydrocarbon deposits in the enhanced performance of direct-oxidation SOFCs, *J. Electrochem. Soc.* **150** (4) (2003), pp. A470–A476.

23. H. He, J. M Vohs and R. J. Gorte, Carbonaceous deposits in direct utilization hydrocarbon SOFC anode, *J. Power Sources* **144** (2005), pp. 135–140.

24. K. M. Walters, A. M. Dean, H. Zhu and R. J. Kee, Homogeneous kinetics and equilibrium predictions of coking propensity in the anode channels of direct oxidation solid-oxide fuel cells using dry natural gas, *J. Power Sources* **123** (2003), pp. 182–189.

25. C. Yang, P. Costamagna, S. Srinivasan, J.Benziger and A. B. Bocarsly, Approaches and technical challenges to high temperature operation of proton exchange membrane fuel cells, *J Power Sources* **103** (2001), pp. 1–9.

26. T.Uda, D. A. Boysen, C. R. I. Chisholm and S. M. Haile, Alcohol fuel cells at optimal temperatures, *Electrochem Solid State Lett.* **9** (2006), pp. A261–A264.

27. N. Sammes, R. Bove and K. Stahl, Phosphoric acid fuel cells: Fundamentals and applications, *Curr. Opin. Solid State Mater. Sci.* **8** (2004), pp. 372–378.

28. J. H. Hirschenhofer, D. B. Stauffer, R. R. Engleman and M. G. Klett, *Fuel Cell Handbook*, 4th Edn. (Parsons Corporation, Pasadena, CA, USA, 1998).

29. Q. F. Li, R. H. He, J. O. Jensen and N. J. Bjerrum, Approaches and recent development of polymer electrolyte membranes for fuel cells operating above 100 degrees C, *Chem. Mater.* **15** (2003), pp. 4896–4915.

30. N. Q. Minh, Ceramic Fuel-Cells, *J. Am. Ceram. Soc.* **76** (1993), p. 563.

31. W. Nernst, Über Die Elektrolytische Leitung Fester Körper Bei Sehr Hohen Temperaturen, *Z. Electrochemistry* **6** (1899), p. 41.

32. E. Baur and H. Z. Preis, Uber brennstoff-ketten mit festLeitern, *Z. Electrochemistry* **43** (1937), p. 727 (in German).

33. F. J. Gardner, M. J. Day, N. P. Brandon, M. N. Pashley and M. Cassidy, SOFC technology development at Rolls-Royce, *J. Power Sources* **86** (2000), p. 122.

34. F. G. E. Jones, P. A. Connor, A. J. Feighery, J. Nairn, J. Rennie and J. T. S. Irvine, SOFC Roll development at St. Andrews fuel cells Ltd., *J. Fuel Cell. Sci. Technol.* **4** (2007), p. 1.

35. R. Stevens, *Magnesium Elektron* (London, UK, 1986)

36. T. H. Etsell and S. N. Flengas, Electrical Properties of Solid Oxide Electrolytes, *Chem. Rev.* **70**, (1970), p. 339.

37. E. C. Subbarao and H. S. Maiti, Solid electrolytes with oxygen ion conduction, *Solid State Ionics* **11** (1984), p. 317.

38. J. B. Goodenough, Oxide-ion electrolytes, *Annu. Rev. Mater. Res.* **33** (2003), p. 91.

39. S. J. Skinner and J. A. Kilner, Oxygen ion conductors, *Mater. Today* **6** (2003), p. 30.

40. M. Mogensen, N. M. Sammes and G. A. Tompsett, Physical, chemical and electrochemical properties of pure and doped ceria, *Solid State Ionics* **129** (2000), p. 63.

41. B. C. H. Steele, Appraisal of Ce1-yGdyO2-y/2 electrolytes for IT-SOFC operation at 500 degrees C, *Solid State Ionics* **129** (2000), p. 95.

42. B. C. H. Steele and A. Heinzel, Materials for fuel-cell technologies, *Nature* **414** (2001), p. 345.

43. S. Spacil, Electrical device including nickel-containing stabilized zirconia electrode, US Patent 3,558,360 (1970).

44. H. Kim, S. Park, J. M. Vohs and R. J. Gorte, Direct oxidation of liquid fuels in a solid oxide fuel cell, *J. Electrochem. Soc.* **148** (2001), p. A693.

45. O. A. Marina, C. Bagger, S. Primdahl and M. Mogensen, A solid oxide fuel cell with a gadolinia-doped ceria anode: preparation and performance, *Solid State Ionics* **123** (1999), p. 199.

46. S. Primdahl, J. R. Hansen, L. Grahl-Madsen and P. H. Larsen, Sr-doped LaCrO(3) anode for solid oxide fuel cells, *J. Electrochem. Soc.* **148** (2001), p. A74.

47. G. Pudmich, B. A. Boukamp, M. Gonzalez-Cuenca, W. Jungen, W. Zipprich and F. Tietz, Chromite/titanate based perovskites for application as anodes in solid oxide fuel cells, *Solid State Ionics* **135** (2000), p. 433.

48. Y. H. Huang, R. I. Dass, Z. L. Xing and J. B. Goodenough, Double perovskites as anode materials for solid-oxide fuel cells, *Science* **312** (2006), p. 254.

49. J. Larminie and A. Dicks, *Fuel Cells Systems Explained*, 2nd Edn. (John Wiley & Sons Ltd, London, 2005).

50. D. Cao, Y. Sun and G. Wang, Direct carbon fuel cell: Fundamentals and recent developments, *J. Power Sources* **167** (2007), pp. 250–257.

51. B. de Boer, SOFC Anodes: Hydrogen oxidation at porous nickel and nickel/yttria-stabilised zirconia cermet electrodes, Ph.D thesis, University of Twente (1998).

Chapter 20

Electrical Grids

Róisin Duignan* and Mark O'Malley
School of Electrical, Electronic and Communications Engineering
University College, Dublin, Ireland
**roisin.duignan@ucd.ie*

As an essential of modern life, electricity is different from other commodities. It cannot be stored economically, and the supply of and demand for electricity must be balanced in real time. Mismatches in supply and demand can threaten grid integrity within seconds. Grid conditions can change significantly from day to day or hour to hour especially with the introduction of renewable energy sources in the electricity generation fuel mix. This chapter introduces the electrical grid, providing details of its operation and it's key building blocks. Market environments that govern the electrical grid operation are discussed along with the engineering challenges faced with the ever increasing demand for renewable energy sources.

1 Introduction

The electrical power grid is recognized by the National Academy of Engineering, US, as the most important engineering accomplishment over the last century.[1] Indeed, the electricity infrastructure, as one of the most fundamental infrastructures of our society, provides electric energy for lighting and heating to billions of people and powers the industry and businesses that serve as the foundation of the world economy. In order to fully understand the electrical power grid, one needs to understand the constituent components together with the engineering and economic factors which play such a vital role in the electrical power grid. This chapter will describe the fundamental building blocks of the electrical power grid, the tools used in its analysis and control and the market dynamics of the electrical power market.

2 Power Grids

The following sections will serve as an introduction to the infrastructure, operation, and planning governing the electrical power grid to provide context for the subsequent sections in this chapter.

2.1 Electric Power Infrastructure

The infrastructure to provide electric energy to the industry and commercial and residential users includes the electric power grids and the associated communication and computer infrastructures. A power grid includes a number of power plants, each with electric power generators and other equipment to support the power generation.[2-4] A **generator** is an energy conversion device. For example, a fossil power plant converts the energy stored in the fossil fuel into the electric energy generated by the generators. Generators operate using a variety of energy sources and the net generation by energy source in the US for the past decade can be seen in Table 1.[5-8]

Load refers to the electrical equipment of the users including electric motors, lighting, heating, and air-conditioning devices. Generators are connected to substations through **transmission lines** (or the **transmission network**). The **distribution network** carries electricity from the transmission system and delivers it to the consumers. The four main parts of a power grid i.e. generation, transmission, distribution, and load, form a large-scale interconnected grid with numerous pieces of power equipment distributed over a large geographic area. Figure 1 is an illustration of an electric power grid.

The communication infrastructure is essential for the operation of the electric power grids. A **control center** monitors the grid conditions and control commands are issued for maintenance and emergency purposes as well as for normal operation. A typical control center setting can be seen in Fig. 2. Computer hardware and software in a control center provide the tools for monitoring and remote control for the grid operators. The electricity markets, where electric energy is traded, are generally operated with computer tools. Suppliers and loads from a power market send their bids to the market through computers. Modern protective relays of the power grids are computer-based; these relays are designed to issue control signals to open circuit breakers when the relays sense a fault such as a short circuit that leads to an excessive amount of electric currents on the grid. A fault can be caused, for example, by a tree that falls on the transmission line during a storm.

2.2 Operation, Planning, and Service Restoration

The operation of an interconnected power grid involves a number of control systems. The **Supervisory Control and Data Acquisition** (SCADA) system in a control center provides the hardware and software tools for monitoring and remote control of the power grid devices. Control centers also have **Energy Management Systems** (EMSs) that consist of software tools for estimating the state of the system as well as performing a security assessment. In a control center, system operators monitor and control the transmission grid and ensure that the grid is secure in the sense that it is able to withstand contingencies such as equipment failures and faults.

Table 1: Net Generation by Energy Source (US) in Thousand Megawatthours.[5-8]

Period	Coal	Petroleum	Natural gas	Other gases[a]	Nuclear	Hydroelectric conventional	Other renewables[b]	Hydroelectric pumped storage	Other[c]	Total
2000	1,966,265	111,221	601,038	13,955	753,893	275,573	80,906	-5,539	4,794	3,802,105
2001	1,903,956	124,880	639,129	9,039	768,826	216,961	70,769	-8,823	11,906	3,736,644
2002	1,933,130	94,567	691,006	11,463	780,064	264,329	79,109	-8,743	13,527	3,858,452
2003	1,973,737	119,406	649,908	15,600	763,733	275,806	79,487	-8,535	14,045	3,883,185
2004	1,978,301	121,145	710,100	15,252	788,528	268,417	83,067	-8,488	14,232	3,970,555
2005	2,012,873	122,225	760,960	13,464	781,986	270,321	87,329	-6,558	12,821	4,055,423
2006	1,990,511	64,166	816,441	14,177	787,219	289,246	96,525	-6,558	12,974	4,064,702
2007	2,016,456	65,739	896,590	13,453	806,425	247,510	105,238	-6,896	12,231	4,156,745
2008	1,985,801	46,243	882,981	11,707	806,208	254,831	126,101	-6,288	11,804	4,119,388
2009	1,755,904	38,937	920,979	10,632	798,855	273,445	144,279	-4,627	11,928	3,950,331
2010	1,847,290	37,061	987,697	11,313	806,968	260,203	167,173	-5,501	12,855	4,125,060

[a]Blast furnace gas, propane gas, and other manufactured and waste gases derived from fossil fuels.
[b]Other renewables represents the summation of the sub-categories of Wind, Solar, Thermal, and Photovoltaic, Wood and Wood Derived Fuels, Geothermal, and Other Biomass.
[c]Small number of electricity-only, non-Combined Heat and Power plants may be included. The numbers in the Pumped Storage column are negative since more energy is used to pump water than is later extracted to the grid.

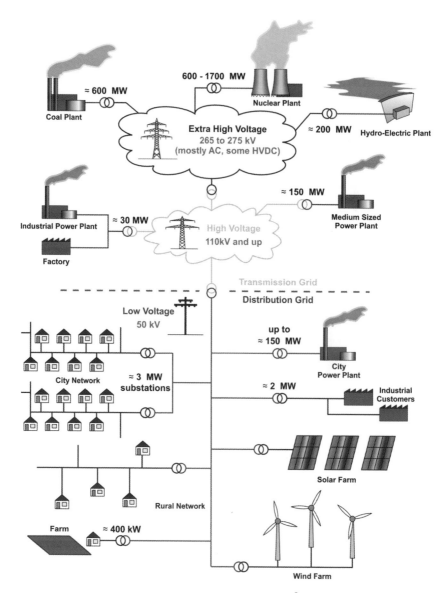

Fig. 1.   Electric grid systems.[9]

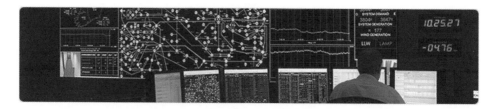

Fig. 2.   Control center of a power grid (© National Control Centre, Eirgrid 2011).[10]

An electrical grid requires careful planning to ensure that the load growth can be met. System planning may be driven by different factors, including load growth, addition/replacement of facilities, and installation of new technologies. Generation planning is intended to ensure that there will be sufficient generation resources to meet the forecasted load while meeting the reliability requirements. Transmission planning involves the expansion or modification of the transmission lines or substations. At a lower voltage level, distribution system planning involves the expansion or reconfiguration of distribution feeders to serve load growth or new developments.

Planning involves engineering studies to ensure that the new system will meet the reliability and security criteria. Commonly adopted reliability and security criteria involve a requirement on the **Loss of Load Probability** (LOLP).[2]

Most power outages do not involve all loads served from a grid. A blackout, however, implies that service to all loads is lost. **Restoration** is the process of bringing electricity back to all customers of a power grid. Some generating units can be started without electricity from outside; they are called **black start units**, examples of which are combustion turbines and hydro units.

The North American Electric Reliability Councils (NERC),[11] record the number and cause of U.S. transmission circuit automatic outages by cause code and by region. The NERC regions are shown in Fig. 3.

Table 2 details the number of transmission outages by cause and by NERC region in 2010.[12] Note: an automatic outage that results from the automatic operation of a switching device, causing an element to change from an in-service state to a not-in-service state. A momentary outage is an automatic outage with an outage

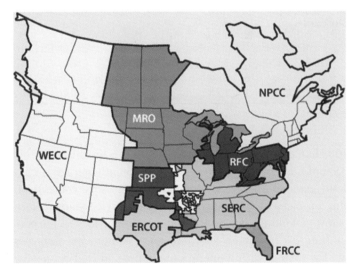

Fig. 3.   North American electric reliability councils (NERC) regions.
Source: US Energy Information Administration. http://www.eia.gov/cneaf/electricity/chg_str_fuel/html/fig02.html

Table 2: US Transmission Circuit Automatic Outage Counts by Cause Code and by NERC Region, 2010.[12]

	FRCC	MRO	NPCC	RFC	SERC	SPP	ERCOT	WECC	Contigious US
Circuit Outage Counts									
Automatic Outages									
Momentary	96	110	35	231	281	115	121	444	1,433
Sustained	136	217	70	264	320	128	71	838	2,044
Non-Automatic Outages									
Operational	110	42	61	136	219	10	24	661	1,263
Planned	2,447	358	505	1,571	1,943	227	548	3,043	10,642
Circuit Outage Counts per 1,000 Circuit Miles									
Automatic Outages									
Momentary	13	6	5	9	8	15	13	7	8
Sustained	19	11	10	10	10	17	7	14	12
Non-Automatic Outages									
Operational	15	2	9	5	7	1	3	11	7
Planned	344	19	75	62	58	29	58	51	63
Circuit Outage Hours									
Automatic Outages									
Sustained	6,864	7,619	14,530	13,341	4,642	1,117	1,089	22,422	71,623
Non-Automatic Outages									
Operational	270	8	84	39	46	1	4	39	493
Planned	350	314	1,467	797	602	96	111	367	4,103
Circuit Outage Hours per Outage Incident									
Automatic Outages									
Sustained	50	35	208	51	15	9	15	27	35
Non-Automatic Outages									
Operational	2	0	1	0.29	0.21	0.09	0.17	0.06	0
Planned	0.14	1	3	1	0.31	0.42	0.20	0.12	0

duration of less than one minute. A sustained outage is an automatic outage with an outage duration of a minute or greater. A non-automatic outage is an outage which results from the manual operation (including supervisory control) of a switching device, causing an element to change from an in-service state to a not in-service state. An operational outage is a non-automatic outage for the purpose of avoiding an emergency (i.e. risk to human life, damage to equipment, damage to property) or to maintain the system within operational limits and that cannot be deferred. A planned outage is a non-automatic outage with advance notice for the purpose of maintenance, construction, inspection, testing, or planned activities by third parties that may be deferred.

3 Electric Power Grid Analysis Tools and Fundamentals

In the previous section, the electrical power infrastructure, its operations, and planning were briefly introduced. To analyze the electrical power grid, these fundamental concepts need to be clearly understand. In this section, a brief introduction to the concept of phasors and phasor notation is presented. The fundamentals of power and three-phase circuits are then introduced. These concepts are very important in the analysis of electric power grids. A more detailed account of these topics can be found at Elgard[2] and Von Meier.[13]

3.1 *Phasors Fundamentals*

A sinusoidal supply voltage can be represented by the equation

$$v(t) = V_m \sin(\omega t).$$

Its phasor form is given by

$$v(t) = V_m \sin(\omega t + \alpha),$$

where V_m is the maximum or peak voltage, ω is the angular frequency of the supply, and t is time (as shown in Fig. 4). A phasor can also be thought of as a quantity that varies sinusoidally with time and for which the amplitude, (angular) frequency, and phase angle (α) are specified.

The Root Mean Square (RMS) value is used as a way of averaging phasors, whose true average value is always zero. The root mean square value of $v(t)$, V_{RMS} is

$$V_m = \sqrt{2} V_{RMS}.$$

Two phasors may have identical amplitudes and frequencies, but may also have a difference in phase, α. The phase of a periodic signal is the angular difference between a point on the signal and a corresponding point on a reference signal.

A convenient way of representing phasor quantities is using complex numbers. The Argand diagram[13] depicting these quantities is called a phasor diagram. To

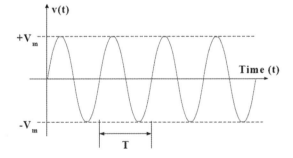

Fig. 4.   Sinusoidal supply voltage with amplitude V_m and varying with time t.

find the impedance of any component, we find the relationship between the voltage across its terminals and the current flowing through it, (which is described by Ohm's Law).[14] **Impedance** (Z) is given by

$$Z = R + jX,$$

where R and X are the resistance and reactance, respectively.

Admittance (Y) is given by

$$Y = \frac{1}{Z} = G + jB,$$

where G and B are the **conductance** and **susceptance**, respectively.

The complex impedance of the inductor is given by

$$Z_{\mathrm{L}} = j\omega L,$$

where ω is the angular frequency of the supply and L is the inductance.

The complex impedance of the capacitor is given by

$$Z_C = \frac{1}{j\omega C} = \frac{-j}{\omega C},$$

where C is the capacitance.

3.2 *Alternating Current and Direct Current*

In the previous subsection, phasor notation was introduced, which can be thought of as a quantity that varies sinusoidally with time and for which the amplitude, (angular) frequency, and phase angle are specified. Throughout this chapter, the notion of **alternating current** (**AC**) and **direct current** (**DC**) will be discussed. AC involves the movement of electric charge for which it periodically reverses direction. For **direct current** (**DC**), the flow of electric charge is only in one direction. That is, the frequency is equal to zero for direct current.

3.3 *Power Fundamentals*

The power, or real power, is given by the voltage multiplied by the current, i.e.

$$P(t) = v(t) \times i(t).$$

Consider a network supplied by a voltage

$$v(t) = V_{\mathrm{m}} \sin \omega t,$$

and whose load impedance, Z, draws a current

$$i(t) = I_{\mathrm{m}} \sin(\omega t + \alpha),$$

where by Ohm's Law

$$v(t) = Z \times i(t).$$

Then instantaneous power to the load is

$$P(t) = \frac{1}{2}V_m I_m [\cos\alpha - \cos(2\omega t + \alpha)].$$

The average real power is given by

$$P_{av} = \frac{1}{T}\int_0^T P(t)\mathrm{dt} = \frac{V_m}{\sqrt{2}}\frac{I_m}{\sqrt{2}}\cos\alpha = V_{rms}I_{rms}\cos\alpha.$$

If the circuit is purely resistive, then angle $\alpha = 0$. This implies that

$$P_{av} = V_{rms}I_{rms},$$

and the term $\cos\varphi$ is known as the power factor for the load.

The efficiency (η) of power transmission can be defined as the proportion of received active power to sent active power

$$\eta = \frac{\text{Received Power}}{\text{SendingPower}} = \frac{P_{Load}}{P_{Load} + P_{Losses}}.$$

The apparent power, S, delivered to a network is given as

$$S = V_{rms} \times I_{rms} \ VA,$$

The reactive power, Q is given as

$$Q = V_{rms} \times I_{rms} \sin\varphi \ VAR.$$

Reactive power is exchanged back and forth between inductive and capacitive parts of a reactive load (one containing inductors and capacitors) during each cycle of the supply, and does zero work. Reactive power refers to the energy storage part of a load where there is no net transfer of energy.

3.4 *Phase Circuit Fundamentals*

Modern electricity supply systems invariably employ three-phase alternating current (A.C.), for economic and efficiency reasons.[15] Three-phase electric power systems have at least three conductors carrying voltage waveforms that are $\frac{2\pi}{3}$ radians offset in time. Consider a single-phase generator as shown in Fig. 5.

If a resistive load is introduced at AA' terminals then the instantaneous power, P, is given by

$$\boldsymbol{P = V}\cos\omega\boldsymbol{t} \times \boldsymbol{I}\cos\omega\boldsymbol{t} = \boldsymbol{VI}\cos^2\omega\boldsymbol{t}.$$

Now consider a two-phase generator as shown in Fig. 6.

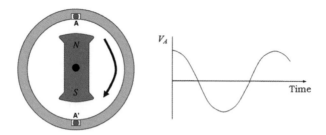

Fig. 5.   Single-phase generator.

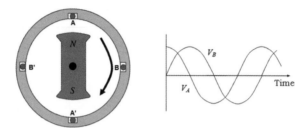

Fig. 6.   Two-phase generator.

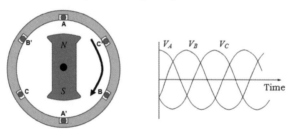

Fig. 7.   Three-phase generator.

With resistive loads on phase A and phase B, (i.e. $\frac{\pi}{2}$ radians out of phase), the instantaneous power, P, is given by

$$P = V \cos \omega t \times I \cos \omega t + V \sin \omega t \times I \sin \omega t = VI.$$

We can see that the two-phase generator produces a constant power output equivalent to VI.

Finally, consider a three-phase generator as shown in Fig. 7.

With a three-phase load on the generator, the instantaneous power, P, is given by

$$P = V \cos \omega t \times I \cos \omega t + V \cos(\omega t - 120°) \times I \cos(\omega t - 120°),$$
$$+ V \cos(\omega t - 240°) \times I \cos(\omega t - 240°) = 1.5 \, VI.$$

We can see from this analysis that a three-phase generator produces a constant output power. Another feature is that the required mechanical input is constant.

We also find that a three-phase generator produces three times the power output. In general, three-phase motors, generators, and transformers are simpler, cheaper, and more efficient. We also find that under balanced conditions, the sum of the currents in a three-phase generator equals zero, i.e.

$$\sum i = I \cos \omega t + I \cos(\omega t - 120°) + I \cos(\omega t - 240°) = 0.$$

We find that:

- No return wire is required in the transmission system.
- All the phase conductors carry the same current and so can be the same size, for a balanced load.
- The power losses are halved, and the voltage drop is halved.

In practice, a single neutral conductor is provided to carry any imbalance current.

3.5 *Electric Power Fundamentals*

Electrical current is provided in the form of Alternating Current (AC) with a certain frequency (e.g. 50 Hz in mainland Europe and 60 Hz in North America). Electricity is generated by electric generators and therefore maintaining the frequency requires that the rotor speed of a generator be controlled at a specified constant.[2] The **load** can also change suddenly e.g. the loss of a substation due to equipment problems can lead to the loss of load in the service area. All these changes have to be "absorbed" by the grid so that the system frequency can be maintained. This can be achieved by **automatic generation control** that monitors the system frequency and controls the generators through a feedback mechanism. The power grid is also installed with **under-frequency load shedding** relays that automatically disconnect loads when the frequency falls below a pre-specified threshold in order to avoid power outages.

Voltage is also an important variable to monitor. Typically voltages are raised to a higher level voltage for transmission to the load centers over the transmission lines to reduce power losses. Electric energy is transported across the countryside with high-voltage lines because the line losses are much smaller than with low-voltage lines. The voltage is then stepped down to the distribution level voltage and typically distributed to the users. The voltages are raised or lowered by transformers in the substations. Under some conditions when the system load is heavy and the available voltage control is weak, the system voltage may fall to unacceptable levels or there may even be a **voltage collapse**. A voltage collapse can cause widespread power outages.

4 Transformers

In this section, the transformer, which is one of the key devices used in the electrical power grid, is discussed.

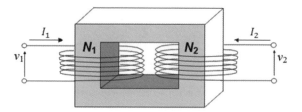

Fig. 8.   Ideal transformer.

4.1 *Transformers: Introduction*

A transformer is a device for transferring electric energy from one circuit to another.[17] Transformers are often used to step up or down supply voltages. A transformer comprises a pair of magnetically coupled coils, such that some of the magnetic flux produced by current in the first coil links the turns of the second, and vice versa. The magnetic coupling can be improved by winding the coils on a common ferromagnetic (high μ) core. The coils are then known as the "windings" of the transformer. Transformers are used in power stations to step up generated voltage to high levels for transmission to reduce current and hence I^2R losses. Voltage is then stepped down at substations for local distribution (which was shown earlier in Fig. 1).

4.2 *Ideal Transformer*

Most transformers are highly efficient and operate almost ideally. The ideal transformer is presented in Fig. 8.

The key properties of an ideal transformer are:

- The primary (N_1) and secondary (N_2) windings have zero resistance, while the iron core has infinite permeability.
- There are no internal losses, and the ideal transformer has an efficiency of 100%.
- The voltage at the input and the voltage at the output can be determined by the number of primary and secondary windings. The relationship is given by the equation (termed "Turns Ratio" or "Voltage Equation"):

$$\frac{v_1}{v_2} = \frac{N_1}{N_2} = \frac{I_2}{I_1}.$$

4.3 *Real Transformer*

In the previous section, the ideal transformer was introduced. However, the ideal transformer model is not sufficiently accurate for many applications.

A practical transformer may be drawn as an ideal device whose windings are in series with external components that account for the losses and voltage drops. In practical power transformers we find that:

- The primary and secondary windings have resistance.

- Not all of the primary flux links the secondary winding, and vice versa. This leakage flux does not contribute to energy transfer.
- The iron core exhibits hysteresis and eddy current losses.

4.4 *Transformer Core Losses*

There are two types of core losses which are of concern in any real transformer. These are:

- Hysteresis Losses
- Eddy Current Losses

Hysteresis losses

This is a nonlinear phenomenon in which the response to a driving force in one direction is different from the response to a force in the opposite direction. Energy is required to accomplish the reversals of domain alignment and field direction. The heat dissipated is proportional to the area under the hysteresis loop.

Eddy current losses

If a closed conductor is placed in an alternating magnetic field, the induced voltage will cause current to flow around the conductor. These are termed eddy currents and are a source of energy loss $(I^2 R)$. Laminations are often used to minimize the effect of eddy currents. The laminations are insulated from each other by iron oxide or special coatings.

Iron losses are conventionally accounted for in equivalent models by placing a resistance in parallel with the magnetizing reactance.

The new current, I_0 , is the no load or secondary open circuited current and is given by

$$I_0 = I_l + I_m,$$

where, I_l and I_m are the core-loss and magnetizing currents.

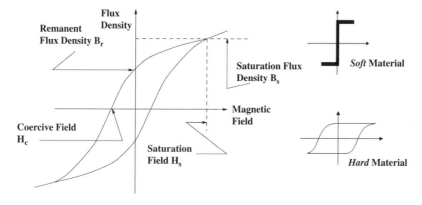

Fig. 9. Typical representation of hystersis losses in a real transformer.

4.5 Determination of Real Transformer Circuit Parameters

The real transformer parameters can be deduced either by:

- Calculations based on the various transformer dimensions and design details, or
- Tests conducted on the transformer after manufacture

Two tests are carried out on a real transformer equivalent circuit to determine the parameters: an open-circuit test and a short-circuit test.

Open circuit test

The Open circuit test, or "no-load test", is used to determine the no load impedance in the excitation branch of a real transformer.

By open-circuiting the secondary and feeding the primary with rated voltage then one measures the input voltage, current, and power.

Short circuit test

The short circuit test is used to determine the series branch parameters of the equivalent circuit. If the secondary terminals are short circuited and the primary is fed with reduced voltage such that I_1 is the rated primary current which also ensures rated copper losses. Then, measure the input voltage, current, and power.

4.6 Transformer Cooling and Winding Connections

Transformer cooling

Transformer losses in the windings and iron core generate heat which must be removed to prevent deterioration of the insulation and the magnetic properties of the core. Since losses are proportional to the volume, while heat dissipation is proportional to surface area, then larger transformers are more difficult to cool. This means that:

- Cooling by natural convection of air only used for small transformers.
- Transformers $> 10\,\text{kVA}$ are most likely oil-cooled, with the transformer contained in a tank of oil.
- In very large devices, pumps may force circulate oil through the windings and external radiators.

Winding connections

There are a multitude of different connections available for three-phase transformers:

- A star connection reduces phase voltage and hence insulation requirements, as well as the number of turns required. It also provides a neutral connection, which can be used for single-phase distribution.
- A delta connection stabilizes unbalanced voltages and removes third harmonics.

Table 3: Uses and Features of Different Transformer Connection Types.

Connection type	Uses and features
Delta/Star	— Power supply transformers — star point is useful for mixed loading i.e. single-phase loads.
	— The closed delta primary damps out third harmonic currents — which stabilizes the star point potential.
Delta/Delta	— Used in moderate voltage systems where there is no insulation problem — since more turns required per phase.
	— Large load imbalances are easily met.
Star/Star	— May be used in high voltage applications — number of turns and insulation is minimum.
	— If a neutral is not provided the phase voltages tend to become severely unbalanced if the load is unbalanced.
	o There are also problems with third harmonics.

The uses and features of the different connection types are given in Table 3.

5 Synchronous Machines

5.1 *Synchronous Machines: Introduction*

Synchronous machines are so called because their speed is directly related to the system frequency.[15] They are characterized by a permanent magnet or DC electromagnet on the rotor. The stator windings are called the armature, while those on the rotor are labeled the field windings. The most important synchronous machines operate as generators (alternators), and are connected in parallel with many others, in large interconnected power systems. As a motor, they have the characteristics of constant-speed operation, power-factor control through adjusting field excitation, and high operating efficiency.

5.2 *Synchronous Motor Operation*

In this section, we describe the operation of the synchronous motor. The stator is fed with three-phase balanced voltages as is shown in Fig. 10:

The phase currents at instants ① to ④ are

$$① : i_A = 0, i_B = -\frac{\sqrt{3}}{2}\hat{I}, i_C = \frac{\sqrt{3}}{2}\hat{I} \quad ② : i_A = i_C = \frac{1}{2}\hat{I}, i_B = -\hat{I},$$

$$③ : i_A = \hat{I}, i_B = i_C = -\frac{1}{2}\hat{I} \quad ④ : i_A = 0, i_B = \frac{\sqrt{3}}{2}\hat{I}, i_C = -\frac{\sqrt{3}}{2}\hat{I}.$$

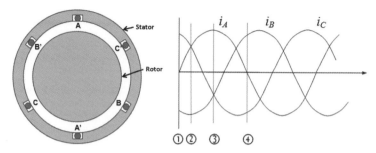

Fig. 10.   Synchronous motor fed with three-phase balanced voltages.

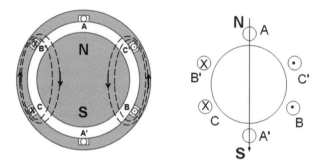

Fig. 11.   Field produced by stator winding at instant ①.

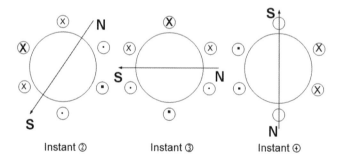

Fig. 12.   Field produced by stator winding at instant ②, ③, and ④.

So, the field produced by the stator winding at instant ① is shown in Fig. 11.

Similarly, the fields produced by the stator winding at instants ②, ③, and ④ are shown in Fig. 12.

The resultant field produced by all three phases rotates in space through the same phase as the currents cycle through in time. Therefore the stator field rotates at a speed of $\omega = 2\pi f$, where f is the frequency.

If the rotor is energized with DC to form an electromagnet, the rotating stator field will interact with the rotor field. The magnetic field produced by the rotor locks in with the stator field, so that they rotate in synchronism. The number of

Generating Mode   Pumping Mode

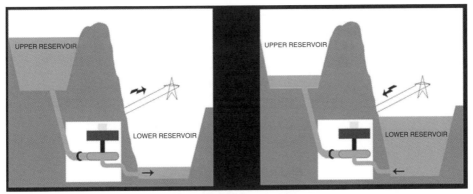

Fig. 13. Typical representation of pumped storage scheme (© Eskom 2011).[16]

poles on the machine rotor and stator can be increased with the resultant stator field rotating at a speed of $\omega = \frac{2\pi f}{p}$ where p equals the number of pole pairs.

5.3 *Synchronous Machines Example — Pumped Storage Schemes*

Like any electrical machine, synchronous machines may be operated as either motors or generators. Some of the largest AC motors are pumped-storage hydroelectricity generators that are operated as synchronous motors to pump water to a reservoir at a higher elevation for later use to generate electricity using the same machinery. Large quantities of water are pumped to a higher elevation, where it can be stored indefinitely. On demand, the water is released through hydraulic turbines, generating electrical energy. This is illustrated in Fig. 13 (see also Chapter 14).

Pumped storage is useful in smoothing out the peaks and troughs of electrical demand.

6 Transmission of Electricity

6.1 *Electricity Transmission Fundamentals*

The main purpose of a transmission system is to carry real power reliably and economically from points of generation to points of consumption.[21] The key elements of a transmission system as illustrated in Fig. 15 are:

- Synchronous machines generate three-phase (11–25 kV).
- High Voltage (HV) transmission required to reduce RI^2 (i.e. Power) losses.
- Within substations, voltage levels are reduced before feeding subtransmission and distribution networks. Substations also control network connectivity.
- Subtransmission and low voltage (LV) networks supply customers.

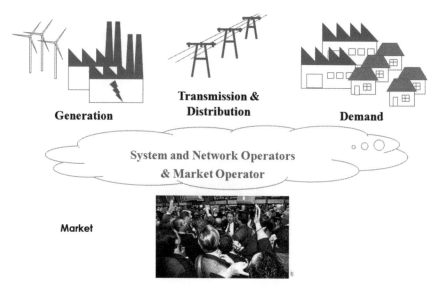

Fig. 14.   Typical representation of an electricity market enviornment (© EirGrid & Soni, 2007).[20]

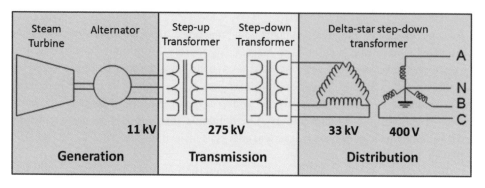

Fig. 15.   Typical representation of electricity transmission system.

6.2 *Transmission Line Fundamentals*

A transmission line model must account for the series resistance and inductance, and shunt capacitance and conductance of each phase. For short lines, shunt parameters can be neglected, leading to a simple, single-phase representation as shown in Fig. 16.

The change in voltage at the receiving end of a line as the load changes from zero to full-load is more commonly known as Voltage Regulation.

$$\text{Voltage Regulation (\%)} = \frac{V_{\text{send}} - V_{\text{receive}}}{V_{\text{receive}}} \times 100\%$$

where V_{send} is assumed constant with load power factor specified.

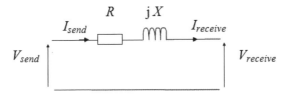

Fig. 16.   Transmission line single-phase representation.

The efficiency is given by:

$$\text{Efficiency}, \eta = \frac{\text{Power output at receiving end}}{\text{Power input at sending end}} \times 100\%.$$

The change in voltage at the receiving end of a line as the load changes from zero to full-load is more commonly known as Voltage Regulation.

6.3 *Power Flows on Transmission Lines*

As an electrical network, the power grid satisfies the associated physical laws which are called Kirchoff's Circuit Laws.[14] Kirchoff's Current Law (KCL) states that all power flows entering a point ("node") of the network from transmission lines will also flow out of the point to other lines. Kirchoff's Voltage Law (KVL) states that the sum of the source voltages around a network closed loop must be equal to the sum of the voltage drops across the transmission lines and the load devices. Power flows on transmission lines are typically measured in the unit of Megawatts (MWs). The power flows on a power grid can be approximated by an analytical model based on linear equations. In Fig. 17, a simple power grid with three substations ("3-bus") is illustrated. Bus 1 and Bus 3 represent generators G_1 and G_2 at the power plants and Bus 2 is a load center. Three transmission lines connect the generators and the load center.

Each transmission line is characterized by a parameter called susceptance, denoted by b_{12}, b_{13}, and b_{23} respectively. The power injection at each node is the

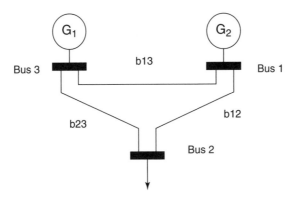

Fig. 17.   A three-bus power grid.

Table 4:   Transmission Line Losses in MW per 100 Miles.[22]

	Line losses – MW/100 miles		
	Resistive	Corona[a]	Total
765 kV Line @1000MW Load			
Original 4-conductor ("Rail") bundle	4.4	6.4	10.8 (1.1%)
Newer 4-conductor ("Dipper") bundle	3.3	3.7	7.0 (0.7%)
Current 6-conductor ("Tern") bundle	3.4	2.3	5.7 (0.6%)
Planned 6-trapezoidal cond. ("Kettle") bundle	3.1	2.3	5.4 (0.5%)
500 kV Line @1000MW Load			
Typical 2-conductor bundle	11.0	1.6	12.6 (1.3%)
345 kV Line @1000MW Load			
Typical 2-conductor bundle	41.9	0.6	42.5 (4.2%)

[a]Yearly average corona loss at sea level based on 20%/2%/78% rain/snow/fair weather conditions respectively.

amount of power that is injected into the transmission system from the node. The bus angles δ_1, δ_2 and δ_3 at buses 1, 2, and 3, respectively, are electrical characteristics from the AC voltages at the buses. The physical significance of the bus angles is that the angle difference across the two sides of the line is proportional to the power flow on that line.

6.4 *Transmission Line Efficiencies*

An energized transmission line carrying load incurs power losses due to heating and so-called "corona" effects. Heating (or resistive) losses increase linearly with line resistance and quadratically with loading. Corona losses result from undesirable discharge of electric energy, which can be visible and/or audible especially during rain, caused by air ionization around line conductors and hardware. Corona losses increase with voltage level and elevation above sea level of the line. The following statistics in Table 4 characterize Extra High Voltage (EHV) transmission lines operating at different voltages (i.e. 765 kV, 500 kV, and 345 kV), in normal weather, carrying 1,000 MW of power.[22]

The markedly superior transmission efficiency of 765 kV transmission is attributable to its higher operating voltage and thermal capacity/low resistance compared to 500 kV and 345 kV.

6.5 *Transmission Interconnection*

Transmission interconnection can be viewed as power systems formed as multiple generation stations and load centers in disparate locations, linked by a network of transmission lines operating at a wide range of voltages. There are many reasons as to why interconnection is advantageous and some of these reasons are presented in Table 5.

Table 5: Advantages of Interconnection.

Key factor	Advantages of interconnection
The initial cost per kVA of power stations decreases with increased generation capacity.	• Economical to run most efficient stations to full capacity and transmit energy considerable distances. • At low loads, instead of running many units at reduced capacity, more expensive stations can be shut down.
Extra generating capacity required to meet sudden increases in load or cope with generator outages by increasing the power output of generators that are already connected to the power system, (i.e. Spinning reserve)	• By interconnecting generators, spinning reserve may be spread across all system units.
Distribution of energy sources does not generally coincide with location of load centers.	• Renewable sources often found in remote locations. • Nuclear stations generally located in isolated areas. (Interconnection allows for these energy sources to be harnessed and utilized).
Consumers require continuity of supply.	• Load centers may be supplied by several generators through multiple paths (see Fig. 1). • Maintenance on individual generators, transmission lines, etc., can be performed without affecting supply.

There are, however, some disadvantages to interconnection. Some of the main issues are:

• During short circuits, interconnected systems have increased fault currents.
• There is a remote possibility that a remote disturbance may lead to blackout.

Interconnection does, however, come with a cost. Table 6 shows the interconnection cost and capacity for new generators, by voltage class, in the U.S. for the years 2009 and 2010.[8]

6.6 *Direct Current Transmission*

Although AC transmission is the most common type of transmission of electricity, **direct current** (or DC) permits an asynchronous or flexible link between two rigid systems.[23]

DC transmission is illustrated in Fig. 18. We see that convertor 1 (used as a rectifier) is used to convert alternating current (AC) to direct current (DC). This

Table 6: Interconnection Cost and Capacity for New Generators, by Voltage Class, in the US for the Years 2009 and 2010.[8]

Voltage group	Units[a]	Nameplate capacity[a] (megawatts)	Cost[a] (thousand dollars)
2009			
Total	**382**	**23,144**	**819,680**
Less than 100 kV	207	1,831	96,452
Between 100 kV and 199 kV	78	6,086	268,834
Greater than 200 kV	97	15,227	454,394
2010			
Total	**418**	**19,661**	**493,909**
Less than 100 kV	287	2,223	66,801
Between 100 kV and 199 kV	69	4,305	145,940
Greater than 200 kV	62	13,133	281,168

[a]Cost is the total cost incurred for the direct, physical interconnection of generators that started commercial operation in the respective years. These generator-specific costs may include costs for transmission or distribution lines, transformers, protective devices, substations, switching stations, and other equipment necessary for interconnection. Units and Nameplate Capacity represent the number of units and associated capacity for which interconnection costs were incurred and reported.

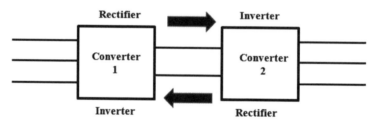

Fig. 18. Typical representation of direct current transmission.

process is known as rectification. Convertor 2 reverses this process and converts DC to AC and is known as an inverter.

DC transmission has advantages and disadvantages, which are summarized in Table 7:

Table 7: Advantages and Disadvantages of DC Transmission.

Advantages	Disadvantages
Only two conductors required, so right of way is reduced.	Limited by high cost of conversion equipment, only becoming economic for long line lengths.
Power can be transmitted in cables over large distances.	DC lines are usually point to point systems (therefore difficult to tap power off at intermediate points).
Power transfer is independent of system frequency, line reactance and phase angle between terminal voltages.	More difficult to interrupt DC supplies, since current does not reduce to zero twice a cycle.

7 Power Systems Operations

7.1 *Unit Commitment*

In any electricity market environment, decisions are made on which electricity generators need to be started up and shut down based on the given load forecasts and the generators available for electric energy production. The problem of Unit Commitment involves finding the least-cost committed set (on/off) of available generation resources to meet the electrical load and the approximate time horizon for which these decisions are taken from a number of hours to 1 week ahead of when the generator is needed.[18,19] In any unit commitment optimization problem, the objective is to minimize the cost where costs can include fuel costs, start-up costs, the expense of the crew and maintenance expenses etc. Typically constraints in this optimization problem are the minimum generation and the maximum generation of a unit, crew constraints, generator units which are "must-run" units or those which are "must not run" units and "spinning reserves".

7.2 *Economic Dispatch*

The calculation of power outputs of the committed generators in an electricity market to minimize costs and meet the load demand is called Economic Dispatch. For M generators, then the economic dispatch problem is to find the mimimum of the total generation cost where the total generation cost, (C_T), is given by

$$C_T = \sum_{i=1}^{M} C_i(P_{Gi}) \quad \text{for all } i = 1, 2, \ldots, M,$$

and C_i is the cost of generator i and P_{Gi} is the power output of generator i.

7.3 *Electricity Market Environment*

An electricity market is a system for effecting purchases, through bids to buy. Bids and offers use supply and demand principles to set the price. Long-term trades are contracts similar to power purchase agreements and generally considered private bilateral transactions between counterparties. Figure 14 gives a typical representation of an electricity market environment.

8 Renewables and the Electrical Power Grid

8.1 *Renewables and the Electrical Power Grid: Introduction*

The demand for carbon-free electricity is driving a growing movement of adding renewable energy to the grid. Renewable sources such as wind and solar are proving the sources contributing to the largest penetration levels for renewable energy.[24–26]

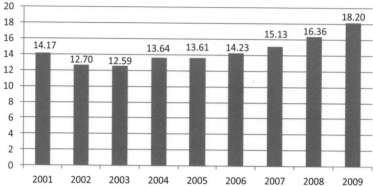

Fig. 19. Electricity generated from renewable sources as a percentage of gross electricity consumption in EU-27 countries.[27]

Electricity generated from renewable sources as a percentage of gross electricity consumption has been steadily increasing in the 27 EU member states over the last number of years as can be seen in Fig. 19.

Renewable energy sources have a number of disadvantages when considered in the context of integration with the electrical power grid:

- Location
- Variability

Renewable resources are typically concentrated far from population centers, requiring additional long-distance, high-capacity transmission to match supply with demand. The variability of renewables due to the characteristics of weather is high, much larger than the few percent uncertainty in load that the grid now accommodates by dispatching conventional resources in response to demand.

8.2 Renewable Example — Wind Turbines

As discussed previously, electricity can be generated from a number of renewable sources. One such renewable source is wind. Generating electricity from wind is a three-step process where the kinetic energy of moving air is converted to mechanical energy and then to electrical energy.

Electricity from wind is typically generated from large, grid-connected wind turbines, deployed either in a great number of smaller wind power plants or a smaller number of much larger plants. For a number of factors including installation costs and maintenance costs then wind power plants are commonly built on land (i.e. "onshore") as opposed to at sea (i.e. offshore). A typical wind turbine can be seen in Fig. 20 (see also Chapter 14).

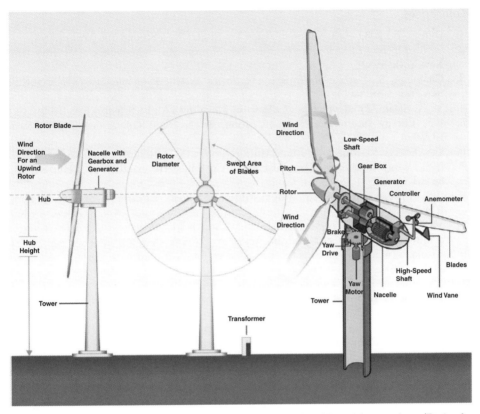

Fig. 20. Basic components of a modern, horizontal-axis wind turbine with a gearbox. (Design by the National Renewable Energy Laboratory (NREL).[28]

The electrical conversion system in a wind turbine is one of the most important features from a system reliability perspective. In the case of large grid-connected turbines, electrical conversion systems come in the following forms:

- Fixed-speed induction generators.
- Variable-speed machines: doubly-fed induction generators with a power electronic converter and synchronous generators with a full power electronic converter, both of which are almost always coupled with pitch-controlled rotors.

The key challenge of these designs when taken in context with the synchronous generators found in most large-scale fossil fuel–powered plants is that they result in no intrinsic inertial response capability. In other words, there is no synchronization between their power output and system power imbalances. This lack of inertial response is an important consideration for electric system planners because less overall inertia in the electric system makes the maintenance of stable system operation more challenging.

References

1. DOE, Grid 2030 — A National Vision for Electricity's Second 100 Years (2003).
2. O. I. Elgerd, *Electric Energy Systems Theory: An Introduction* (McGraw-Hill, New York, 1982).
3. Chapman, S. *Electric Machinery and Power System Fundamentals* (McGraw-Hill, San Francisco, 2002).
4. M. A. Salam, *Fundamentals of Electrical Machines* (Alpha Science, UK, 2005).
5. U.S. Energy Information Administration, Power Plant Operations Report, form EIA-923.
6. U.S. Energy Information Administration, Power Plant Report, form EIA-906.
7. U.S. Energy Information Administration, Combined Heat and Power Plant Report, form EIA-920.
8. U.S. Energy Information Administration, Annual Electric Generator Report, EIA-860.
9. My Solar Energy/Renergy. Available at: http://mienergiasolar.es/en/energy_cummunities/energy_systems.htm. Accessed 15 September 2012.
10. Available at: http://www.eirgrid.com/operations/nationalcontrolcentre/. Accessed 21 May 2012.
11. Available at: http://www.nerc.com/. Accessed 21 May 2012.
12. U.S. Energy Information Administration, Coordinated Bulk Power Supply Program Report, form EIA-411.
13. A. Von Meier, *Electric Power Systems: A Conceptual Introduction* (1996).
14. J. D. Irwin, *Basic Engineering Circuit Analysis* (John Wiley and Sons, New Jersey, 2005).
15. M. G. Say, *Alternating Current Machines* (John Wiley and Sons, New York, 1976).
16. Available at: http://www.eskom.co.za/. Accessed 21 May 2012.
17. J. A. Fleming, *The Alternate Current Transformer in Theory and Practice* ("The Electrician" Printing And Publishing Company Ltd., London, 1896–1900).
18. A. J. Wood and B. F. Wollenberg, *Power Generation, Operation and Control* (Wiley, NJ, 1996).
19. C. Harris, *Electricity Markets: Pricing, Structures and Economics* (Wiley, NJ, 1996).
20. Available at: http://www.sem-o.com/. Accessed 21 May 2012.
21. C. Dannatt and J. W. Dalgleish, *Electrical Power Transmission and Interconnection* (Sir I. Pitman and Sons, Ltd., UK, 1930).
22. Available at: http://www.aep.com/about/transmission/docs/transmission-facts.pdf. Accessed 21 May 2012.
23. W. Breuer, V. Hartmann, D. Povh, D. Retzmann and E. Teltsch, Application of HVDC for large power system interconnections, CIGRE, Paris, 2004.
24. T. Burton, D. Sharpe, N. Jenkins and E. Bossanyi, *Wind Energy: Handbook* (John Wiley, NJ, 2001).
25. EASAC, Transforming Europe's Electricity Supply–An Infrastructure Strategy for a Reliable, Renewable and Secure Power System, Policy Report 11, 2009.
26. APS Panel on Public Affairs, Integrating Renewable Electricity on the Grid, Report, 2003.
27. Available at: http://epp.eurostat.ec.europa.eu/portal/. Accessed 21 May 2012.
28. Available at: http://www.nrel.gov/. Accessed 21 May 2012.

Chapter 21

Energy Use and Energy Conservation

V. Ismet Ugursal

Department of Mechanical Engineering, Dalhousie University
P.O. Box 1000 Halifax, NS, Canada B3J 2X4
Ismet.Ugursal@dal.ca

This chapter presents the evolution of energy use and related economic indicators over the past 30 years. The data presented illustrates the relationship between energy use and human development and wealth. The opportunities and methods to reduce energy consumption are discussed to provide an overview of energy conservation techniques in various sectors of the economy.

1 Energy Use: Trends and Implications

Energy is one of the key commodities required to sustain civilization and one of the largest components of the world economy. While no definitive figures have been published, it is estimated that energy expenditures constitute about 8% of the global GDP.[1] In the US, where records have been maintained since 1970, the share of energy expenditures in the GDP varied from a low of 6% to a high of close to 14%, as shown in Fig. 1.[2]

Owing to the various ways in which energy is converted and used, there are different methods to quantify and report energy use.[a] "Primary energy" refers to energy extracted or captured directly from natural resources such as crude oil, coal, and natural gas, whereas "secondary energy" refers to energy that is produced from primary energy by conversion processes, such as generation of electrical energy by combustion of a fuel. The amount of energy that is available for use as energy, feedstock, or for conversion to other forms of energy is referred to as "energy supply" and includes the energy that is produced and recovered from stocks and net of imports/exports. "Energy consumption" (also referred to as "end-use energy consumption" or "final energy consumption") refers to the energy delivered to consumers to satisfy their energy needs and does not include the energy losses due to

---

[a]A comprehensive discussion of definitions, units and methodologies used in energy statistics is presented in "Energy Statistics Manual" published jointly by the IEA, OECD, and EUROSTAT.[3]

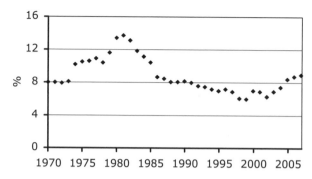

Fig. 1.   Energy expenditures as a percentage of GDP in the US.

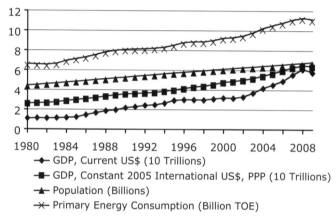

Fig. 2.   Evolution of GDP, population, and primary energy consumption (TOE: tonnes of oil equivalent).

conversion and distribution. Commonly, energy statistics are reported on an annual basis, as done here.

Global energy consumption has been increasing due to the increasing population and GDP as shown in Fig. 2.[4–6] However, the evolution of population, GDP, and energy consumption has been substantially different in different parts of the world as shown in Figs. 3–5.[7] The drop in the global energy consumption and GDP in 2009 is due to the economic recession that started in the US in the last quarter of 2008 with the mortgage crisis, and expanded into most countries. While the effects of the recession were felt widely, high-income countries[b] (such as EU countries and US) were hit most severely, as shown in Figs. 4 and 6.[7] However, most economies returned to a growth mode by the end of 2009.

---

[b]The World Bank classifies national economies according to the annual gross national income (GNI) per capita in US\$, calculated using the World Bank Atlas method.[7] The classification in 2009 was: low income, \$995 or less; lower middle income, \$996–\$3945; upper middle income, \$3946–\$12,195; and high income, \$12,196 or more.

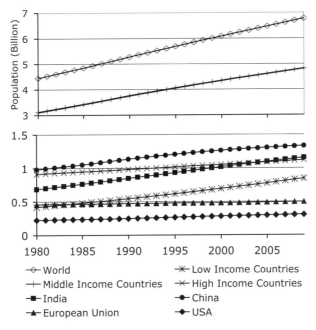

Fig. 3. Evolution of population.

DEFINITIONS: Gross Domestic Product (GDP) is the sum of gross value added by all producers in the economy plus any product taxes and minus any subsidies not included in the value of the products. It is calculated without making deductions for depreciation of fabricated assets or for depletion and degradation of natural resources. **Purchasing Power Parity (PPP) GDP** is GDP converted to international dollars using PPP conversion factor. **PPP conversion factor** is the number of units of a country's currency required to buy the same amounts of goods and services in the domestic market as US dollar would buy in the US. An **international dollar** has the same purchasing power over GDP as the US dollar has in the US.

There is a direct correlation between energy consumption and wealth, as shown in Fig. 7, where per-capita energy and electricity consumption in countries is plotted against their per-capita GDP. Owing to this relationship, as the per-capita GDP increases over the years, so does the energy and electricity consumption, as shown in Figs. 6, 8 and 9.[7] The vast difference in the level of wealth (measured in GDP per capita) among the countries is reflected in the amount of per capita energy and electricity consumption.

Energy consumption is also related to socio-economic development, measured in terms of the Human Development Index (HDI).[8] The HDI is measured in three basic dimensions: longevity, educational level (literacy index and registration combined

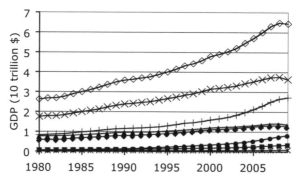

Fig. 4. Evolution of GDP, PPP (constant 2005 international US$10 trillion). (Legend same as for Fig. 3).

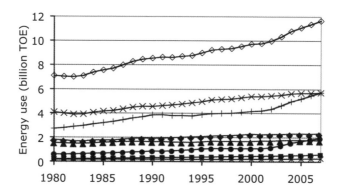

Fig. 5.   Evolution of energy consumption (Legend same as for Fig. 3).

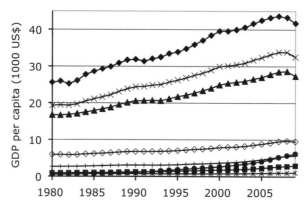

Fig. 6.   Evolution of GDP per capita (PPP, constant 2005 international 1000 US$). (Legend same as for Fig. 3)

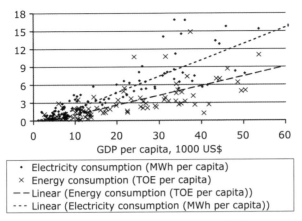

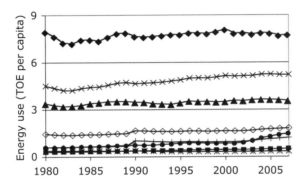

Fig. 7. Energy/electricity consumption versus GDP per capita (PPP, constant 2005 international 1000 US$).

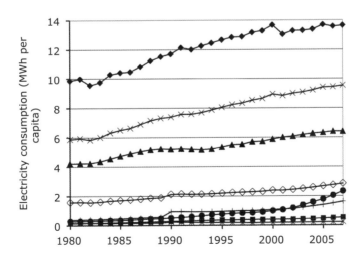

Fig. 8. Evolution of per capita energy use. (Legend same as for Fig. 3).

Fig. 9. Evolution of per capita electricity consumption. (Legend same as for Fig. 3).

index), and economic performance through GDP in dollar PPP, and spans a range of 0 to 1 with the interpretation:

High human development:

$$1 \geq \text{HDI} \geq 0.8.$$

Moderate human development:

$$0.799 \geq \text{HDI} \geq 0.5.$$

Low human development:

$$0.499 \geq \text{HDI} \geq 0.0.$$

The HDI, GDP, population, and energy consumption data for countries are presented in Table 1.[7,8] As shown in Fig. 10,[7,8,13] in the lower end of the HDI range (0–0.5), the energy consumption per capita is negligibly small and has no impact on the value of HDI. However, in the moderate HDI range, small increases in energy consumption result in substantial increases in HDI, while once HDI exceeds 0.8, large increases in energy consumption result in small HDI increases. This indicates that (i) in highly developed countries, energy consumption can be reduced without a significant reduction in socio-economic level, (ii) medium developed countries would greatly benefit from small increases in energy consumption, and (iii) a minimum level of energy consumption must be reached in developing countries before improvement in socio-economic level is possible. A comparison of energy production, consumption, and GDP in different economies is presented in Fig. 11.[10]

The global energy supply comes from three sources: fossil fuels (oil, natural gas and coal), renewables (hydro, solar, wind, geothermal, and combustible renewables such as wood and biomass) and nuclear. Toward the end of the 19th century, coal started to replace wood as the primary energy source, and reached a peak early in the 20th century.[9] Fifty years after its discovery in Pennsylvania, oil became the dominant source of energy globally, and its peak reached in the early 1970s, coinciding with the first oil crisis of 1973. Since then, with a decrease in the share of oil and coal in the global energy supply, and with increases in the share of natural gas, nuclear, hydro, and combustible renewables and waste, the sources of energy supply became more diverse, as shown in Fig. 12.[10] Also, although currently miniscule (less than half of 1% of the total), the use of wind, solar, and geothermal energy is increasing, primarily due to government subsidies in the form of feed-in tariffs in many OECD countries.

While there are global efforts to reduce the dependency on fossil fuels, especially coal, due to concerns associated with carbon dioxide emissions, fossil fuels still supply more than 80% of the global energy need, as shown in Fig. 12. The use of coal is mostly for production of electricity. As shown in Figs. 12 and 13,[7] in spite of efforts to the contrary, the use of coal in total as well as for the production of electricity is steadily increasing worldwide due to its diverse global availability,

Table 1: Socioeconomic and Energy Data for Countries.

Country	Population (million)	HDI	GDP, PPP (const. 2005 int'l billion US$)	Energy use (1000 TOE)	GDP per capita, PPP (const. 2005 int'l	Energy use (kgoe) per capita	Electricity consumption (kWh per capita)	Energy use (kgoe) per US$1,000
Albania	3.13	0.709	21.34	2,173	6,811	694	1,186	102
Algeria	33.86	0.662	247.69	36,863	7,316	1,089	902	149
Angola	17.55	0.392	85.58	10,630	4,875	606	185	124
Argentina	39.49	0.764	493.86	73,065	12,506	1,850	2,659	148
Armenia	3.07	0.697	16.17	2,844	5,261	926	1,692	176
Australia	21.07	0.931	713.26	124,068	33,848	5,888	11,249	174
Austria	8.30	0.846	295.31	33,177	35,576	3,997	8,033	113
Azerbaijan	8.58	0.691	63.46	11,910	7,395	1,388	2,394	188
Bahrain	0.76	0.806	23.52	8,774	30,962	11,551	14,153	373
Bangladesh	157.75	0.449	185.80	25,759	1,178	163	144	139
Belarus	9.70	0.720	101.34	28,047	10,446	2,891	3,345	281
Belgium	10.63	0.864	355.81	57,022	33,486	5,366	8,614	160
Benin	8.39	0.423	11.21	2,883	1,336	343	72	257
Bolivia	9.52	0.625	36.08	5,438	3,788	571	515	151
Bosnia Herzegovina	3.78	0.706	26.74	5,604	7,078	1,483	2,381	207
Botswana	1.89	0.614	23.85	2,022	12,604	1,068	1,435	85
Brazil	190.12	0.685	1,745.46	235,556	9,181	1,239	2,171	135
Brunei Darussalam	0.38	0.804	18.45	2,767	47,949	7,190	8,393	150
Bulgaria	7.66	0.736	86.08	20,231	11,238	2,641	4,456	250
Cambodia	14.32	0.484	24.59	5,134	1,717	358	94	209
Cameroon	18.66	0.446	37.24	7,294	1,996	391	265	196
Canada	32.98	0.885	1,189.56	269,369	36,074	8,169	16,995	226
Chile	16.64	0.773	217.01	30,790	13,045	1,851	3,318	142
China	1,317.89	0.639	6,903.97	1,955,766	5,239	1,484	2,332	283
Colombia	44.36	0.676	358.62	29,480	8,084	665	977	83
Congo	3.55	0.472	12.48	1,266	3,514	357	135	101

(*Continued*)

Table 1: (Continued)

Country	Population (million)	HDI	GDP, PPP (const. 2005 int'l billion $)	Energy use (1000 TOE)	GDP per capita, PPP (const. 2005 int'l)	Energy use (kgoe) per capita	Electricity consumption (kWh per capita)	Energy use (kgoe) per $1,000
Congo	62.52	0.235	17.54	18,089	281	289	97	1,031
Costa Rica	4.46	0.719	45.69	4,769	10,247	1,070	1,863	104
Côte d'Ivoire	20.12	0.387	30.75	9,978	1,528	496	178	325
Croatia	4.44	0.763	74.59	9,318	16,814	2,101	3,738	125
Cyprus	0.85	0.804	20.12	2,437	25,662	2,854	5,441	121
Czech Republic	10.33	0.843	236.26	45,755	22,862	4,428	6,496	194
Denmark	5.46	0.864	189.14	19,649	34,633	3,598	6,670	104
Dominican Republic	9.81	0.651	70.96	7,892	7,231	804	1,378	111
Ecuador	13.34	0.685	95.04	11,805	7,123	885	788	124
Egypt	80.06	0.601	381.26	67,246	4,762	840	1,384	176
El Salvador	6.11	0.653	37.55	4,884	6,149	800	939	130
Estonia	1.34	0.816	26.53	5,633	19,773	4,198	6,273	212
Ethiopia	78.65	0.309	58.35	22,805	742	290	40	391
Finland	5.29	0.870	177.03	36,467	33,474	6,895	17,162	207
France	61.94	0.864	1,956.20	263,718	30,651	4,258	7,772	135
Gabon	1.42	0.635	19.05	1,848	13,399	1,300	1,066	97
Georgia	4.36	0.698	19.35	3,343	4,409	767	1,620	173
Germany	82.27	0.883	2,744.72	331,257	33,364	4,027	7,184	121
Ghana	22.87	0.459	29.61	9,502	1,295	415	259	323
Greece	11.19	0.847	298.77	32,180	26,693	2,875	5,628	108
Guatemala	13.35	0.550	57.86	8,284	4,333	620	558	143
Haiti	9.72	0.404	10.11	2,777	1,040	286	30	275
Honduras	7.17	0.594	25.54	4,745	3,560	661	692	186
Hong Kong, China	6.93	0.855	276.74	13,745	39,958	1,985	5,899	50
Hungary	10.06	0.803	179.65	26,728	17,865	2,658	3,977	149
Iceland	0.31	0.888	11.48	4,894	36,860	15,708	36,853	426
India	1,124.79	0.500	3,031.64	594,913	2,695	529	542	196

(Continued)

Table 1: (*Continued*)

Country	Population (million)	HDI	GDP, PPP (const. 2005 int'l billion $)	Energy use (1000 TOE)	GDP per capita, PPP (const. 2005 int'l)	Energy use (kgoe) per capita	Electricity consumption (kWh per capita)	Energy use (kgoe) per $1,000
Indonesia	224.67	0.580	791.15	190,647	3,521	849	566	241
Iran	71.02	0.684	734.75	184,935	10,346	2,604	2,325	252
Ireland	4.36	0.896	179.23	15,060	41,136	3,457	6,263	84
Israel	7.18	0.869	180.44	21,963	25,130	3,059	7,002	122
Italy	59.38	0.848	1,707.97	178,163	28,766	3,001	5,713	104
Jamaica	2.68	0.682	19.40	4,956	7,252	1,852	2,542	255
Japan	127.77	0.880	4,045.20	513,519	31,660	4,019	8,474	127
Jordan	5.68	0.665	27.47	7,201	4,840	1,269	1,971	260
Kazakhstan	15.48	0.707	158.85	66,462	10,259	4,292	4,448	418
Kenya	37.75	0.456	54.55	18,305	1,445	485	151	335
Korea (Rep. of)	48.46	0.865	1,212.44	222,197	25,021	4,586	8,502	183
Kuwait	2.66	0.767	121.27	25,200	45,539	9,463	16,198	208
Kyrgyzstan	5.23	0.584	9.95	2,913	1,900	556	1,772	293
Latvia	2.28	0.777	37.03	4,670	16,268	2,052	3,064	126
Libya	6.17	0.741	90.77	17,823	14,715	2,889	3,871	196
Lithuania	3.38	0.785	57.42	9,250	17,010	2,740	3,414	161
Luxembourg	0.48	0.861	35.72	4,219	74,422	8,790	16,315	118
Malaysia	26.56	0.735	339.60	72,589	12,788	2,733	3,667	214
Malta	0.41	0.809	9.02	867	22,053	2,120	4,845	96
Mexico	105.28	0.742	1,407.69	184,262	13,371	1,750	2,036	131
Moldova	3.67	0.616	9.17	3,337	2,564	910	1,319	364
Mongolia	2.61	0.605	7.97	3,087	3,053	1,182	1,369	387
Morocco	31.22	0.551	119.72	14,361	3,776	460	707	120
Mozambique	21.87	0.270	16.21	9,150	741	418	472	564
Namibia	2.09	0.593	12.22	1,556	5,849	745	1,541	127
Nepal	28.29	0.411	27.89	9,554	986	338	80	343
Netherlands	16.38	0.886	613.75	80,423	37,466	4,909	7,097	131

(*Continued*)

Table 1: (Continued)

Country	Population (million)	HDI	GDP, PPP (const. 2005 int'l billion $)	Energy use (1000 TOE)	GDP per capita, PPP (const. 2005 int'l)	Energy use (kgoe) per capita	Electricity consumption (kWh per capita)	Energy use (kgoe) per $1,000
New Zealand	4.23	0.903	108.61	16,771	25,686	3,966	9,622	155
Nicaragua	5.60	0.555	13.58	3,474	2,427	621	446	256
Nigeria	147.72	0.412	276.57	106,683	1,872	722	137	386
Norway	4.71	0.937	229.81	26,859	48,800	5,704	24,980	117
Pakistan	162.59	0.481	381.81	83,271	2,348	512	474	218
Panama	3.34	0.737	36.12	2,824	10,803	845	1,592	78
Paraguay	6.13	0.631	25.62	4,202	4,182	686	958	164
Peru	28.51	0.707	206.46	14,079	7,242	494	961	68
Philippines	88.72	0.628	282.19	39,980	3,181	451	586	142
Poland	38.12	0.784	596.76	97,111	15,655	2,547	3,662	163
Portugal	10.61	0.785	233.31	25,065	21,993	2,363	4,860	111
Qatar	1.14	0.800	85.79	22,187	75,415	19,504	12,915	259
Romania	21.55	0.754	231.63	38,908	10,750	1,806	2,452	168
Russian Federation	142.10	0.708	1,991.70	672,139	14,016	4,730	6,317	340
Saudi Arabia	24.24	0.741	516.28	150,326	21,302	6,202	7,224	291
Senegal	11.89	0.399	19.58	2,673	1,646	225	128	137
Serbia	7.38	0.729	71.29	15,806	9,658	2,141	4,155	222
Singapore	4.59	0.836	228.23	26,754	49,739	5,831	8,514	117
Slovakia	5.40	0.811	104.47	17,847	19,356	3,307	5,250	171
Slovenia	2.02	0.825	53.12	7,329	26,321	3,632	7,138	138
South Africa	48.26	0.590	452.00	134,337	9,366	2,784	4,944	297
Spain	44.88	0.857	1,279.89	143,950	28,519	3,208	6,296	112
Sri Lanka	20.01	0.646	80.19	9,284	4,008	464	417	116
Sudan	40.43	0.369	75.96	14,675	1,879	363	90	193
Sweden	9.15	0.885	318.19	50,422	34,782	5,512	15,238	160
Switzerland	7.55	0.876	285.84	25,718	37,854	3,406	8,164	90
Syria	20.08	0.582	82.80	19,639	4,123	978	1,469	237

(Continued)

Table 1:   (Continued)

Country	Population (million)	HDI	GDP, PPP (const. 2005 int'l billion $)	Energy use (1000 TOE)	GDP per capita, PPP (const. 2005 int'l)	Energy use (kgoe) per capita	Electricity consumption (kWh per capita)	Energy use (kgoe) per $1,000
Tajikistan	6.73	0.563	11.16	3,900	1,659	580	2,176	349
Tanzania	41.28	0.379	46.15	18,278	1,151	443	82	396
Thailand	66.98	0.642	491.19	103,991	7,333	1,553	2,055	212
Togo	6.30	0.419	4.90	2,458	777	390	96	502
Trinidad Tobago	1.33	0.727	31.32	15,282	23,583	11,506	5,642	485
Tunisia	10.23	0.665	72.62	8,837	7,102	864	1,248	122
Turkey	73.00	0.672	874.09	100,005	11,973	1,370	2,238	114
Turkmenistan	4.98	0.652	28.16	18,073	5,657	3,631	2,279	642
Ukraine	46.51	0.710	304.50	137,342	6,547	2,953	3,529	451
United Arab Emirates	4.36	0.806	231.04	51,636	52,944	11,832	16,165	223
United Kingdom	60.98	0.845	2,079.37	211,308	34,099	3,465	6,123	102
US	301.58	0.899	13,167.63	2,339,942	43,662	7,759	13,638	177
Uruguay	3.32	0.749	35.88	3,167	10,796	953	2,197	88
Uzbekistan	26.87	0.600	61.52	48,682	2,290	1,812	1,658	791
Venezuela	27.48	0.689	313.43	63,745	11,404	2,319	3,077	203
Viet Nam	85.15	0.554	209.03	55,787	2,455	655	728	267
Yemen	22.27	0.424	49.17	7,212	2,208	324	202	147
Zambia	12.31	0.370	14.94	7,442	1,213	604	720	498
WORLD	6,620.50		63,128.26	11,664,890	9,535	1,819	2,846	185

Source: With the exception of HDI and sectoral energy consumption, all data in this table are from World Bank Databank, and are for year 2007 (http://databank.worldbank.org/). Human Development Index (HDI) data are from "Human Development Reports (HDR) Statistics" published by United Nations Development Program (http://hdr.undp.org/en/statistics/), and are for the year 2007.

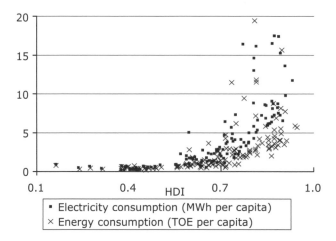

Fig. 10.   Energy/electricity consumption per capita versus HDI of countries (2007 data).

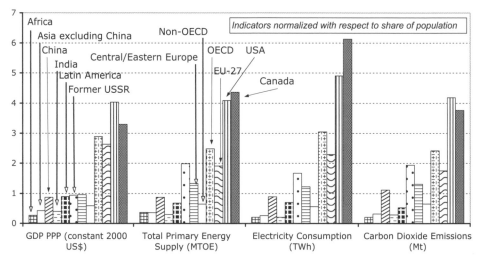

Fig. 11.   Comparison of economies — all indicators are normalized with respect to share of population (2008 data).

relatively low price and well-developed technology for utilization. In the expanding middle-income economies, the growth of coal usage is faster due to the vast appetite for energy. The reduction in the importance of coal in the EU is due to the large-scale switch in countries such as the UK, France, and Germany from coal to natural gas,[11] which is largely imported from Russia and other former Soviet Republics.

Energy consumption in the poorest economies is a fraction of that in the middle- and high-income economies as seen in Figs. 5, 7–9, and 11. In addition to differences

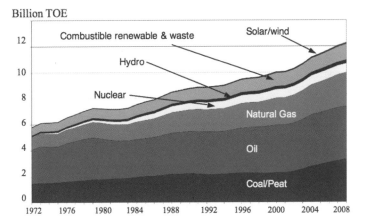

Fig. 12. Evolution of fuel mix in total primary energy supply.[10]

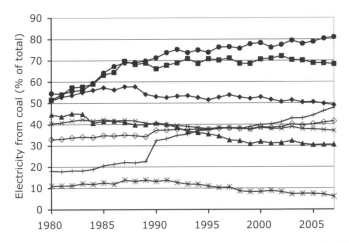

Fig. 13. Evolution of electricity generation from coal. (Same legend as for Fig. 3).

in magnitude, there is also a marked difference in the energy mix of the poorest economies compared to that of middle- and high-income economies. As shown in Fig. 14,[7] a large part of the energy consumption in the poorest economies is from traditional biomass, which is mostly in the form of wood and charcoal. In these economies, the minimal energy consumption is primarily for cooking, and to some extent, lighting. The choice of wood and charcoal is due to their low cost and local availability. However, both fuels are highly problematic because of their contribution to deforestation, air pollution, and respiratory illnesses due to cooking indoors on open fire.

While the global energy use is increasing with population and GDP, the efficiency at which energy is used is also increasing as shown in Fig. 15[7] due to advances

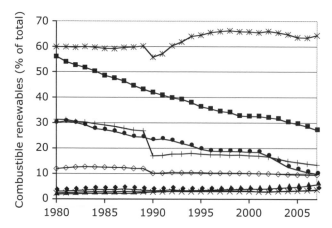

Fig. 14.   Evolution of the use of combustible renewables and waste as percent of total energy use. (Legend same as for Fig. 3).

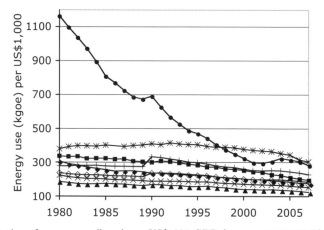

Fig. 15.   Evolution of energy use (kgoe) per US$1000 GDP (constant 2005 PPP). (Legend same as for Fig. 3).

in technology and energy conservation efforts. Due to the vast difference in access to technology, the energy use efficiency is highest in developed countries and lowest in low-income countries. The remarkable improvement in energy efficiency in China is due to the rapid technological advances the Chinese economy achieved in the past 30 years.

Commonly, energy use in a country is broadly categorized in five end-use sectors: industry, transportation, agriculture, commercial and public, and residential. The distribution of energy use amongst the end-use sectors varies substantially from one country to another, depending on parameters such as wealth, level of energy use and industrialization, climate and land mass, as shown in Table 1. In addition

Table 2:   Sectoral Energy Consumption in Selected Countries.

Country	Industry	Transportation	Agriculture	Commerce & Public service	Residential	Non energy
Angola	12.8	13.9	0	3.8	66.5	2.9
Australia	35	38.7	2.2	6.9	12.7	4.4
Austria	25.7	28.5	2.7	9.4	28.3	5.3
Bangladesh	27.8	8.6	3.5	1	56	3.2
Brazil	42.7	31.4	5	5	12.9	2.9
Canada	35.2	28.5	2.1	14.8	16.2	3.3
China	40	9.6	4.2	3.7	38.2	4.3
DR Congo	21.2	2	0	0	74.9	1.8
Côte d'Ivoire	6.6	10.2	1.4	11.8	58.2	11.7
Ethiopia	1.7	3.5	0	0.2	2	92.5
Germany	29.8	26.8	1.1	9.8	28.6	3.8
Haiti	15.3	16.3	0	2.3	65.4	0.7
Iceland	32.7	14.7	12.2	5.1	28.7	6.7
India	26.2	11.7	2	0.9	56.5	2.7
Indonesia	19.8	19.7	1.7	1.8	56.1	1
Ireland	23	37.2	2.3	13.2	22.5	1.7
Japan	35.9	28.1	1.9	16.9	13.6	3.5
Mexico	32.6	40.5	2.9	3.8	18.6	1.6
Mozambique	14.1	4.6	0.1	0.4	80.5	0.3
Netherlands	32.7	24.2	7	13.3	17.3	5.4
New Zealand	42.2	36.5	2.5	7.1	10.1	1.6
Nigeria	10.8	9.8	0	0.3	78.3	0.7
Norway	40.5	22.1	3.8	10.7	19.9	3.1
Pakistan	26.7	17.3	1.4	2.5	51.6	0.6
Philippines	29.9	33.1	1.7	9.3	22	3.9
Romania	41.9	17.3	1.2	4.8	29.5	5.4
Russian Fed.	34.7	19.3	3	4.6	34.1	4.2
Senegal	19.2	21.5	2.7	0.7	55	0.8
Sudan	24	14.4	1	14	45.4	1.3
Sweden	38	23.8	1.3	12.7	21.9	2.3
Tanzania	10.2	4.6	4.6	0	77	3.6
Turkey	31.6	23.2	5.7	4.1	30.8	4.6
US	26.2	39.6	1	12.2	16.6	4.5
Zambia	22.8	6	0.4	1.4	68.8	0.7

Source: *Energy Balances of OECD Countries* (2003 Edition) and *Energy Balances of non-OECD Countries* (2003 Edition), published by the IEA (2001 data).

to their use as a primary source of energy in the end-use sectors of the economy, fossil fuels are also used as a chemical feedstock (non-energy use). The magnitude of the energy consumed for non-energy uses is less than 10% of the total energy consumption in most countries as shown in Table 2.

Energy is an essential input to all processes related to the production of goods and services that contribute to human development, and as the data presented so far in this chapter show, advances in human development require increased energy consumption. However, the increasing cost of energy, the depletion of energy resources and the degradation of the environment as a result of the extraction, conversion and use of energy are substantial and escalating problems that challenge human development. One of the effective approaches to reduce the economic cost of development and the impact of energy consumption on the environment is energy conservation i.e. achieving more with less energy. The methods and approaches for energy conservation in the various sectors of the economy are reviewed in the Sec. 2.

2 Energy Management and Conservation

Energy engineering is a profession that combines the skills of engineering and scientific knowledge to identify, formulate and solve problems associated with the conversion and utilization of energy and to design, build, operate, maintain, and decommission systems for these purposes. Energy engineering requires a "systems approach" and is multidisciplinary in nature. Thus, an energy engineer needs engineering skills in electrical, mechanical, and process engineering, as well as good management knowledge.

Energy management and conservation (EM&C), a component of energy engineering, has as its primary objective the judicious and effective use of energy to minimize waste and costs while maximizing output. Thus, EM&C seeks to improve *energy efficiency.*

Energy Efficiency

In the most general sense, efficiency is the ratio of desired output to required input:

$$\text{Efficiency} = \frac{\text{Desired output}}{\text{Required input}}.$$

Based on this general concept of "efficiency," "energy efficiency" can be defined in a variety of ways, depending on the application. For example:

Energy efficiency of a hot water boiler

$$= \frac{\text{Energy received by the water heated in the boiler}}{\text{Energy input to the boiler}}.$$

$$\text{Energy efficiency of a production facility} = \frac{\text{Units produced}}{\text{Energy input to the facility}}.$$

These two examples demonstrate that there can be a variety of ways to quantify "energy efficiency" depending on the objective. EM&C is about increasing "energy efficiency" no matter how it is defined.

Increasing energy efficiency can be accomplished by (i) improving the efficiency of the individual components of the system, (ii) improving the system integration to produce an energy efficient system. Both approaches should be applied to maximize energy efficiency since inefficient components in an integrated system, as well as efficient components in a poorly integrated system result in energy waste.

Examples

Automobile: Efficient engine and drive train (efficient components), but brick-shaped body (inefficient system)

Chemical plant: Efficient boilers, furnaces, heaters, refrigeration system (efficient components), but no heat recovery (inefficient system)

The impetus for energy conservation is often to reduce energy expenditures. Thus, as cost of energy increases due to scarcity or anticipation of scarcity, energy conservation and/or energy substitution efforts gain focus. It is possible to reduce energy consumption and costs by lowering standards and/or production, but this is not a desirable approach. Alternatively, energy consumption can be reduced by increasing energy efficiency without compromising standards and/or production through EM&C measures. Energy expenditures can also be reduced by inter-fuel substitution to permit a less expensive alternative fuel to replace the scarce fuel.

Energy Efficiency and Jevons Paradox/Khazzoom–Brookes postulate

The Jevons paradox points out that in some cases, contrary to intuition, energy efficiency increases result in an increase of energy consumption rather than a decrease. The argument is as follows:

— When energy efficiency improves, energy consumed to carry out a given task decreases, resulting in reduced energy costs.
— The savings realized are then used to increase energy consumption for other purposes.

It is argued that increased energy efficiency increases energy consumption by two mechanisms:

(i) increased energy efficiency makes the use of energy cheaper and promotes increased energy use (the direct rebound effect), and
(ii) increased energy efficiency leads to increased economic growth, which increases energy use by the whole economy.

EM&C provides the following benefits to an organization:

• reduced capital and operating costs, and consequently, higher profits;
• reduced vulnerability to energy price and availability fluctuations; and

- increased "output" as well as increased understanding of the operation, which will likely lead to improved productivity.

EM&C can be achieved by:

- developing and maintaining a corporate culture that is committed to the efficient and effective use of energy through communication and education;
- establishing an energy policy that includes an iterative monitoring, reporting and management system to develop, review and revise measurable performance goals and action plans and to monitor results;
- adopting an integrated approach to EM&C across all aspects of the organization with professionals in-charge; and
- developing and maintaining an adoption, research and development program for energy-efficient technologies and practices.

To be effective, energy conservation efforts should be undertaken in four stages:

1. *Identification of the need and establishment of the objectives*
   - Need for a global or a sectional study
   - Selection of the study team — in house or independent consultants
   - Setting the financial ground rules, such as available capital, maximum allowable payback period and minimum acceptable return on investment

2. *Energy audit*
   - Data collection
   - Analysis of data
   - Process and energy flow diagrams
   - Identification of opportunities
   - Technical evaluation of each potential measure to determine savings
   - Economic evaluation — costs and savings
   - Integration of various proposals to form a viable project proposal

3. *Engineering design and implementation of proposals*
4. *Follow-up: Verification, evaluation and modifications*

2.1 *Energy Audit*

To improve energy efficiency, the first step is to carry out an energy audit (also known as an "energy study"). An energy audit is an organized survey to identify and measure all energy uses to determine probable sources of energy losses and wastes and identify potential energy conservation measures. This effort is usually performed by a team of specialists under the direction of an experienced energy engineer.

The energy audit serves to identify all of the energy streams into a facility and quantify energy use according to discrete functions. Thus, the main purpose of an energy audit is to identify quantitatively how and where a facility, such as a building or an industrial plant, uses energy, and the opportunities to reduce energy consumption.

The cost and effectiveness of an energy audit is proportional to the magnitude of the effort. To conduct a comprehensive energy audit that can identify a wide range of energy-saving opportunities, detailed energy and operational data have to be collected and analyzed. An energy audit consists of several phases that are interrelated with each other as summarized below.

1. Data collection and processing

In this phase, data on all sources and uses of energy, as well as design and operation of mechanical, electrical, and process systems and equipment and building envelope are collected. To identify patterns, energy consumption data for preferably the last three to five years are necessary. The sources of data include energy bills, design and as-built drawings and specifications as well as operational data records on set-points, occupancy and production characteristics. Operational data are then checked against the design parameters and verified, while discrepancies are identified and justified. The data are reduced to more practical formats such as graphs and tables, and historical energy consumption data and energy-use profiles for all systems are developed. When possible, energy consumption data is normalized with respect to parameters such as degree-days, building area, or number of units produced to provide indicators that can be used to compare energy consumption during different periods and between different facilities and processes. Table 3 provides a list of energy indices that are used for comparative evaluation of energy consumption figures. Table 4 lists data requirements and documentation needed to conduct an in-depth energy audit.

Table 3: Energy Indices Used for Comparative Evaluation of Energy Consumption.

Energy consumption and/or energy expenditure:
— Per degree day
— Per unit floor area
— Per unit envelope area
— Per (unit floor area × degree day)
— Per (unit envelope area × degree day)
— Per unit output
— Per unit income
— Per unit profit

Table 4:   Data Requirements and Documentation Needed to Conduct an In-Depth Energy Audit.

— Fuel and electricity cost and consumption history (3- to 5-year data)
— Mechanical, electrical, and architectural drawings and specifications of buildings (preferably "as built" drawings)
— Lighting levels, wattage, and state of lighting fixtures
— Building tightness regarding infiltration/exfiltration
— Material properties of the building envelope components (R-values, density, color, etc.)
— Process flow diagrams, process design drawings, and specifications
— Operational data from processes
— Operation, occupancy, temperature, and relative humidity schedules
— Measured data on flow rates, temperatures, pressures
— Name-plate specifications and operational data of major energy using equipment, including heat exchangers, coils, heaters
— Pump and fan capacity and operational data, flow control equipment, and strategies
— Automatic control strategies, equipment and operation
— Electrical equipment, motors, demand control, power factor, and power factor correction
— Estimates of incomplete information

Degree-day

A degree-day (DD) is the difference of one degree between the mean outdoor temperature on a certain day and a reference temperature. Commonly 18°C is used as the reference temperature. The number of *annual heating degree-days* (HDD) is determined by cumulating, over an entire year, the DDs for the days with average temperature below 18°C:

$$\text{Annual heating degree-days} = \sum_{i=1}^{365}(T_{\text{reference}} - T_{\text{average}})$$

Similarly, *annual cooling degree-day* (CDD) is calculated by cumulating the DDs with average temperature above the reference temperature:

$$\text{Annual cooling degree-days} = \sum_{i=1}^{365}(T_{\text{average}} - T_{\text{reference}})$$

where T_{average} is the daily average outdoor temperature.

The annual HDD and annual CDD provide a measure of the severity of the heating and cooling seasons, respectively. The higher the DD, the higher would be the space heating and space cooling energy consumption of a building. Thus, annual degree days are used to monitor, target, and normalize space heating and space cooling energy consumption. For example, if a building consumes 1000 L of heating oil per year in a location where HDD = 4500, the same building would be expected to consume about 2000 L per year in a location where HDD = 9000.

   DD data for many locations are available at http://www.degreedays.net/

Building Energy Simulation Software — Capabilities, Uses, Availability

To determine the energy-saving potential in an existing building as a result of energy efficiency improvements and renewable/alternative energy applications, it is necessary to model the building using building energy simulation software. Building simulation software simulates the thermal performance of the building and estimates the energy consumption over a given period or an entire year based on typical weather data of the location. There are a wide range of software packages varying in complexity, capability, and accuracy. Some are capable of only estimating the heating energy requirement, whereas others can estimate both heating and cooling requirements. Similarly, while some software packages are capable of simulating the operation of complex HVAC systems and plants, others can only simulate simple systems.

A survey of the capabilities and availability of building energy simulation software is given in the *Building Energy Software Tools Directory* of the US Department of Energy (available at http://apps1.eere.energy.gov/buildings/tools_directory/).

Generally, building simulation software utilizes "long-term climatic normal" weather conditions averaged over periods ranging from 5 to 30 years depending on the location and require the following input data:

• Schedules for lighting, occupancy, internal loads, and equipment operation
• Wall, window, roof, building mass, internal load, or infiltration characteristics
• HVAC system characteristics, zoning, and controls
• Cooling and heating plant characteristics and control
• Utility rate structures

The approach that is commonly used in building energy conservation studies that make use of building energy simulation can be summarized as follows:

Step 1. Architectural, mechanical, and electrical drawings, specifications and operational data as well as building energy (fuel and electrical) consumption data are collected.

Step 2. Using the architectural, mechanical, and electrical data, and the energy simulation software, the energy engineer models the building to reflect the normal operation practices. This is usually called the "baseline model." The objective of the baseline model is to match the energy consumption predicted by the baseline model simulation with the actual energy consumption data. A close agreement (within 10%) between the simulation results and the actual data indicates that the building model is accurate and can be considered to represent the actual operation of the building. If, however, the simulation results do not match the actual data, further refinement of the model is necessary to improve the accuracy of the baseline model.

Building Energy Simulation Software — Capabilities, Uses, Availability (*Continued*)

Step 3. Once the baseline simulation is accurately accomplished, the baseline model is updated to include the energy conservation measures that the engineer wishes to test. After modifying the baseline model to incorporate an energy-saving measure, a simulation run is done, and the result is compared with the baseline simulation results to determine the expected savings. This procedure is repeated as many times as necessary to estimate the energy savings possible with each one of the energy-saving methods the engineer wishes to consider. It may also be necessary to do simulation runs for groups of energy-saving measures since one may have an effect on the other (for example, using more efficient reduced wattage lights will reduce cooling energy consumption since heat gains will be reduced; however, heating energy consumption will increase).

Step 4. Once the energy savings are estimated based on the simulation results, the necessary modifications are identified for each energy-saving measure, and the associated costs are estimated. Using the estimated costs and savings, an economic analysis is carried out to make a comparative evaluation of energy-saving measures.

---

2. Study and analysis of systems

Systems in a building or an industrial facility can be categorized as non-energized systems and energized systems. Non-energized systems are those that require no outside energy sources such as electricity and fuel, whereas energized systems require outside energy sources. Both energized and non-energized systems are divided into sub-systems defined by function. For example, HVAC, DHW and lighting systems are sub-systems of energized systems, whereas building site and envelope components are non-energized sub-systems. A list of energized and non-energized systems encountered in buildings and industrial facilities is given in Table 5. Often, energy-saving measures have complex interrelationships. For example, a reduction in lighting will reduce the cooling load in a building, but will increase the heating load during winter. In addition to the energy and operation related data, an industrial energy audit requires the following:

- Process flow diagram
- Process limitations (such as safety, operability, technological, etc.)
- Flow rates, specific heats and entering and leaving temperatures for all processes and process streams
- Production data

Based on the collected design, operation, and energy billing data, the operational characteristics and the energy consumption of each system are determined and the

Table 5:   Examples of Energized and Non-Energized Systems.

Non-energized systems:

— Building site
— Walls, windows, exterior doors, skylights, roof
— Filters, strainers
— Tanks, piping, ducts, coils

Energized systems:

— HVAC systems and components (pumps, fans, blowers, humidifiers)
— Appliances (including computers, copiers, printers, etc.)
— Indoor and outdoor lighting
— Service and process hot water heaters
— Furnaces, boilers, fired heaters, kilns, dryers
— Refrigeration equipment, chillers, heat pumps, cooling towers
— Electric motors
— Compressors, turbo-expanders, turbines

energy consumption is tabulated. This allows identifying the systems that consume the most energy so that high priority can be assigned to these.

Many energy auditors use specially prepared software for quick and efficient data analysis. The advantage of using software is that uniform results and summaries can be obtained in a form suitable for review or further analysis. Computer analysis also provides easy modification of the results to reflect specific reporting requirements of management or to present desired comparisons, such as for different energy uses and types of equipment. Some software packages include savings and cost-estimation routines to conduct an economic analysis.

Energy Efficiency Analysis using RETScreen

The *RETScreen Clean Energy Project Analysis Software* is a unique decision support tool developed with the contribution of numerous experts from the Canadian government, industry and academia. This free software can be used worldwide to evaluate the energy production and savings, costs, emission reductions, financial viability and risk for various types of Renewable-energy and Energy-efficient Technologies (RETs). The software (available in 36 languages) also includes product, project, hydrology and climate databases, a detailed user manual and a case study based college/university level training course, including an engineering e-textbook. RETScreen is available at www.retscreen.net.

---

3. Identification of potential energy-saving measures

With the help of the understanding of systems, energy-use profiles and operation characteristics, potential energy-saving measures are identified. A list of commonly adopted energy-saving measures is given in Table 6.

When considering energy-saving measures, a concept that is useful to remember is the concept of "losses breed losses," which asserts that energy losses that occur

Table 6:   Commonly Adopted Energy-Saving Measures.

— Improve building envelope (e.g. increase insulation, replace windows and doors with high
  efficiency units, add weather-stripping/caulking)
— Apply temperature set-back and shut-off fresh air systems during unoccupied periods
— Improve efficiency of HVAC systems (e.g. retrofit constant volume reheat systems with
  variable air volume systems, incorporate controls for free cooling, use reverse cycle heat pump
  systems)
— Recover heat and/or power from waste heat streams (such as exhaust air, stack and exhaust
  gas streams, liquid effluents, cooling, and refrigeration systems)
— Improve efficiency of energy conversion equipment or switch to high-efficiency equipment
  (boilers, furnaces, fired heaters, lights, electric motors, and drives)
— Upgrade process control systems to computer-based systems
— Eliminate compressed air and steam leaks
— Maintain all equipment to maintain efficient operation

Table 7:   Losses Breed Losses.

Consider a processing plant:

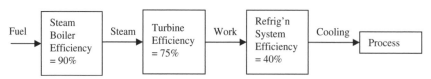

Cooling requirement by the process: 100 units Fuel requirement to produce 100 units of cooling:

$$(100)/(0.9 \times 0.75 \times 0.4) = 370 \text{ units.}$$

Improve the efficiency of the steam boiler by 5%:

$$\text{Savings: } 370 - (100)/(0.95 \times 0.75 \times 0.4) = 19 \text{ units.}$$

Improve the efficiency of the turbine by 5%:

$$\text{Savings: } 370 - (100)/(0.9 \times 0.80 \times 0.4) = 23 \text{ units.}$$

Improve the efficiency of the refrigeration system by 5%:

$$\text{Savings: } 370 - (100)/(0.9 \times 0.75 \times 0.45) = 41 \text{ units.}$$

further from the source of energy result in larger losses at the source due to the
compounding of losses through the system, as shown in Table 7.

4. Preliminary engineering design of potential energy-saving measures

Preliminary engineering designs for the potential energy-saving measures are proposed. The effects of these measures on issues such as production, building use,
operations as well as on other systems are studied and evaluated. The level of
detail of the preliminary design is such that cost estimates can be developed within

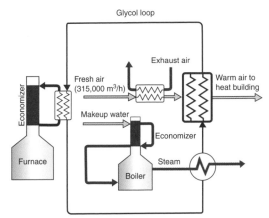

Glycol loop

Exhaust air

Fresh air
(315,000 m³/h)

Makeup water

Economizer

Furnace

Boiler

Steam

Warm air to
heat building

Economizer

Sample energy saving measure:
Industrial heat recovery for fresh
air and boiler make-up water
preheating.

Base case: An industrial facility
uses steam produced in a boiler to
heat fresh air introduced into the
facility for ventilation.

Energy saving measures: (i) Heat
recovery from the exhaust air and
furnance stack gas to preheat fresh
air, (ii) Heat recovery from the
boiler stack gas to preheat make-
up water.

Fig. 16. Industrial heat recovery for fresh air and boiler make-up water preheating.
Source: www.retscreen.net.

15%–20% accuracy. A sample preliminary engineering design of an energy saving measure is shown in Fig. 16.

5. Economic evaluation of potential energy-saving measures

The equipment required to implement an energy conservation measure cost money to purchase and install (initial capital cost) as well as to maintain (annual cost), and result in a reduced energy cost (annual cost). Often, there are a multitude of competing energy-saving measures that have different costs and savings. To select the option that will provide the best long-term economic benefit, an economic analysis has to be conducted. The first step in the economic analysis is the estimation of the capital and annual operating costs for implementing the potential energy-saving measure, as well as the annual amount of energy savings. Then an economic analysis, considering capital and annual operating costs, and annual cost savings is carried out for each potential energy-saving measure. A listing of commonly used economic analysis methods is presented in Table 8. There is a plethora of cost-estimating software packages, such as the RSMeans construction cost data (available at http://rsmeans.reedconstructiondata.com/), Cost Data On line (available at www.CostDataOnLine.com) and Aspen Economic Evaluation Family, which contains software for both cost estimation and economic analysis (available at www.aspentech.com/).

6. Identification of feasible energy-saving measures and final report

Based on the economic feasibility criteria developed by the owners (such as acceptable payback period, minimum return on investment), feasible energy-saving measures are identified and rank ordered. A number of energy-saving measures could be grouped together and proposed as a package.

A comprehensive report that includes all data and analysis is prepared, and submitted to the owner who decides on the implementation phase. The final report

Table 8:   Commonly Used Economic Analysis Methods and Cost Estimation Tools.

Life Cycle Cost analysis

For each energy-saving option considered as well as for the current state, the Life Cycle Cost (LCC) analysis sums the capital cost with the annual costs (i.e. cost of money, energy and maintenance) over the life of the investment. The preferred option is the one that has the lowest total LCC.*

Benefit–Cost analysis

For each energy-saving option considered as well as for the current state, the Benefit–Cost (B–C) analysis finds the difference between the net benefits (net monetary savings) and the costs over the life of the investment, expressed in present value or as an annuity. The preferred option is the one that produces the highest difference between benefit and cost.

Savings to Investment Ratio analysis

The Savings to Investment Ratio (SIR) calculates the ratio of lifetime savings to lifetime cost, expressed in present value or as an annuity, of each energy-saving option considered as well as for the current state. The preferred option is the one that produces the largest ratio. An SIR greater than one indicates that the option is cost-effective, whereas an SIR less than one is not cost-effective and an SIR of one is neutral.

Internal Rate of Return analysis

The Internal Rate of Return (IRR) is the interest rate for which the lifetime monetary savings associated with an energy-saving option are just equal to its lifetime cost. The preferred option is the one that produces the highest IRR. Also, the IRR has to be higher than the investor's minimum acceptable rate of return. The IRR is found by trial and error.

Discounted Payback analysis

The Discounted Payback (DPB) analysis determines the amount of time it takes for an energy-saving measure to accumulate monetary savings that equals the capital cost.

*Further reading on LCC analysis: http://www.wbdg.org/resources/lcca.php?r=use_analysis

of an energy audit presents the findings, results and recommendations of the energy audit.

2.2 *Energy Conservation Opportunities — Examples*

Energy conservation opportunities exist in every sector and facet of the economy. It is not possible to discuss or introduce the wide spectrum of energy conservation opportunities within the limited scope of this chapter. There is a wide range of excellent resources, many of them freely available on the internet, on energy conservation. To provide a starting point for the interested reader, a selection of resources is presented under "Further suggested readings." In the next few paragraphs, a brief introduction to energy conservation opportunities in lighting and personal transportation are presented as examples.

1. Energy conservation opportunities in lighting

Substantial energy-saving opportunities are possible by replacing inefficient lights with efficient ones. The output of a light fixture (luminous flux) is measured in lumens (lm). The "efficiency" of a light, called *luminous efficacy*, is the ratio of

Table 9: Luminous Efficacy and Life Expectancy of Light Sources.

Type of light	Luminous efficacy (lumens/watt)	Life expectancy (hours)
Incandescent	10 to 35	1,000 to 4,000
Mercury Vapor	20 to 60	24,000+
Light Emitting Diode (LED)[a]	100 to 140+	100,000+
Fluorescent	40 to 100	6,000 to 24,000
Metal Halide	50 to 110	6,000 to 20,000
High Pressure Sodium	50 to 140	24,000 to 40,000
Low Pressure Sodium	100 to 180	16,000

[a]The LED technology is constantly improving, and higher levels of luminous efficacy are expected in the future.

the light output to the energy input and is measured in lumens per watt (lm/W). The luminous efficacy as well as the life expectancy of different light sources vary dramatically, as shown in Table 9.

From an economic standpoint, the process of providing the desired lighting involves the expenditure of money for goods (lamps, luminaries, wire) and services (labor, electric energy), which leads to the idea of cost per unit of lighting delivered. The "cost of light" is commonly expressed as cost per lumen-hour. Because the cost per lumen-hour of typical general lighting systems is very small, the unit of dollars per million lumen-hours is usually used.

Cost-of-light data may be calculated for a given light fixture (luminaire) by multiplying the average light output of the luminaire (lumens) by the life of the lamps (hours) and dividing the result into the total costs (owning and operating) of the luminaire for the same time period. The following relationship provides a simplified approach[14]:

$$U = \frac{10}{Q} \left(\frac{p+h}{L} + WR \right).$$

where,

U = Unit cost-of-light for a lamp in dollars per million lumen-hours
Q = Mean lamp lumens
P = Lamp price in cents
h = Labor cost to replace one lamp in cents
L = Average rated lamp life in thousands of hours
W = Mean luminaire input watts (lamp + ballast)
R = Energy cost in cent per kilowatt-hour

2. Energy conservation opportunities in personal transportation

Energy conservation in personal transportation can be achieved through careful selection of the vehicle to be purchased, proper maintenance and improved driving habits. Energy conservation opportunities for each category are summarized in the following.

Table 10:   Fuel Consumption Ratings of a Sample of Passenger Cars.

	Engine size (L)	Transmission	Fuel consumption (L/100 km)	
			City	Highway
Audi A4	2.0	Automatic	8.9	6.5
Audi A8	4.2	Automatic	12.5	7.8
BMW 323	2.5	Automatic	11.2	6.7
Cadillac CTS-V	6.2	Automatic	17.7	10.7
Chevrolet Aveo	1.6	Manual	7.6	5.7
Chevrolet Impala	3.9	Automatic	12.0	7.4
Dodge Avenger	2.4	Automatic	9.9	6.7
Dodge Charger	5.7	Automatic	13.5	8.0
Ford Fiesta	1.6	Manual	7.1	5.3
Ford Taurus	3.5	Automatic	11.6	7.2
Honda Insight	1.3	Automatic	4.8	4.6
Honda Civic Hybrid	1.3	Automatic	4.7	4.3
Honda Civic	2.0	Manual	10.2	6.8
Lincoln Towncar	4.6	Automatic	13.4	8.8
Mazda 6	2.5	Manual	9.8	6.6
Toyota Corolla	1.8	Automatic	7.4	5.6
Volkswagen Golf Diesel	2.0	Manual	6.7	4.6

Source: http://oee.nrcan.gc.ca/transportation/tools/fuelratings/fuel-consumption-guide-2009.pdf.

Energy conservation through selection of vehicle: Fuel consumption varies widely from one vehicle to another, depending on the size and weight of the vehicle; the size and type of its engine; the type of its transmission (standard or automatic); and whether a vehicle is front-wheel drive, rear-wheel drive or four-wheel drive. Fuel consumption ratings of a sample of passenger vehicles given in Table 10 show the wide range. Thus, to minimize fuel consumption, vehicle selection should be made to avoid buying more vehicle than one needs. Many countries publish fuel consumption guides that provide the fuel consumption ratings of cars sold in the country (see, for example, the Fuel Consumption Guide published by Natural Resources Canada: http://oee.nrcan.gc.ca/transportation/tools/fuelratings/fuel-consumption-guide-2009.pdf).

Energy conservation through proper maintenance: Keeping a vehicle in top operating condition will save fuel and reduce operating and maintenance costs as well as exhaust emissions. Using the least viscous oil within the range of oils recommended by the manufacturer will improve the fuel efficiency of the engine, particularly when starting it cold. Maintaining proper air pressure in tires is also critical: operating a vehicle with just one tire underinflated by 56 kPa (8 psi) can increase the vehicle's fuel consumption by four percent and reduce the life of the tire by 15,000 km.

Energy conservation through improved driving habits: High speed and aggressive driving increases fuel consumption. Driving at 120 km/h requires 20% more fuel compared to driving at 100 km/h. Aggressive driving — speeding, quick acceleration, and hard stops — can increase fuel consumption by 25% or more. Also, use of air-conditioning can increase fuel consumption by up to 20%. Carrying roof or bicycle racks, or heavy items in the trunk when they are not needed increases fuel consumption. Typically, idling for more than 10 sec uses more fuel than it takes to restart a vehicle.

More comprehensive energy saving tips for driving can be found at http://oee. nrcan.gc.ca.

References

1. Energy's large share of the economy requires caution in determining policies that affect it, Institute for Energy Research, 16 February 16 2010. Available at: http://www.instituteforenergyresearch.org/2010/02/16/a-primer-on-energy-and-the-economy-energys-large-share-of-the-economy-requires-caution-in-determining-policies-that-affect-it/. Accessed 4 January 2011.
2. Energy Consumption, Expenditures, and Emissions Indicators, 1949–2009 in Annual Energy Review, Table 1.5, U.S. Energy Information Administration, 19 August 2010. Available at: http://www.eia.gov/aer/overview.html. Accessed 4 January 2010.
3. Energy Statistics Manual, IEA, OECD and EUROSTAT, 2005. Available at: http://www.iea.org/textbase/nppdf/free/2005/statistics_manual.pdf. Accessed 5 January 2011.
4. World Population Prospects: The 2008 Revision (Updated 21 May 2010), United Nations, Dept. of Economic and Social Affairs. Available at http://esa.un.org/unpd/wpp2008/all-wpp-indicators_components.htm. Accessed 14 January 2011.
5. The World Bank. Available at: http://www.worldbank.org/. Accessed 14 January 2011.
6. *Statistical Review of World Energy 2010*, BP. Available at: http://www.bp.com/productlanding.do?categoryId=6929&contentId=7044622. Accessed 14 January 2011.
7. *World Bank Databank, 2011.* Available at: http://databank.worldbank.org/. Accessed 16 January 2011.
8. Human Development Reports, United Nations Development Program (UNDP). Available at: http://hdr.undp.org/en/statistics/hdi/. Accessed 17 January 2011.
9. H. Crane, E. Kinderman and R. Malhotra, *A Cubic Mile of Oil — Realities and Options for Averting the Looming Global Energy Crisis* (Oxford University Press, USA, 2010).
10. 2010 Key World Energy Statistics, International Energy Agency (IEA). Available at: http://www.iea.org/textbase/nppdf/free/2010/key_stats_2010.pdf. Accessed 4 January 2011).
11. *BP Statistical Review of World Energy 2010.* Available at: http://www.bp.com/productlanding.do?categoryId=6929&contentId=7044622. Accessed 17 January 2011.
12. Energy Poverty; How to make modern energy access universal, International Energy Agency, 2010. Available at: http://www.worldenergyoutlook.org/docs/weo2010/weo2010_poverty.pdf. Accessed 17 January 2011.
13. Energy Balances of OECD Countries (2003 Edition) and Energy Balances of non-OECD Countries (2003 Edition), International Energy Agency (IEA).

14. M. S. Rea (ed.), *IESNA Lighting Handbook*, IESNA Illuminating Engineering Society of North America, annual publication.

Further Suggested Readings

B. L. Capehart, W. C. Turner and W. J. Kennedy, *Guide to Energy Management* (CRC Press, Boca Raton, FL, 2008).

Energy Efficiency Guide for Industry in Asia. Available at: www.energyefficiency-asia.org.

Energy Efficiency Planning and Management Guide (Office of Energy Efficiency, Natural Resources Canada. Available at: http://oee.nrcan.gc.ca/publications/infosource/pub/cipec/efficiency/toc.cfm?attr=20.

Energy Use and Management, *2007 ASHRAE Handbook* — Applications, Ch. 35 (ASHRAE, 2007).

Lighting Reference Guide, Natural Resources Canada (2005). Available at http://oee.nrcan.gc.ca/publications/equipment/lighting/doc/LightningReferenceGuide-NRCAN-E.pdf.

M. D. Lyberg (ed.), *Source Book for Energy Auditors*, Vols. 1 and 2 (IEA, 1987). Available at: http://www.ecbcs.org/docs/annex_11_source_book_vol1.pdf; http://www.ecbcs.org/docs/annex_11_source_book_vol2.pdf.

A. Thumann and S. Dunning, *Plant Engineers and Managers Guide to Energy Conservation* (The Fairmont Press, Lilburn, GA, USA, 2008).

A. Thumann and D. P. Mehta, *Handbook of Energy Engineering* (The Fairmont Press, Lilburn, GA, USA, 2008).

A. Thumann, W. J. Younger, T. Niehus, *Handbook of Energy Audits* (The Fairmont Press, Lilburn, GA, USA, 2010).

W. C. Turner and S. Doty, *Energy Management Handbook* (The Fairmont Press, Lilburn, GA, USA, 2009).

Chapter 22

The Earth's Energy Balance

Gerard M. Crawley

Marcus Enterprises, Columbia, SC 29206, USA

marcusenterprise@gmail.com

The temperature of the earth is computed by the basic radiation laws both at the upper atmosphere and at the earth's surface. The importance of the atmosphere in providing a living environment is demonstrated, and the composition of the atmosphere leading to the "greenhouse" effect is shown. Data on the history and current levels of greenhouse gases are given. The history of the global temperature and current measurements both on land and sea are provided, as are data on sea levels and ice cover. Finally, predictions of various climate models are shown.

1 Introduction

The temperature of the earth is governed principally by incident solar radiation, although there is a small heat flux from the earth's core.[a] Some of this solar radiation is reflected (about 30%) and the remainder is absorbed. The absorbed radiation is re-emitted as long-wave radiation, and the balance of the absorbed and emitted radiation determines the equilibrium temperature of the earth. The atmosphere plays a critical role in maintaining the surface of the planet at a considerably higher temperature than that would exist in the absence of an atmosphere. This chapter outlines the basic physical principles underlying the earth's energy balance and explains the effect of the atmosphere on the equilibrium surface temperature.

Data showing the recent dramatic increases in atmospheric "greenhouse" gases will be given, as will measurements of global temperatures as a function of time. Finally, predictions using various models to predict future global temperatures and sea level changes will be discussed.

---

[a]The geothermal heat flux to the earth's surface is about 3.9×10^{15} kWh per year compared to the total solar energy incident of about 1.5×10^{18} kWh per year (see also Chapters 9, 10, and 12 of this book).

2 Black Body Radiation

All objects that are not at temperatures of absolute zero ($T = 0$ K) emit electro-magnetic radiation, whether they be the sun, the earth, or a piece of cheese. The total amount of radiation emitted and the shape of the spectrum of the emitted radiation critically depend on the temperature of the object. When electromagnetic radiation falls on any object, it can be reflected, transmitted, or absorbed. In equi-librium i.e. when the temperature has stabilized, the amount of radiation absorbed must equal the amount of radiation emitted.

The total amount of energy emitted per second by an object at an absolute temperature T, is given by the Stefan–Boltzmann law[b]:

$$\Phi_{\text{tot}} = \sigma.AT^4, \tag{1}$$

where Φ_{tot} is the total energy emitted per second in joules per second or watts, A is the total area in square meters, T is the absolute temperature in Kelvin, and σ is a constant.

$$\sigma = 5.6705 \times 10^{-8}\,\text{W}/(\text{m}^2\text{K}^4).$$

The dependence of the radiation emitted on frequency and temperature is given by Planck's radiation law[1]:

$$u_f(f, T) = \frac{8\pi f^2}{c^3} \cdot \frac{hf}{e^{hf/(kT)} - 1}, \tag{2}$$

where $u_f(f, T)$ is the spectral radiant energy density in Joule sec/m^3, f is the frequency in Hertz, T the temperature in Kelvin, c the speed of light in meters/sec ($c = 2.998 \times 10^8$ m/s), h is Planck's constant ($h = 6.626 \times 10^{-34}$ J sec), and k is Boltzmann's constant ($k = 1.381 \times 10^{-23}$ J/K).

A useful additional law governing the frequency of the emitted radiation is given by the Wien displacement law:

$$\lambda_{\text{max}} = \frac{2.8978 \times 10^{-3}}{T}, \tag{3}$$

where λ_{max} is the wavelength in meters at which maximum energy is emitted and T is the absolute temperature in Kelvin.

The energy emitted by the sun and its distance from the earth are required to estimate the temperature of the earth. A list of some of the properties of the sun[2] is given in Table 1.

The total energy emitted per second by the sun and the distance from the sun to the earth can be used to calculate the average amount of radiation falling on a square meter at the top of the earth's atmosphere, viz. 1365 W/m^2. This value agrees to within 0.1% of the average value[3] from satellite measurements of

---

[b]Strictly this applies only to an ideal "blackbody" where the emissivity, $\varepsilon = 1$. This is a very good approximation for the sun and most stars.

Table 1: Properties of the Sun.

Mass	1.99×10^{30} kg
Radius	6.96×10^{8} m
Total energy emitted per second (Luminosity)	3.839×10^{26} W (J/sec)
Mean distance from the Earth	1.496×10^{11} m
Surface temperature	5778 K (5505° C)
Wavelength at which maximum energy is emitted	5.02×10^{-7} m (502 nm)

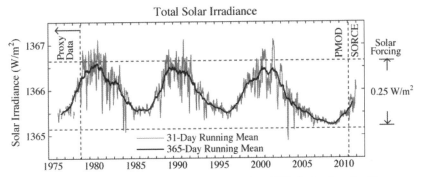

Fig. 1. Sun's energy output at the distance of the earth in W/m^2 showing the 11-year sunspot cycle.

$1366 \, W/m^2$ at the edge of the earth's atmosphere. This number, termed the solar constant, will vary slightly (by about 7%) as the earth's distance from the sun varies over the course of a year. In addition, the sun has an 11-year cycle of activity, which manifests itself as an increase in the number of sun spots and an increase in the temperature of the sun's surface. This 11-year cycle is clearly visible in Fig. 1, where the total solar irradiance is plotted as a function of time from 1975 to the present.[3] Note than in 2012, the sun is emerging from a minimum in the total irradiance. However the range of variability is small compared to the total irradiance of around $1366 \, W/m^2$.

This total energy is distributed over a range of wavelengths, although the wavelength at which peak energy is available is about 500 nm (see Table 1). This is, perhaps not surprisingly, in the middle of what we humans call the "visible" spectrum.

The actual distribution of solar energy is not exactly that of an ideal blackbody at 5778 K because of absorption in the solar atmosphere[4] (see Fig. 2). Even more important, the radiation reaching the surface of the earth is modified by absorption in the earth's atmosphere where ozone, water vapor, and CO_2 remove specific wavelengths from the solar spectrum, as shown in Fig. 2.

Another important factor that greatly influences the amount of radiation reaching a specific location on the earth's surface is the tilt of the earth on its axis as it orbits the sun. This gives rise to the seasons each year. The earth is a sphere (approximately) so that the radiation from the sun is not perpendicular to the

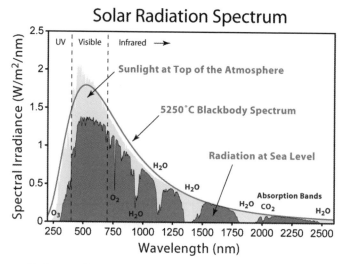

Fig. 2. Solar radiation reaching the earth's surface showing the effect of the atmospheric absorption.

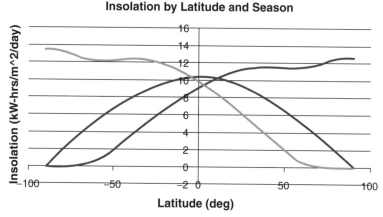

Fig. 3. The medium dark line is for June 21, the Northern Hemisphere summer solstice, and the pale line is for the Southern Hemisphere summer solstice, December 21. The darkest line with the peak near 0° is for insolation for both equinoxes, March 21 and September 21.

surface but depends on the latitude. A summary of this effect is presented in Fig. 3, which gives the amount of solar radiation falling on the earth's surface (insolation) as a function of latitude.[5]

3 Albedo

Not all the solar energy incident on the earth is absorbed. Some fraction of this short wavelength radiation is reflected from either the atmosphere (clouds) or the surface particularly by snow. The percentage of the radiation reflected by a

Table 2: Albedos for Various Surfaces (%).

Type	Surface	Range	Typical value
Water	Deep water, low wind	5–10	7
	Deep water, high wind	10–20	12
Bare surface	Moist dark soil	5–15	10
	Dry soil, desert	20–35	30
	Dry light sand	30–40	35
	Asphalt pavement	5–10	7
	Concrete pavement	15–35	20
Vegetation	Short green vegetation	10–20	17
	Coniferous forest	10–15	12
	Deciduous forest	15–25	17
Snow, ice	Sea ice, no snow cover	25–40	30
	Old melting snow	35–65	50
	Dry cold snow	60–75	70
	Fresh dry snow	70–90	80
Clouds	Cirrus (high)		21
	Cumulus (medium)		48
	Stratus (low)		69

particular surface is termed the "albedo" of the surface. A list of the albedos[2] of different surfaces is shown in Table 2. The albedo depends, to some extent, on the wavelength of the radiation falling on the surface as well as the detailed condition of the surface in the same way that a mirror's reflectance will be affected by flaws, cracks, or grime on its surface. Table 2 also shows that as sea ice and snow decreases, the albedo of the earth will decrease and more solar energy will be absorbed.

The overall albedo of the earth as measured by the Big Bear Solar Observatory[6] is 0.297 ± 0.005 (29.7%) but varies by about 5% on a daily basis, mainly dependent on cloud cover.

4 Calculation of the Earth's Temperature at the Upper Atmosphere

Given the earth's albedo, it is now possible to calculate the temperature of the earth at least at the upper edge of the atmosphere. The basic concept follows from the conservation of energy and can be expressed as:

Solar energy absorbed = energy radiated by the earth.

If more solar energy were absorbed than was radiated by the earth, the temperature of the earth would increase, and more radiation would be emitted by the earth until a new equilibrium was established at a higher temperature.

The earth is a sphere but presents a circular cross section to the incoming essentially parallel beam of radiation from the distant sun. So the total solar radiation absorbed is given by:

$$\text{Solar energy absorbed} = S_0 \cdot (1 - a)\pi R_E^2,$$

where S_0 is the solar constant $(1365\,\text{W/m}^2)$, a is the earth's albedo (0.297), and R_E is the radius of the earth.

Likewise for the earth, the total radiation emitted per square meter of surface area at a temperature T Kelvin is given by:

$$\text{Energy radiated by the earth} = \text{Surface area of earth} \times \sigma T^4$$

$$= 4\pi R_E^2 \sigma T^4.$$

Therefore by equating the energy absorbed to the energy radiated, the temperature is obtained as:

$$T^4 = \frac{S_0(1 - a)}{4\sigma}.$$

This gives a temperature at the edge of the earth's upper atmosphere of only $255\,\text{K}$ or $(-18°\text{C})$.

This calculation shows clearly that the earth's atmosphere has a very important effect on the temperature at the surface of the earth. Because of the atmosphere, the earth's surface is at a more benign temperature for life.

5 Effect of the Atmosphere on the Earth's Surface Temperature

Fortunately for life as we know it, the earth's atmosphere acts as a gaseous "blanket" which increases the temperature of the surface of our planet. The Wein displacement law, given in Eq. (3), implies that the wavelength at which the earth radiates the maximum energy is $11.36 \times 10^{-6}\,\text{m}$ (11.36 micron or $11.36\,\mu\text{m}$), which is a much longer wavelength (infra red) than that at which the sun radiates its maximum energy ($502\,\text{nm}$). The blanket effect arises because some of the gas molecules in the atmosphere, particularly water vapor and CO_2, are transparent to the short-wavelength radiation from the sun but absorb and re-emit the longer-wavelength radiation from the earth's surface. This is often termed the "greenhouse" effect because it was believed that a greenhouse heats up because its glass walls are transparent to sunlight but reflect the long-wavelength radiation (infra-red) emitted from objects within the greenhouse.[c]

5.1 Composition of the Earth's Atmosphere

Almost 99% of the earth's atmosphere consists of two gases, nitrogen and oxygen. The remaining 1% is made up of a number of trace gases, the most important of

---

[c]It has been shown that the main effect in warming of a glass greenhouse is actually because of the trapping of warm air within the enclosure, thus reducing convective heat losses.

Table 3: Composition of Earth's Atmosphere.

Gas	Amount in parts per million by volume (ppmv)
Nitrogen (N_2)	780,840 (78.08%)
Oxygen (O_2)	209,460 (20.95%)
Argon (Ar)	9340 (0.934%)
Carbon Dioxide (CO_2)	394.45
Neon (Ne)	18.18
Helium (He)	5.24
Methane (CH_4)	1.79
Krypton (Kr)	1.14
Water vapor (H_2O)	Variable[a]

Note: [a]Water vapor is about 0.4% overall in the atmosphere but varies between 1% and 4% near the earth's surface.

which are CO_2 and water vapor.[2] Table 3 lists the various gases in the atmosphere which are present at a level greater than one part per million by volume (ppmv).

The density of the atmosphere decreases with increasing height above the earth's surface. The atmosphere can be conveniently divided into various regions. The lowest region, up to about 14 km above the surface, is the **troposphere**, where most of the weather phenomena takes place. The atmospheric pressure at the top of the troposphere is only 10% of the pressure at the surface, and the temperature within the troposphere decreases with height above the surface. Next, there is a shallow layer, the tropopause, only 4 km in height before reaching the **stratosphere**, which stretches from 18 km to a height of 50 km. At the top of the stratosphere is a narrow region called the ozone layer, containing the highly reactive gas, ozone (O_3). Air movement in the stratosphere is mainly horizontal. There are two other layers above the stratosphere, the mesosphere and the ionosphere, but they contain very small amounts of gas and will not be relevant to the discussion of the greenhouse effect.

5.2 *Radiation Balance for Short- and Long-Wavelength Radiation*

What happens to the radiation from the sun (short-wavelength radiation) when it enters the earth's atmosphere is illustrated in Fig. 4. This figure shows the disposition of the short-wavelength radiation as a percentage of the incident radiation, taken as 100 units. A small amount is absorbed in the stratosphere (3 units) and a greater percentage (17 units) is absorbed lower in the troposphere. Some of the radiation is scattered by clouds and molecules in the atmosphere and reaches the surface as diffuse radiation (25 units) while some reaches the surface directly (25 units) and is absorbed. Thus the total amount absorbed at the surface is 50 units (25 + 25 units) and the total absorbed in the atmosphere is 20 units (3 + 17 units) making a total amount absorbed of 70 units, consistent with the earth's albedo of 30% ($a = 0.297$). The 30 units of short-wavelength radiation that are reflected

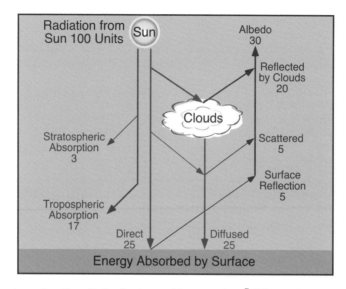

Fig. 4. Short-wavelength radiation in the earth's atmosphere[7] (*Copyright*: Michael Pidwirny).

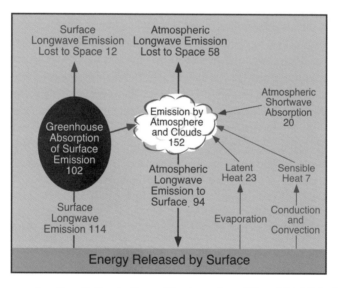

Fig. 5. Long wavelength radiation in the earth's atmosphere (*Copyright*: Michael Pidwirny).

arise from reflection from clouds (20 units), reflection from the surface (5 units) and scattering upward from the atmosphere (5 units).[7]

Similarly the balance sheet for the long-wavelength radiation emitted from the earth's surface is shown in Fig. 5. The earth's surface emits 114 units of long-wavelength radiation, 102 units of which are absorbed by the atmosphere and the remaining 12 units are emitted to space from the top of the atmosphere. The 102 units absorbed by molecules in the atmosphere like CO_2 plus other additions to atmospheric energy from conduction and convection of water vapor from the surface (7 units) plus latent heat from water condensation in the atmosphere (23 units) and the energy from the short wavelength radiation absorbed in the atmosphere (20 units) make a total of 152 units of energy that is contained in the atmosphere.[7]

This energy is then transferred in two directions, one to the surface, where 94 units of this energy are absorbed by the surface, leaving the remaining 58 units to be emitted to space as long-wavelength radiation.

Using the surface long-wavelength radiation emission value of 114 units, it is now possible to simply calculate the average surface temperature of the earth.

Here the surface temperature T_S will be given as:

$$T_S^4 = \frac{1.14(S_0)}{4\sigma},$$

where T_S is the surface temperature in Kelvin and the other terms are as before. This leads to a surface temperature of 288 K or $+15°C$, a much more livable temperature for life.

This overview for the earth as a whole neglects the transfer of heat from lower latitudes to higher latitudes by the atmosphere and ocean. Atmospheric circulation moves warm air from the tropics toward the polar regions and cold air from higher latitudes toward the tropics. Similarly, warm water from the tropics is moved to higher latitudes by ocean currents such as the Gulf Stream. Without these effects there would be a much greater difference in surface temperature between the tropics and the polar regions. More details of both short- and long-wave emission and absorption for 24-year averages for different regions of the globe using satellite data are available[8] from the US National Aeronautics and Space Administration (NASA).

5.3 *Abundances of Greenhouse Gases*

Section 5.2 illustrated the importance of the atmosphere in determining the surface temperature of the earth. The greenhouse gases, including water vapor, CO_2, methane (CH_4) and nitrous oxide (N_2O), which transmit a large fraction of the short-wavelength radiation but absorb and re-emit longer wave radiation, play an important role in atmospheric warming of the surface.

There is strong evidence that the amount of these gases in the atmosphere, particularly CO_2, has been increasing rapidly in recent years. The most widely quoted data is from the observatory on Mauna Loa in Hawaii,[9] as shown in Fig. 6.

A more detailed plot of the month by month variations of the CO_2 levels from Mauna Loa,[9-11] is shown in Fig. 7. The cyclic nature of this variation is explained by the seasonal variation in the northern hemisphere, where vegetation takes up CO_2 during the growing months so that the level decreases in the spring/summer months and then increases again during the fall/winter months as the vegetation

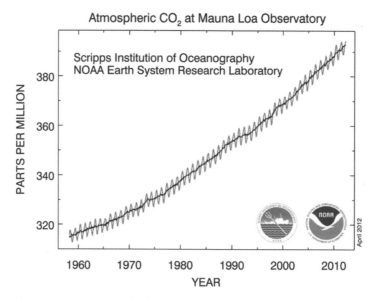

Fig. 6.   Measurements of atmospheric CO_2. The jagged curve shows the annual variations.

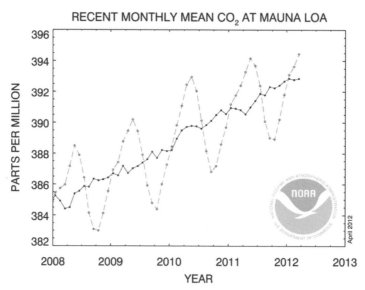

Fig. 7. Monthly variation of CO_2 levels at Mauna Loa. The oscillating curve with diamond symbols represents the monthly mean values, centered on the middle of each month. The smoothed line with the square symbols represents the same, after correction for the average seasonal cycle.

becomes dormant. This cyclic variation also gives an indication of the importance of vegetative growth on the levels of CO_2 in the atmosphere.

There have been many measurements made of CO_2 concentrations in the atmosphere over many years in both the northern and southern hemispheres and the

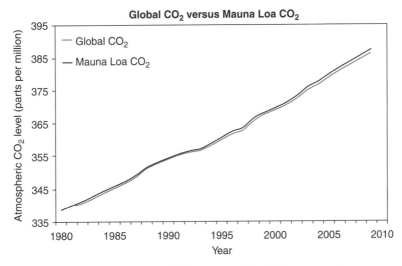

Fig. 8. Comparison of global averages of CO_2 concentrations in the atmosphere compared to the measurements from Mauna Loa.

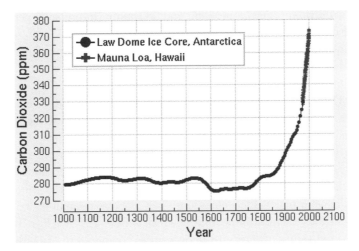

Fig. 9. Measurements of CO_2 concentrations using ice core from Antarctica dating back 1,100 years compared to results from Mauna Loa. Note that the plot has an off-set zero of about 270 ppmv which makes the recent increase appear even more dramatic.

results are quite consistent with a steady increase in CO_2 levels. Figure 8 shows results from a global summary compared to the results from Mauna Loa, and the measurements are in very good agreement.[9,12]

There have also been measurements of CO_2 concentrations from ice cores that allow CO_2 estimates extending thousands of years before the present.[12,13] Figure 9 shows CO_2 concentrations back to 1000 AD, which link in well to the more recent measurements. These results indicate that the amount of CO_2 in the

Table 4:   Current Concentration of Important Greenhouse Gases.[10]

Gas	Pre-1,750 concentration (ppmv)	Recent concentration (ppmv)	Lifetime in atmosphere (Years)	Increase in radiative forcing (W/m^2)
Carbon dioxide	280	390.5	100	1.79
Methane	0.70	1.871[a] 1.750[b]	12	0.50
Nitrous oxide	0.270	0.323[a] 0.322[b]	114	0.18
CFC-12	zero	0.534[a] 0.532[b]	100	0.17

[a]Measurements taken at Mace Head, Ireland.
[b]Measurements taken at Cape Grim, Tasmania, Australia.

atmosphere remained reasonably stable at a concentration of around 280 ± 10 ppmv from 1000 AD until the late 19th century. Since then there has been a steady increase in CO_2 levels to current values of above 390 ppmv. In fact, the concentration of CO_2 at a value of 280 ppmv represents the high end of the naturally occurring CO_2 cycle of the last one million years. During this time, the CO_2 concentration has fluctuated between 180 ppmv (glacial periods) and 280 ppmv (interglacial periods) over a 100,000-year cycle.

A current measurement (April 2012) from Scripps Mauna Loa Laboratory indicates[14] that the concentration of CO_2 in the atmosphere is now 396.18 parts per million.

While CO_2 is the most important greenhouse gas after water vapor, there are contributions from other gases such as methane (CH_4) and nitrous oxide (N_2O) which are also present in the atmosphere. Methane arises as a byproduct of biological processes in low oxygen environments such as swamps and from digestion. The increase in the number of domestic farm animals is expected to be a factor in the increase of the level of methane. Nitrous oxide results from the use of fertilizers containing nitrogen, plus vehicle emissions. There is also a small contribution from chlorofluorocarbons the most important of which is CFC-12 (CCl_2F_2). Table 4 shows current measured values of these greenhouse gases compared to pre-industrial levels obtained from ice core measurements.[10] The table also lists the atmospheric lifetime of these gases and an indication of their relative contribution to the increase in radiation emitted from the surface of the earth, called radiative forcing. Carbon dioxide is approximately three times more important than the next most important gas, methane, in increasing the radiative forcing, and therefore the temperature.

6 Measurements of Earth's Temperature

Obviously the temperature at various points on the earth's surface varies considerably depending on the latitude, the time of the day, and the season of the year. However for the same place on the earth, it is possible to track temperature over the course of many years and determine trends. And it is also possible to measure the average temperatures for the earth as a whole using a large number of measurements over a wide geographical distribution on the earth's surface. Such measurements have been made over the last few centuries, and reasonably reliable data sets are available over this time frame. In order to reach further back in time to estimate temperatures, more indirect methods are used. These include paleoclimatic proxies such as tree rings or ratios of the isotopes of oxygen O^{18}/O^{16} in ice cores and coral reefs. These measurements, when combined into large-scale syntheses, allow the estimates of global temperatures back many centuries, although the accuracy of the methods decreases as one moves further back in time.

An extensive summary of much of this data is contained in the Fourth Assessment report[15] of Working Group I of the International Panel on Climate Change (IPCC) and one set of data from this report is shown in Fig. 10. Temperature differences relative to the period 1961–1990 are shown for both Northern and Southern Hemispheres separately and also for the Global combination of these results. The temperatures shown in Fig. 10 are a combination of land and sea measurements.

Similar measurements[17] from the US National Oceanic and Atmospheric Administration (NOAA) shown in Fig. 11 extend these temperature measurements to 2010. These results are in excellent agreement with the summary results in Fig. 10, although a slightly different period is used as the standard period (1901–2000).

Likewise, measurements from NASA Goddard[18,19] (see Fig. 12) but using a different base are also quite consistent. This gives some confidence in the accuracy of the measurements.

It is interesting to compare the temperature measurements[20] for land and ocean separately as seen in Fig. 13. There is a significantly greater increase in land temperatures than that of the ocean. A number of possible reasons can explain this variation. The ocean has a larger heat capacity than land, and there is also the large effect of evaporative cooling, which is always present for the ocean but only applies to wet or at least moist surfaces on land. Finally, there may be a time lag between land and ocean temperatures because of mixing of the surface water with deeper levels in the oceans.

Another method of examining temperature records is to observe the rate of temperature changes over different periods of time.[15] Such results from different groups are shown in Table 5, where temperature increases over three time periods are given. Temperature increases over the recent past (1979–2005) are significantly greater than averages over longer periods (1850–2005). The agreement between the different measurements gives some confidence in these conclusions. The trends observed in other data shown in Figs 10–13 are confirmed by the averages over time periods. For instance, there is a consistently greater increase in Northern than in Southern latitudes. There is also a greater temperature increase on land than the measurements of sea surface temperatures.

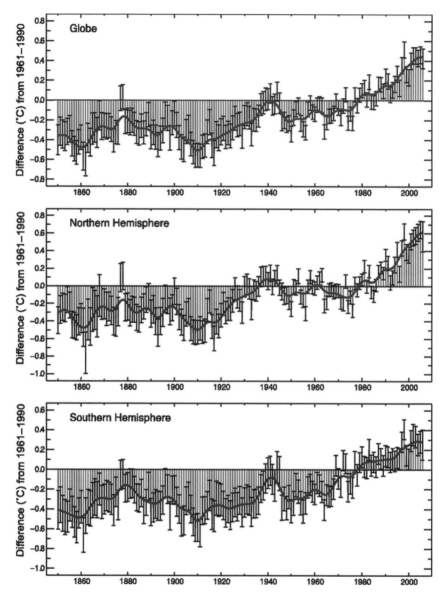

Fig. 10. Global and hemispheric annual combined land-surface air temperature and sea-surface temperature anomalies (°C) (red) for 1850 to 2006 relative to the 1961 to 1990 mean, along with 5 to 95% error bar ranges, adapted from Brohan et al. (2006).[15,16] The smooth curves show decadal variations.

7 Sea Ice Extent

In addition to the temperature increases noted earlier, there is an indication that the sea ice extent in the Arctic is also decreasing with time.[26] Figure 14 shows the percentage change in the sea ice in the Arctic in the month of March each

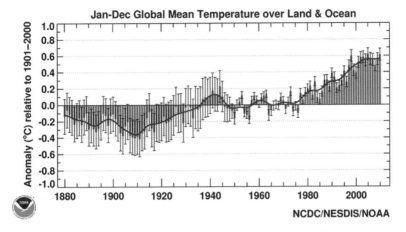

Fig. 11.   Global mean temperature over land and ocean till 2010.

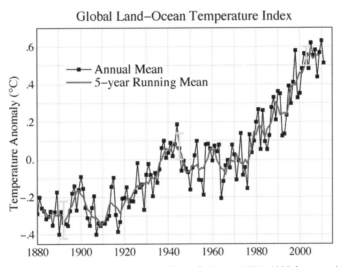

Fig. 12.   Global surface air temperature anomalies relative to 1951–1980 base period for annual and five-year running means. Vertical bars are 2σ error estimates.[18,19]

year from 1980 to 2012. While there are fluctuations from year to year, the trend observed is decreasing sea ice coverage at a rate of 2.6 % per decade. The situation in Antarctica is less clear, although recent evidence[27] suggests that warmer water around the Antarctic coastline is causing undermining of the ice shelf in some parts.

8 Sea Level Rise

Increasing global temperatures would also increase the sea level both because warmer water expands and because melting ice adds to the total amount of water in

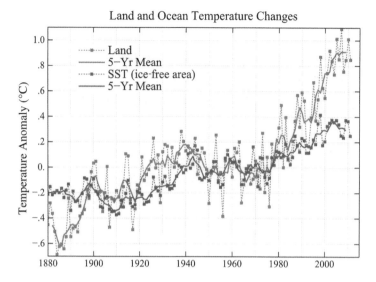

Fig. 13.   Comparison of land and ocean temperature changes.

Table 5:   Temperature Trends (°C per Decade).

Data set	1850–2005	1901–2005	1979–2005
Land: Northern Hemisphere			
CRU (Brohan et al.[16])	0.063 ± 0.015	0.089 ± 0.025	0.328 ± 0.087
NCDC (Smith and Reynol[21])		0.072 ± 0.026	0.344 ± 0.096
GISS (Hansen et al., 2001[22])		0.083 ± 0.025	0.294 ± 0.07
Lugina et al. (2005)[23]		0.079 ± 0.029	0.301 ± 0.075
Land: Southern Hemisphere			
CRU (Brohan et al.[16])	0.036 ± 0.024	0.077 ± 0.029	0.134 ± 0.070
NCDC (Smith and Reynolds[21])		0.057 ± 0.017	0.220 ± 0.093
GISS (Hansen et al., 2001[22])		0.056 ± 0.012	0.085 ± 0.055
Lugina et al. (2005) [23]		0.058 ± 0.011	0.091 ± 0.048
Land: Globe			
CRU (Brohan et al.[16])	0.054 ± 0.016	0.084 ± 0.021	0.268 ± 0.069
NCDC (Smith and Reynolds[21])		0.068 ± 0.024	0.315 ± 0.088
GISS (Hansen et al., 2001[22])		0.069 ± 0.017	0.188 ± 0.069
Lugina et al. (2005)[23]		0.069 ± 0.020	0.203 ± 0.058
Ocean: Northern Hemisphere			
UKMO HadSST2 (Rayner, 2006)[24]	0.042 ± 0.016	0.071 ± 0.029	0.190 ± 0.134
UKMO HadMAT1 (Rayner, 2003)[25]	0.038 ± 0.011	0.065 ± 0.020	0.186 ± 0.060
Ocean: Southern Hemisphere			
UKMO HadSST2 (Rayner, 2006)[24]	0.036 ± 0.013	0.068 ± 0.015	0.089 ± 0.041
UKMO HadMAT1 (Rayner, 2003)[25]	0.040 ± 0.012	0.069 ± 0.011	0.092 ± 0.050
Ocean: Globe			
UKMO HadSST2 (Rayner, 2006)[24]	0.038 ± 0.011	0.067 ± 0.015	0.133 ± 0.047

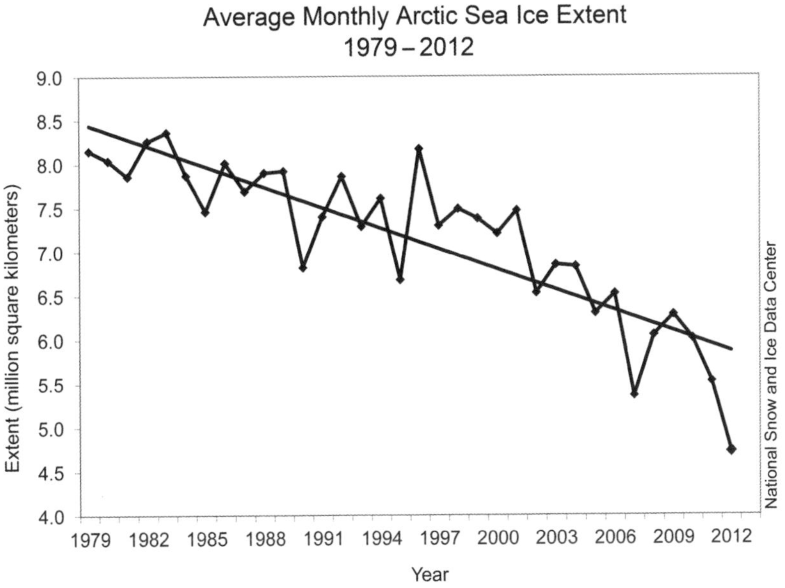

Average Monthly Arctic Sea Ice Extent
1979 – 2012

Fig. 14.   Monthly mean sea ice extent and area, in millions of square kilometers, for the Northern Hemisphere by year for March each year. Anomalies and median extent are calculated using a reference period of 1979 through 2000.[26]

the oceans. Historically, sea level has been measured by tide-gauge measurements in many locations around the world. Since the 1980s, with the advent of satellites, tide-gauge measurements have been supplemented by satellite measurements to measure sea level changes.[28] Figure 15 provides data on sea levels back to 1880. During this period, global-averaged sea level rose about 21 cm, with an average rate of rise of about 1.6 mm/yr between 1900 and 2000.

More recent sea-level data, using satellite altimeter measurements, are shown between 1992 and 2011 in Fig. 16. The trend line for this data indicates an increase of 3.1 mm/year, almost twice the value for the period between 1900 and 2000.

9 Climate Predictions

Predicting the future is always difficult, and the longer the time scale, the more difficult it is to make accurate predictions. The prediction of the earth's future climate is particularly difficult because of the many variables that play a role and the interplay between these variables. Possibly important feedback mechanisms can make the climate system quite unstable and therefore even more difficult to predict. An example of such a positive feedback mechanism is where the increase in carbon dioxide levels in the atmosphere causes a temperature increase of the oceans, which then release more dissolved CO_2 into the atmosphere, thus increasing the temperature and so on.

The amount of CO_2 in the atmosphere is one of the important determinants of future climate so that it is important to establish some indication of the amount

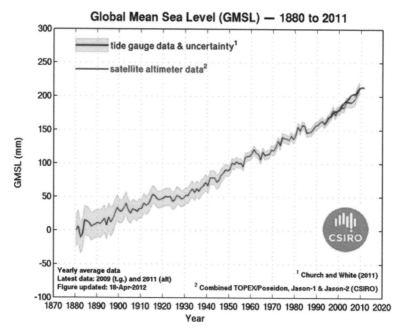

Fig. 15.   Global mean sea level rise from 1870 to 2010. The shading shows the uncertainty in the measurements.

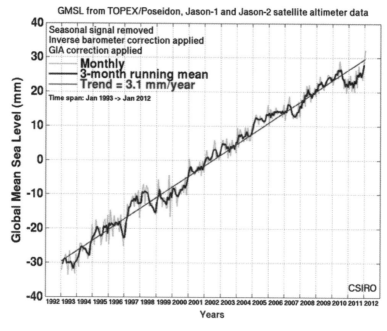

Fig. 16.   Global mean sea levels till the end of 2011 using satellite altimeter data.

Table 6:   Prediction of CO_2 Emissions[29] in Millions of Metric Tons per Year from Major Producers through 2035.

Region/Country	2005	2010	2015	2020	2025	2030	2035
US	5,996	5,644	5,680	5,777	5,938	6,108	6,311
Canada	620	569	569	582	608	635	679
Mexico/Chile	463	480	524	565	623	688	782
OECD Europe	4,400	4,094	4,115	4,147	4,156	4,198	4,257
Japan	1,241	1,090	1,125	1,142	1,136	1,110	1,087
South Korea	494	528	553	562	597	634	678
Australia/New Zealand	437	456	466	477	492	509	528
Russia	1,645	1,632	1,648	1,607	1,603	1,659	1,747
China	5,513	8,262	9,386	10,128	11,492	12,626	13,441
India	1,182	1,602	1,802	2,056	2,398	2,728	3,036
Middle East	1,400	1,692	1,889	2,019	2,199	2,435	2,659
Africa	978	1,107	1,209	1,311	1,430	1,568	1,735
Brazil	365	440	528	579	644	739	874
Total World	**28,181**	**31,305**	**33,457**	**35,210**	**37,341**	**40,160**	**42,220**

of CO_2 which will be emitted in the future. Predictions[29] for the emissions by a number of the important countries emitting CO_2 through 2035 are shown in Table 6. The main reason for the continued growth rate of CO_2 emissions in China is its strong dependence on coal-fired power plants. The US also depends to a large extent on coal-fired power plants and, in addition, is a large emitter of CO_2 in the transportation sector. The 24 European countries of the Organization for Economic Cooperation and Development (OECD) are predicted to have a reasonably constant level of production of CO_2.

9.1 *Global Temperature Projections*

There are many predictions of the future global temperature.[15,30] This is particularly controversial since there is a belief that limiting greenhouse gas emission is likely to have a deleterious effect on the economy of a country. The countries where the emissions are largest have so far been unwilling to restrict CO_2 emissions significantly. The most recent report[15] from the IPCC makes a number of predictions of future global temperature increases through 2100 using various assumptions about the change of greenhouse gas emissions. These predictions are shown in Fig. 17.

It will be some time before these different predictions can be tested. Even with the assumption that greenhouse gas emissions will remain constant at the levels of year 2000, (a very unlikely scenario) the model predicts that the global temperature will increase by about 0.6°C. For the high-growth scenario, the prediction is for the global temperature to increase by more than 3°C by 2100.

It should also be noted that the temperature increase will not be uniform across the globe. One of the outcomes of the model predictions is that northern latitudes will experience even greater temperature increases compared to the tropical regions.

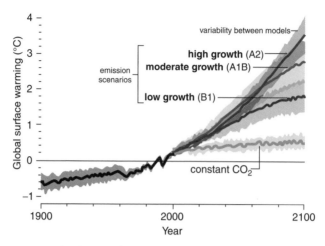

Fig. 17. Temperature projections to the year 2100, based on a range of emission scenarios and global climate models. Scenarios that assume the highest growth in greenhouse gas emissions provide the estimates in the top end of the temperature range. The "constant CO_2" line projects global temperatures with greenhouse gas concentrations stabilized at year 2000 levels.[15,30]

There have been numerous criticisms of the models used in the IPCC report but the consensus of informed scientific opinion is that the projections are reasonable. If no steps are taken to limit greenhouse gas emissions then the accuracy of the projections will become clearer in 10–20 years.

9.2 Sea-Level Projections

As noted already in Figs. 15 and 16, the global sea level has been rising at an increasing rate in the past century. This is obviously affected by the rise in temperature which causes expansion of the ocean volume and the increasing rate of melting of ice at the poles. The IPCC report[15] uses various models to predict future sea level rise through 2100 (Fig. 18). The average projected rise in sea level is 36 ± 14 cm by 2100.

A recent paper by Rahmstorf[32] presents a summary of the results from a number of models, including the IPCC report[15], which are used to predict global sea level increase as shown in Fig. 19. All of the other predictions[33–37] are larger than that of the IPCC report. The large uncertainty in the predictions is clearly shown. However, the average of these models suggests a possible rise in global sea level by more than 1 meter by 2100.

A recent study by Wageningen University and Research Center and the Royal Netherlands Meteorological Institute on behalf of the Dutch government[38] lists the various contributions to sea level rise along the Dutch coastline as shown in Table 7.

This table shows that sea rise in 2100 is dominated by two factors, the thermal expansion due to warming and the melting of the Antarctic Ice field. Note the large uncertainties in the numbers. However, because Holland is a low-lying country,

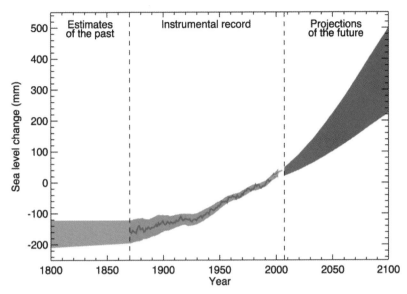

Fig. 18.   Past and projected global average sea level. The gray shaded area shows the estimates of sea level change from 1800 to 1870 when measurements are not available. The region between 1870 and 2010 is a reconstruction of sea level change measured by tide gauges with the surrounding shaded area depicting the uncertainty. The upper shaded area beyond 2010 represents the range of model projections conducted by IPCC[15] for a medium growth emissions scenario (IPCC SRES A1B) and excludes the additional rise in sea level if Greenland or Antarctica contribute more ice to the oceans in the future. *Source*: IPCC (2007).[31]

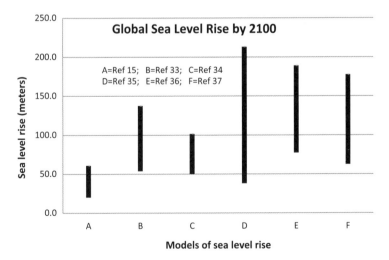

Fig. 19.   Global sea level rises (in meters) by the year 2100 predicted using various models.[32]

Table 7: Overview of All Estimated Contributions and the Total High-End Projections A&B for Local Sea Level Rise Along the Dutch Coast for 2100 (in Meters) and the Corresponding Contributions Reported in KNMI'06 for the Warm Scenario (4°C Temperature Rise).

Component	Delta committee	Delta committee	KNMI (2006)
	High End A	High End B	Warm scenario
Global mean thermal expansion	+0.12 to +0.49	+0.12 to +0.49	+0.27 to +0.35
Local expansion	−0.05 to +0.2	−0.05 to +0.2	−0.04 to +0.15
Small glaciers	+0.06 to +0.14	+0.06 to +0.14	+0.06 to +0.15
Antarctic ice sheet	−0.01 to +0.45	−0.03 to +1.07	−0.02 to +0.33
Greenland ice sheet	+0.03 to +0.04	−0.55 to −0.33	Not separated
Terrestrial water storage	0.0 to 0.04	0.0 to 0.04	0.0 to 0.04
Total	0.40 to 1.05	−0.05 to 1.15	
Total without elasto-gravity	(+0.55 to +1.2)	(+0.55 to +1.2)	0.4 to 0.85

Note: Numbers in brackets result from disregarding the elasto-gravity effect completely.

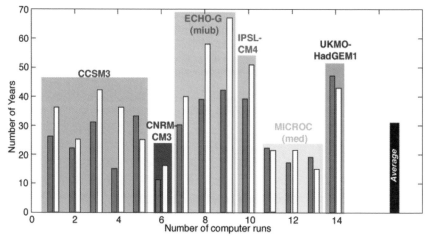

These six IPCC models, which most faithfully reproduced observed features of the recent climate, used the 2007-2008 September ice extents as a starting point for projections into the future.

Fig. 20.   Vertical bars indicate estimated number of years for sea ice extent to decline from the current value to a summer ice free Arctic. Runs by each of the six IPCC models are shown in the six boxes.[15,33]

which is particularly vulnerable to sea level rise, the Dutch are looking ahead to possible worst-case scenarios.

9.3 *Glacier and Ice Sheet Projections*

The ice sheets in Greenland and Antarctica show a seasonal variation of decreasing in mass during summer and replenishing during snowfall in winter. The determination of the net effect is difficult because the global models typically have too large a cell size to accurately describe the effects on smaller scale near the coastlines where much of the accumulation takes place. It does appear however that situation in the northern hemisphere will continue the trend of decreasing ice cover. One illustration is given in Fig. 20, where six models[15,33] are used to predict the number of years before the Arctic Ocean is fully ice-free in summer.

The average value from the models is about 30 years but it is clear that the uncertainty in this number is quite large. The situation with the Antarctic ice sheet continues to be the subject of considerable debate. A recent paper[27] suggests that wind-driven warmer water near the Antarctic coastline may be producing greater melting of the ice below the water surface than had previously been expected.

References

1. W. Benenson, J. W. Harris, H. Stocker and H. Lutz, *Handbook of Physics* (Springer, New York, 2002).
2. D. L. Hartmann, *Global Physical Climatology* (Academic Press, San Diego, 1994).
3. US National Aeronautics and Space Administration, Langley Research Center Atmospheric Sciences data Centere NASA/GEWEX SRB project. Available at: http://gewex-srb.larc.nasa.gov/common/php/SRB_data_graphics.php. Accessed 15 April 2012.
4. Wikipedia, *Greenhouse Effect.* Available at: http://en.wikipedia.org/wiki/Greenhouse_effect. Accessed 12 April 2012.
5. San Jose State University, Available at: http://www.applet-magic.com/insolation.htm. Accessed 22 April 2012.
6. J. Qiu, P. R. Goode, E. Pallé, V. Yurchyshyn, J. Hickey, P. Montañés, Rodriguez, M.-C. Chu, E. Kolbe, C. T. Brown and S. E. Koonin, Earthshine and the earth's albedo: 1. Earthshine observations and measurements of the lunar phase function for accurate measurements of the Earth's Bond albedo, *J. Geophys. Res.* **108** (4709) (2003), p. 22.
7. M. J. Pidwirny (2006). Net radiation and the planetary energy balance. In *Fundamentals of Physical Geography*, 2nd edn. Available at: http://www.physicalgeography.net/fundamentals/7i.html. Accessed 5 May 2012.
8. National Aeronautics and Space Adminstration, Goddard Institute for Space Studies, *GISS Surface Temperature Analysis.* Available at: http://data.giss.nasa.gov/gistemp/2011/. Accessed 2 May 2012.
9. P. Tans and R. Keeling, NOAA/ESRL, Scripps Institute of Oceanography, National Aeronautics and Atmospheric Administration, Earth System Research Laboratory. Available at: http://www.esrl.noaa.gov/gmd/ccgg/trends/. Accessed 15 April 2012.
10. Carbon Dioxide Information Analysis Center, Oak Ridge National Lab. Available at: http://cdiac.ornl.gov/pns/current_ghg.html. Accessed 15 March 2012.
11. National Oceanic and Atmospheric Administration, Earth System Research Laboratory. Available at: http://www.esrl.noaa.gov/gmd/outreach/faq_cat-3.html. Accessed 2 April 2012.
12. Skeptical Science, How reliable are CO_2 measurements. Available at: http://www.skepticalscience.com/co2-measurements-uncertainty.htm. Accessed 2 May 2012.
13. V. R. Gray, Greenhouse Bulletin No. 120 Feb 1999. Available at: http://www.johndaly.com/bull120.htm. Accessed 10 April 2012.
14. CO2Now.org. Available at: http://co2now.org/Current-CO2/CO2-Now/. Accessed 2 May 2012.
15. Contribution of Working Group I to the Fourth Assessment Report of the Intergovernmental Panel on Climate Change, 2007. S. Solomon, D. Qin, M. Manning, Z. Chen, M. Marquis, K. B. Averyt, M. Tignor and H. L. Miller (eds.) (Cambridge

University Press, Cambridge, UK, New York, USA). Available at: http://www.ipcc. ch/publications_and_data/ar4/wg1/en/ch3s3-2.html. Accessed 15 March 2012.

16. P. Brohan *et al.*, Uncertainty estimates in regional and global observed temperature changes: A new dataset from 1850, *J. Geophys. Res.* **111** (2006), D12106.

17. National Oceanic and Atmospheric Administration, National Climate Data Center, *Global Surface Temperature Anomalies.* Available at: http://www.ncdc.noaa.gov/cmb-faq/anomalies.php. Accessed 15 March 2012.

18. National Aeronautics and Space Administration, Goddard Institute for Space Studies. Available at: http://www.columbia.edu/~mhs119/UpdatedFigures/LOTI.pdf. Accessed 14 April 2012.

19. J. Hansen, R. Ruddy, M. Sato and K. Lo, Global surface temperature change, *Rev. Geophys.* **48** (2010), p. rg4004.

20. National Oceanic and Atmospheric Administration, May 2010 Global State of the Climate. Available at: http://www.noaanews.noaa.gov/stories2010/20100615_globalstats_sup.html. Accessed 15 March 2012.

21. T. M. Smith and R. W. Reynolds , A global merged land and sea surface temperature reconstruction based on historical observations (1880–1997), *J. Clim.* **18** (2005), pp. 2021–2036.

22. J. Hansen *et al.*, A closer look at United States and global surface temperature change, *J. Geophys. Res.* **106** (2001), pp. 23947–23963.

23. K. M. Lugina *et al.*, Monthly surface air temperature time series area-averaged over the *30*-degree latitudinal belts of the globe, 1881–2004. In *Trends: A Compendium of Data on Global Change.* (Carbon Dioxide Information Analysis Center, Oak Ridge National Laboratory, US Department of Energy, Oak Ridge, TN, 2005). Available at: http://cdiac.esd.ornl.gov/trends/temp/lugina/lugina.html. Accessed 20 May 2012.

24. N. A. Rayner *et al.*, Improved analyses of changes and uncertainties in sea surface temperature measured in situ since the mid-nineteenth century: The HadSST2 dataset, *J. Clim.* **19** (2006), pp. 446–469. Available at: http://www.cru.uea.ac.uk/cru/data/temperature/. Accessed 22 May 2012.

25. N. A. Rayner *et al.*, Global analyses of sea surface temperature, sea ice, and night marine air temperature since the late nineteenth century, *J. Geophys. Res.* **108** (2003), p. 4407.

26. National Snow and Ice Data Center, Boulder, CO. Available at: http://nsidc.org/data/. Accessed 10 May 2012.

27. S. S. Jacobs, A. Jenkins, C. F. Giulivi and P. Dutrieux, *Nat. Geosci.* **4** (2011), pp. 519–523.

28. Commonwealth Scientific and Industrial Research Organization, Australia. Available at: http://www.cmar.csiro.au/sealevel/sl_hist_few_hundred.html. Accessed 15 April 2012.

29. International Energy Agency, International Energy Outlook (2011). Available at: http://www.eia.gov/forecasts/ieo/index.cfm. Accessed 10 May 2012.

30. US Environmental Protection Agency, *Future Temperature Changes.* Available at: http://www.epa.gov/climatechange/science/futuretc.html. Accessed 10 May 2012.

31. Greenhouse effect and sea level rise: "Future sea level rise". Available at: http://papers.risingsea.net/future-sea-level-rise.html. Accessed 15 July 2012.

32. S. Rahmstorf, A new view on sea level rise. Nature Reports: Climate Change. 6 April 2010. Available at: http://www.nature.com/climate/2010/1004/full/climate.2010.29.html#B2. Accessed 15 July 2012.

33. S. Rahmstorf, A semi-empirical approach to projecting future sea-level rise, *Science* 315, pp. 368–370 (2007).

34. R Horton *et al.* Sea level rise projections for current generation CGCMs based on the semi-empirical method, *Geophys. Res. Lett.* **35** (2008), p. L02715.

35. A. Grimstead, J.C. Moore and S. Jevrejeva, Reconstructing sea level from paleo and projected temperatures 200 to 2100AD, *Climate Dynamics* **34** (2009), pp. 461–472.

36. M. Vermeer and S. Rahmstorf, Global sea level linked to global temperature, *Proc. Natl. Acad. Sci. USA* **106** (2009), pp. 21527–21532.

37. S. Jevrejeva, J.C. Moore and A Grimstead, How will sea level respond to changes in natural and anthropogenic forcings by 2100? *Geophys. Res. Lett.* **37** (2010), p. L07703.

38. Exploring high-end climate change scenarios for flood protection of the Netherlands, September 2008. Available at: http://www.knmi.nl/bibliotheek/knmipubWR/WR2009-05.pdf. Accessed 20 April 2012.

39. National Oceanic and Atmospheric Administration, Computer Model Predictions of Sea Ice. Available at: http://www.arctic.noaa.gov/future/sea_ice_models.html. Accessed 20 April 2012.

Index

About the Contributors

Bertrand Barré is presently Scientific Advisor to the Chairperson of AREVA and Professor Emeritus of Nuclear Engineering at the French National Institute for Nuclear Sciences and Technologies (INSTN). From 1967 to 2003, he spent most of his career in the French Atomic Energy Commission (CEA), alternating between scientific and managerial positions. He was notably Nuclear Attaché at the French Embassy in Washington DC, Delegate Director for Innovation in the CEA, Director of Engineering in Technicatome (now AREVA TA), Head of the Nuclear Reactors Directorate of the CEA, Vice-President in charge of R&D at Cogema (now AREVA NC) and Vice-President for Scientific Communications in AREVA. A Fellow and member of the Board of the American Nuclear Society, Bertrand Barré was successively Chairman of the French Nuclear Energy Society (SFEN), Chairman of the European Nuclear Society (ENS), Chairman of the International Nuclear Societies Council (INSC) and lately, Chairman of the International Nuclear Energy Academy (INEA). He was also Chairman of the Standing Advisory Group on Nuclear Energy (SAGNE) to the Director General of the International Atomic Energy Agency, French Governor of the European Joint Research Center, member of the French Conseil Supérieur de la Recherche et de la Technologie and Vice-Chairman of the EURATOM Scientific and Technical Committee.

H. J. M. (Jos) Beurskens headed the Renewable Energy and Wind Energy Units of the Energy Research Centre of the Netherlands (ECN) from 1989 to 2005 and was a scientific advisor at ECN from 2005 to 2012. At present, he has his own consulting company. He is the scientific director of the We@Sea Foundation, which focuses on offshore wind energy research and transfer of knowledge. Throughout his career, Jos Beurskens has been extensively involved with various industry associations involved in renewable energy. He was a founding member of the European Wind Energy Association (EWEA), on the Board of which he still serves as a member. He was one of the founders of the International Meeting of Test Stations (IMTS), the European Academy of Wind Energy (EAWE), and the European Renewable Energy Centres Agency (EUREC). He has chaired the Executive Committee of the Wind Energy program of the International Energy Association (IEA) and has been retained as an advisor to the European Commission and several national governments on R&D programs in the field of wind energy. He is Chairman of the Scientific Advisory Board of ForWind (the joint wind energy research institute of the universities of Oldenburg, Bremen, and Hannover, Germany) and a member of the Scientific Advisory Council of the Hanse Wissenschaftskolleg, Delmenhorst, Germany. He serves on the Steering Committee of the European Wind Energy Technology Platform and is Chairman of WG Wind Power Systems. Mr. Beurskens received the honor of Wind Energy Pioneer of the British Wind Energy Association. In 2008, he was awarded the Poul la Cour Prize, which was presented to him by Mr. Janez Potočnik, EU Commissioner for Science and Research. In November 2009, Mr. Beurskens was conferred an Honorary Doctor's Degree from the University of Oldenburg, Germany, for his work on initiating European research in the field of wind energy.

 **R. Gordon Bloomquist** has been an independent consultant since 2007, specializing in geothermal development with a special focus on East Africa. Dr. Bloomquist provides consulting services on geothermal, legal, and institutional issues; power generation; power purchasing; and project planning and financing. His clients include the World Bank, US AID, US TDA, KfW, and a number of private companies and utilities. He has most recently specialized in the design and implementation of geothermal exploration and drilling risk mitigation programs (World Bank GeoFund and ARGeo and the KfW/AUC Geothermal Risk Mitigation Fund). He holds a Ph.D. in Geochemistry from the University of Stockholm, Sweden. Prior to starting his own consulting business, Dr. Bloomquist served as Senior Scientist and Director of the Center for Distributed Generation and Thermal Distribution with Washington State University (WSU). As such, he was responsible for state policy, technical assistance to resource developers and investigations related to geothermal energy development. While at WSU, Dr. Bloomquist also managed the Bonneville Power Administration (BPA) Four-State Geothermal Program. This included developing case studies of numerous geothermal power generation facilities, developing criteria for geothermal project financing, setting up a Four-State GIS system and providing overall geothermal project advice to BPA management. Dr. Bloomquist has served on the Board and as President of the GRC and on the Board and as Chairman of the Finance Committee of the International Geothermal Association. He also serves on the Board of Nevada Geothermal Power and chairs the Technical Committee.

Arno J. Brand is a Research Scientist at the Wind Energy Unit of the Energy Research Centre of the Netherlands (ECN). He received his M.S. degree in Applied Physics and Ph.D. degree in Mechanical Engineering from the Delft University of Technology in 1987 and 1992, respectively. His dissertation work was on the structure of the turbulent boundary layer. Dr. Brand has been employed in wind energy at ECN since 1991. Until 1999, he was mainly involved in experimental and computational wind turbine rotor aerodynamics. In this context, he participated in four projects under the European Non-Nuclear Energy Program and two tasks of the International Energy Agency. In 1998, he became involved in wind energy meteorology, with emphasis on short-term prediction and wind resource assessment. Recent projects were on renewable energy forecasting, the wind (power) resource of the North Sea, integration of wind power into the grid, the effect of wind farms on the wind resource, and the flow modeling aspects of wind farm control. Currently, he is involved in modeling and control of clusters of wind farms. Dr. Brand was Secretary of the Netherlands Wind Energy Association (NEWIN) from 2000 to 2006, and has been a member of the Steering Committee Wind Technology of KiviNiria since 2003. His current research interests include fluid dynamics (experimental as well as numerical), aerodynamics, turbulence, and statistics.

Ian Bryden is a Professor of Renewable Energy at the University of Edinburgh, Scotland, where he is Head of the Institute for Energy Systems (IES). The IES is the university body with responsibility for research into (i) Low- and zero-carbon generation of electricity, (ii) Integration of intermittent and variable generation into electrical distribution grids, and (iii) Alternative vectors for the export of energy. Professor Bryden obtained his Ph.D. degree in the dynamics of flexible floating wave energy devices from the University of Edinburgh in 1984 and has worked in marine energy research, both wave and tidal currents, for more than 30 years in both commercial and academic environments. Professor Bryden is a Fellow of the Royal Society of Edinburgh; the Institution of Mechanical Engineers; the Institute for Marine Engineers, Scientists and Technologists, and the Institute of Physics. In addition, he is a non-executive director of the European Marine Energy Centre and a director of FloWave TT Ltd., which is the company responsible for the construction and operation of the university-owned All Waters Current and Wave Test Facility, a unique test facility capable of 1/10th scale simulation of wave and current conditions for device array testing.

Mo-Yuen Chow is the Founder and Director of the Advanced Diagnosis, Automation and Control (ADAC) Laboratory at North Carolina State University. His recent research focuses on distributed control and fault management, with applications on smart grids, PHEVs, batteries, and robotic systems. He has served as a Principal Investigator in projects supported by various federal agencies and private companies. He earned his B.S. degree in Electrical and Computer Engineering from the University of Wisconsin-Madison in 1982 and his M.Eng. and Ph.D. degrees from Cornell University in 1983 and 1987, respectively. Upon completion of his Ph.D. degree, Dr. Chow joined the Department of Electrical and Computer Engineering at North Carolina State University as an Assistant Professor in 1987, was promoted to Associate Professor in 1993, and to Professor in 1999. He spent his sabbatical leave as a Visiting Scientist in 1995 in ABB Automated Distribution Division. He has published 1 book, contributed 7 book chapters, and published over 200 journal and conference articles. Dr. Chow is an IEEE Fellow, the Editor-in-Chief of *IEEE Transactions on Industrial Electronics*, was a Co-Editor-in-Chief of *IEEE Transactions on Industrial Electronics*, a past Technical Editor of *IEEE Transactions on Mechatronics* and a past Associate Editor of the *IEEE Transactions on Industrial Electronics*. He was the Vice President for Publication of IEEE Industrial Electronics Society in 2006–2007, and the Vice President for Membership of IEEE Industrial Electronics Society in 2000–2001. He was the General Chair of IEEE-IECON05 and the General Co-Chair of IEEE-IECON10 and IEEE-ISIE12. He was a Senior Fellow of the Japan Society for the Promotion of Science in 2003. He has received the IEEE Eastern North Carolina Section Outstanding Engineering Educator Award in 2004, the IEEE Region-3 Joseph M. Biedenbach Outstanding Engineering Educator Award in 2005 and the IEEE Eastern North Carolina Section Outstanding Service Award in 2007. Dr. Chow is a Changjiang Scholar and a Visiting Professor at Zhejiang University. He worked in the US Army TACOM TARDEC Division as a Senior Research Scientist during the summer of 2003.

Gael O. G. Corre received a chemical engineering degree from the Institut National des Sciences Appliques de Toulouse, France, in 2003. His main research interest is the development of fuel cells that can efficiently operate on biofuels. After having worked on computational fluid dynamics projects addressing the physicochemical modeling of environmental processes, he joined Prof. John Irvine's research group at the University of St. Andrews in 2005 and received a Ph.D. degree in 2009. His Ph.D. work addressed Solid Oxide Fuel Cell anode fuel flexibility, with a strong focus on the development of oxide anodes for Solid Oxide Fuel Cells (SOFC) and their direct operation with ethanol and methane fuels. During his Ph.D., Dr. Corre actively collaborated with the research groups of Profs. Gorte and Vohs at the University of Pennsylvania, leading to the development of nickel-free oxide anodes offering performance levels comparable to nickel anodes and able to operate on undiluted methane. After receiving his Ph.D. degree, Dr. Corre undertook a more engineering-focused project, working on the development of short stacks for the Hybrid Direct Carbon Fuel Cell Technology. In 2010, Dr. Corre joined the Hydrogen Centre at the University of Glamorgan, where he is currently a Research Fellow. His research addresses the direct use of biogas in SOFCs.

Gerard "Gary" M. Crawley is the President of Marcus Enterprises LLC., based in South Carolina. He currently serves as the Senior Advisor to the Director of the National Research Foundation of the UAE and as the Vice-Chair of the International Experts Council of the Republic of Kazakhstan. Previously, Prof. Crawley served as the Director of the Frontiers Engineering and Science Directorate of Science Foundation Ireland from 2004 to 2007. Prior to this, Prof. Crawley was the Dean of the College of Science and Mathematics at the University of South Carolina from 1998 to 2004. At Michigan State University, he was Dean of the Graduate School from 1994 to 1998 and earlier Chair of the Department of Physics and Astronomy from 1988 to 1994. Professor Crawley served two terms at the US National Science Foundation, one as the Director of the Physics Division, 1987–1988, and earlier as a Program Officer in the Nuclear Physics Program. He has also served as the Chair of the Nuclear Physics Division of the American Physical Society during 1991–1992. Professor Crawley was born in Scotland but his first degree is from the University of Melbourne in Australia. He obtained his Ph.D. in Physics from Princeton University in 1965. He is the author of over 150 articles in refereed journals and wrote a textbook *Energy* published in 1975 by Macmillan.

John B. Curtis is Professor of Geology and Geological Engineering and Director, Potential Gas Agency at the Colorado School of Mines. He received his B.A. and M.Sc. degrees in Geology from Miami University in 1970 and 1972, respectively. He received his Ph.D. degree in Geology from The Ohio State University in 1989. He is a licensed Professional Geologist (Wyoming). He was an officer in the US Air Force from 1972 to 1975. Dr. Curtis has been at the Colorado School of Mines since July 1990. He had 15 years' prior experience in the petroleum industry with Texaco, Inc., SAIC, Columbia Gas, and Brown & Ruth Laboratories/Baker-Hughes. He serves on and has chaired several professional society and natural gas industry committees, which previously included the Supply Panel, Research

Coordination Council and the Science and Technology Committee of the Gas Technology Institute (Gas Research Institute). He co-chaired the American Association of Petroleum Geologists (AAPG) Committee on Unconventional Petroleum Systems from 1999 to 2004 and is an invited member of the AAPG Committee on Resource Evaluation. He was a Counselor to the Rocky Mountain Association of Geologists from 2002 to 2004. Professor Curtis was an Associate Editor of the *AAPG Bulletin* from 1998 to 2010. He has published studies and given numerous invited talks concerning hydrocarbon source rocks, exploration for unconventional reservoirs, and the size and distribution of US, Canadian and Mexican natural gas resources and comparisons of resource assessment methodologies. As Director of the Potential Gas Agency, he directs a team of 100 geologists, geophysicists and petroleum engineers in their biennial assessment of the remaining US natural gas resources. He teaches petroleum geology, petroleum geochemistry, and petroleum design at the Colorado School of Mines, where he also supervises graduate student research.

Mark Downing is an Agricultural Economist and Senior Scientist in the Environmental Sciences Division at Oak Ridge National Laboratory (1992–present). He received a Ph.D. in Agricultural Economics from Texas A&M University in 1992 and has been working in the field of biomass for energy since then. He has background and commercial experience in forestry, horticulture, and botany. His current research is understanding the agricultural and forest commodities used as inputs to the commercial biorefinery, and the markets for multiple products produced as outputs. He has been involved in the Department of Energy's research, development, and demonstration of biomass to electric power facilities and ethanol production from lignocellulosics and has supported transportation technologies such as renewable fuels, and alternative propulsion systems for on-road and off-road vehicles. He was a contributor to the 2011 Billion Ton Update Report. He now acts as Group Leader for the Energy and Environmental Systems Science area at ORNL, and the Science Team Lead for Bioenergy Resource and Engineering Systems Team.

Róisín Duignan is a Lecturer in University College Dublin (UCD), where she also received her B.E. and Ph.D. degrees in Electrical & Electronic Engineering. In June 2011, Dr. Duignan became a member of the Electricity Research Centre (ucd.ie/erc) in UCD. Her research interests include nonlinear systems stability and control, dynamic modeling, and optimization and design and control of switched systems. She has significant experience in the development of control algorithms and numerical methods, including convex optimization and linear programming. Her recent research has focused on the frequency stability and control of an increasingly distributed and stochastic grid, demand response strategies in commercial buildings and domestic dwellings, dynamic modeling and control of buildings, and responsive load modeling.

William Ellis graduated from the University of Pittsburgh in 1976 with a B.S. degree in Electrical Engineering and received his MBA degree from Waynesburg College in 2002. Throughout his career, he has worked in energy and energy-related fields. Mr. Ellis served in a variety of positions in power plants in the Pittsburgh area, PA, for a local utility before taking a position as a support contractor at the U.S. Department of Energy National Energy Technology Laboratory (NETL). At NETL, he has utilized his extensive power plant experience to support DOE projects ranging from external project communications products to leading a team performing Life Cycle Assessment work for power generation facilities. Mr Ellis is employed by KeyLogic Systems.

William L. Fisher was a Recipient of the 2002 UT Presidential Citation and holder of the Leonidas T. Barrow Centennial Chair in Mineral Resources in the Department of Geological Sciences. He was the inaugural Dean of the Jackson School of Geosciences (2005–2006) and has had a significant impact on the University of Texas at Austin for the past 42 years. He served as director of the Bureau of Economic Geology from 1970–1994 and again in 1999, creating the current facility with the largest core library in the US and transforming the Bureau into an internationally acclaimed research institution. He has also served the White House and the US Departments of the Interior and Energy, with appointments as Deputy Assistant Secretary of Energy (1975–1976) and Assistant Secretary of Energy and Minerals (1976–1977). Under his direction, UT's Geology Foundation increased its endowment fivefold, including a leadership gift from John A. and Katherine G. Jackson that established the Jackson School of Geosciences. The author of 14 books and 200 articles, Dr. Fisher has served as department Chairman, first Director of the Jackson School, inaugural Dean of the school, member of the UT faculty council, and President of the American Association of Petroleum Geologists. Professor Fisher is a member of the National Academy of Engineering and the recipient of several awards and medals for his pioneering contributions in the field of geology.

Magnus Gehringer has been the Senior Energy Specialist at the World Bank/ESMAP since January 2009, specializing in the field of geothermal energy and power generation. His responsibilities include (i) development of cutting-edge knowledge products and (ii) support to WBG project teams on geothermal projects. Since joining the World Bank, Mr. Gehringer has focused entirely on enhancing the WBG understanding and involvement in power generation from geothermal resources in developing countries. He has assisted country teams in initiating new geothermal projects in Djibouti, Ethiopia, and Malawi, while assessing the options and potential of several other East African, Central American and Asian countries. Prior to joining the Bank, Mr. Gehringer served as Executive Director of Business Development at Landsvirkjun Power Ltd., the private power generation development subsidiary of the National Power Company of Iceland. Until 2007, he was Chief Executive Officer of Exorka Ltd., an Icelandic engineering company that specializes in the design, construction and operation of "Kalina" binary power plants for geothermal, biomass and industrial waste heat recovery applications. Mr. Gehringer holds an M.Sc. degree in Business and degrees in Fisheries Technologies and Project Management, all from Iceland.

Robert J. Goldston is a Professor in the Department of Astrophysical Sciences at Princeton University. He was Director of the DOE Princeton Plasma Physics Laboratory from 1997 to 2009. He has made numerous experimental and theoretical contributions to fusion plasma physics over the last 40 years, exploring plasma heating, energy confinement, and — since stepping down as Director of PPPL — the physics of plasma power and particle exhaust. He is Associated Faculty with the Princeton Environmental Institute and Affiliated Faculty with Princeton's Program in Science and Global Security, and has been examining socio-economic aspects of fusion power. He is a fellow of the American Physical Society and recipient of its "Excellence in Plasma Physics" award. He chaired the American Physical Society's Physics Policy Committee. He is the author or co-author of many peer-reviewed publications, and co-author of the textbook *Introduction to Plasma Physics* (Taylor & Francis, 1995).

John Irvine is a Professor of Chemistry at the University of St. Andrews and a Royal Society Wolfson Merit award holder. His first degree is in Chemical Physics from Edinburgh University and he obtained a D. Phil from the University of Ulster in Photoelectrochemistry. He performed his postdoctoral studies, working with Anthony West in Aberdeen, and was subsequently appointed to a BP/RSE fellowship, lectureship, and senior lectureship at Aberdeen University. In 1994 he was Visiting

Professor at Northwestern University and then moved to the University of St Andrews as Reader and then Professor of Inorganic Chemistry. In 1999, he was awarded the IOM/RSC/SCI Beilby medal for his research in Materials and Energy Research. In 2003 he was awarded the RSC Francis Bacon medal for fuel cell research and was the IPS Bourner Lecturer. In 2005 he was elected as a Fellow of the Royal Society of Edinburgh and in 2008 received the Royal Society of Chemistry Materials Chemistry Award. He has over 300 publications in refereed scientific journals. His research interests are in solid state ionics, new materials, ceramic processing, electrochemistry, fuel cell technology, hydrogen, photoelectrochemistry, electrochemical conversion and heterogeneous catalysis. He is also Chairman of the Scottish Hydrogen and Fuel Cells Association that seeks to promote these technologies to the industry and the public in Scotland.

 **John W. Lund** has recently retired from the National Renewable Energy Laboratory in Golden, CO, USA, in their Low Temperature Geothermal Program. He is the former Director of the Geo-Heat Center at the Oregon Institute of Technology (OIT) and has worked in the direct utilization of geothermal energy for over 37 years. He received his B.S. and Ph.D. degrees from the University of Colorado, and his M.E. degree from the University of California, Berkeley, all in Civil Engineering. He is an Emeritus Professor of Civil Engineering at OIT, where he taught for 32 years. He has also taught at the University of Alaska, Fairbanks, and at Oregon State University. He is a registered Professional Engineer (PE) in California, Colorado, and Oregon. While at Oregon Institute of Technology, he was a Professor of Civil Engineering, Department Chair of Civil Engineering Technology, and Dean of Engineering and Industrial Technologies. He has been a member of the Advisory Committee on Geothermal Energy for the United Stated Department of Energy, and a past president of the Geothermal Resources Council (GRC) (2001–2002). He received the Geothermal Pioneer Award for "Outstanding Achievement in the Development of Geothermal Resources" from GRC in 1997 and the Joseph W. Aidlin Award in 2008 for "outstanding contributions to the GRC and to the development of geothermal resources". He has also been an invited lecturer at the geothermal schools in New Zealand and Iceland. He was the President (2004–2007) and the Past President of the International Geothermal Association (2007–2010). In 2002, he received the Medal of Honor from the International Summer School, and in 2005 an award from the Italian Geothermal Union. He received the Patricius Medal from the German Geothermal Association in 2008 for "his pioneer work in the direct use of geothermal energy, as well as his activities with international geothermal committees for 30 years". His recent interest is in geothermal energy related to direct utilization, geothermal heat pumps, and small-scale and low-temperature power generations. He has written several hundred papers on the utilization of geothermal energy and has

participated in numerous workshops, seminars, conferences, and training sessions in 35 countries. He has been responsible for collecting, editing, and summarizing the country update papers for the World Geothermal Conferences in 2000, 2005, and 2010. He was the Editor of the Geo-Heat Center Quarterly Bulletin, which presents papers and case studies on the direct utilization of geothermal energy. He lives in a geothermally heated home and works as an Emeritus Professor on a geothermally heated campus. He was also involved with the installation of a 280 kW geothermal binary power plant on the Oregon Institute of Technology campus, and the drilling of a 5,300 ft (1,600 m) deep geothermal well for a 1 MW geothermal power plant, the first campus in the world to receive both heat and electrical energy needs from geothermal energy.

Gérard C. Nihous graduated from the École Centrale in Paris in 1979 and from the University of California at Berkeley in 1983. His doctorate thesis in Ocean Engineering dealt with wave power extraction. After moving to Hawaii in 1987, he became involved in research on Ocean Thermal Energy Conversion (OTEC) for more than a decade. He taught a graduate course on renewable energy at Hiroshima University in 1996 and 1997. He also worked extensively on the ocean sequestration of CO_2 until 2002 before joining the University of Hawaii in 2003, primarily to do research on methane hydrates at the Hawaii Natural Energy Institute. He has been with the Department of Ocean and Resources Engineering since 2009, where the focus of his activities is, once again, the promotion and development of marine renewable energy.

Torbjørn K. Nielsen graduated in Mechanical Engineering from Møre og Romsdal State Technical College in Norway in 1972. He received his M.Sc and Ph.D. degrees in Mechanical Engineering from the Norwegian University of Science and Technology (NTNU) in 1976 and 1991, respectively. He began his career as a Research Assistant at the Waterpower Laboratory at NTNU and carried on as a research engineer at the Norwegian Hydrotechnical Laboratory. After a period as consultant at Nybro Hansen AS and as Head of Section of Unsteady Flow at the Norwegian Hydrotechnical Laboratory, he joined SINTEF Fluid Machinery as Head of the Section of Fluid Machinery Systems in 1985. In 1995, he was employed by Kværner Energy and became Head of the Turbine Laboratory in Trondheim, taken over by General Electric in 1997. In 2002, he obtained a professorship at NTNU. Professor Nielsen is currently the head of the Waterpower Laboratory at NTNU. His main interest is the behavior of hydraulic machinery in interaction with the connected hydraulic system in steady state, transient and oscillatory conditions. Regarding Hydro Power Plants, he has worked as consultant with surges and pressure transients, governor stability and flow induced vibrations, as well as optimisation of power output from multi aggregate systems.

Mark O'Malley graduated with B.E. and Ph.D. degrees in Electrical Engineering from the National University of Ireland in 1983 and 1987, respectively. He is a Professor of Electrical Engineering at University College Dublin (UCD), and the Director of the Electricity Research Centre (ucd.ie/erc), an industry supported research group, chaired by the Irish energy regulator. His teaching and research interests are in Grid Integration of Renewable Energy. He has spent sabbaticals at the University of Virginia, the University of Washington, and the National Renewable Energy Laboratory, Colorado. He has received two Fulbright awards (1994 and 1999). In 2007, he was elected a Fellow of the Institute of Electrical and Electronic Engineers (IEEE) and in 2008, he was elected a member of the Royal Irish Academy. He is a Fellow of the Institution of Engineering and Technology (IET) and a Fellow of the Institute of Engineers of Ireland (IEI). He has authored over 200 academic papers and supervised 17 Ph.D. students to completion. He is a member of the Engineering Sciences Panel of the European Research Council, a member of the European Academy of Sciences Advisory Council Energy Panel, the Chair of the IEEE Power and Energy Society Task Force on Capacity Value of Wind, a member of the Editorial Board of the IEEE Transactions on Power Systems, a member of the North American Electric Reliability Corporation Task Force on Integrating Variable Generation, and an IEEE Distinguished Lecturer in the grid integration of renewable energy. He is the Irish representative on the International Energy Agency Research Task 25: Design and Operation of Power Systems with Large Amounts of Wind Power and is a lead author on the International Panel on Climate Change Special Report on Renewable Energy Sources and Climate Change Mitigation. In 2008, he received the Utility Wind Interest Group (UWIG) Achievement Award.

Adrian Radziwon received his B.S. degree in Chemical Engineering from Pennsylvania State University and an M.S. degree in Chemical Engineering from the University of Pittsburgh. He is a Registered Professional Engineer in Pennsylvania. Prior to his employment as a support contractor, currently with KeyLogic Systems, at the US Department of Energy's National Energy Technology Laboratory (NETL), he held positions in the steel industry, a regulatory agency and several positions in chemical process design. The bulk of his experience has been related to energy, primarily coal. His prior coal-related experience includes metallurgical coke manufacturing and byproduct processing, combustion, pyrolysis, direct liquefaction, beneficiation, and gasification for both power generation and synthetic natural gas production. His prior energy related work also includes crude oil desalination. At NETL, he has primarily worked providing engineering support to the Clean Coal Technology Programs as well as the former Combustion and Flue Gas Cleanup Divisions.

Habiballah Rahimi-Eichi received his B.Sc. and M.Sc. degrees in Electrical Engineering from Isfahan University of Technology and Khaje-Nasir University of Technology, Iran, respectively with a specialty in adaptive and robust control. He is currently working toward his Ph.D. degree in the Advanced Diagnosis Automation and Control Laboratory at North Carolina State University, Raleigh, with Prof. M. Y. Chow. His research focus is online state of charge and state of health monitoring of batteries, especially in electric vehicles and smart grid applications. He has several publications on adaptive modeling, parameter identification and state estimation of different battery types.

Ignacio Rey-Stolle has held the position of Professor at the Technical University of Madrid (Madrid, Spain) since 2003, where he lectures on Electronics and Photovoltaic Solar Energy. He obtained his Ph.D. degree in 2001 from the Technical University of Madrid working on high concentrator GaAs solar cells. Since receiving his Ph.D., Dr. Rey-Stolle has spent more than 14 years working in the field of high concentrator III-V multijunction solar cells doing device design and simulation, epitaxial growth, characterization and reliability studies. This work has resulted in several world-record efficiency solar cell devices that have been produced in Prof. Rey-Stolle's research group, the latest being the best concentrator dual-junction solar cell with an efficiency of 32.6% at 1,026 suns. Professor Rey-Stolle has co-authored more than 100 scientific papers, two book chapters and one patent. He has been principal investigator on 7 research projects, has participated as an associate researcher in more than 20 projects, and has been deeply engaged in two technology transfer projects to photovoltaic industries. In 2007, Prof. Rey-Stolle received the "Best Young Professor Award" of the Technical University of Madrid for outstanding scientific research and innovation.

Thomas A. Sarkus has 33 years of technical and managerial experience, including 26 years in project management at the US Department of Energy's (US DOE's) National Energy Technology Laboratory (NETL). He is currently the Division Director of NETL's Project Financing & Technology Deployment Division, Office of Major Demonstrations, Strategic Center for Coal, and provides expert guidance and consultation to major DOE-funded clean coal technology and carbon sequestration demonstration projects. As a key manager within the major demonstration programs administered by DOE's Office Fossil Energy, he has worked on a variety of coal preparation, flue gas cleanup (e.g. SO_2, NOx, and Hg emissions control)

and advanced power generation (e.g. CFB, IGCC) technologies. Major technology demonstration projects have included advanced SO_2 scrubbers at Chesterton, IN, Newnan, GA, & Lansing, NY; a 265 MWe (net) Circulating Fluidized-Bed (CFB) power plant in Jacksonville, FL; and coal-based Integrated Gasification Combined-Cycle (IGCC) power plants near Terre Haute, IN & Tampa, FL. These and other major projects have garnered national and international attention, plus numerous awards, including the National Society of Professional Engineers' Outstanding Engineering Achievement Award, *Power*'s Annual Powerplant Award, *Power Engineering*'s Coal Project of the Year, and *R&D*'s R&D 100 Award. He has hosted over 20 technical workshops on subjects ranging from unburned carbon in fly ash to SO_3, reburning technology, and selective catalytic and non-catalytic reduction technologies for NO_x control. He speaks frequently on topics related to energy, the environment, and cleaner coal technologies. A native of Donora, Pennsylvania, his efforts have contributed to the capture or avoidance of millions of tons of air pollutant emissions at significantly reduced compliance costs. Mr. Sarkus holds degrees in Chemistry, Geology, Earth Science, and Law and is a licensed attorney in Pennsylvania.

Wes Stein is the Manager of the National Solar Energy Centre and Leader of the Solar Power Program at the Commonwealth Scientific and Industrial Research Organization (CSIRO), Australia. He holds a B. Eng degree in Mechanical Engineering with first class honors and a Master's degree in Engineering. His present responsibilities include developing and delivering next-generation solar thermal power technologies. Currently he leads a team of 30 research engineers and scientists delivering leading Concentrating Solar Power (CSP) research with over AU$40 million worth of CSP research projects presently underway. He is also a member of the Research Advisory Committee for the Australian Solar Institute, a co-author of a study for the World Bank on the global status and opportunities for Concentrating Solar Power (2006) and a lead author for the IPCC Special Report into Renewable Energy (2011). He is responsible for initiating and leading the development of Australia's two high-density solar tower arrays for research. He was awarded an Australian Institution of Engineers scholarship to develop high-temperature solar cycles, working at each of the major solar thermal research institutions in the world. He also won the UK's Rendel Award for his paper on "Global Energy Issues for the next 30 years". He is Australia's member on the Executive Committee of the International Energy Agency's SolarPACES (Solar Power and Chemical Energy Systems), the predominant CSP international body.

Erik Storm is internationally recognized as an expert in the physics and technology of Inertial Confinement Fusion (ICF). He received his B.S., M.S., and Ph.D. degrees in Aeronautical Engineering and Physics of Fluids from California Institute of Technology in 1967, 1968 and 1972, respectively. He began his career at Lawrence Livermore National Laboratory (LLNL) in 1974 and, during the next 20 years, worked on and directed the ICF campaigns on the Janus, Argus, Shiva, and Nova facilities at LLNL, becoming the Deputy Associate Director for Lasers in 1985. Over the next eight years, he was responsible for executing a series of dedicated underground nuclear tests that led to rest the question of the scientific feasibility of ICF and led the LLNL ICF Program through two National Academy of Sciences (NAS) reviews that culminated in the NAS recommendation to build the National Igition Facility (NIF). In 1994, he was asked to negotiate and put in place a Government-to-Government agreement between the US Department of Energy and the French Commissariat de l'Energie Atomique to collaborate on High Energy/High Power lasers. This agreement allowed an extremely fruitful collaboration between the two countries to develop the technology for, and subsequently build, the NIF at LLNL, and the Laser Megajoule (LMJ) at the CEA laboratory near Bordeaux. Dr. Storm returned to LLNL as Deputy Associate Director for the Defense Nuclear Technologies program from 1997 to 1998, before once again taking on the responsibility of coordinating the joint US/French NIF/LMJ technology and physics development programs. From 2006, Dr. Storm has been responsible for developing Advanced Inertial Fusion Energy options at LLNL.

Anthony F. Turhollow Jr. is a Staff Economist in the Environmental Sciences Division of Oak Ridge National Laboratory (1982–1993, 1998–present). He received his Ph.D. degree in Agricultural Economics from Iowa State University in 1982 and has been working in the field of biomass for energy since 1980. His particular expertise is in the economics (costs), logistics, energy and greenhouse balances, and resource assessments of biomass energy crops. From 1983 to 1993, he managed herbaceous energy crop screening and selection research and rapeseed breeding research. He was one of the main authors on the 2005 Billion Ton Report and 2011 Billion Ton Update. Economics analyses include estimating the costs of producing herbaceous and short rotation woody energy crops. He has estimated costs associated with the harvest, handling, transport, and storage of energy crops and crop residues. He has served on a number of US Department of Energy and US Department of Agriculture panels to review and select biomass-related research projects for funding.

V. Ismet Ugursal is a Professor of Mechanical Engineering at Dalhousie University in Halifax, Nova Scotia, Canada. He holds a B.Sc in Mechanical Engineering, an M.Eng in Industrial Engineering, and a Ph.D. in Chemical Engineering. He teaches courses on thermodynamics, energy management, energy conversion, internal combustion engines, and heating, ventilating and air-conditioning. His research and consulting areas include energy conversion systems for electric power and heat generation, building thermal systems, modeling of energy consumption in the residential sector, and energy management and conservation. Dr. Ugursal has worked in various areas of energy engineering as a consultant, project director, researcher, and teacher. He has worked on energy and international development projects in Canada, Turkey, Sierra Leone, Russia, Columbia, Gambia, Trinidad and Tobago, the Philippines, and Thailand. He has received more than 35 research contracts and grants as the project director/principal investigator. Dr. Ugursal served as the Associate Dean, Graduate Studies and Research, of the Faculty of Engineering, at Dalhousie University (2001–2003); the Chair, Department of Mechanical Engineering at the University of Victoria (2003–2005); and the Head of the Department of Mechanical Engineering at Dalhousie University (2007–2009). He has published extensively in his areas of expertise and graduated 30 Doctorate and Master's students. Currently, he is an Associate Editor of *Journal of Energy Engineering*, published by the American Society of Civil Engineers, and a member of the editorial board of *Buildings*, an open-access journal for the built environment. He is a Fellow of the Canadian Society of Mechanical Engineering.

Lu Wei is a visiting Ph.D. student in the School of Materials Science and Engineering at Georgia Institute of Technology, USA, with Prof. Gleb Yushin as her advisor, effective September 2008. She received her B.S. degree in Materials Science and Engineering from Zhengzhou University, China, in June 2005. Upon graduation, she joined Northwestern Polytechnical University (China) as a graduate student in the National Key Laboratory of Thermostructural Composite Materials, where she worked on the structural design and preparation techniques of high-temperature oxidation protective coatings for carbon/carbon composites. At Georgia Institute of Technology, she has been engaged in the synthesis of innovative nanostructured materials for energy storage applications, such as supercapacitors and Li-batteries. Her research interests are focused on the investigation of low-cost nanoporous carbon electrodes with controlled pore size from environmentally friendly precursors, and the manufacturing of on-chip micro supercapacitors consisting of single particle electrodes with precisely defined dimensions. Ms. Wei's recent work has been published in *Advanced Functional Materials*, *Advanced Energy Materials*, *Carbon*, and *Journal of Power Sources*.

Xiaoxiang Xu received his Bachelor's degree in Materials Science and Engineering (MSE) from the University of Science & Technology of China (USTC) in 2006, where he worked on lithium rechargeable batteries under Prof. Chunhua Chen's supervision. He then moved to University of St. Andrews in the same year to start research on proton conductors under the supervision of Prof. John Irvine and Prof. Shanwen Tao. After receiving his Ph.D. degree in Chemistry in 2009, he continued his research as a postdoctoral research fellow in Prof. John Irvine's group. His current research project focuses on the development of new materials as photocatalysts for hydrogen generation and reduction, which is part of the collaboration with Prof. Michael Hoffmann from California Institute of Technology (Caltech) and Prof. Peter Robertson from Robert Gordon's University (RGU). Dr. Xu's research interests include the design and synthesis of new protonic conductors, and new photo-catalysts that have potential applications for energy conversions, particularly in solar energy conversions, such as water splitting into H_2 and O_2, as well as CO_2 reduction into fuels under solar irradiation.

Gleb Yushin is an Associate Professor in the School of Materials Science & Engineering and Director of the Center for Nanostructured Materials for Energy Storage at Georgia Institute of Technology, USA. The current research activities of his laboratory are focused on energy storage technologies with emphasis on nanostructured and nanocomposite materials for use in advanced lithium-ion batteries, high-energy, high-power supercapacitors, and lightweight structural materials. For his contributions to these and other areas, he has received numerous awards and recognition, including the *R&D*'s R&D 100 Award, National Science Foundation Faculty Early Career Development Award, Air Force Office of Scientific Research Young Investigator Award, Honda Initiation Award, Petroleum Research Fund Doctoral New Investigator Award, NASA Nano 50 Award, Roland B. Snow Award from the American Ceramic Society, Micromeritics' Instrument Award, and Creating and Energy Options Award from the Strategic Energy Institute. Professor Yushin's work has been covered by *Science, Nature, Nature Materials, Advanced Functional Materials, Advanced Energy Materials, ACS Nano, Nano Letters, Physical Review Letters*, and *Journal of the American Chemical Society*, among other leading scientific journals. He has co-authored over 80 journal articles and 20 US and international patents and patent applications. He and his students won multiple best poster awards from the *Electrochemical Society, Materials Research Society, Carbon Society*, and *North American Membrane Society* meetings, and multiple awards at ceramographic competitions. Since 2010, he has served as an Associate Editor of the *Journal of Nano Energy and Power Research* (JNEPR). Professor Yushin received his M.S. degree in Physics from Saint-Petersburg Technical University, Russia, and Ph.D. degree in Materials Science from North Carolina State University, Raleigh, NC, USA.

Michael Zarnstorff is the Deputy Director for Research at the Princeton Plasma Physics Laboratory. He is an experimental plasma physicist with interests in the basic physics of plasma confinement and configuration optimization. Dr. Zarnstorff received his Ph.D. in Physics from the University of Wisconsin-Madison in 1984 and was named a Distinguished Research Fellow by the laboratory in 1995. In 2008, he received the APS Dawson Award for Excellence in Plasma Physics Research. His research included the first observation and systematic study of the bootstrap current, investigations of neoclassical and turbulent transport, transport barriers and the confinement and stability of different magnetic field configurations. He led the National Compact Stellarator Experiment physics group and was one of the leaders of the TFTR research program. He has collaborated on experiments across the US and in Germany, Japan, and the UK. Dr. Zarnstorff is a fellow of the American Physical Society and has served as a Division of Plasma Physics Distinguished Lecturer and on the Executive Committee, Fellowship Committee, and Program Committee. He was a member of the NRC Plasma Science Committee, Burning Plasma Panel, and the Committee to Review the US ITER-Science Participation Planning. He served as a member of the DOE Fusion Energy Science Advisory Committee, and a number of sub-committees. He also served as Vice-Chair of the Council of the US Burning Plasma Organization, as Chair of the BPO International Collaboration Task Group, and on numerous advisory and review committees.